Water, Earth, and Man

First published in 1969, *Water, Earth, and Man*, was written to demonstrate the advantages of adopting a unified view of the earth and social sciences.

The book considers the connection between an understanding of physical environments and an understanding of social environments. It explores the hydrologic cycle and highlights the significance of the relationship between natural environments and the activities of humankind, drawing together physical and human geography to produce a highly detailed study.

T0133534

Water, Earth, and Man

A Synthesis of Hydrology, Geomorphology, and Socio-Economic Geography

Edited by Richard J. Chorley

Routledge
Taylor & Francis Group

First published in 1969
by Methuen & Co. Ltd.

This edition first published in 2021 by Routledge
2 Park Square, Milton Park, Abingdon, Oxon, OX14 4RN
and by Routledge
605 Third Avenue, New York, NY 10017

Routledge is an imprint of the Taylor & Francis Group, an informa business

© 1969 Methuen & Co. Ltd.

Publisher's Note
The publisher has gone to great lengths to ensure the quality of this reprint but points out that some imperfections in the original copies may be apparent.

Disclaimer
The publisher has made every effort to trace copyright holders and welcomes correspondence from those they have been unable to contact.

A Library of Congress record exists under LCCN: 78410024

ISBN 13: 978-0-367-77195-9 (hbk)
ISBN 13: 978-1-003-17018-1 (ebk)
ISBN 13: 978-0-367-77194-2 (pbk)

Book DOI: 10.4324/9781003170181

Water, Earth, and Man

A SYNTHESIS OF HYDROLOGY, GEOMORPHOLOGY,
AND SOCIO-ECONOMIC GEOGRAPHY

EDITED BY

Richard J. Chorley

CONTRIBUTORS

*R. G. Barry, R. P. Beckinsale, M. A. Carson, R. J. Chorley,
G. H. Dury, I. S. Evans, R. W. Kates, B. A. Kennedy,
A. N. Kirkby, M. J. Kirkby, M. G. Marcus, R. J. More,
M. A. Morgan, R. L. Nace, T. O'Riordan, J. C. Rodda, J. F. Rooney,
S. A. Schumm, W. R. D. Sewell, D. B. Simons, M. Simons,
C. T. Smith, D. R. Stoddart, J. P. Waltz, and P. W. Williams.*

LONDON

METHUEN & CO LTD

First published 1969 by
Methuen & Co Ltd,
11 New Fetter Lane,
London EC4
Reprinted 1974 and 1977

© *1969 Methuen & Co Ltd*

Printed in Great Britain by
Fletcher & Son Ltd, Norwich

ISBN 0 416 12030 X

Distributed in the USA by
Harper & Row Publishers, Inc
Barnes & Noble Import Division

Acknowledgements

The editor and contributors would like to thank the following editors, publishers and individuals for permission to reproduce figures and tables:

Editors

American Journal of Science for figs. 2.II.8 and 7.II.6; *Annals of the Association of American Geographers* for figs. 10.II.2 and 10.II.3; *Bulletin of the American Association of Petroleum Geologists* for fig. 2.II.15; *Bulletin of the Geological Society of America* for figs. 2.II.4, 2.II.11, 2.II.17, and 5.I.6(*a*); *Bulletin of the International Association of Scientific Hydrology* for figs. 4.II.3, 4.II.5 and 5.I.3; *Canadian Geographer* for fig. 7.III(ii).3; *Department of Civil Engineering Technical Reports, Stanford University* for figs. 2.1.2 and 9.I.5; *Geographical Bulletin* for fig. 8.1.5(*a–c*); *International Science and Technology* for fig. 6.III.2; *Journal of Applied Meteorology* for fig. 3.I(i).3; *Journal of Geology* for figs. 2.II.18 and 3.II.6; *Journal of Glaciology* for fig. 8.I.3; *Marine Geology* for fig. 1.III.8; *Petermann's Geographische Mitteilungen* for fig. 1.1.4; *Proceedings of the Institution of Civil Engineers* for fig. 9.I.3; *Professional Geographer* for fig. 10.II.1; *Transactions of the American Geophysical Union* for figs. 2.1.3, 2.II.20, and 5.II.3.

Publishers

American Association of Petroleum Geologists for figs. 1.III.6 and 1.III.7 from *Recent Sediments, Northwest Gulf of Mexico* by F. P. Shepard *et al.*; Edward Arnold Ltd., London for fig. 4.1.4 from *Irrigation and Climate* by H. Oliver; W. H. Freeman and Co., San Francisco, for fig. 3.II.10 from *Fluvial Processes in Geomorphology* by L. B. Leopold, M. G. Wolman, and J. P. Miller; Heinemann, London, for fig. 3.II.1 from *Water, The Mirror of Science* by K. S. Davis and J. A. Day, and fig. 4.III(i).1 from *The Plant in Relation to Water* by R. O. Knight; Macmillan, London, for fig. 12.1.5 from *Techniques for Design of Water Resource Systems* by M. M. Hufschmidt and M. B. Fiering; The Macmillan Co., New York for fig. 4.III(i).5 from *The Nature and Properties of Soils* by H. O. Buckman and N. C. Brady, and fig. 11.1.1 from *Rainfall and Runoff* by E. E. Foster; McGraw-Hill Book Co., New York for figs. 2.II.3, 2.II.6, and 2.II.14 from *Handbook of Applied Hydrology* by Ven Te Chow (Ed.); Merrill Books, Inc., Columbus, for fig. 3.II.2 from *Oceanography* by M. G. Gross; Methuen and Co. Ltd., London, for fig. 1.1.1 from *Models in Geography* by

R. J. Chorley and P. Haggett (Eds.); Thos. Nelson and Sons, London, for figs. 3.I(i).4 and 11.1.5 from *The British Isles* by J. W. Watson and J. B. Sissons (Eds.); Oliver and Boyd, Edinburgh for fig. 1.III.3 from *Principles of Lithogenesis* by N. M. Strakhov; Prentice-Hall, Inc., New York, for figs. 3.II.3 and 3.II.9 from *Physical Geology* by L. D. Leet and S. Judson; Presses Universitaires de France for figs. 1.III.1 and 1.III.2 from *Climat et Erosion* by F. Fournier; Princeton University Press for figs. 1.II.1, 11.II.2, and 11.II.3 from *The Quaternary of the United States* by H. E. Wright and D. G. Frey (Eds.); University of Chicago Press for figs. 1.1.5, 1.1.9, and 4.1.7 from *Physical Climatology* by W. D. Sellers, and figs. 12.1.3 and 12.1.4 from *Readings in Resource Management and Conservation* by I. Burton and R. W. Kates (Eds.); John Wiley and Sons, Inc., New York, for fig. 1.III.9 from *The Sea* (Vol. 3) by M. N. Hill (Ed.), fig. 2.II.13 from *Geohydrology* by R. J. M. DeWiest, fig. 4.III(i).3 from *Irrigation Principles and Practices* by O. W. Israelson and V. E. Hansen, and fig. 4.III(i).4 from *Soil and Water Conservation Engineering* by G. O. Schwab *et al.* U.N.E.S.C.O. for fig. 8.1.3.

Individuals

Chief Engineer, U.S. Bureau of Reclamation for figs 3.III.2 and 3.III.3; The Controller, Her Majesty's Stationary Office (Crown Copyright Reserved) for figs. 1.1.3 and 12.1.7; The Director, Department of Lands and Forests, British Columbia, for the base maps for fig. 8.II(i).2; The Director, Geographical Branch, Department of Mines and Technical Surveys, Ottawa, for figs. 8.II(ii).2 and 8.II(ii).3 from the *Memoir* by J. R. Mackay; The Director, Geographical Branch, Office of Naval Research, Washington, for fig. 2.II.16 from the *Technical Report* by A. Broscoe and figs. 2.II.21 and 5.II.4 from the *Technical Report* by M. A. Melton; The Director, Irrigation Districts Association of California for fig. 12.1.8; The Director, Military Engineering Experiment Establishment, Christchurch, Hampshire, for fig. 2.II.1; The Director, Road Research Laboratory, Watford for fig. 5.III.5; The Director, Tennessee Valley Authority, Office of Tributary Development, Knoxville, Tennessee, for fig. 5.1.5 from the *Research Paper* by R. P. Betson; The Director, U.S. Army Engineer Experiment Station, Vicksburg, Mississippi, for fig. 2.II.1; The Director, U.S. Geological Survey for figs. 2.II.12, 2.II.21, 3.II.7, and for the use of the originals from which a number of figures in Chapters 7.II and 9.II were prepared; Professor K. M. King, Ontario Agricultural College for figs. 4.1.4 and 4.1.5; R. P. Matthews of Portsmouth College of Technology for part of Table 3.I(i).2; Ralph Parsons Co., New York, for fig. 11.III.2 of the N.A.W.A.P.A. Scheme; Superintendent of Documents, U.S. Government Printing Office for the Frontispiece and fig. 7.III(i).1.

Finally, the following thanks are also due:

Mr A. Burn, Mr R. Smith, Mrs B. Human, and Miss L. Thorne of the Drawing Office, Department of Geography, Southampton University, for drawing the figures for Chapters 1.1., 3.1(i), 4.1 and 11.1; Members of the City Engineer's Department, Corporation of Bristol for valuable assistance with Chapter 5.III; Mr M. A. Church for data on Banks Island slope angles employed in Chapter 8.II(ii); Dr R. M. Holmes, Inland Waterways Branch, Calgary, for valuable comments on Chapter 4.1; Professor J. N. Jennings, Department of Biogeography and Geomorphology, Australian National University for valuable comments on Chapter 6.II; Dr Mark Meier, U.S. Geological Survey for material used in Chapter 8.1; Dr Gunnar Østrem, Glaciology Section, Vassdragsvesenet for material used in Chapter 8.1; Mr R. W. Robertson of the University of Victoria for drawing the figures for Chapter 9.III; Mr M. Young, Miss R. King, and Mr M. J. Ampleford of the Drawing Office, Department of Geography, Cambridge University, for drawing figures for Chapters 2.II, 3.II, 3.III, 4.III(i), and 12.1; Dr K. A. Edwards and Members of the Institute of Hydrology, Wallingford, for their kind assistance with Chapters 3.1(ii) and 9.1, Mr V. J. Nash for preparing the illustrations.

The Editor and Publishers would like to thank Mrs D. M. Beckinsale for her painstaking and authoritative preparation of the Index, which has contributed greatly to the value of this volume.

Contents

1. THE WORLD

2. THE BASIN

3. PRECIPITATION

4. EVAPOTRANSPIRATION

7. CHANNEL FLOW

8. SNOW AND ICE

9. SHORT-TERM RUNOFF PATTERNS

10. ANNUAL RUNOFF CHARACTERISTICS

11. LONG-TERM TRENDS

12. CHOICE IN WATER USE

Introduction

R. J. CHORLEY and R. W. KATES

Department of Geography, Cambridge University and Graduate School of Geography, Clark University

Who would not choose to follow the sound of running waters? Its attraction for the normal man is of a natural sympathetic sort. For man is water's child, nine-tenths of our body consists of it, and at a certain stage the foetus possesses gills. For my part I freely admit that the sight of water in whatever form or shape is my most lively and immediate form of natural enjoyment: yes, I would even say that only in contemplation of it do I achieve true self-forgetfulness and feel my own limited individuality merge into the universal.

(Thomas Mann: *Man and his Dog*)

1. 'Physical' and 'human' geography

Perhaps it is of the nature of scholarship that all scholars should think themselves to be living at a time of intellectual revolution. Judged on the basis of the references which they have cited (Stoddart, 1967, pp. 12–13), geographers have long had the impression that they were the immediate heirs of a surge of worthwhile and quotable research. There is good reason to suppose, however, that geography has just passed through a major revolution (Burton, 1963), one of the features of which has been profoundly to affect the traditional relationships between 'physical' and 'human' geography.

Ever since the end of the Second World War drastic changes have been going on in those disciplines which compose physical geography. This has been especially apparent in geomorphology (Chorley, 1965a), where these changes have had the general effect of focusing attention on the relationships between process and form, as distinct from the development of landforms through time. In the early 1950s geomorphologists, especially in Britain, were able to look patronizingly at the social and economic branches of geography and dismiss them as non-scientific, poorly organized, slowly developing, starved of research facilities, dealing with subject matter not amenable to precise statement, and denied the powerful tool of experimentation (Wooldridge and East, 1951, pp. 39–40). It is true that by this time most geographers had long rejected the dictum that physical geography 'controlled' human geography, but most orthodox practitioners at least paid lip service to the idea that there was a physical *basis* to the subject. This view was retained even though traditional geomorphology had little or nothing to contribute to the increasingly urban and industrial preoccupations of human geographers (Chorley, 1965b, p. 35), and its place in the subject as a whole was maintained either as a conditioned reflex or as increasingly embarrassing grafts on to new geographical shoots. American geographers, who had largely abandoned geomorphology to the geologists even before the war, tended to look more to climatology for their physical basis. However, despite the important researches of Thornthwaite and of more recent work exemplified by that of Curry [1952] and Hewes [1965], the proportion of articles relating to weather and climate appearing in major American geographical journals fell more or less steadily from some 37% in 1916 to less than 5% in 1967 (Sewell, Kates, and Phillips, 1968). Even in the middle of the last decade Leighly (1955, p. 317) was drawing attention to the paradox that instructors in physical geography might be required to teach material quite unrelated to their normal objects of research.

The problems of the relationships between physical and human geography facing Leighly were small, however, compared with those which confront us today. Little more than a decade has been sufficient to transform the leading edge of human geography into a 'scientific subject', equipped with all the quantitative and statistical tools the possession of which had previously given some physical geographers such feelings of superiority. Today human geography is not directed towards some unique areally-demarcated assemblage of information which can be viewed either as a mystical *gestalt* expressive of some 'regional personality' or simply as half-digested trivia, depending on one's viewpoint. In contrast, most of the more attractive current work in human geography is aimed at more limited and intellectually viable syntheses of the pattern of human activity over space possessing physical inhomogeneities, leading to the disentangling of universal generalizations from local 'noise' (Haggett, 1965). Today it is human geography which seems to be moving ahead faster, to have the more stimulating intellectual challenges, and to be directing the more imaginative quantitative techniques to their solution.

One immediate result of this revolution has been the demonstration, if this were further needed, that the whole of geomorphology and climatology is not coincident with physical geography, and that the professional aims of the former are quite distinct from those of the latter. This drawing apart of traditional physical and human geography has permitted their needs and distinctions, which had previously been obscure, to emerge more clearly. Perhaps the distinctions may have become too stark, as evidenced by current geographical preoccupations with a rootless regional science and with socio-economic games played out on featureless plains or within the urban sprawl. Perhaps this is what the future holds for geography, but it is clear that without some dialogue between man and the physical environment within a spatial context geography will cease to exist as a discipline.

There is no doubt that the major branches of what was previously called physical geography can exist, and in some cases already are existing, under the umbrella of the earth sciences, quite happily outside geography, and that they are probably the better for it. It is also possible that this will be better for geography in the long run, despite the relevance to it of many of the data and certain of the techniques and philosophical attitudes of the earth sciences. In their place a more meaningful and relevant physical geography may emerge as the product of a new generation of physical geographers who are willing and able to face up to the contemporary needs of the whole subject, and who are prepared to concentrate on the areas of physical reality which are especially relevant to the modern man-oriented geography. It is in the extinction of the traditional division between physical and human geography that new types of collaborative synthesis can arise. Such collaborations will undoubtedly come about in a number of ways, the existence of some of which is already a reality. One way is to take a philosophical attitude implied by an integrated body of techniques or models (commonly spatially oriented) and demonstrate their analogous application to both human and physical phenomena (Woldenberg and Berry, 1967; Haggett and Chorley, In press). Another way is to assume that the stuff of the physical world with which geographers are concerned are its resources – resources in the widest sense; not just coal and iron, but water, ease of movement, and even available space itself. In one sense the present volume represents both these approaches to integration by its concentration on the physical resource of water in all its spatial and temporal inequalities of occurrence, and by its conceptualization of the many systems subsumed under the hydrological cycle (Kates, 1967). In the development of water as a focus of geographical interest the evolution of a human-oriented physical geography and an environmentally sensitive human geography closely related to resource management is well under way.

2. Water as a focus of geographical interest

Water, Earth, and Man, both in organization and content, reflects the foregoing attitudes by illustrating the advantages inherent in adopting a unified view of the earth and social sciences. The theme of this book is that the study of water provides a logical link between an understanding of physical and social environments. Each chapter develops this theme by proceeding from the many aspects of water occurrence to a deeper understanding of natural environments and their fusion with the activities of man in society. In this way water is viewed as a highly variable and mobile resource in the widest sense. Not only is it a commodity which is directly used by man but it is often the mainspring for extensive economic development, commonly an essential element in man's aesthetic experience, and always a major formative factor of the physical and biological environment which provides the stage for his activities. The reader of this volume is thus confronted by one of the great systems of the natural world, the hydrologic cycle, following water through its myriad paths and assessing its impact on earth and man. The hydrologic cycle is a great natural system, but it should become apparent that it is increasingly a technological and social system as well. It has been estimated that 10% of the national wealth of the United States is found in capital structures designed to alter the hydrologic cycle: to collect, divert, and store about a quarter of the available surface water, distribute it where needed, cleanse it, carry it away, and return it to the natural system. The technical structures are omnipresent: dams, reservoirs, aqueducts, canals, tanks, and sewers, and they become increasingly sophisticated in the form of reclamation plants, cooling towers, or nuclear desalinization plants. The social and political system is also pervasive and equally complex, when one reflects on the number of major decision makers involved in the allocation and use of the water resources. White has estimated that for the United States the major decision makers involved in the allocation and use of water include at least 3,700,000 farmers, and the managers of 8,700 irrigation districts, 8,400 drainage districts, 1,600 hydroelectric power plants, 18,100 municipal water-supply systems, 7,700 industrial water-supply systems, 11,400 municipal sewer systems, and 6,600 industrial-waste disposal systems.

This coming together of natural potential and of human need and aspiration provides a unique focus for geographic study. In no other major area of geographic concern has there been such a coalescence of physical and human geography, nor has there developed a dialogue comparable to that which exists between geographers and the many disciplines interested in water. How these events developed is somewhat speculative. First, there is the hydrologic cycle itself, a natural manifestation of great pervasiveness, power, and beauty, that transcends man's territorial and intellectual boundaries. Equally important is that in the human use of water there is clear acknowledgement of man's dependence on environment. This theme, developed by many great teachers and scholars, (e.g. Ackerman, Barrows, Brunhes, Davis, Gilbert, Lewis, Lvovich, Marts, Powell, Thornthwaite, Tricart, and White), is still an important geographic concern, despite the counter trends previously described. Finally, there is no gainsaying the universal appeal of water itself, arising partly from necessity, but also from myth, symbol, and even primitive instinct.

The emergence of water as a field of study has been paralleled in other fields. In the application of this knowledge to water-resource development, a growing consensus emerges as to what constitutes a proper assessment of such development: the estimation of physical potential, the determination of technical and economic feasibility, and the evaluation of social desirability. For each of these there exists a body of standard techniques, new methods of analysis still undergoing development, and a roster of difficult and unsolved problems. Geographers have made varying contributions to these questions, and White reviewed them in 1963. Five years later, what appear to be the major geographic concerns in each area?

Under the heading of resource estimates, White cites two types of estimates of physical potential with particular geographic significance. The first is 'the generalized knowledge of distributions of major resources . . . directly relevant to engineering or social design'. While specific detailed work, he suggests, may be in the province of the pedologist, geologist, or hydrologist, there is urgent need for integrative measures of land and water potential capable of being applied broadly over large areas. The need for such measures has not diminished, but rather would seem enhanced by developments in aerial and satellite reconnaissance that provide new tools of observation, and by the widespread use of computers that provide new capability for data storage and analysis. In the developing world the need is for low-cost appraisal specific to region or project.

A second sort of estimate of potential that calls upon the skills of both the physical and human geographer is to illuminate what White calls 'the problem of the contrast between perception of environment by scientists . . . (and) others who make practical decisions in managing resources of land and water'. These studies of environmental perception have grown rapidly in number, method, and content. They suggest generally that the ways in which water and land resources receive technical appraisal rarely coincide with the appraisals of resource users. This contrast in perception is reflected in turn by the divergence between the planners' or technicians' expectation for development and the actual course of development. There are many concrete examples: the increase in flood damages despite flood-control investment, the almost universal lag in the use of available irrigation water, the widespread rejection of methods of soil conservation and erosion control, and the waves of invasion and retreat into the margins of the arid lands. Thus a geography that seeks to characterize environment as its inhabitants see it provides valued insight for the understanding of resource use.

In 1963 White differentiated between studies of the technology of water management and studies of economic efficiency. Today one can suggest that, increasingly, technical and economic feasibility are seen as related questions. The distinction between these areas, one seen as the province of the engineer and hydrologist, the other as belonging to the economist and economic geographer, is disappearing, encouraged by the impressive results of programmes of collaborative teaching and research between engineering and economics (e.g. at Stanford and Harvard Universities). In this view, the choice of technology and of scale is seen as a problem of cost. The choice of dam site, construction material, and height depends on a comparison of the incremental costs and of the incremental benefits arising from a range of sites, materials, and heights. This decision can be simultaneously related through systems analysis to the potential outputs of the water-resource system.

The methodology for making such determinations has probably outrun our understandings of the actual relationships. The costs and benefits of certain technologies are not always apparent, nor are all the technologies yet known. Geographic research on a broadened range of resource use and specific inquiry into the spatial and ecological linkages (with ensuing costs) of various technologies appears to be required. Indeed, as the new technologies of weather forecasting and modification, desalinization, and cross-basin transport of water and power expand, the need for such study takes on a special urgency.

Finally, there appears to be a growing recognition that much of what may be socially important in assessing the desirability of water-resource development will escape our present techniques of feasibility analysis for much time to come. The need for a wider basis of choice to account for the social desirability of water-resource development persists and deepens as the number of water-related values increase and the means for achieving them multiply. A framework for assessing social desirability still needs devising, but it could be hastened by careful assessment of what actually follows water-resource development. There is much to be learned from the extensive developments planned or already constructed. However, studies such as Wolman's [1967] attempt to measure the impact of dam construction on downstream river morphology or the concerted effort to assess the biological and social changes induced by the man-made lakes in Africa are few and far between. Studies built on the tradition of geographic field research but employing a rigorous research design over an extended period of observation are required. Geographers, freed from the traditional distinction between human and physical geography and with their special sensitivity towards water, earth, and man, have in these both opportunity and challenge.

REFERENCES

ACKERMAN, E. A. [1965], The general relation of technology change to efficiency in water development and water management; In Burton I. and Kates, R., Editors, *Readings in Resource Management and Conservation* (Chicago), pp. 450–67.

BURTON, I. [1963], The quantitative revolution and theoretical geography; *The Canadian Geographer*, **7**, 151–62.

BURTON, I. and KATES, R. [1964], The perception of natural hazards in resource management; *Natural Resources Journal*, **3**, 412–41.

CHORLEY, R. J. [1965a], The application of quantitative methods to geomorphology; In Chorley, R. J. and Haggett, P., Editors, *Frontiers in Geographical Teaching* (Methuen, London), pp. 147–63.

CHORLEY, R. J. [1965b], A re-evaluation of the geomorphic system of W. M. Davis; In Chorley, R. J. and Haggett, P., Editors, *Frontiers in Geographical Teaching* (Methuen, London), pp. 21–38.

CURRY, L. [1952], Climate and economic life: A new approach with examples from the United States; *Geographical Review*, **42**, 367–83.

HAGGETT, P. [1965], *Locational Analysis in Human Geography* (Arnold, London), 339 p.

HAGGETT, P. and CHORLEY, R. J. [1969], *Network Models in Geography* (Arnold, London).

HEWES, L. [1965], Causes of wheat failure in the dry farming region, Central Great Plains, 1939–57; *Economic Geography*, **41**, 313–30.

HUFSCHMIDT, M. [1965], The methodology of water-resource system design; In Burton, I. and Kates, R., Editors, *Readings in Resource Management and Conservation* (Chicago), pp. 558–70.

KATES, R. W. [1967], Links between Physical and Human geography; In *Introductory Geography: Viewpoints and Themes* (Washington), pp. 23–31.

LEIGHLY, J. [1955], What has happened to physical geography?; *Annals of the Association of American Geographers*, **45**, 309–18.

SEWELL, W. R. D., Editor [1966], Human Dimensions of Weather Modification; *University of Chicago, Department of Geography, Research Paper* **105**, 423 p.

SEWELL, W. R. D., KATES, R. W., and PHILLIPS, L. E. [1968], Human response to weather and climate; *Geographical Review*, **58**, 262–80.

STODDART, D. R. [1967], Growth and structure of geography; *Transactions of the Institute of British Geographers*, No. 41, 1–19.

WHITE, G. F. [1963], Contribution of geographical analysis to river basin development; *Geographical Journal*, **129**, 412–36.

WHITE, G. F. [1968], *Strategies of American Water Management* (Ann Arbor).

WOLDENBERG, M. J. and BERRY, B. J. L. [1967], Rivers and central places: Analogous systems?; *Journal of Regional Science*, **7** (2), 129–39.

WOLMAN, M. G. [1967], Two problems involving river channel changes and background observations; In Garrison, W. L. and Marble, D. F., Editors, *Quantitative Geography: Part II Physical and Cartographic Topics* (Northwestern University), pp. 67–107.

WOOLDRIDGE, S. W. and EAST, W. G. [1951], *The Spirit and Purpose of Geography* (Hutchinson, London), 176 p.

CHAPTER I

The World

I. The World Hydrological Cycle

R. G. BARRY

Department of Geography, University of Colorado

1. Global water and the components of the hydrological cycle

We begin our consideration of water in the global context with some figures to illustrate the storage capacity of the earth–atmosphere system. The oceans, with a mean depth of 3·8 km and covering 71% of the earth's surface, hold 97% of *all* the earth's water ($1·31 \times 10^{24}$ cm³). 75% of the total *fresh* water is locked up in glaciers and ice sheets, while almost all of the remainder is ground water. It is an astonishing fact that at any instant rivers and lakes hold only 0·33% of all fresh water and the atmosphere a mere 0·035% (about 12×10^{18} cm³).

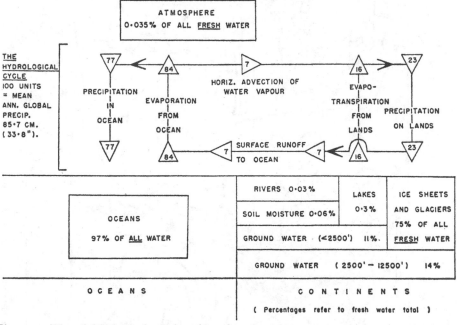

Fig. 1.1.1 The global hydrological cycle and water storage.
The exchanges in the cycle are referred to 100 units which equal the mean annual global precipitation of 85·7 cm (33·8 in.). The storage figures for atmospheric and continental water are percentages of all *fresh* water. The saline oceans make up 97% of *all* water (From More, 1967).

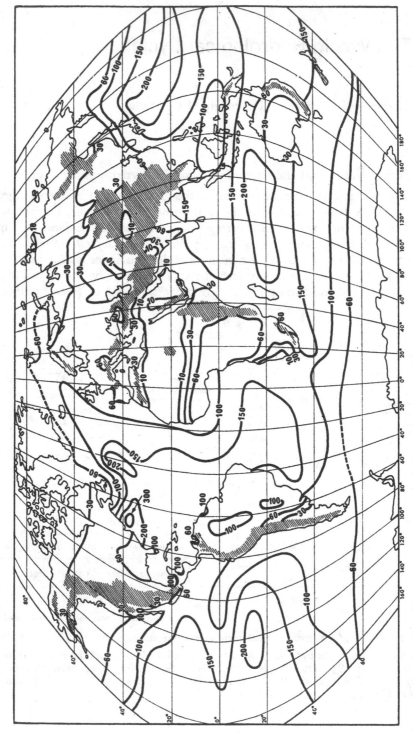

Fig. 1.1.2 Annual evaporation in cm (After Budyko *et al.*, 1962).

In hydrological studies the primary focus of interest is the transfer of water *between* these stores (fig. 1.1.1). The exchanges of water involved in the various stages of the hydrological cycle are evaporation, moisture transport, condensation, precipitation, and run-off. The global characteristics of these components will now be examined to provide a framework for the discussion in subsequent chapters.

2. Evaporation

Evaporation (including transpiration) provides the moisture input into the atmospheric part of the hydrological cycle and may be taken as our starting-point. The oceans provide 84% of the annual total and the continents 16%. Figure 1.1.2 shows the general pattern, although the magnitudes are only to be regarded as approximate in view of our present limited knowledge concerning evaporation. The highest annual losses, exceeding 200 cm, occur in the sub-tropics of the western North Atlantic and North Pacific, where evaporation over the respective Gulf Stream and Kuro Shio Currents is very pronounced in winter, and in the trade-wind zones of the southern oceans. The land maximum occurs primarily in equatorial regions in response to high solar radiation receipts and the growth of luxuriant vegetation. It is noticeable that amounts over land are two–three times less than over the oceans in equivalent latitudes. The factors which determine evaporation rates are discussed fully in Chapter 4.1.

3. Atmospheric moisture

The atmospheric moisture content, comprising water vapour and water droplets and ice crystals in clouds, is determined by local evaporation, air temperature, and the horizontal atmospheric transport of moisture. The cloud water may be ignored on a global scale, since it amounts to only 4% of atmospheric moisture.

Air temperature sets an upper limit to water-vapour pressure – the saturation value (i.e. 100% relative humidity) – consequently we may expect the distribution of mean vapour content to reflect this control (fig. 1.1.3). In January minimum values of 0·1–0·2 cm (equivalent depth of water) occur in continental interiors and high latitudes, with secondary minima of 0·5–1 cm in tropical desert areas. Maximum vapour contents of 5–6 cm are over southern Asia during the summer monsoon and over equatorial latitudes of Africa and South America.

The average water content of the atmosphere is about 2·5 cm (1 in.), which is sufficient only for some ten days' supply of rainfall over the earth as a whole. Clearly, a frequent and intensive turnover of moisture through evaporation, condensation, and precipitation must occur. While atmospheric moisture is essential for precipitation, the relationship between these two items is determined by the efficiency of rain-producing weather systems (Chapter 3.1.2) in any particular climatic region. For example, observations show that on average only 5% of the water vapour crossing Illinois is precipitated there, and in the case of the Mississippi basin only about 20%.

B

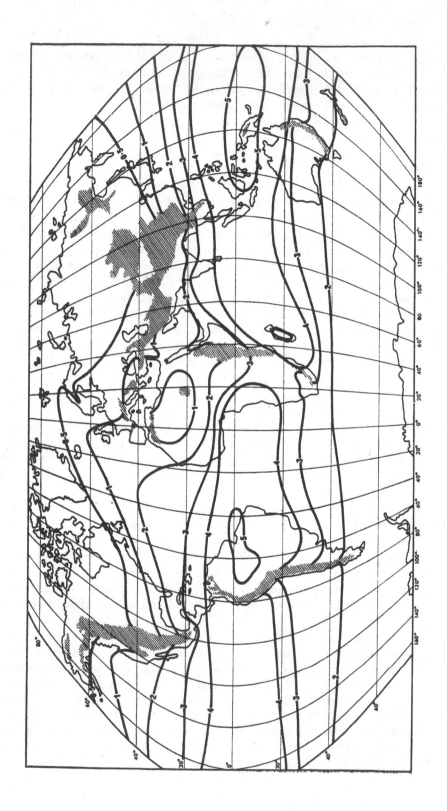

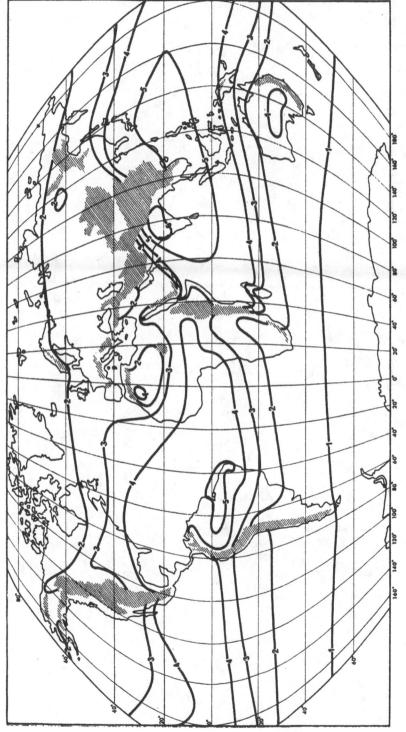

Fig. 1.1.3 Mean atmospheric water vapour content in January (*above*) and July (*below*) 1951–5, in cm of precipitable water (After Bannon and Steele, 1960). (Crown Copyright Reserved).

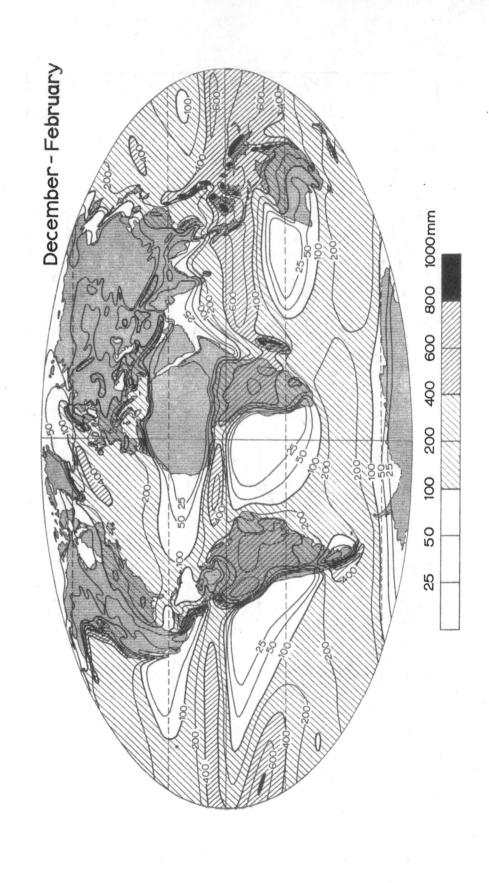

December ~ February

25 50 100 200 400 600 800 1000mm

June – August

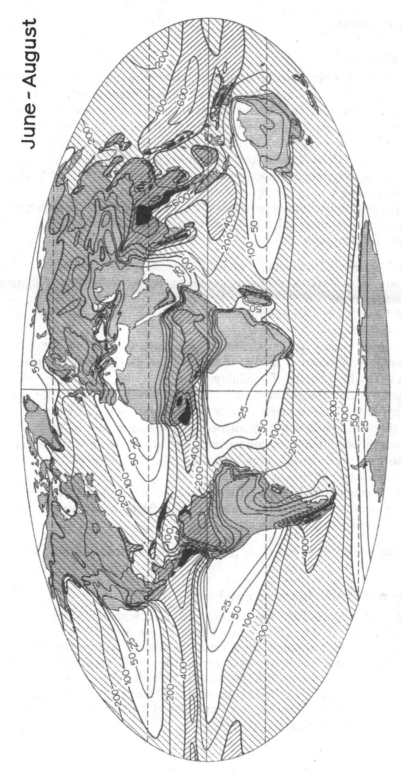

Fig. 1.1.4 Global precipitation for December–February (*above*) and June–August (*below*) in mm (From Möller, 1951).

4. Precipitation

The major types of precipitation are drizzle, rain, snow, and hail, although dew, fog drip, hoar frost, and rime may also make significant contributions to the total (see Chapter 3.1). The distribution of summer and winter precipitation is shown in fig. 1.1.4. The least reliable parts of the maps are the oceans, especially in the southern hemisphere. Analysis of precipitation frequency at North Atlantic weather ships suggests that previous estimates of annual totals in the north-western sector are 20–50 cm too low, while around 45° N, 15° W they are 40–50 cm too high.

The patterns reflect many complex weather factors and geographical influences, such as topography and the land–sea distribution, but the most significant features are:

1. The 'equatorial' maximum, which is displaced into the northern hemisphere. This is related primarily to the converging trade-wind systems and monsoon regimes of the summer hemisphere, particularly in southern Asia and West Africa. Annual totals over large areas are of the order of 200–250 cm (80–100 in.) or more.
2. The west coast maxima of middle latitudes associated with the belt of travelling disturbances in the westerlies. The precipitation in these areas has a very high degree of reliability.
3. The dry areas of the subtropical high-pressure cells, which include not only many of the world's major deserts but also vast oceanic expanses. In the northern hemisphere the remoteness of the continental interiors extends these dry conditions into middle latitudes. In addition to very low average annual totals, often less than 15 cm (6 in.), these regions are subject to considerable year-to-year variability.
4. Low precipitation in high latitudes and in winter over the continental interiors of the northern hemisphere. This reflects the low vapour contents of the extremely cold air.

Types of seasonal regime and other precipitation characteristics are examined in Chapter 3.1.

5. Water circulation in the atmosphere

The previous sections have been concerned only with the static aspects of moisture transfer at the surface and storage in the air, but the atmospheric transport of moisture is an important aspect of climatic differentiation over the earth. Comparison of annual average precipitation and evaporation totals for latitude zones shows that in low and middle latitudes $P > E$, whereas in the subtropics $P < E$ (fig. 1.1.5). These regional imbalances are maintained by net moisture transport into (convergence) and out of (divergence) the respective zones (ΔD, where divergence is positive).

$$E - P = \Delta D \qquad (1)$$

For the year 1949 Benton and Estoque investigated the moisture inflow or outflow over the coasts of North America (Table 1.1.1). The major inflow in winter is across the Gulf Coast, while in summer and especially in autumn the Pacific Coast is more important. High evaporation rates allow a net *export* of moisture from the continent in summer, and a similar result has been obtained by other workers for eastern Asia. These findings have necessitated the revision of earlier views about the role of oceanic moisture sources in summer in both areas. The total annual 'balance' in Table 1.1.1 represents the amount of water which must be discharged by the rivers. Rasmusson [1967] has recently carried out a more detailed study for vapour transport over North America during 1961–3 which essentially confirms the previous results.

The spatial distribution of the horizontal convergence and divergence of water-vapour flux is less reliably known, although maps have been prepared for the

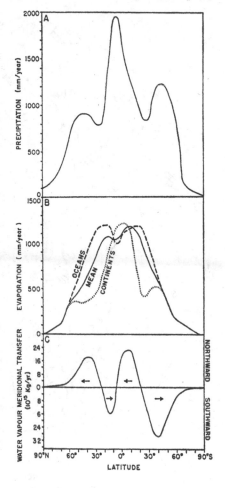

Fig. 1.1.5 Mean precipitation and evaporation for latitudinal zones and meridional transport of water vapour (After Sellers, 1965).

TABLE 1.1.1 Moisture inflow across the coasts of North America during 1949 (after Benton and Estoque, 1954)

	Winter	Spring	Summer	Autumn	Year
			Units: 10^6 *kg/sec*		
Gulf Coast	244	167	168	84	157
Pacific Coast	190	181	197	311	220
Labrador Coast	—78	—64	—167	—148	—114
Atlantic Coast	—249	—248	—307	—190	—248
All coastal sections*	206	83	—79	79	72

* This table omits the smaller fluxes across the Arctic and Alaskan coasts and the south-western border of the United States.

A minus sign indicates outflow.

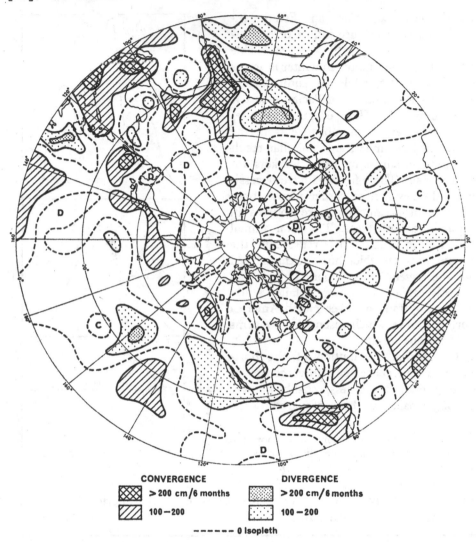

CONVERGENCE ☒ >200 cm/6 months ▨ 100–200

DIVERGENCE ⊡ >200 cm/6months ⊡ 100–200

------ 0 Isopleth

Fig. 1.1.6 The horizontal divergence of water-vapour transport for the six summer months, 1958 (After Peixoto and Crisi, 1965).

Divergence is positive, convergence negative. The units are cm of water/6 months.

northern hemisphere for 1950 and 1958, as well as for continental and other areas. Figure 1.1.6 shows a map for the six summer months of 1958 prepared by Peixoto and Crisi. In studying such maps it is essential to remember that only the *balance* of precipitation and evaporation is shown. Prominent features are the divergent zones, i.e. $P - E < 0$, of the oceanic subtropical high-pressure systems and the convergent areas of monsoon regime in India and Malaysia. The large divergence, implying a moisture source, east of the Persian Gulf is probably a spurious result due to the shortcomings in the available data.

6. Water circulation in the lithosphere

After precipitation has reached the ground it is distributed in three ways; some is re-evaporated, some runs off to be discharged into the oceans, and the remainder percolates into the soil. The runoff and percolation stages may be delayed for days or even months where the precipitation falls as snow. According to the time of year, snow accumulation occurs on 14–24% of the earth's surface (⅔ on land, ⅓ on sea-ice). The global detention of water on land (i.e. snow accumulation, stream runoff, and soil water) reaches a maximum in March–April, when there is extensive snow-cover in the northern hemisphere and lakes, rivers and the soil are frozen over vast areas of Eurasia and North America.

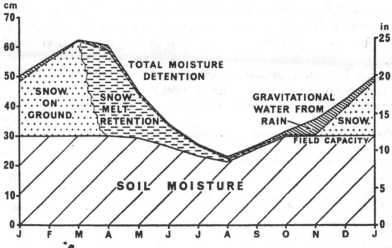

Fig. 1.1.7 The four components of total moisture at Sapporo, Japan (43° N.) (Based on data in Hylckama, 1956).

Figure 1.1.7 illustrates the annual march of water detention at Sapporo, Japan. In the tropics there is a late-summer maximum resulting from the summer rainfall and especially the monsoons.

The seasonal variation in global detention is matched by an inverse pattern of storage in the oceans. In October the seas are estimated to hold $7 \cdot 5 \times 10^{18}$ cm³ more water than in March, although this is equivalent to a sea-level change of only 1–2 cm.

In the above situation the residence time of water on the land is comparatively short, approximately $10^0–10^2$ days. In the case of glaciers and ice-caps, however, the storage time is of the order of years. In the extreme case of central Antarctica it is estimated by Shumskiy and his co-workers [1964] that the maximum storage time of ice is about 200,000 years.

Ground water is a similarly stable component of the hydrological cycle. Most ground water represents precipitation which has percolated through the soil layers into the zone of saturation, where all interstices are water-filled. Water of this origin is termed *meteoric*. Minor sources of water are located in the earth's crust. They are *connate* water, representing water trapped during the formation of sedimentary rocks, and *juvenile* water. The latter, earlier considered to be reaching the surface for the first time in connection with volcanic activity, is now thought to be mainly connate. In many arid areas with internal drainage the major source of ground water is seepage from stream runoff and lakes. Near the water-table the cycling-time of water is a year or less, but in deep aquifers it is of the order of thousands of years (see Chapter 1.11). A similar storage-time applies in the case of ground ice in the permafrost regions. In addition, there are deep-seated brines (connate water), which are effectively isolated from the hydrological cycle, where any circulation has a time-scale of geological epochs.

Ground water contributes on average approximately 30% of total runoff, although within different geographical zones this proportion varies considerably. Some illustrations based on calculations by L'vovich for river basins in the

TABLE I.I.2 Run-off relationships for selected river basins in different geographical zones of the U.S.S.R. (after L'vovich, 1961)

Zone	Precipitation (cm)	Evaporation (cm)	Runoff / Precipitation (%)	Surface Runoff / Total Runoff (%)
Tundra	45	11	76	97
Taiga with permafrost	40	21	49	95
Taiga	50	16	69	73
Mixed forest	58	37	36	76
Mixed forest	61	43	30	60
Forest steppe	41	33	19	79
Steppe	35	32	9	87
Semi-desert	20	19	5	90

U.S.S.R. are shown in Table 1.1.2. The proportion of precipitation going into runoff decreases as the heat available for evaporation increases. The effect of frozen ground on ground-water discharge is pronounced in the tundra and Siberian sections of the taiga. In the drier areas precipitation is mainly evaporated after moistening the soil, and consequently the contribution of ground water to streamflow is negligible. The volume of surface runoff going into streamflow reaches a maximum in the tundra and taiga zones.

Average runoff from the land masses can be estimated from precipitation and evaporation data if allowance is made for areas of inland drainage. Such areas, predominantly in Asia, Africa, and Australia, account for 25% of stream runoff. For all continents the average annual runoff is 26·7 cm (10·5 in.), although amounts far exceed this in South America and the Malayan Archipelago. The latter region provides 12% of the total runoff from only 2% of the land surface

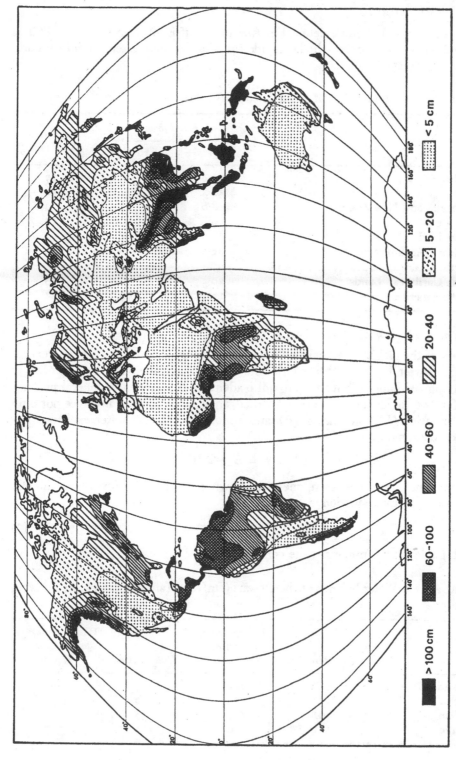

Fig. 1.1.8 Annual runoff in cm. Information is not available for areas left blank (After L'vovich, 1964).

(see Table 1.1.3). A more detailed picture of runoff is given in fig. 1.1.8. This is based on Soviet sources, which provide some of the most extensive information on the subject.

TABLE 1.1.3 Annual Runoff (after L'vovich)

	Land area		% of total global runoff
	$10^6 \ km^2$	$10^6 \ ml^2$	
Africa	29·8	11·5	17
Asia	42·2	16·3	20
Australia and New Zealand	8·0	3·1	2
Europe	9·6	3·7	7
Greenland	3·8	1·5	2
Malayan Archipelago	2·6	1·0	12
North America, Central America, and the Caribbean Lands	20·5	7·9	18
South America	17·9	6·9	22

7. The global water balance

Study of the global water balance has been made by Budyko and his co-workers in the Soviet Union (Table 1.1.4). The figures must obviously be regarded as first approximations, but they are quite adequate to convey the general picture. As far as individual oceans are concerned, their water balance involves not only precipitation (P), evaporation (E), and runoff (r) but also water exchange between oceans (ΔW).

$$P + r = E \pm \Delta W \tag{2}$$

The magnitude of these components is shown in Table 1.1.4. For the continents, the water balance equation is

$$P = E + r \tag{3}$$

and Table 1.1.5 summarizes the computations.

TABLE 1.1.4 Water balance of the Oceans (cm/year) (after Zubenok, in Budyko, 1956)

	Precipitation	Runoff from adjoining land areas	Evaporation	Water exchange with other oceans
Atlantic Ocean	78	20	104	—6
Arctic	24	23	12	35
Indian	101	7	138	—30
Pacific	121	6	114	13

TABLE 1.1.5 Water balance of the continents (cm/year)

	Precipitation	Evaporation	Runoff
Africa	67	51	16
Asia	61	39	22
Australia	47	41	6
Europe	60	36	24
North America	67	40	27
South América	135	86	49

Combining these data shows that for the earth as a whole precipitation and evaporation are of the order of 100 cm. The difference in the runoff figures in Table 1.1.6 is due, of course, to the respective surface areas of the oceans and continents. It is worth noting at this point that for each latitude zone around the globe the net convergence or divergence (equation (1)) is equivalent to the total runoff in that zone.

TABLE 1.1.6 The global water balance (cm/year) (after Budyko *et al.*, 1962)

	Precipitation	Evaporation	Run-off
Oceans	112	125	−13
Continents	72	41	31
Whole Earth	100*	100	0

* The computed value is actually 102 cm.

8. Global water circulation in relation to the energy budget

Water circulation in the atmosphere and oceans is intimately linked with the global energy budget. This is not the place for a detailed account, but it is appropriate to stress the interdependence of the moisture and energy budgets. Fuller details may be found in Sellers [1965, p. 100].

The annual input of solar radiation into the earth–atmosphere system and the net loss of terrestrial radiation produce a positive energy budget in low latitudes and a negative one in middle and higher latitudes. For annual averages the budget is balanced at about 35° latitude. Consequently, poleward heat transport is essential if the higher latitudes are not to become progressively colder and lower latitudes hotter. This transport occurs in three forms – atmospheric transport of sensible heat and of latent heat (i.e. water vapour which subsequently condenses), and the transport of warm water by ocean currents. The role of these three components is shown in fig. 1.1.9. Some 80% of the poleward heat transport takes place in the atmosphere. Sensible heat transport, by warm air masses, need not concern us here, although it accounts for the majority of the heat transport. The transport of latent heat reflects the pattern of the global wind belts on either side of the subtropical high-pressure zones, so that about 10° N and 10° S the flux is equatorward as a result of the transport of moisture into the equatorial low-pressure trough by the trade winds. Ocean currents, predominantly of course the Gulf Stream and Kuro Shio in the northern hemisphere and the Brazil Current and the poleward branches of the Equatorial Currents in the south-west Pacific and south-west Indian Ocean in the southern hemisphere, are most important about 35°–40° latitude.

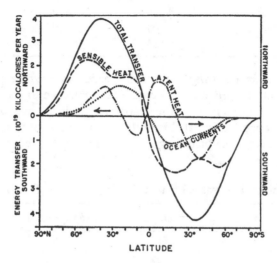

Fig. 1.1.9 Components of the poleward energy flux in the earth–atmosphere system (From Sellers, 1965).

9. Climatic change and the global water budget

The interdependence of the various components of the hydrological cycle makes it apparent that a change in any climatic parameter may have far-reaching repercussions. For instance, a 1% increase in evaporation from the tropical oceans would cool a 200-m layer by 3° C in fifty years (Malkus, 1962). Quantitative assessment of actual changes is obviously uncertain owing to observational limitations, but in the case of sea-level fluctuations, at least, the picture is now reasonably complete for late and post-glacial times.

It is estimated that at the Quaternary glacial maximum – the Illinoian, Riss, or Saale glacial – ice affected an area 3 × that of the present ice cover with 5 × the present mass. As a result of this long-term storage of water on the land, and the consequent diminution of run-off, a eustatic lowering of sea-level took place. Crary estimates the eustatic lowering as follows:

Glacial phase	Eustatic lowering	
Classical Wisconsin	105·5 m	348 ft
Early Wisconsin	114·5	378
Illinoian	137·4	453

Different calculations of ice volume by Novikov indicate a lowering of 159 m during the Illinoian maximum. The actual sea-level change in the glacierized areas was, of course, complicated by the isostatic depression of the continents due to their ice load.

One possible result of a falling sea-level is a decrease in evaporation and precipitation because a greater land area is exposed. Conversely, a sea-level rise, by reducing the land area, may promote warm inter-glacial conditions with higher average evaporation and precipitation. Fairbridge estimates that since the end of the Pliocene period sea-level has fallen by 200 m, and this should favour colder, drier conditions. Indeed, some authorities consider that the Last Glacial phase was the coldest and driest. This would be consistent with the suggested relationship, but the present evidence is too uncertain for confirmation

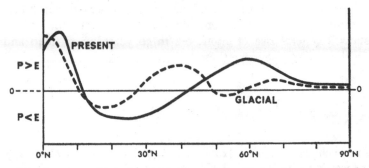

Fig. 1.1.10 The location of humid and arid zones in glacial times and the present day (From Flohn, 1953).

of this hypothesis. There is, in any case, little doubt that more important modifications of the precipitation–evaporation cycle took place due to the global cooling during glacial phases. In lower-middle latitudes evaporation, and consequently precipitation, was probably reduced by 20% at the full glacial stage, creating arid conditions in subtropical areas such as the Mediterranean. However, in the *early* stages of the glacial phases, before the oceans had begun to cool, the subtropics experienced 'pluvial' conditions (Butzer, 1957). In the south-west United States Pleistocene lake levels provide further evidence of subtropical pluvials. It is estimated that in order to maintain lakes at their maximum level, annual precipitation must have been 25 cm (10 in.) greater than now in north-central New Mexico, and 20 cm (8 in.) greater in east-central Nevada, with reduced evaporation due to cooler summers.

A model of the shift of the moisture budget zones under glacial conditions compared with the present has been outlined by Flohn [1953] as shown in fig. 1.1.10. Considerable work is required to refine such concepts and extend their application to the various phases of the glacial–interglacial cycle, and even when this has been satisfactorily achieved it will be necessary to establish appropriate correlations between climatic and hydrological parameters. The runoff during pluvial phases must have increased proportionately more in semi-arid than sub-humid climates (Schumm, 1965), but only tentative estimates of the magnitude of these changes can be made. Indeed, in some cases the sign of the change is still a matter of controversy. The implications of such changes are discussed in Chapter 11.II.

As climate started to ameliorate towards the end of the last glacial period the increased runoff from the melting ice-caps caused an eustatic rise in sea-level. This is considered to have begun about 18,000 years ago, when sea-level was some 105–120 m below the present level, and proceeded rapidly until about 6,000 years ago. Authorities disagree as to whether or not the level at that time exceeded present mean sea-level. One view is that it was approximately 5 m higher, while another school of thought considers that the post-glacial rise has been continuous, but diminished markedly over the last few thousand years. Whichever is correct, the relative stability during recent times indicates that the ice-caps of Antarctica and Greenland are more or less in equilibrium with climatic conditions. Direct measurements of sea-level since the end of the last century show a general rise of about 0·2 m/50 years up to 1940 and a 40% decrease in the rate of rise since then.

REFERENCES

BANNON, J. K. and STEELE, L. P. [1960], *Average Water Vapour Content of the Air; Geophysical Memoir No. 102* (H.M.S.O., London), 38 p.

BENTON, G. S. and ESTOQUE, M. A. [1954], Water-vapor transfer over the North American Continent; *Journal of Meteorology*, 11, 462–77.

BUTZER, K. W. [1957], The recent climatic fluctuation in lower latitudes and the general circulation of the Pleistocene; *Geografiska Annaler*, 39, 105–13.

BUDYKO, M. I. [1956], *The Heat Balance of the Earth's Surface* (Leningrad), translation by N. A. Stepanova (Washington), 1958, 259 p.

BUDYKO, M. I., et al. [1962], The heat balance of the surface of the earth; *Soviet Geography*, 3, No. 5, 3–16.

CHOW, VEN TE, Editor [1964] *Handbook of Applied Hydrology* (New York).

DONN, W. L., FARRAND, W. R., and EWING, M. [1962], Pleistocene sea volumes and sea-level lowering; *Journal of Geology*, 70, 206–14.

FAIRBRIDGE, R. W. [1961], Eustatic changes in sea level; *Physics and Chemistry of the Earth*, 4, 99–185.

FLOHN, H. [1953], Studen über die atmosphärische Zirkulation in der letzten Eiszeit; *Erdkunde*, 7, 266–75.

HYLCKAMA, T. E. A. VAN [1956], *The Water Balance of the Earth*, Publications in Climatology, 9, No. 2, Drexel Institute of Technology (Centerton, New Jersey), 117 p.

JELGERSMA, S. [1966], Sea-level changes during the last 10,000 years; In Sawyer, J. S., Editor, *World Climate from 8000 to 0 B.C.*, Royal Meteorological Society (London), pp. 54–69.

L'VOVICH, M. I. [1961], The water balance of the land; *Soviet Geography*, 2, No. 4, 14–28.

L'VOVICH, M. I. [1962], The water balance and its zonal characteristics; *Soviet Geography*, **3**, No. 10, 37–50.

MALKUS, J. S. [1962], Inter-change of properties between sea and air. Large-scale interactions; in Hill, M. N., Editor, *The Sea*, volume 1 (New York), pp. 88–294.

MILLER, D. H. [1965], The heat and water budget of the earth's surface; *Advances in Geophysics*, **11**, 175–302.

MÖLLER, F. [1951], Vierteljahrskarten des Niederschlags für die ganze Erde; *Petermann's Geographische Mitteilungen*, **95**, 1–7.

PEIXOTO, J. P. and CRISI, A. E. [1965], *Hemispheric Humidity Conditions during the IGY*, Scientific Report No. 6, Meteorology Department, Massachusetts Institute of Technology (Cambridge, Mass.).

RASMUSSON, E. M. [1967], Atmospheric water vapor transport and the water balance of North America; *Monthly Weather Review*, **95**, 403–26.

SCHUMM, S. A. [1965], Quaternary palaeohydrology; In Wright, H. E., Jr., and Frey, D. G., Editors, *The Quaternary of the United States* (Princeton, New Jersey), pp. 783–94.

SELLERS, W. D. [1965], *Physical Climatology* (Chicago), 272 p.

SHUMSKIY, P. A., KRENKE, A. N., and ZOTIKOV, I. A. [1964], Ice and its changes; In Odishaw, H., Editor, *Research in Geophysics*, volume 2 (Cambridge, Massachusetts), pp. 425–60.

SUTCLIFFE, R. C. [1956], Water balance and the general circulation of the atmosphere; *Quarterly Journal of the Royal Meteorological Society*, **82**, 385–95.

TUCKER, G. B. [1961], Precipitation over the North Atlantic Ocean; *Quarterly Journal of the Royal Meteorological Society*, **87**, 147–58.

II. World Water Inventory and Control [1]

R. L. NACE

U.S. Geological Survey

The total amount of water in the earth system and its partition and movement among major earth realms have been topics of speculation and investigation during more than a century. Nevertheless, quantitative data are scarce, and the hydrology of the earth as a complete system is still poorly known. Only approximate values can be assigned to most components of the system. Table 1.II.1 is a summary estimate of water in the world exclusive of water of composition and crystallization in rocks and of pore water in sediments beneath the floor of the sea. Most of the water is salty, and much of the fresh water is frozen assets in the cold-storage lockers of Antarctica and Greenland. Most studies of water concern its occurrence and availability in specific areas. However, the global situation has more than intellectual interest, and it is necessary to put local conditions in perspective within that situation.

World globes usually are set up with a prominent land mass to the fore, so the earth looks quite earthy. A more realistic orientation would have Jarvis Island – one of the Line Islands in the Pacific Ocean – in the centre of the field of view. In this perspective the earth is very watery indeed. The preponderance of water area (71% of the earth's surface) and the great extent of the ice-caps are unfortunate in the eyes of people who would prefer more land to accommodate more people (as though the world needed more people!). The effects of reduced ocean area or of melted ice-caps, however, would be far-reaching and generally unfavourable to man.

1. Significance of world inventory

In scientific parlance a system is any region in space together with the things and processes that operate in that region. A closed system is self-contained – nothing enters and nothing leaves. In many ways closed systems are preferable for study to open systems, because once the parameters within the system have been identified their interactions can be studied independently of outside phenomena.

No natural hydrological system is closed. Even the global system is open, because radiant solar energy enters, and reflected and re-radiated energy leaves. However, the solar constant is known more accurately than perhaps any other factor in the hydrological cycle except the physical and chemical properties of water itself. Solar energy is the main driving force of the hydrological cycle.

[1] Publication authorized by the Director, U.S. Geological Survey.

TABLE I.II.I World supply and volume of annually cycled water*

Item	Area (km$^2 \times$ 10^{-3})	Volume (km$^3 \times$ 10^{-3})	% of total water
Atmospheric vapour (water equivalent)	510,000 (at sea-level)	13	0·0001
World ocean	362,033	1,350,400	97·6
Water in land areas:	148,067†	(124,000)‡	—
Rivers (average channel storage)	—	1·7	0·0001
Fresh-water lakes	825	125	0·0094
Saline lakes; inland seas	700	105	0·0076
Soil moisture; vadose water	131,000	150	0·0108
Biological water	131,000	(Negligible)	—
Ground water	131,000	7,000	0·5060
Ice-caps and glaciers	17,000	26,000	1·9250
Total in land areas (rounded)		33,900	2·4590
Total water, all realms (rounded)		1,384,000	100
Cyclic water:			
Annual evaporation –§			
From world ocean		445	0·0320
From land areas		71	0·0050
Total		516	0·0370
Annual precipitation –			
On world ocean		412	0·0291
On land areas		104	0·0075
Total		516	0·0370
Annual outflow from land to sea –			
River outflow		29·5	0·0021
Calving, melting, and deflation from ice-caps		2·5	0·0002
Ground-water outflow¶		1·5	0·0001
Total		33·5	0·0024

* Values are approximations, computed on data from many sources which are not mutually consistent. None of the values is precise. Data on evaporation and precipitation modified from L'vovich [1945, p. 54].
  † Total land area, including inland waters.
  ‡ Continental mass above sea-level.
  § Evaporation is a measure of total water participating annually in the water cycle.
  ¶ Arbitrarily set equal to about 5% of runoff.

Ocean basins, the atmosphere, and the outer crust of the earth form a single gigantic plumbing system, all of whose parts communicate directly or indirectly with all other parts. The occurrence and movement of water in one part of the system are related to its occurrence and movement in all other parts. Furthermore, the whole is more than the sum of its parts, because it includes the interactions of the parts. Therefore the complete system requires study and synoptic observation before men can realistically hope to modify climate or increase water supply predictably and safely. Further, man is the only species which is capable of destroying the habitability of the world. Reckless exploitation of the earth has brought many unwanted side effects that are clearly undesirable and often not understood. It behooves us to understand the entire system in which we live. No other is attainable.

2. Oceanic water

Ocean volume is equivalent to oceanic evaporation during about 3,000 years, which might be taken as the average residence time of a water molecule in the ocean. Some molecules reside only for an instant, however, and water in great ocean deeps may be out of the water cycle during many thousands of years.

Oceanic waters comprise only about 0·023% of the earth's total mass. However, the hydrological cycle is largely an external phenomenon in which oceans have the major role. The system ocean–atmosphere–continents is a great heat engine that drives the water cycle, and oceans are the principal heat reservoir.

3. Water aloft

Evaporation from the huge oceanic reservoir is continual, but an average column of atmosphere contains vapour equivalent only to about 25 mm of liquid water. Estimates of annual precipitation averaged for the whole earth range from 700 to more than 1,000 mm. About 1,000 mm is an acceptable value, equivalent to about 2·7 mm da^{-1} if precipitated uniformly in time and space. The apparent average residence time of a water molecule in the atmosphere is about 10 days, and in any case, water in the atmosphere obviously undergoes rapid flux and reflux.

Strong variability of atmospheric vapour is worth noting. Estimates indicate, for example, 0·6–1·5 mm (water equivalent) over Antarctica (Loewe, 1962, p. 5175) and 50–70 mm in typhoon air masses over Japan (Arakawa, 1959). Using a conservatively rounded value of 50 km hr^{-1} for the typhoon wind speed and 60 mm of water, vapour transport over Japan in a belt 1 km wide in this situation would be equivalent to about 800 m^3 s^{-1}. Considering that major moving air masses are hundreds of kilometres in width, it is apparent that unseen rivers aloft are equivalent to great rivers aground, as they must be to maintain the water cycle.

4. Rivers

The Amazon, mightiest of all rivers, was first seen by Europeans in A.D. 1500. Commanding four caravelles under the Spanish flag, Vicente Yañez Pinzon reached the estuary of the Amazon on 7 February. Four hundred and sixty-three years elapsed before the flow of the river was accurately measured. In 1963–4 a joint Brazilian–United States expedition aboard a Brazilian Navy corvette measured the Amazon's discharge three times, once each at high, low, and intermediate stages.

The average of the Amazon flow at Obidos, 800 km above the mouth (the upstream limit of tidal effects), is 157,000 m³ s⁻¹. The computed average flow at the mouth is 175,000 m³ s⁻¹ (Oltman, 1968). The annual volume of discharge is thus a little more than 5,500 km³, or nearly 20% of the total discharge of all rivers of the world and 4½ times the discharge of the Congo, next largest river.

A. Total discharge

Table 1.11.2 summarizes estimated averaged discharge from all the world's rivers into the sea, amounting to 29,500 km³ yearly. L'vovich [1945, p. 54] estimated a larger total discharge of 36,300 km³, including 1,100 km³ of ice discharge, which is estimated separately herein. Other authors have estimated smaller values, but new data, including that from the International Hydrological Decade, permits improved estimation.

TABLE 1.11.2 Summary of river discharges from land areas to the sea*

Land region	Area† (km² × 10⁻³)	Mean discharge (m³ s⁻¹ × 10⁻³)
Europe, including Iceland	7,960	75·0
Asia and East Indies	31,500	226·0
Africa	18,700	105·4
North America	21,400	151·4
South America	17,000	353·0
Australia, including Tasmania and New Zealand	5,380	13·3
Greenland	2,180	12·4
Totals	104,120	936·5
Equivalent annual volume	29, 500 km³	

* Based on data from manuscript (in preparation) by W. H. Durum, R. B. Vice, and S. G. Heidel.

† Includes effective area only; excludes areas that yield no external drainage.

Many hundreds of rivers discharge to the sea, but most of the flow occurs in a few large streams. Sixteen great rivers – those discharging 10,000 m³ s⁻¹ or more – discharge 13,600 km³ annually, or about 45% of the world total. Fifty additional rivers with individual discharges of 500 m³ s⁻¹ or more bring the total to 17,600 km³, or about 60% of all discharge to the sea. Many hundreds of small rivers have not been measured accurately, but individually they contribute little to the total. Aggregate ungauged flow from each continent can be estimated on the basis of climate, topography, vegetation, river characteristics, and other factors. The total discharge estimated in the table is probably within 10% of the true value.

B. Channel storage

Engineers regularly calculate channel storage and storage changes for many river segments as an aid to water management. Although these calculations represent only a small sample of the millions of miles of river channels around the world, they suffice to show that total channel storage, important though it may be locally, is insignificant on the world scale. L'vovich's classical study included an approximate calculation of the isochronal volume of water in river channels. He derived a value of $1,200$ km^3.

The Amazon channel below Obidos illustrates the relative insignificance of channel storage. Assuming a channel length of 800 km, an average width of 2,500 m, and an average depth of 60 m, channel storage would be about 120 km^3, or enough to maintain the flow at the mouth during about eight days.

At median flow the width of the Amazon at Obidos is 2,260 m, and the mean depth is about 46 m (Oltman, 1968). Channel configuration above Obidos is virtually unknown, but it might be assumed that the average cross-section is midway between the dimensions at Obidos and zero at the Peruvian border, 3,000 km distant. If so, upstream storage would be about 160 km^3, which brings total storage to 280 km^3. This is enough to maintain flow at the mouth during about nineteen days.

Among the hundreds of tributary channels in the vast Amazon river basin, none approach the main stream in capacity. Total storage of 1,000 km^3 in the basin might be assumed, equivalent to flow at the mouth during about two months. This is probably excessive.

The Amazon basin covers about 7% of the non-arid, non-glaciated, run-off-producing land area of the world, but other regions do not yield water in the same proportion to area. The Congo basin, equal to 70% of the area of the Amazon, yields only a fourth as much runoff. The Mississippi river basin, half the size of the Amazon basin, yields only a tenth as much runoff. The average daily discharge of all rivers of the world is 80 km^3. Inasmuch as 90% of the run-off-producing area is much less copiously supplied with water, and hence with much less channel storage than the Amazon, it may be assumed that total channel storage is 1,700 km^3, equivalent to total flow during about three weeks. This value is only a guess, but it illustrates that isochronal channel storage is a vanishingly small percentage of water in the hydrological cycle. At a slow velocity of $1 \cdot 5$ m s^{-1}, in twenty days water could flow from head to mouth of a river 2,600 km in length, so the guess seems reasonable.

The sustained flow of rivers is truly remarkable, considering that precipitation is an unusual event in most areas of the earth. Localization of precipitation in time and space is striking. At Paris, France, a reasonably typical temperate-zone location, total duration of precipitation during a forty-five-year period averaged 577 hours per year (1 year = 8,766 hours), or about 7% of the time (Péguy, 1961, p. 189). Few storms last more than a few hours, so even storm days are mainly rainless. Yet rivers flow throughout the year. The sustaining source of flow is effluent ground water, about which more will be said later.

5. Wide places in rivers

Lakes have been called wide places in rivers. This is true of many small lakes that are impounded by relatively minor and geologically temporary obstructions across river channels. But no single over-simplified metaphor accurately describes all lakes, which vary widely in their physical characteristics and in the geological circumstances under which they occur. The handsome little tarn occupying an ice-scooped basin in the glaciated Alps differs radically from the deep and limpid Crater Lake of Oregon, occupying the collapsed crater of an extinct volcano. The North American Great Lakes occupy huge basins formed in a complex manner by isostatic subsidence of that whole region of the earth's crust, glacial excavation, moraine and outwash deposition, and other factors. These lakes have no resemblance to Lake Tanganyika in the great Rift Valley of Africa, where geological processes created the rift by literally pulling two sections of the earth's crust apart, opening a deep gash, part of which is occupied by the lake. The world contains many other spectacular examples of genetically different lakes. |

The earth's land areas are dotted with hundreds of thousands of lakes. Areas like Wisconsin–Minnesota and Finland each contain some tens of thousands. But these small lakes, important though they may be locally, contain only a minor amount of the world supply of fresh surface water, most of which is in a relatively few large lakes on three continents.

The aggregate volume of all fresh-water lakes in the world is about 125,000 km^3, and their surface area is about 825,000 km^2. About 80% of the water is in forty large lakes. For the purpose of this chapter a lake is called large if its contents are 10 km^3 or more. Thus the group excludes water bodies such as the Zürichsee of Switzerland (about 4 km^3). The range of volume among the 'large' lakes is enormous, from the lower limit of about 10 km^3 in Lake Okeechobee (Florida) to an upper limit of nearly 22,000 km^3 in Lake Baikal (central Asia), the world's bulkiest and deepest single body of fresh water. The latter contains nearly as much water as the five Great Lakes of North America. The latter are large in surface area, but their average depth is very much less than that of Baikal.

The Great Lakes and other large lakes in North America contain about 32,000 km^3 of water, which is a fourth of all the liquid fresh surface water in existence. Large lakes of Africa contain 36,000 km^3, or nearly 30% of the earth's total. Asia's Lake Baikal alone contains about 18% of the total.

Lakes on these three continents account for more than 70% of the world's fresh surface water. Large lakes on other continents – Europe, South America, and Australia – contain a comparatively small amount, about 3,000 km^3, or roughly 2% of the total. About a fourth of the total fresh surface-water supply occurs in the hundreds of thousands of rivers and lesser lakes throughout the world.

6. Inland seas

Inland seas and saline lakes of the world are equivalent in volume to fresh-water lakes. Their aggregate area is 700,000 km², and their volume is about 105,000 km³. The distribution, however, is quite different from that of fresh-water lakes. About 80,000 k³ (76% of the total saline volume) is in the Caspian Sea, and most of the remainder is in lesser saline lakes of Asia. North America's shallow Great Salt Lake is relatively insignificant, with less than 30 km³.

7. Life fluid of the vegetable kingdom

Aside from plants that grow directly in water or marshy ground, the great mass of useful vegetation lives on so-called dry land. But the dryness is only relative, and even dust may contain a few per cent of water by weight. Soil holds small amounts of water so tenaciously that plant roots cannot extract it. Desert plants are adapted to low water supply, but most vegetation flourishes only where the soil contains extractable moisture most of the time. A quite ordinary tree may extract and transpire 200 litres of water per day. Frequent replenishment of soil moisture, therefore, is essential for vegetation.

It may be assumed for illustration that the soil zone, except in arid and semi-arid climates, contains on the average about 10% by volume of water within a depth of 2 m beneath the land surface, or about 20 cm of water. In 82×10^6 km² of land area that is not arid to semi-arid and not under permanent ice the estimated total of soil water is about 16,500 km³. Some soil moisture is present in dry areas, and the value actually is larger. From data in an exhaustive study by Van Hylckama [1956], it may be calculated that the average of soil-moisture values for all land areas totals about 25,400 km³. This includes unmelted snow, but snow should be included in the water budget. Therefore Van Hylckama's data are acceptable. The amount of soil water is about fifteen times the amount in channel storage in rivers.

A zone of non-saturation generally occurs between the soil belt and the underlying water table. By non-saturated flow, water percolates below the root zone and migrates towards the water-table. The amount of migrating water is appreciable even in areas of igneous, metamorphic, and other crystalline rocks. The average thickness of this zone of vadose water is unknown, because average depth to the zone of saturation is indeterminate.

Again, crude estimates indicate the order of magnitude of the water volume. Assuming an average thickness of 15 m for the vadose zone, and assuming that it is perennially wetted at an average field capacity of 20%, the moisture content would be 246,000 km³ in humid to sub-humid areas not covered by ice. This estimate probably is excessive because of the vast areas of crystalline rocks in which the moisture content is much smaller, perhaps on the order of 3–5% as a maximum and much less on the average. It seems reasonable to halve the estimate to obtain an order-of-magnitude value. The assumed rounded value for soil moisture and vadose water together is 150,000 km³.

8. Biological water

Animal bodies are largely water, and plant tissue also contains much water. Plants have an important role in the water cycle through the process of transpiration. In an average temperate area having annual precipitation of 750 mm about 70% may be dissipated by evaporation and transpiration. The two processes cannot be measured separately, and their proportions vary widely in different environments.

Van Hylckama [1956] estimated that water seasonally (ephemerally) stored in vegetation is about 5.3×10^{15} cm^3 [= 5.3 km^3]. This does not include water perennially stored in vegetation nor total annual circulation of water through vegetation. Lotka [1956, p. 217] estimated that about 120 mi^3 [500 km^3] of water annually passes through the organic cycle. According to Furon [1967, p. 10], the amount of water needed annually for photosynthesis is about 65×10^{10} tons [= 650 km^3]. Both of the latter two values seem unreasonably low.

Precipitation on the world land areas is about 107,000 km^3 annually. If only 25% [26,000 km^3] passes through plants the volume is equal to about 90% of the annual discharge of the world's rivers. Evaporation from land areas is about 71,000 km^3. If only a third of this is vegetative transpiration the annual volume is about 24,000 km^3. This is no indication of the amount of water permanently stored in vegetation, which cannot be large.

9. Unseen reservoirs

The pores in granular geological formations, fissures and joints in hard rocks, and solution cavities and channels in limestone are examples of voids within masses of rock and sediment that can store and transmit water. The storage capacities and transmissivities of rock types range enormously. Storage and transmission are virtually nil in massive dense rock. A clean coarse gravel may have porosity of 25–35%, and rates of flow through the gravel may achieve velocities of 15–20 m da^{-1}. On the other hand, a clay bed may have a porosity of 60%, but will hold water so tenaciously that it moves only in response to osmotic pressure or to compressive stress.

The tremendous volume of ground water in storage is comprehended by relatively few people. Many believe that water occurs underground as lakes and rivers. It is true that limestone caverns contain pools and rivers, but these are exceptional. In general, ground water permeates and fills the pores of underground masses of rock and sediment quite like those we see at the land surface. On the average, ground-water movement is so slow that it would be imperceptible even if the water were visible. The storage capacity of aquifers will be illustrated with a few examples in Chapter 6.III.

Imbeaux [1930, p. 37] cited various estimates of ground-water volume made during the previous hundred years. The smallest of these was 15 km^3 and the largest was 1,175. Even the larger value is absurdly small, because a single major aquifer may contain hundreds of cubic kilometres of water. It is not possible to estimate accurately the total amount of fresh ground water. In many areas deep water is saline, and few regional determinations of the lower limit of fresh water have been made. An order-of-magnitude estimate, however, is permissible and necessary.

Porosity ranges from a small fraction of 1% in massive dense rock to perhaps 35% in highly permeable sediment. With an average effective porosity of only 1%, the upper 1,000 m of the world's land areas exclusive of ice-cap areas (131×10^6 km^2) could contain $1\cdot31 \times 10^6$ km^3 of water. The actual amount is probably at least five times that value, or nearly 7×10^6 km^3. Much of this water participates in the hydrological cycle, but an undetermined amount is immobilized in the $9\cdot5 \times 10^6$ km^2 of permafrost area. The residence time of water in aquifers ranges from a few minutes or hours to hundreds of years in most aquifers, but the residence time in some aquifers ranges up to tens of thousands of years.

A great deal of water occurs at depths greater than 1,000 m, but much of it is saline and much is so-called fossil water, not participating in the hydrological cycle. However, it seems safe to round off the estimated volume of recoverable fresh ground water to 7×10^6 km^3. Probably an additional equal amount is present but not recoverable for use.

Most circulating ground water enters stream channels within the continents, but some discharges directly into the sea by diffuse percolation and through submarine springs. Submarine springs have been noted in many parts of the world, but few quantitative data on their discharge are available. Owing to low water-table gradients in most coastal areas, the aggregate amount of water thus discharged cannot be a large percentage of cycled water. Crude calculations for the conterminous United States indicate that the amount may be equal to about 5% of streamflow, and this is acceptable as an average value for land areas of the world. The derived total volume is 1,500 km^3 per year.

10. Global refrigeration system

The largest share of the earth's fresh water is locked up in the deep-freeze systems of Antarctica and Greenland. Alpine and valley glaciers and small ice-caps are locally important, containing in the aggregate about 80,000 km^3 of water equivalent, or about the same amount as in large fresh-water lakes.

The main Greenland ice-cap, nearly $1\cdot73 \times 10^6$ km^2 in area and averaging about 1,500 m in thickness, contains $2\cdot6 \times 10^6$ km^3 of ice. Outlying ice masses and glaciers covering 60,000 km^2 increase the total to about $2\cdot65 \times 10^6$ km^3, equivalent to a water volume of $2\cdot3 \times 10^6$ km^3. The Antarctic ice-cap, however, is the largest single item in the world's water budget outside the oceans. The area of the main ice sheet in East Antarctica is about $12\cdot1 \times 10^6$ km^2, its average thickness is 2,200 m, and the ice volume is about $26\cdot6 \times 10^6$ km^3. No good estimate for West Antarctica has been found, but it seems reasonable to round the estimated total to 27×10^6 km^3. The equivalent water volume is about 24×10^6 km^3. This does not include shelf ice.

The water equivalent of ice-caps and glaciers, in rounded sum, is 26×10^6 km^3. These volumes of ice and water are difficult to conceive. The ice-caps and glaciers contain enough water to feed all the rivers of the world at their present discharge rates for nearly 900 years. They could feed the Mississippi River for 46,000 years or the Amazon for somewhat more than a tenth of that time.

A glacier may be regarded as a peculiar kind of river and an ice-cap as a special kind of lake. The great ice-caps and some glaciers discharge water to the sea by melting, by calving, and by wind transport of drifting snow. Only crude data are available, but they seem to indicate that the aggregate water equivalent of discharge by these processes is about 2,500 km³ yr⁻¹.

The average turnover time of water in alpine, piedmont, and valley glaciers is in the range of a few decades to a few centuries. The turnover time for great ice-caps has an extreme range. Today's snowfall at the edge of a cap may return to the sea tomorrow. But snow that forms ice in the heart of the cap may remain in residence during hundreds of thousands of years, and perhaps millions.

In addition to the great ice-caps and lesser glaciers, vast areas in the northern parts of North America, Siberia, and Europe (about $8 \cdot 5 \times 10^6$ km²) are locked in permafrost – permanently frozen ground. Little of the water in this ground participates in the hydrological cycle. In all, nearly 26×10^6 km² of the land area is under ice or otherwise frozen.

11. Possibilities for control

With 97% of the world's water in the sea and 2% in deep freeze, the world evidently is a fine place for whales and penguins, but it has its shortcomings for man. In addition, 17% of the land area is under ice or frozen, and 32% is arid to semi-arid. Small wonder that man, throughout his history, has sought ways to interfere with the water cycle!

Men have managed, mismanaged, and tried with varying degrees of success to control water since the dawn of civilization some 6,000 years ago. Hydraulic engineering has a long and preponderantly successful history. Technology of water control has advanced continually, but the specific means have not changed in hundreds of years, consisting of dams, diversion structures, canals, and the like. The principal change has been in the size and design characteristics of structures and hydraulic systems.

Ever more ambitious plans have now evolved to the stage where armchair planners propose alteration of natural water systems on a continental scale in North and South America and Asia. No doubt engineering skill is equal to the proposed tasks. It is possible also that such projects would be economically feasible from the standpoint of conventional benefit/cost ratios. But engineering and conventional economic feasibility are totally irrelevant in the present state of knowledge about what effects vast projects would have on the ecological systems of the continents.

Human activity has already scarred the face of the earth and upset many ecological systems, but men have not yet learned to predict or forestall the un-wanted side effects of landscape alterations. With characteristic heedlessness, however, we continue to plan ever larger alterations, ignoring the fact that construction of monumental projects entails the risk of monumental blunders.

Men have turned their eyes to the skies also and have sought a cloud-wringer to get more water down to earth. Competent analysts say that in at least some situations cloud-seeding has been successful. Completely missing, however, is information whether rain-making at one place diminishes precipitation at another place.

More ambitious is the dream of some men to control world climate. In the present state of knowledge this is beyond possibility. Ability to predict is a good measure of understanding. Weather prediction has not yet advanced from an art to a quantitative science. Until forecasters can make accurate long-range forecasts of natural weather events they can hardly predict the results of efforts at climate modification. It would be foolhardy therefore to try to modify climate, lest the attempt backfire and produce an unwanted change.

Adequate global study of water and weather would be impossible by conventional means, and prohibitively expensive even if possible. Fortunately, new technology offers hope for economical data collection on a global scale and for synoptic communication. So-called remote-sensing techniques are under extensive trial at present. Remote-sensing is any process for determining the nature of an object or phenomenon without direct contact. The human eye is a remote-sensing organ.

Conventional and colour photography with airborne cameras revolutionized geologic and topographic mapping during recent decades. Cameras and other instruments in orbiting or synchronous satellites may, in the future, not only revolutionize mapping but resources surveys in general. Among the more promising techniques are radar and far-infra-red scanning and near-infra-red photography.[1] These have proved useful for mapping topography despite heavy forest cover, for mapping surface drainage, differentiating soils and vegetation, mapping diseased vegetation, detecting geological structures, detecting submarine springs, mapping ocean currents, and a variety of other uses. The whole subject is too complex for treatment in these few paragraphs. The techniques are usable with aeroplanes, but satellites have the advantage that they scan large segments of the earth and its atmosphere during short periods, and they can do this repeatedly.[2]

Acknowledgements. I am grateful to the following colleagues for critical review of the manuscript for this chapter and suggestions for its improvement: P. H. Jones, J. T. Barraclough, J. B. Robertson, and Alfonso Wilson.

[1] See, for example, Colwell, R. N. [1968], Remote sensing of natural resources; *Scientific American*, **21**, No. 1, 54–69.

[2] The United States Government has published an atlas-type volume of remote imagery, consisting of 250 pages in colour. See *Earth Photographs from Gemini III, IV and V* (National Aeronautics and Space Administration SP-129, 1967), $7.00.

REFERENCES

ARAKAWA, H. [1959], Cosmic-ray intensities and liquid-water content in the atmosphere; *Journal of Geophysical Research*, **64**, 625–9.

AUSTRALIAN WATER RESOURCES COUNCIL [1963], *Review of Australia's Water Resources, 1963* (Canberra), 107 p.

CENTRO DE ESTUDIOS HIDROGRÁFICOS [1965], *Anuario de Aforos, 1961–62, v. 9 – Cuencas del Ebro*, Ministerio de Obras Públicas, Madrid [pages not numbered (about 200)].

IMBEAUX, EDOUARD [1930], *Essai d'Hydrologie. Recherche, étude et captage des eaux souterraines* (Paris), 704 p.

LOEWE, F. [1962], On the mass economy of the interior of the Antarctic icecap; *Journal of Geophysical Research*, **67**, 5171–7.

LOTKA, A. J. [1956], *Elements of Mathematical Biology* (New York), 465 p.

L'VOVICH, M. I. [1945], Elementy vodnogo rezhima rek zemnogo shara; *Glavnoe Upravlenie Gidrometeorologischeskoe Sluzhby CCP, Trudy Nauchno – Issledovatel' skikh Uchrezhdenii, Seriya IV, Gidrologiya Sushi*, Bypusk **18**, 109 p.

OLTMAN, R. E. [1968], Reconnaissance investigations of the discharge and water quality of the Amazon; *U.S. Geological Survey Circular 552*.

PÉGUY, C. P. [1961], *Précis de climatologie* (Paris), 347 p.

U.S.S.R. COMMITTEE FOR THE INTERNATIONAL HYDROLOGICAL DECADE [1967], *Water Resources and Water Balance of the Area of the Soviet Union* [in Russian with Russian and English titles] (Hydrometeorological Publishing House, Leningrad), 199 p.

VAN HYLCKAMA, T. E. A. [1956], The water balance of the earth; *Drexel Institute of Technology, Laboratory of Climatology*, Publications in Climatology, **9**, 58–117.

III. World Erosion and Sedimentation

D. R. STODDART

Department of Geography, Cambridge University

The mean elevation of the continents is +840 m and that of the oceans −3,800 m. Extreme elevations are, respectively, +8,882 m and −11,500 m. The mean elevation of the earth's surface, with respect to present sea-level, is −2,440 m. Erosion processes are constantly transferring material from the lands to the sea in the way that James Hutton described in 1795 in his *Theory of the Earth*. This section considers the magnitudes and areal variations of both erosion and sedimentation on a world scale.

1. Contemporary erosion: magnitude and patterns

Gross rates of denudation of the land surface are given by measuring the sediment load of rivers at their mouths. In practice, because of the difficulty of measuring bed load, rates are based on suspended sediment load and in some cases also solution load. Compilation of data from rivers throughout the world can thus be used to identify patterns of present denudation, especially when reduced to some common unit, such as metres per thousand years (m/10^3 yr). Because of the multivariate controls of rate of erosion, including relief, lithology, climate, and human use, these studies can yield only a first-order approximation, and it is not surprising that published syntheses are often mutually inconsistent, especially when based on sediment yields from basins of widely differing size.

Corbel studied total erosion for different temperature zones, in terms of three humidity and two relief categories. He found that erosion rates vary inversely with temperature, being lowest in the tropics; within each temperature zone they vary directly with humidity. Erosion is in all cases greater in mountainous areas than in plains, but the disparity between the two relief types is least in the tropics (a factor of 2), greatest in the temperate regions (a factor of 4–5), and intermediate in cold regions (a factor of 3–4). Figures for erosion in each category can be multiplied by area to give total erosion, and then averaged to give a mean world figure for unglaciated lands of 28·3 $m^3/km^2/yr$, or roughly 0·03 m/10^3 yr. This agrees remarkably with Dole and Stabler's estimate in 1909 of a rate of 1 ft in 9,000 yr (0·034 m/10^3 yr) for rivers in the continental United States. According to Corbel, all tropical areas, except for mountainous humid areas, are being eroded more slowly than this, and the temperate and cold lands, except for arid temperate plains, more rapidly. This pattern would presumably reflect more rapid mechanical weathering in cold lands, and would discount the effect of rapid chemical weathering and high rainfall in the tropics.

Other studies on a similar scale, using comparable data, yield rather different conclusions. Fournier studied suspended sediment yield in seventy-eight basins ranging in size from 0·0025 to 1·06 × 10⁶ km², correlating yield with a climatic parameter p^2/P, where p is the rainfall of the month with greatest rainfall and P the mean annual rainfall. The scatter of basins in terms of sediment yield and p^2/P, while generally confirming the expected increase of sediment yield with increase in rainfall, showed a grouping in terms of relief, into: (a) basins with

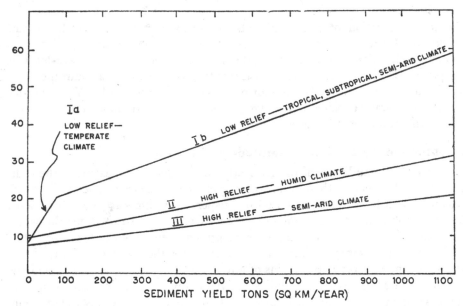

Fig. 1.III.1 Relationship between climate and suspended sediment yield. Vertical scale is p²/P (After Fournier, 1960).

low relief; (b) basins with high relief and a humid climate; and (c) basins with high relief and a semi-arid climate. It is thus impossible to relate rainfall to erosion without taking relief into account. Fournier derives a general empirical equation which fits his data, and which can be used for predicting sediment yield when climate and relief are known:

$$\log E \text{ (tons/km}^2\text{/yr)} = 2\text{·}65 \log (p^2/P)(\text{mm}) + 0\text{·}46 \log \bar{H} \cdot \tan \phi - 1\text{·}56$$

where E is suspended sediment yield, $\bar{H}$ mean height, and ϕ mean slope in a drainage basin. If $\bar{H} \cdot \tan \phi$ is not readily available, basins may be grouped into one of the three classes already distinguished, and erosion calculated from the regression equations for each group (fig. 1.III.1): using this approximate method, Fournier has mapped the world distribution of erosion based on suspended sediment yield (fig. 1.III.2). He finds maximum rates in the seasonally

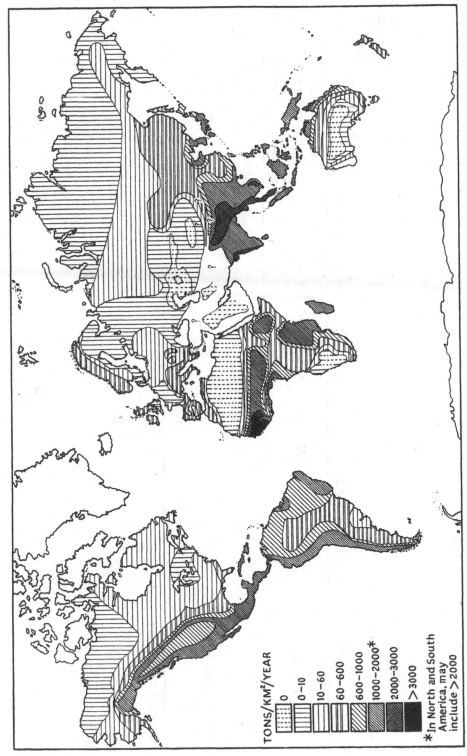

TONS/KM²/YEAR

0
0–10
10–60
60–600
600–1000
1000–2000*
2000–3000
>3000

*In North and South America, may include >2000

Fig. I.III.2 World distribution of erosion (After Fournier, 1960).

c

TABLE I.III.I Rates of erosion of the continents

Continent	Area (km² × 10⁶)	Suspended sediment yield (tons × 10⁶)	Dissolved load yield (tons × 10⁶)	Demudation rate (tons/km²)	
				Mechanical	Chemical
Europe	9·67	420	305	43·0	32·0
Asia	44·89	7,445	1,916	166·0	42·0
Africa	29·81	1,395	757	47·0	25·2
North and Central America	20·44	1,503	809	73·0	40·0
South America	17·98	1,676	993	93·0	55·0
Australia	7·96	257	88	32·1	11·3

Source: Strakhov [1967]

humid tropics, declining in the equatorial regions where the seasonal effect is lacking and in the arid regions where the total amount of runoff is low. In the deserts long-distance transport of sediment, except by wind, is nil. The rate of erosion rises again in the seasonally wet Mediterranean lands, but over the temperate and cold regions it is low except in mountainous areas. While both Fournier and Corbel agree on the importance of climatic controls on erosion rates, in practice they invert the trend.

Fournier's conclusions are generally supported by Strakhov, using similar data on sediment yield in rivers. Strakhov's sample of rivers vary in drainage area from 13·4 to 7,050 × 10³ km², in discharge from 11 to 3,187·5 km³/yr, and in sediment yield from 0·82 to 1,000 × 10⁶ tons/yr. The tropical areas of intense chemical weathering are being eroded most rapidly, with sustained rates in the high-rainfall, high-relief areas of south-east Asia of 390 tons/km²/yr. Strakhov's rates (fig. 1.III.3 and Table 1.III.1) are in general less than Fournier's, sometimes by an order of magnitude; and this may reflect abnormally high rates, caused by human interference and accelerated soil erosion during historic time, in Fournier's sample. Thus Strakhov's rates may be geologically more 'normal' (Douglas, 1967).

Because of the problems of standardizing data from basins of differing magnitudes, it is possible to proceed by taking a series of small basins and extrapolating from them. This has been done in the western United States by Langbein and Schumm, using both stream-gauging and reservoir-fill data. Schumm found the following relationship between sediment yield and rainfall (standardized to a mean annual temperature of 50° F), sediment mass having been converted to erosion rates by assuming 1 ton of sediment to be equivalent to $4·34 \times 10^{-7}$ ft/yr:

TABLE I.III.2

Gauging station data		Reservoir-fill data	
Effective precipitation	m/10³ yr	Effective precipitation	m/10³ yr
10	0·088	8–9	0·186
10–15	0·104	10	0·155
15–20	0·073	11	0·198
20–30	0·073	14–25	0·149
30–40	0·052	25–30	0·189
40–60	0·030	30–38	0·104
		38–40	0·073
		40–55	0·064
		55–100	0·058

Source: Schumm [1963]

These data demonstrate a maximum sediment yield with a rainfall of 250–350 mm/yr: with a lower rainfall runoff is inadequate, and with a higher rainfall vegetation growth hinders the entrainment of sediment. Much higher rates than

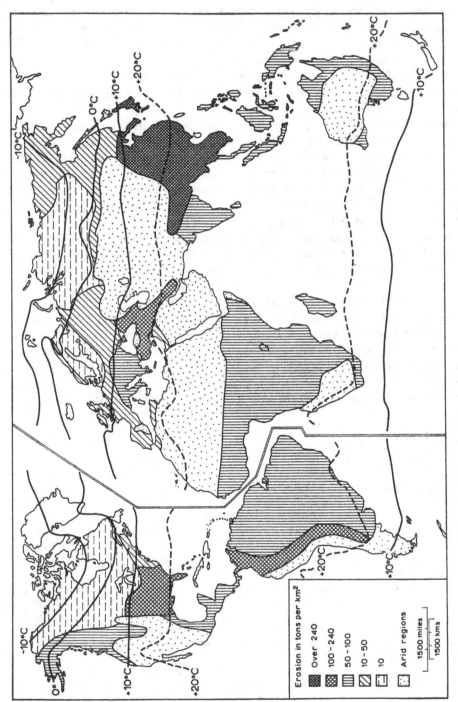

Fig. 1.III.3 World distribution of erosion (After Strakhov, 1967).

those listed in Table 1.III.2 are found in small basins in unconsolidated material (up to 12·8 m/10³ yr in loess badlands), and there is a general relationship between sediment yield and basin size, yield decreasing with the —0·15 power of basin area. Over the whole Mississippi basin, for example, erosion rate based on suspended load is only 0·04 m/10³ yr.

In addition to variation in sediment yield with rainfall, Schumm found sediment yield to be an exponential function of relief:

$$\log S \text{ (acre-ft/sq. mile)} = 27\cdot35R - 1\cdot187$$

where R is the relief–length ratio. Similarly, denudation rates, standardized for a basin area of 3,885 km², are also a function of relief:

$$\log D \text{ (ft/10}^3\text{ yr)} = 26\cdot866H - 1\cdot7238$$

where H is the relief–length ratio. Schumm's data indicate an average *maximum* rate of denudation of 0·9 m/10³ yr, and it is also clear that, though sediment

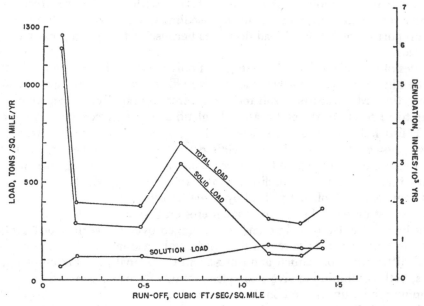

Fig. 1.III.4 Relationship between erosion rates and runoff (After Judson and Ritter, 1964).

yield is not a linear function of rainfall, meaningful distinctions can be made between arid, semi-arid, and humid areas. It is not clear from these data what happens in the hot wet lands of the tropics: Fournier's work would suggest that at rainfalls greater than 1,500 mm sediment yield rises to a new peak, and a similar inference can be drawn from Judson and Ritter's survey of regional rates of erosion in the United States (fig. 1.III.4).

In view of the problems raised in this survey of the world patterning of erosion by rivers, it is clearly premature to attempt to estimate the net sediment input in the erosion–depositional system. Confining attention to river sediments, moreover, neglects dissolved load, and also other modes of erosion, particularly by wind and sea. Dissolved load, which Clarke estimated to be 1.5 km^3/yr over the earth, is important both in the denudation of the lands and also in the supply of minerals to the sea, for direct precipitation and for conversion to plant and animal skeletons and subsequent accumulation. Thus silica is being supplied to the sea at the rate of 320×10^6 tons/yr and calcium carbonate at the rate of 560×10^6 tons/yr. Rates of chemical denudation published by Strakhov vary from 3.9 to 290 tons/km^2, whereas those of mechanical denudation range from 3.9 to $2,000$ tons/km^2. Mechanical denudation adjusts much more rapidly than chemical to changes in flow conditions, for above a threshold value chemical denudation rates are a function of the availability of solutes rather than of discharge. Hence, while in a given basin the concentration of suspended load in rivers increases with increasing discharge, that of dissolved load decreases. In the United States the maximum dissolved load of $125–150$ tons/mile^2/yr is reached with a runoff of about 250 mm, i.e. with an effective precipitation greater than that giving maximum solid sediment yield. With increasing relief the proportion of dissolved load decreases because of the much greater yield of solid load.

Rates of coastal erosion are of interest, if only because of the prolonged debate in the nineteenth century over the relative efficiency of marine and sub-aerial denudation. Many coasts in soft rocks are retreating rapidly: in the Sea of Azov at up to 12 m/yr, on the east coast of England at $1.5–5$ (maximum 11) m/yr, on the Polish Baltic coast and on Cape Cod at 1 m/yr. Harder rocks retreat more slowly, however (Normandy chalk cliffs 0.3 m/yr; tidal notches on coral limestone coasts 0.001 m/yr), and many coasts are actively aggrading. Kuenen estimates that the world's shorelines, 370×10^3 km long, yield 0.12 km^3/yr of sediment, only 1% of that yielded by fluvial erosion.

These estimates of fluvial and coastal erosion give a world total annual sediment loss to the lands of 13.6 km^3, or, averaged over the surface of the globe, 27×10^{-9} m/yr. This figure can have little real meaning, both because of the zonal distribution of rainfall and runoff, and the azonal distribution of upland areas, both combining to give a complex mosaic of erosion rates which are only beginning to be understood on the basin level. In terms of the azonal controls, a fundamental distinction must be made between the shield areas of the globe, covering 105×10^6 km^2 with mean elevation 0.75 km, and the fold mountain belts, with half the area and twice the mean height (42×10^6 km^2; 1.25 km). It should be noted also that the proportions of different kinds of rock vary in these different provinces, often in ways not indicated by conventional geological mapping, and that even on a continental scale lithology provides a powerful control of erosion rates (contrast, for example, Andean America with the Brazilian shield; 82% of the suspended sediment at the mouth of the Amazon comes from the Andes, according to Gibbs [1967]).

2. Contemporary sedimentation

Sediment loads from the continents are being transferred to ocean basins of differing sizes and shapes, and which differ widely in the proportions of land surface run-off they receive. Lyman has calculated that 68·5% of the world land area drains into the Atlantic, 44% in the Arctic, 26% into the Indian, and only

TABLE 1.III.3 Magnitudes of world's oceans

Ocean	Area, $km^2 \times 10^6$	Volume, $km^3 \times 10^6$	Mean depth, m	% world ocean area
Pacific	166·24	696·19	4,188	
With adjacent seas	181·34	714·41	3,940	50·1
Atlantic	86·56	323·37	3,736	
With adjacent seas	94·31	337·21	3,575	26·0
Indian	73·43	284·35	3,872	
With adjacent seas	74·12	284·61	3,840	20·5
Arctic	9·48	12·61	1,330	
With adjacent seas	12·26	13·70	1,117	3·4
World ocean	362·03	1,349·93	3,729	

Source: Menard and Smith [1966]

11% into the largest ocean, the Pacific. As a result, deep-sea sedimentation rates are almost an order of magnitude greater in the Atlantic than in the Pacific, averaging in the Holocene 0·088 $m/10^3$ yr in the former and 0·01 $m/10^3$ yr in the latter. As a first-order generalization, an erosion rate of 0·03 $m/10^3$ yr in the great river basins (i.e. not including the deserts and other areas of low erosion) may be compared with a mean deposition rate of 0·01 $m/10^3$ yr over the water 70% of the surface of the globe. Table 1.III.3 shows the area, volume, and depth of the world's oceans, and Table 1.III.4 the proportions of each ocean in different physiographic provinces.

A. Deltas

The 14 km^3 yielded every year by erosion of the continents is provided mainly by areal (slope) and linear (rivers, coasts) processes: its pattern of deposition is by contrast primarily punctiform. Figure 1.III.5 shows the world's twenty largest drainage basins, all with an area greater than 9×10^5 km^2: in these, erosion products from almost 30% (437×10^5 km^2) of the land area of the earth are being delivered to the twenty deltas and estuaries mapped. Apart from dissolved load which is added to the oceans, an average of 60% of the solid load of rivers is deposited at their mouths. Figure 1.III.5 also demonstrates the wide range of solid discharge of these major rivers, from 10 to $1,800 \times 10^6$ tons/yr: some rivers with smaller basins, notably the Mekong and the Irrawaddy, have much higher sediment loads than even a river like the Congo with three times their discharge. Holeman [1968] has listed the major rivers of the world in terms of

TABLE 1.III.4 Physiographic provinces of world's oceans

Ocean	Continental shelf and slope	Continental rise	Ocean basin	Mid-ocean ridge	Trench	Other
Pacific	13·1	2·7	43·0	35·9	2·7	2·5
Atlantic	17·7	8·0	39·3	32·3	0·7	2·0
Indian	9·1	5·7	49·2	30·2	0·3	5·4
Arctic	68·2	20·8	0	4·2	0	6·8
World ocean	15·3	5·3	41·8	32·7	1·7	3·1

Figures as percentages.
Source: Menard and Smith [1966]

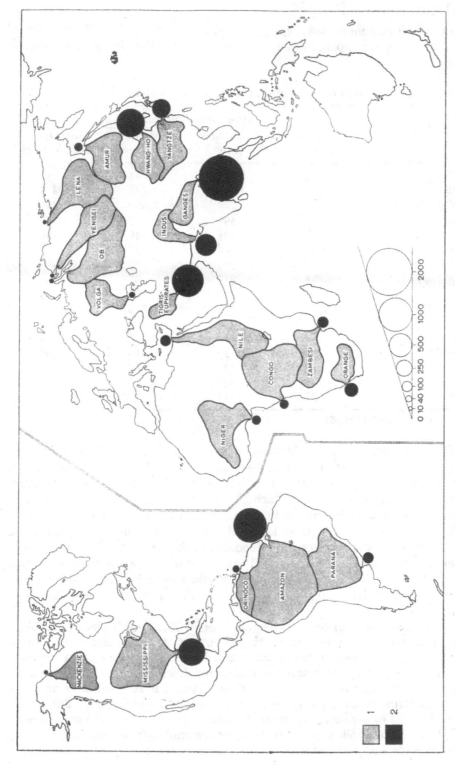

Fig. 1.III.5 World's largest drainage basins (**1**) and magnitude of solid load (**2**) (Data from Strakhov, 1967).

mean annual suspended sediment yield, rather than basin area (Table 1.III.5). It is instructive to compare the data in Holeman's list with that mapped in Figure 1.III.5.

TABLE 1.III.5 Rivers of the world ranked by sediment yield

| River | Drainage basin 10^3 km² | Average annual suspended load | | Average discharge at mouth 10^3 cfs |
		Metric tons $\times 10^6$	Metric tons/km²	
Yellow	673	1,887	2,804	53
Ganges	956	1,451	1,518	415
Brahmaputra	666	726	1,090	430
Yangtze	1,942	499	257	770
Indus	969	435	449	196
Ching	57	408	7,158	2
Amazon	5,776	363	63	6,400
Mississippi	3,222	312	97	630
Irrawaddy	430	299	695	479
Missouri	1,370	218	159	69
Lo	26	190	7,308	—
Kosi	62	172	2,774	64
Mekong	795	170	214	390
Colorado	637	135	212	5·5
Red	119	130	1,092	138
Nile	2,978	111	37	100

Source: Holeman [1968]

The Mississippi River, with a drainage basin area of 29×10^5 km², discharges 590 km³ of water a year. The present sediment load is about 450×10^6 tons/yr, containing 40% silt, 50% clay, and 2% sand. The late Pleistocene and Recent delta covers 114×10^3 km², of which 28·5 are in the present deltaic plain, 45·3 on the continental shelf, 22·0 on the continental slope, and 18·1 in a submarine bulge built during the last glacial low level of the sea. This late Quaternary delta is estimated to contain 33,400 km³, deposited at the rate of $1,570 \times 10^6$ tons/yr when the river was entrenching and of 635×10^6 tons/yr when it was aggrading as sea-level rose. The present distinctive birdfoot delta has been built entirely since A.D. 1500, during the dumping of a further 110 km³ of sediment (fig. 1.III.6).

These rates are so great that they cause easily measurable changes in topography round delta mouths, in striking contrast to the difficulty of measuring erosion distributed over much greater areas. In the Mississippi Delta accretion rates of more than 0·3 m/yr have been measured down to depths of 200 m where the delta is building out on to the continental slope. Maximum rates on delta-front platforms reach 0·3–0·45 m/yr over the last seventy years, falling away from the delta to less than 0·03 m/yr. These are figures for wet sloppy sediment, and should be multiplied by 0·5 to give a net accretion rate (see fig. 1.III.7). They may be compared with rates of 0·12–0·3 m/yr $\times 10^3$ measured in the coastal lagoons along much of the rest of the Gulf of Mexico coast, and are similar to rates measured for other great deltas of the world. Minor coastal accumulation features, such as salt marshes, have typical vertical growth rates of 0·01 m/yr, but are quantitatively of little significance in the planetary sediment budget.

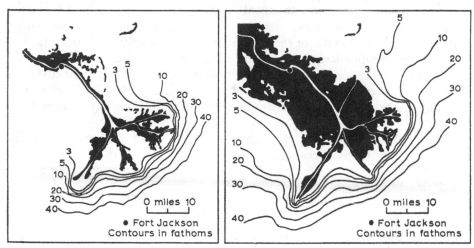

Fig. 1.III.6 Growth of the birdfoot delta of the Mississippi River 1874 (*left*)—1940 (*right*) (After Scruton, 1960).

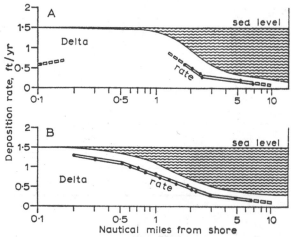

Fig. 1.III.7 Rates of accretion in the Mississippi Delta. A—Main Pass; B—North Pass (After Scruton, 1960).

B. Continental shelves and slopes

Continental shelves have a mean width of 75 km, mean edge depth of 130 m, and mean slope of 0° 07′; Kuenen calculates their area to be 30 × 10⁶ km² or 8% that of the oceans. Being adjacent to the lands, they collect most of the terrigenous solid sediments yielded by them; being relatively shallow, these deposits are subject to considerable reworking by marine action. Above all, the complexity of sedimentation on the shelves results from the fact that they lie within the range of Pleistocene eustatic sea-level shifts, which extended down to at least −150 m.

The sediment pattern on modern shelves consists therefore of deposits laid down when the shelves were emergent in the Würm, partly covered by transgressive marine deposits (transgressive from sea to land) and by terrigenous deltaic and coastal deposits of the present stillstand (transgressive from land to sea). Modern sedimentation is small on the outer parts of continental shelves, and is rapid on the inner parts and in nearshore basins. Hayes has studied the distribution of sediment types on the inner continental shelves in detail (Table I.III.6). He finds that mud is most abundant off areas with high temperature

TABLE I.III.6 Distribution of sediments on the inner continental shelf in terms of coastal climates

Coastal climatic zone	Rocky	Known bottom sediments (%)					Unknown (%)
		Gravel	Coral	Shell	Sand	Mud	
Rainy tropical	3·2	0·3	12·3	4·4	31·4	48·5	21·2
Sub-humid tropical	5·2	1·4	13·5	4·5	38·4	37·0	15·8
Warm semi-arid	3·7	—	8·6	4·2	59·5	24·0	8·5
Warm arid	4·4	—	7·3	4·8	52·1	31·4	24·2
Hyper-arid	9·1	0·6	20·9	12·0	44·5	12·8	33·6
Rainy subtropical	11·7	0·4	3·8	6·6	54·3	23·2	5·4
Summer-dry subtropical	26·1	4·1	2·1	2·7	37·3	27·7	11·9
Rainy marine	26·2	—	2·4	2·4	63·3	5·7	30·0
Wet-winter temperate	29·7	6·3	—	1·6	53·6	8·9	8·6
Rainy temperate	18·6	9·1	—	4·8	48·2	19·2	4·8
Cool semi-arid	—	—	—	7·1	92·9	—	—
Cool arid	20·2	4·8	2·4	3·6	52·8	16·3	20·8
Subpolar	30·8	14·9	—	3·5	39·3	11·5	7·0
Polar	20·8	16·2	—	4·8	43·1	15·1	9·2
Total	13·3	4·1	6·4	4·8	43·5	28·0	—

Source: Hayes [1967]

and high rainfall; sand is everywhere abundant, but with a maximum in areas of moderate temperature and rainfall and in all arid (except extremely cold) areas; gravel is most common off areas of low temperature; rock is most frequent in cold areas; coral is most abundant in areas with high temperature; and the distribution of shell is not related to climate. Figure I.III.8, from Hayes, shows the climatic control in terms of precipitation and temperature of mud, sand, and gravel distribution; these diagrams are comparable to those used by Peltier in his delimitation of world morphoclimatic zones (see Chapter 10.II). In view of the limitations of the data, the known differences in shelf topography,

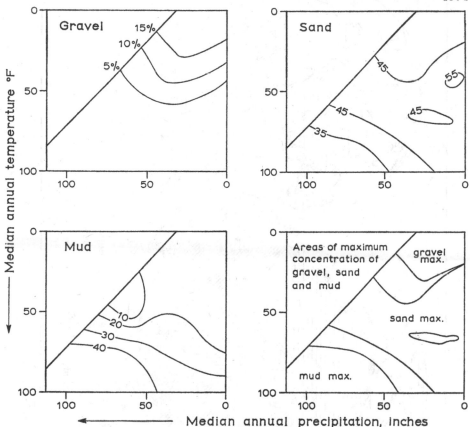

Fig. 1.III.8 Distribution of sediments on the inner continental shelves in terms of climatic controls (After Hayes, 1967).

and the importance of Pleistocene relict sediments, the correlation between sediment type and present coastal climate is remarkable.

Continental slopes have a mean width of 20–100 km, and form a transition zone between the shelf edge and the deep sea floor. Mean slope (contrary to many impressions) is only 4° 17′ for the first 1,800 m (2° 55′ in the Indian Ocean, 3° 05′ in the Atlantic, 5° 20′ in the Pacific), but the topography is fairly rugged because of the incision of submarine canyons in the slopes, cutting back into the shelves. Continental slopes received terrigenous sediment during Pleistocene low sea-levels: aggradation rates have been higher on the slopes than on the outer shelves. The slope is not simply an aggradation surface, however, comparable to deltaic foresets; it is too irregular, and Cretaceous and Tertiary sediments outcrop on it.

Submarine canyons are important as channels through which sediments from the lands are funnelled to the deep sea floor. Daly in 1936 first proposed that they were cut by turbidity currents, dense suspensions of sediment flowing down-slope, and Kuenen in 1950 experimentally demonstrated the reality of such

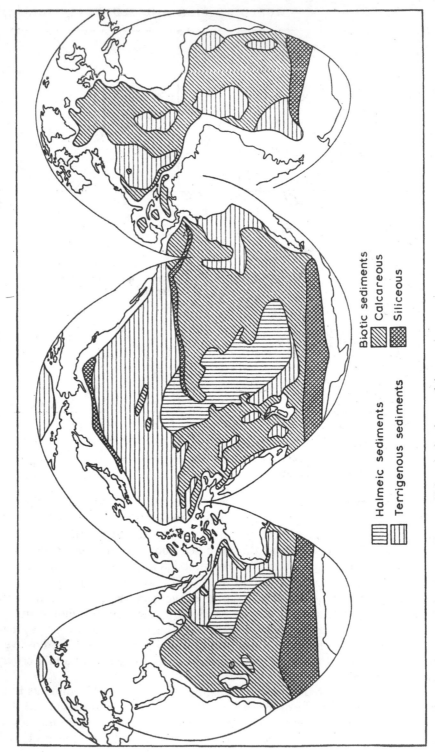

Biotic sediments
Calcareous
Siliceous

Halmeic sediments
Terrigenous sediments

Fig. 1.III.9 Distribution of deep-sea sediments (After Arrhenius, 1963).

currents. Whether turbidity currents can actually erode canyons remains open to doubt, but fine sands and other sediments are certainly transported by them through canyons and deposited in cones at their mouths. Large canyons have been identified off the mouths of many large rivers, including the Hudson and the Congo.

C. Deep ocean floor

Our knowledge of the deep ocean floor and its sediments began with the voyage of H.M.S. *Challenger* in 1872–6; recent work has demonstrated the diversity of deep-sea topography. The mid-ocean ridges occupy 32·7% of the oceans, basins proper 41·8%, and the continental rise at the foot of the continental slopes 5·3%; proportions differ between different oceans (Table 1.III.3). Sedimentation on the floor is of two main types: pelagic, formed by the slow settling of deposits far from land, and terrigenous.

(a) Pelagic deposition

Pelagic deposits consist of either biogenic material, diagenetic deposits, or clays. Table 1.III.7, from Kuenen, shows the proportions of different types in each

TABLE 1.III.7 Distribution of pelagic sediments

Sediment type	Area $km^2 \times 10^3$	% of deep ocean	% of total sea	Relative rate of deposition	% of total sediment volume
Calcareous sands and oozes	127·9	47·7	35·4	3	71·9
Brown (red) clay	102·2	38·1	28·3	1	19·2
Siliceous oozes	38·0	14·2	10·5	1·25	8·9
Total	268·1	100·0	74·2		100·0

Source: Kuenen [1950]

ocean, and Figure 1.III.9 their distribution. Calcareous oozes cover 128×10^6 km^2 of floor: they are formed from the tests of foraminifera and other organisms settling to the ocean floor. At deeper levels calcium carbonate is dissolved by sea-water, and the 'compensation depth', below which calcareous deposits are rare, is 4,500–5,000 m. Present rates of supply of dissolved carbonates from the continents would provide 0·34 $g/cm^2/10^3$ yr of calcareous deposits over the entire ocean floor: two-thirds of this enters the Atlantic, however, and as a result the average $CaCO_3$ content of Atlantic sediments is 41%, compared with 19% in the Pacific. In the equatorial Pacific high-carbonate belt, accretion is at the rate of 0·01 $m/10^3$ yr, compared with 0·0004–0·0005 $m/10^3$ yr in the south-east Pacific. To understand what these rates mean, Menard has calculated that to cover the floor of the Pacific with a layer of beer cans 1 mm (0·001 m) thick over 10^3 yr would require a supply of 10^9 beer cans per day, equivalent to the world's entire output of steel.

Siliceous oozes, consisting of radiolarian tests and diatom frustules, cover 38×10^6 km². They accumulate in areas where terrigenous sediment supply is low, and where carbonates have been dissolved. Accumulation rates are about one-third of those for calcareous deposits.

Brown pelagic clays of mean particle size 1μ accumulate far from land, in the absence of both calcareous and siliceous material. Dating of the clay minerals has shown that the sediment source is terrigenous, including wind-transported desert dust and volcanic ash, together with meteoritic material. There is therefore some correlation with regional continental weathering patterns, and also a probability that accumulation rates were higher in the glacial periods. Brown clays cover 102×10^6 km², mostly deeper than 4,500 m, and they are accumulating at the rate of $0.0005-0.0006$ m/10^3 yr; rates are higher in the north Pacific than in the south, and on the abyssal floor near Baja California reach $0.009-0.02$ m/10^3 yr.

The main diagenetic deposit on the deep sea floor consists of manganese nodules, composed of 16% Mn and 17.5% Fe. Mero finds the mean of 100 Pacific floor measurements to be 11 kg/m² of nodules, with a total volume on the floor estimated to be 1.6×10^{12} metric tons: manganese nodules are thus one of the most common rocks of the surface of the lithosphere. It is generally believed that nodule accumulation is taking place extremely slowly by accumulation of authigenic manganese at a rate of 0.0003 m/10^6 yr, but in this case nodules should be soon buried by pelagic deposits; recently it has been suggested that nodules form very rapidly following sporadic underwater volcanic effusion. At the slow rate nodules are being produced at the rate of 6×10^6 metric tons/yr.

Assuming a mean rate of pelagic deposition of 0.003 m/10^3 yr over a total area of 268×10^6 km², Poldervaart has calculated a total sediment accretion of 0.8 km³ or 22×10^8 tons/yr. This is equivalent to 6% of the present sediment load of rivers.

(b) Terrigenous deposition

Terrigenous deposits consist of muds (silty clays), which are green, black, or red, depending on chemistry and depositional conditions; glacial marine ice-rafted deposits in higher latitudes; and turbidites and slide deposits. Turbidite deposits of graded sands have been found over wide areas, related to local topography. They are also important in deep ocean trenches. Trenches more than 6,000 m deep (maximum depth 11,500 m) are concentrated in the Pacific, where they parallel continental shores and thus serve as sediment traps for terrigenous material. In the Puerto Rico trench the top 10 m of sediments consists of fine sands and silts, interpreted as turbidity current deposits: one such flow 3 m thick covers 10^4 km² and contains 3×10^7 m³ of sediment. In this case it is suggested that the entire $1-1.7$ km of sediments on the trench floor may be Pleistocene turbidites. Other trenches contain little sediment, perhaps because they have formed so recently.

Other terrigenous deposits are important in smaller seas and basins such as the

Mediterranean and the Black Sea. Under euxinic conditions on the Black Sea floor seasonally varved clays and calcareous muds have accumulated at the rate of 1 m clay and 0·1–0·2 m calcareous mud in 5×10^3 yr. Rates in such nearshore basins tend to be highly variable and difficult to generalize.

(c) Biohermal deposits

Since the sea is approximately saturated with calcium carbonate, deposition of carbonates must be occurring in the ocean basins at the rate at which they are being supplied from the lands. This occurs mainly in the slow accumulation of pelagic carbonates over vast areas, and quantitatively the amount contained in bioherms (mainly in atolls, barrier, and fringing reefs) is minor. From borings on open-ocean atolls, we know that reef limestones have accumulated at the rate of $0·023$ m/10^3 yr over the last 70×10^6 yr.

3. Erosion and sedimentation in the geologic record

The evidence of unconformities in the geologic record was used by James Hutton and both Powell and Dutton to infer prolonged periods of erosion, and such data have been used more recently to calculate the time required for peneplanation. This assumes that erosion rates have been invariant through time, a conclusion which stratigraphical geologists have long tended to dispute. Table 1.III.8 gives mean rates of sedimentation since the early Paleozoic calculated by Joseph Barrell; these show an apparent doubling in this time.

TABLE 1.III.8 Sedimentation rates in geologic time

Period	Maximum known thickness of strata (m × 10³)	Approximate duration (yrs × 10⁶)	Rate (m/10³ yr)
Cenozoic	23·2	70	0·375
Mesozoic	33·2	120	0·299
Late Palaeozoic	27·4	130	0·187
Early Palaeozoic	29·0	180	0·161

Source: Joseph Barrell and A. Holmes, from Pettijohn [1957]

Menard has studied rates of sedimentation based on the geometry of deposits in three geosynclinal areas (the Appalachians, the Mississippi basin, and the Himalayas), in all of which the limits of the closed denudation system are inadequately known. His data, and calculated rates of denudation (which ignore dissolved load), are given in Table 1.III.9. Menard concludes that over 10^7–10^8 yr past erosion rates have varied by only one order of magnitude, compared with much greater variability at the present day. Whether such rates are representative of past conditions must depend on the extent and synchroneity of geosynclinal conditions, and hence on tectonic history. Gilluly has made a more detailed study of the Atlantic coast sediments of North America. He finds the Triassic and younger rocks to have an area of 490×10^3 km^2 and volume of $1,170 \times 10^3$ km^3, representing a volume of bedrock eroded of $1,023 \times 10^9$ km^3. The source area is delimited rather arbitrarily with an area of $1,320 \times 10^3$ km^2. Over the 225×10^6 yr of post-Triassic time the mean erosion rate has thus been

TABLE 1.III.9 Past erosion rates from depositional records

Area	Area eroded (km$^2 \times 10^6$)	Volume of deposits (km$^3 \times 10^6$)	Age (yr $\times 10^6$)	Total denudation (km)	Rate (m/10^3 yr)
Appalachian	1	7·8	125	7·8	0·062
Mississippi	3·2	11·1	150	6·9	0·046
Himalayas	1	8·5	40	8·5	0·21

Source: Menard [1961]

0.0034 m/10^3 yr. Taking the present suspended sediment load and dissolved load of the St Lawrence and other North Atlantic rivers, and making allowance for bed load, Gilluly calculated the present erosion rate to be 0.0218 m/10^3 yr, an order of magnitude greater. Present rates projected into the past would have supplied 60% more sediment than actually is found. These figures must be highly approximate, because of difficulties in delimiting the system; they also ignore reworking of sediments; but they do suggest that present rates are more rapid than those in the past, at least as recorded in basins of sedimentation.

Since long-continued sedimentation only occurs in subsiding depositional basins, inference from such data to past world erosion conditions must depend on the frequency, periodicity, and magnitude of basin development in the past. Thus the modern Mississippi delta is only the top member of a pile of sediments in the Gulf Coast geosyncline, accumulating since the Jurassic. During the Tertiary, contemporary with orogeny in western North America, 7,600 m of sediments formed in Louisiana and Texas, with a volume in the Paleogene of $1,000 \times 10^3$ km^3 and in the Neogene of 420×10^3 km^3. One deep well in south Louisiana, penetrating 6,880 m, reached only the Miocene. Subsidence is so rapid, partly from compaction, partly from deeper causes, that when rapid sedimentation ceases at the present delta, as the locus of deposition swings from side to side, channels are carried beneath the sea: dead deltas are revealed by the parallel islands of drowning levees. It was formerly thought that depression was caused by sediment-loading on the crust, and this certainly occurs, but to a limited extent. Thus at Lake Mead, where 40×10^9 tons of water and (so far) 2×10^9 tons of sediment have been added to 600 km^2 of crust, subsidence is taking place at the rate of 12.2 m/10^3 yr, but will total only 0.25 m.

World-wide geologic rates will also depend on the periodicity of orogeny: whether episodic, as argued by Stille and Umbgrove, or random in space and time, and hence essentially continuous, as argued by Shepard and Gilluly. In this connection present erosion rates may be inflated because of current tectonic activity: surprisingly high rates of earth movement have been revealed by geodetic survey, not only in mobile belts (up to 75 m/10^3 yr) and isostatic up-warps (5–15 m/10^3 yr) but in so-called stable areas like the Russian plains (up to 10 m/10^3 yr). Since we do not yet know tectonic norms for the world, we cannot assume that present conditions represent an erosion norm, especially when we take into account the major factors of Pleistocene glaciation, and the great amounts of unconsolidated sediments it left for later stream removal, and man's influence, in clearing the woodlands and cultivating the soil. These two factors alone could have inflated present river erosion rates by perhaps two orders of magnitude by comparison with a hypothetical norm.

Perhaps the most remarkable fact about sedimentation in the oceans is, over vast areas, its extreme slowness, and the concentration of most sediment into a few depositional basins. However, if present rates of 0·001–0·01 m/10^3 yr are extrapolated into the past, assuming the ocean basins to be 2 × 10^9 yr old, sediment thicknesses should be 1–10 km, nearer the latter than the former. Seismic work in the deep oceans has shown that the mean depth of unconsolidated material on the floors is about 0·5 km in the Pacific, 0·1–3 km in the Arctic, 0·6 km in the Indian Ocean, and 0·3–0·6 km in the Atlantic (down to zero on the mid-ocean ridge). No deep ocean sediments known are older than the Cretaceous. This suggests that ocean basins may be younger than was thought. Alternatively, the second layer in the crust (1·7 km thick) may be consolidated sediment rather than basalt. This problem of the absence in the ocean basins of sediments supplied from the continents has been called 'the great paradox of marine sedimentation' by Menard. Several workers, including Dietz, have speculated that ocean-floor sediments are re-incorporated in the continents, pelagic sediments floating on the tops of convection cells across the ocean floors from the mid-ocean ridges to the continents, where they sink beneath the sial layer.

Such speculations serve to illustrate the paucity of our knowledge on the long-term equilibrium of the world erosion–sedimentation system. At least it can be said that, with James Hutton, we see 'no vestige of a beginning, no prospect of an end'.

REFERENCES

ARRHENIUS, G. [1963], Pelagic sediments; In M. N. Hill, Editor: *The Sea*, volume 3 (New York), pp. 655–727.

CONALLY, J. R. and EWING, M. [1967], Sedimentation in the Puerto Rico Trench; *Journal of Sedimentary Petrology*, **37**, 44–59.

CORBEL, J. [1964], L'érosion terrestre, étude quantitative (Méthodes – techniques – résultats); *Annales de Géographie*, **73**, 385–412.

DOLE, R. B. and STABLER, H. [1905], Denudation; *United States Geological Survey, Water Supply Paper*, **294**, 78–93.

DOUGLAS, I. [1967], Man, vegetation and the sediment yield of rivers; *Nature*, **215**, 925–8.

FOURNIER, F. [1960], *Climat et érosion: la relation entre l'érosion du sol par l'eau et les précipitations atmosphériques* (Paris), 201 p.

GIBBS, R. J. [1967], The geochemistry of the Amazon River system, Part I; *Bulletin of the Geological Society of America*, **78**, 1203–32.

GILLULY, J. [1949], Distribution of mountain building in geologic time; *Bulletin of the Geological Society of America*, **60**, 561–90.

GILLULY, J. [1964], Atlantic sediments, erosion rates, and the evolution of the continental shelf: some speculations; *Bulletin of the Geological Society of America*, **75**, 483–92.

HAYES, M. O. [1967], Relationship between coastal climate and bottom sediment type on the inner continental shelf; *Marine Geology*, **5**, 111–32.

HOLEMAN, J. N. [1968], The sediment yield of major rivers of the world; *Water Resources Research*, **4**, 737–47.

JUDSON, S. and RITTER, D. F. [1964], Rates of regional denudation in the United States; *Journal of Geophysical Research*, **69**, 3395–401.

KUENEN, P. H. [1946], Rate and mass of deep-sea sedimentation; *American Journal of Science*, **244**, 563–72.

KUENEN, P. H. [1950], *Marine Geology* (New York), 568 p.

MENARD, H. W. [1961], Some rates of regional erosion; *Journal of Geology*, **69**, 154–61.

MENARD, H. W. [1964], *Marine Geology of the Pacific* (New York), 271 p.

MENARD, H. W. and SMITH, S. M. [1966], Hypsometry of ocean basin provinces; *Journal of Geophysical Research*, **71**, 4305–25.

PETTIJOHN, F. J. [1957], *Sedimentary Rocks* (New York), 718 p.

POLDERVAART, A., Editor [1955], Crust of the earth; *Geological Society of America Special Paper*, **62**, 1–762.

SCHUMM, S. A. [1963], The disparity between present rates of denudation and orogeny; United States Geological Survey; *Professional Paper*, 454–H, 1–13.

SCRUTON, P. C. [1960], Delta building and the deltaic sequence; In Shepard, F. P., Phleger, F. B., and Van Andel, Tj. H., Editors, *Recent Sediments, Northwest Gulf of Mexico* (Tulsa), pp. 82–102.

SHEPARD, F. P. [1963], *Submarine Geology*; 2nd edn. (New York), 557 p.

SHEPARD, F. P., PHLEGER, F. B., and VAN ANDEL, TJ. H., Editors, *Recent Sediments, Northwest Gulf of Mexico* (Tulsa), 394 p.

STRAKHOV, N. M. [1967], *Principles of Lithogenesis*; volume 1 (London), 245 p.

CHAPTER 2

The Basin

I. The Basin Hydrological Cycle

ROSEMARY J. MORE

Formerly of Department of Civil Engineering, Imperial College, London University

1. Introduction

The river basin, bounded by its drainage divide and subject to surface and sub-surface drainage under gravity to the ocean or to interior lakes, forms the logical areal unit for hydrological studies (figs. 2.1.1 (*a* and *b*)). Within this framework one can conveniently, for example, draw up a water balance and assess water resources; estimate the probability of the occurrence of extreme events, such as floods and droughts, particularly as they affect reservoir storage and water use by man; and mobilize hydrological information to enable man to manage his water resources more efficiently by knowing when and in what ways it is to his advantage to intervene locally in the hydrological cycle. It is the purpose of this section to single out the basin as a proper basis for such studies; to identify basin inputs, storages, transfers, and outputs; and to indicate some of the different methods of studying the relationships between these parameters.

The basin cycle can be viewed simply as inputs of precipitation (p) being distributed through a number of storages by a series of transfers, leading to outputs of basin channel runoff (q), evapotranspiration (e), and deep outflow of ground water (b) (fig. 2.1.1(*c*)). For all practical purposes, the last output is difficult to measure and, except under special geological conditions, is usually assumed to be of such small relative importance as to be commonly ignored in basin input/output studies. Thus the gross operation of the basin hydrological cycle may be simply approximated as:

Precipitation = Basin channel runoff + Evapotranspiration +
$$\text{Changes in storage} \quad (1)$$

or
$$p = q + e + \Delta(I, R, M, L, G, S)$$

The operation of the basin cycle may now be considered in more detail. Precipitation (p) (in the form of rain, sleet, hail, and dew – the delayed hydrological effects of snow will be considered in Chapter 10.1) falls on vegetation, bare rock, debris, and soil surfaces, as well as directly into bodies of standing water and stream channels. Water in transit is stored on the vegetation leaf and stem surfaces as interception storage (I), which either evaporates (e_i) or reaches the ground by stem flow and drip (i). The drainage of water from vegetation, together with direct precipitation on to the ground surface and surface water,

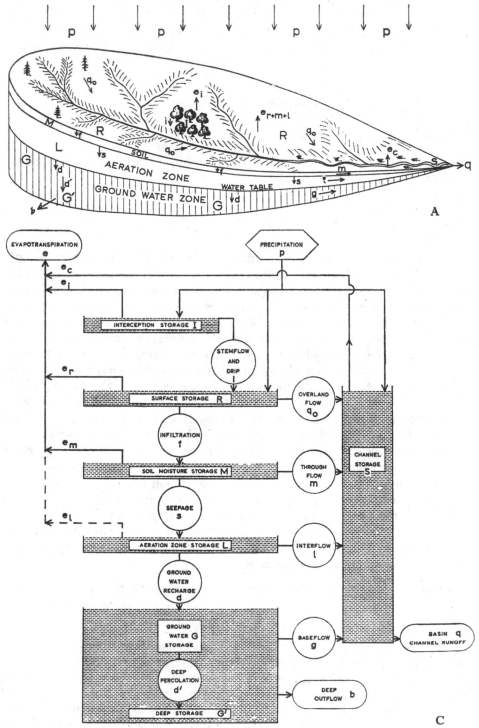

Fig. 2.1.1 The components of the basin hydrological cycle.
A. Block diagram of the basin.
C. Schematic inter-relationships of the basin components.

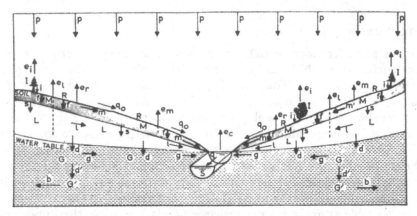

B

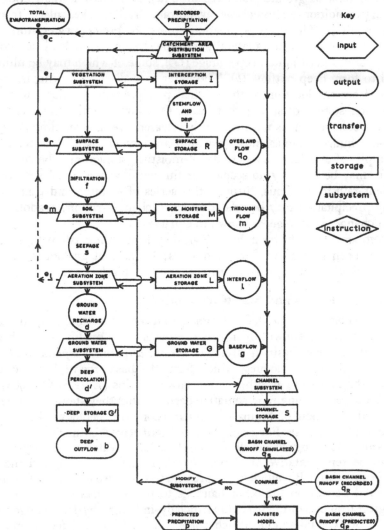

D

B. Cross-section of the basin.
D. Flow diagram of the basin components.

contributes to surface storage (R), which either evaporates directly (e_r), flows over the surface to reach the adjacent stream channels as overland flow (q_0), or infiltrates into the soil (f). The water in the soil (soil moisture storage – M) is similarly depleted by the transpiration of plants (e_m), by throughflow (m) of water downslope within the soil profile to augment the channel storage (S) (itself depleted by evaporation –e_c), or by vertical seepage (s) into the aeration zone. Water in the rock pores and fissures between the base of surface weathering and the water table (the aeration zone storage – L) is depleted by interflow (l), that reaching the adjacent stream channels by flow subparallel to the surface slopes without becoming ground water, and by deeper percolation downwards to the water table as ground-water recharge (d). It is possible that, under pro-longed dry conditions, appreciable evapotranspiration takes place from the aeration zone (e_l). The deep percolation enters the ground-water storage (G), from which water either flows laterally into the stream channels as baseflow (g), or slowly percolates (d') into deep storage (G'), some of which may be ultimately destined to form deep outflow (b). The latter may discharge at depth into the ocean far from the location of the original precipitation, or may augment the ground-water storage of an adjacent basin. Of course, all the water entering a given basin storage zone is not necessarily released over a short time period and basin water-budget studies must take into account changes in water storage. For example, after a long dry period soil-moisture storage may be so depleted that there may be virtually no seepage or throughflow associated with the first subsequent rainstorm. Thus, through this series of storages and transfers, the incoming precipitation may be related to basin channel runoff, evapotranspira-tion, and changes in storage, as shown in equation (1).

The components of the basin hydrological cycle and the relationships between them have been studied in five main ways: by natural analogues, hardware models, synthetic systems, partial systems, and the 'black box' approach.

2. Natural analogues and hardware models

A natural analogue may be defined as some analogous natural system believed to be simpler, better known, or in some respects more readily observable than the original. In river-basin hydrology *representative* and *experimental* basins provide such natural analogues, which form the bases for understanding and predicting the behaviour of other, often larger, basins (UNESCO, 1964).

The aim of establishing representative basins is that they should represent the hydrological operation of basins of similar geometry, geology, climate, soils, vegetation, and land use. For example, the Trent River Authority has eighteen such basins, including Bradwell Brook–Peakshole Water in Derbyshire, in-tended to be representative of Carboniferous Limestone areas, and the River Rea basin in central Birmingham, representative of urban areas. The role of experimental basins, on the other hand, is to provide data for a study of the hydrological response of the basin to artificially-induced physical changes during the period of experimentation. Such changes might include, for example, de-forestation or other variations of land use. Of course, the proper interpretation of these data requires that the whole basin output can be rationalized in terms of the input, and this feature of experimental, as well as of representative, basins of 'needing to know the answer before you can do the sum' is one of the most severe restrictions on their use. There are other difficulties in deciding whether the natural analogies are valid ones; both in that the storm precipitation inputs into a small basin used as an analogue have effects different from similar inputs into larger basins (More, 1967, p. 166) and because the large number of basin parameters means that one is never quite sure that there is complete accordance between the natural analogue and the other basins, the behaviour of which one wishes to predict with its aid.

One obvious way of trying to avoid this last difficulty is to attack the basin hydrological cycle by means of *hardware models*, wherein important structural elements of the basin are physically constructed either as scale models (involving such real world materials as sand and water) or analogue models (where there is a radical change in the media used to represent the basin elements, e.g. the representation of the flow of water by the flow of heat or electricity). A laboratory catchment usually partakes of the features of both scale and analogue models in an attempt to predict the runoff output resulting from a given precipitation input, controlled in space and time. Such catchments have been constructed by the U.S. Department of Agriculture (Chery, 1966) and in the Hydraulics Laboratory at Imperial College, London. The advantages of hardware models are that the basin hydrological cycle can be simplified by only considering the controlled variables pertinent to a particular problem and by the ability of the operator to compress natural time sequences into comparatively short experiments. However, many hydrologists (e.g. Amorocho and Hart, 1965) emphasize the difficulties in attempting to simulate natural basin behaviour from a hardware model, pointing to the impossibility of establishing complete dynamic similitude between the model and the prototype, and to the impossibility of operating the many simulated variables over a sufficiently wide range for adequate simulation. To such workers the electronic computer represents the only acceptable hardware model.

3. Synthetic systems

Nash [1967], for example, considers that progress in understanding the basin hydrological cycle will be made most rapidly by the use of overall *synthetic systems*, analysed by computers. In approaching the operation of the basin cycle viewed as a synthetic system, the investigator attempts to describe the operation of its hydrological cycle by a linkage or combination of components, the presence of which is assumed to exist in the system and whose functions are known and predictable (Amorocho and Hart, 1964). The qualitative specification of the basin components and their relationships can be organized into a flow chart, such as is shown in fig. 2.1.1(d), from which computer programmes can be prepared.

The input into the system may be in the form, for example, of hourly rainfall amounts recorded for a number of stations throughout the basin. Each of these rainfall recorders is strategically situated in a segment of the basin so as to

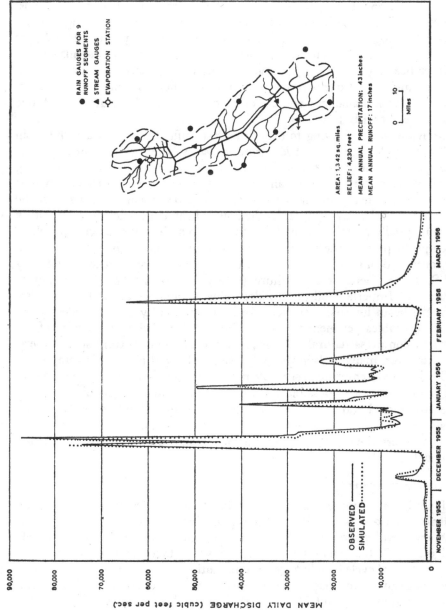

Fig. 2.1.2 Mean daily flow of the Russian River basin at Guerneville, California, November 1955–March 1956, recorded and simulated by the Stanford Watershed Model IV (After Crawford, N. H. and Linsley, R. K., *Dept. of Engineering Stanford University, Technical Rept. 39, 1966*).

represent the input for that segment, and the subsequent basin analysis is largely based on the assumed behaviour of the rainfall input into and through these segments (fig. 2.1.2).

Before the basin system is ready to receive the time sequence of segment inputs from a real basin it must be programmed. This is done by:

1. Inserting estimates of the initial segment storages in each subsystem at the start of the rainfall input sequence.
2. Feeding in an appropriate channel time-delay histogram giving estimated times of surface flow from each of the contributing basin segments along the channel system to the outflow gauging station.
3. Specifying a number of parameters (i.e. mathematical constants) which control and allocate the distribution of precipitation inputs into and between the basin subsystems through time. Some of these parameters are simple invariates (e.g. the measured areas of the basin segments), others can be virtually specified by rule of thumb (e.g. the maximum interception storage for a given vegetation), but many of the parameters are complex time relationships (e.g. the rate of infiltration; the rate of baseflow from ground water). The optimizing of these last parameters is one of the most complex parts of the whole operation.

As the sequences of precipitation inputs are fed into the computerized system the programming described above allows them to filter through the system mathematically, contributing to storage changes, evapotranspiration, and basin channel runoff. (Deep outflow is usually ignored, except in rare geological circumstances where it is significant). The main output from this system is a simulated channel runoff sequence from the basin outlet, and this is compared with the actual recorded channel runoff sequence for the appropriate period. If the two fail to reach some acceptable level of agreement the computer is programmed so that it will automatically modify the subsystem parameters in a predetermined manner, re-run the inputs, and continue this process until the required level of agreement between the simulated and actual basin channel runoff sequences is achieved (fig. 2.1.1(d)). Figure 2.1.2 shows actual and simulated runoffs by the Stanford Watershed Model IV for the Russian River basin above Guerneville, California, for the period November 1955 to March 1956. (For the computer flow diagram and further description see fig. 9.1.5.)

Once the model is working well its potential uses are most important. For example, expected future rainstorm inputs can be used to yield simulated predicted runoff outputs of significance in terms of flood-control planning, and the system parameters can be varied at will to anticipate the hydrological consequences of man-made changes to the basin, such as extensive forest clearance, cultivation, or building construction.

4. Partial systems and the 'black box'

It is clear that the example just given represents research into the basic operation of all the assumed major components of the basin hydrological cycle, in order to

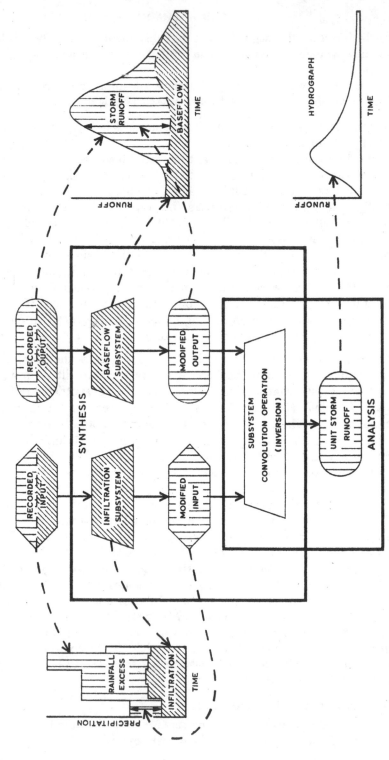

Fig. 2.1.3 Flow chart for partial system synthesis involving the prediction of runoff from precipitation characteristics, with a minimum know-ledge of the internal operations of the system components (From Amorocho and Hart, 1964).

gain an understanding of their mechanism and interactions. For some practical purposes, however, it is unnecessary to attempt this degree of analysis, and the *partial-systems* approach depends for its operation on more limited assumptions about the basin cycle. An example of the partial-systems approach, which is also computer based, has been given by Amorocho and Hart [1964] (fig. 2.1.3). It is assumed that a knowledge of only one system parameter is required for the precipitation input (i.e. the proportion of rainfall which infiltrates during the given storm) and one parameter for the output (the proportion of basin discharge contributed by surface runoff, as distinct from baseflow). The operation of the partial system is based on the assumption that rainfall excess over infiltration can be equated with the storm runoff component of the hydrograph, and that the infiltration makes up the baseflow. A given recorded storm input is thus treated in the infiltration subsystem to disentangle the assumed rainfall excess from the infiltration, and the recorded runoff output is similarly treated in the baseflow subsystem to disentangle the assumed surface runoff from the base-flow. The resulting calculated rainfall excess and calculated surface runoff are then compared in the convolution subsystem and the initial assumptions (i.e. rainfall excess/infiltration and surface runoff/baseflow relationships) are modified until a required measure of accordance has been reached. It is then hoped that any storm precipitation pattern through time can be input into the programme to yield very important information relating to the possible river runoff resulting from it.

The last method of approach to understanding the basin cycle is by the use of 'black-box' techniques. The basin cycle is considered to be a black box, where little or no detailed knowledge is assumed regarding the components or relationships within the cycle and interest is entirely focused on inputs (e.g. rainfall) and outputs (e.g. runoff) and in establishing some direct functional link between them. Because of the complexity of the cycle and the variations of its components from time to time and from area to area it is very difficult to find general mathematical expressions relating rainfall and runoff for a given basin, and little progress has yet been made in this approach.

These five methods of approach to the hydrological cycle are interrelated, and progress in any branch is usually of value in another branch of research, the method chosen being generally dependent on the aims of the investigator.

REFERENCES

AMOROCHO, J. and HART, W. E. [1964], A critique of current methods in hydrologic systems investigation; *Transactions of the American Geophysical Union*, 45, 307–21.

AMOROCHO, J. and HART, W. E. [1965], The use of laboratory catchments in the study of hydrological systems; *Journal of Hydrology*, 3, 106–23.

CHERY, D. L. [1966], Design and tests of a physical watershed model; *Journal of Hydrology*, 4, 224–35.

DAWDY, D. R. and O'DONNELL, T. [1965], Mathematical models of catchment behavior; *Proceedings of the American Society of Civil Engineers, Journal of the Hydraulics Division*, **91**, No. H 74, Part I, 123–37.

MORE, R. J. [1967], Hydrological models and geography; In Chorley, R. J. and Haggett, P., Editors, *Models in Geography* (Methuen and Co., London), pp. 145–85.

NASH, J. E. [1967], The role of parametric hydrology; *Journal of the Institution of Water Engineers*, **21** (5), 435–56.

UNESCO [1964], *Document NS/188, International Hydrological Decade, Inter-Governmental Meeting of Experts, Final Report*.

WOLF, P. O. [1966], Comparison of methods of flood estimation; *The Institution of Civil Engineers, Proceedings of the Symposium on River Flood Hydrology*, 1–23.

II. The Drainage Basin as the Fundamental Geomorphic Unit

R. J. CHORLEY

Department of Geography, Cambridge University

1. Morphometric units

The need for the precise description of the geometry of landforms, particularly those of dominantly fluvial erosive origin, has been a recurring theme in geomorphology, and one of the most important aspects of this has been the search for the basic areal unit within which these data could be collected, organized, and analysed. The conceptions of the nature of these units have been very much a product of the broader methodological approaches to geography and earth science in general, and to geomorphology in particular, and can be grouped into three categories. The first important approach (Fenneman, 1914) sprang from the interest of geographers a half century ago in regional delimitation. The physiographic regions so delimited for the United States were based largely upon considerations of structural geology (e.g. the Ridge and Valley province), although certain gross morphometric attributes, notably relief and degree of dissection, were also used. The modern equivalent of this approach is provided by the terrain analogues of the U.S. Corps of Engineers, who used four terrain factors (characteristic slope, characteristic relief, occurrence of steep slopes greater than 26·5°, and the characteristic plan profile involving the 'peakedness', areal extent, elongation, and orientation of topographic highs) to divide up the *gross landscape* of a region into *component landscapes* in a simple taxonomic manner (fig. 2.II.1). In contrast with this basis, the second approach was concerned to identify 'the physiographic atoms out of which the matter of regions is built' (Wooldridge, 1932, p. 33). These 'atoms', however, were defined as the *facets* of 'flats' and 'slopes' forming the intersecting surfaces characteristic of polycyclic landscapes (Wooldridge, 1932, pp. 31–3), and, although this doctrinaire definition has been relaxed to include *segments* of smoothly curved surface (Savigear, 1965) and to allow the grouping of facets into *landscape patterns*, such as a 'mature river valley' (Beckett and Webster, 1962) (fig. 2.II.1), the genetic overtones and subjective character of this morphometric division limits its usefulness (Gregory and Brown, 1966). The third basis for morphometric division results from the obvious unitary features both of geometry and process exhibited by the erosional drainage basin, as recognized long ago by Playfair (Chorley, Dunn and Beckinsale, 1964, pp. 61–3) and by Davis [1899], who wrote:

D

> Although the river and the hill-side waste sheet do not resemble each other at first sight, they are only the extreme members of a continuous series, and when this generalization is appreciated, one may fairly extend the 'river' all over its basin and up to its very divides. Ordinarily treated, the river is like the veins of a leaf; broadly viewed, it is like the entire leaf.

This topographic, hydraulic, and hydrological unity of the basin provided the basis for the morphometric system of R. E. Horton [1945], as elaborated by Strahler [1964], and it is now employed as a basic erosional landscape element because it is:

1. A limited, convenient, and usually clearly defined and unambiguous topographic unit, available in a nested hierarchy of sizes on the basis of stream ordering.
2. An open physical system in terms of inputs of precipitation and solar radiation, and outputs of discharge, evaporation, and reradiation (Lee, 1964).

2. Linear aspects of the basin

The defining of the perimeter of a drainage basin in the above terms is not difficult, especially as in the majority of instances the ground-water divides are coincident with the topographic ones, but more problems are presented in the definition of the stream-channel network. Definition is especially difficult for the fingertip tributaries in regions of deep soil and plentiful vegetation, whereas in arid shale badlands it is also difficult to distinguish between permanent channels infrequently occupied by runoff and ephemeral rills. The definition of a stream segment, either from the map or in the field, involves five considerations: for a given region or map scale it must have a lower limiting size; it must be connected with the main stream network; it must be 'permanent', as distinct from seasonal; it must form part of a distinctly bifurcating channel pattern; and it must conduct laterally concentrated surface runoff from a well-defined drainage area. A further problem is that the heads of distinct channels are constantly migrating in response to storm excavation or prolonged infill of slope debris (Kirkby and Chorley, 1967).

The linear aspects of stream networks can be analysed from two main viewpoints:

(a) the *topological*, which considers the interconnections of the system and yields some scheme of stream ordering; and
(b) the *geometrical*, having to do with the lengths, shapes, and orientations of the constituent parts of the network.

The recognition of a hierarchy of stream segments is important because of the different morphometric and hydrologic features associated with each. The most widely used ordering scheme was adapted by Strahler [see, for example, 1964] from Horton [1945], in which fingertip channels are specified as order (U) 1, and where two first-order tributaries join, a channel segment of order 2 is formed,

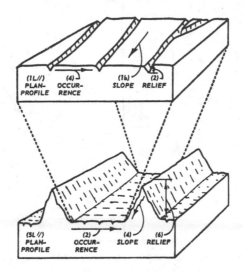

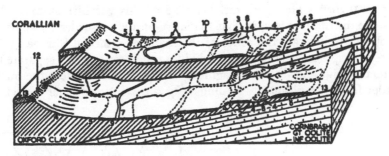

Fig. 2.11.1 Landscape units and geomorphic regions.

Above: Example of a component landscape defined in terms of four terrain factors, and the relation between a component (*top*) and a gross landscape (From Van Lopik and Kolb, 1959).

Below: The pattern of a mature river valley developed by the Upper Thames on the Oxford Clay, illustrating the facets and their relation to each other in the landscape (From Beckett and Webster, 1962).

Facets of river valley and clay

1. High gravel terrace.
3. Clay crest.
5. Clay footslope.
7. River and banks.
9. Flood plain alluvium.

2. Spring line.
4. Clay slope.
6. Unbedded glacial drift.
8. Local bottomland.
10. Old alluvium, not flooded.

Facets of scarplands bounding the river valley pattern

12 (11) Scarp slope.   13. Dipslope.

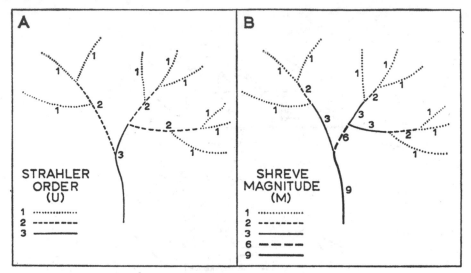

Fig. 2.11.2 Two methods of stream network ordering: (A) Stream segment orders (After Strahler); (B) Stream link magnitudes (After Shreve).

etc. (fig. 2.11.2(*a*)). The main disadvantage of this Strahler system is that it violates the distributive law, in that the entry of a lower-order tributary stream does not always increase the order of the main stream, and Shreve [1966] has proposed a simple remedy for this by dividing the network into separate links at each junction and allowing the magnitude of each link to reflect the number of first-order fingertips ultimately feeding it (fig. 2.11.2(*b*)), and other more involved

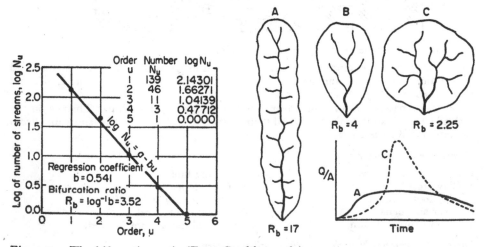

· Fig. 2.11.3 The bifurcation ratio (From Strahler, 1964).
Left: Plot of number of stream segments versus order, with a fitted regression.
Right: Hypothetical drainage basins of differing bifurcation ratios, together with their extreme effects on the runoff hydrograph.

schemes have been suggested. However, the simpler unambiguous Strahler system is now firmly established, and this ordering system provides sequences of stream order numbers (N_1, N_2, ... N_K) which approximate an inverse geometric series for a given basin with the degree of branching, or bifurcation ratio (R_b), given by the ratios N_1/N_2, N_2/N_3, etc., or the antilog of

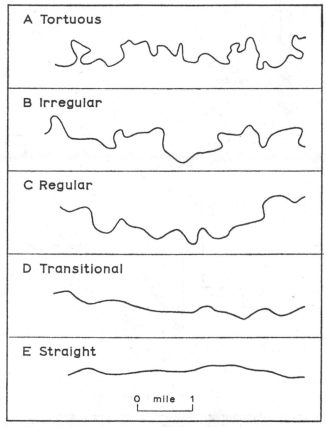

Fig. 2.11.4 Examples of channel patterns (From Schumm, S.A., 1963, *Bulletin of the Geological Society of America*). (A) White River near Whitney, Neb. ($P = 2\cdot1$); (B) Solomon River near Niles, Kan. ($P = 1\cdot7$); (C) South Loup River near St. Michael, Neb. ($P = 1\cdot5$); (D) North Fork Republican River near Benkleman, Neb. ($P = 1\cdot2$); (E) Niobrara River near Hay Springs, Neb. ($P = 1\cdot0$).

the regression coefficient (b) (fig. 2.11.3). The bifurcation ratio, for a given density of drainage lines, is very much controlled by basin shape and shows very little variation (ranging between 3 and 5) in homogeneous bedrock from one area to another. Where structural effects cause basin elongation, however, this value may increase appreciably. Besides influencing the landscape morphometry, the bifurcation ratio is an important control over the 'peakedness' of the runoff hydrograph (fig. 2.11.3) (see Chapter 9.1).

The ratio between the measured length of a stream channel and that of the

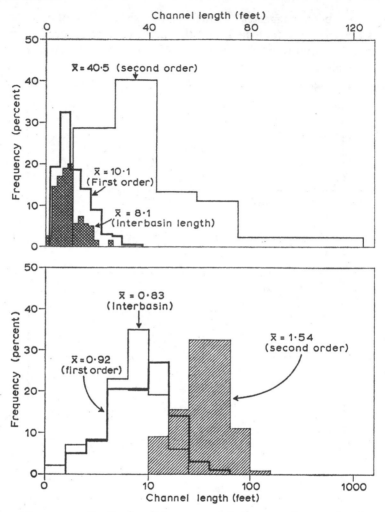

Fig. 2.II.5 Frequency-distribution histograms of first- and second-order channel lengths and maximum interbasin lengths for the Perth Amboy Badlands, New Jersey (From Schumm, 1956).

Above: Actual stream lengths.
Below: Logarithms of stream lengths.

thalweg of its valley is a measure of its sinuosity (fig. 2.II.4). Distributions of lengths of streams of each order in a drainage basin are characteristically right-skewed (log-normal) (fig. 2.II.5), and the plot of mean stream lengths of each order ($L_1, L_2, L_3, \ldots L_K$) in a basin produces an approximation to a direct geometric series (fig. 2.II.6), where the antilog of the regression coefficient is the length ration (R_l).

Obviously the absolute length of the channel system exercises a strong control over the basin lag time (the time difference between rainfall and the resulting

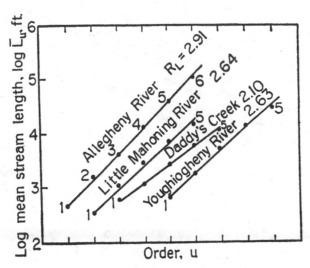

Fig. 2.11.6 Regression of logarithm of mean stream segment length versus order for four drainage basins in the Appalachian Plateau Province (After Morisawa; From Strahler, 1964).

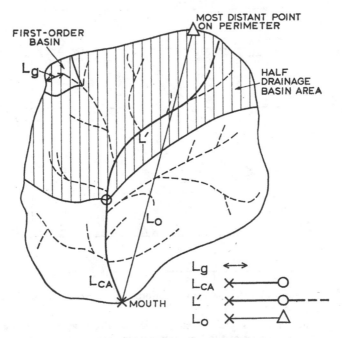

Fig. 2.11.7 Some common drainage basin length parameters.

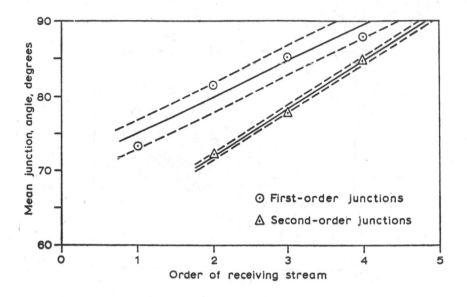

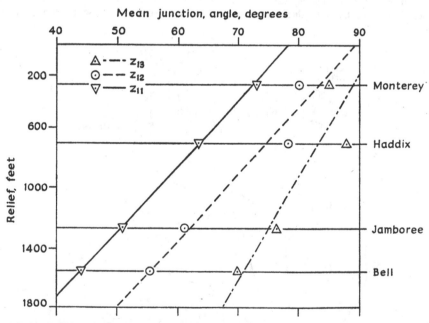

Fig. 2.11.8 Relationships of mean stream junction angles (From Lubowe, J. K., 1964, *American Journal of Science*).

Above: Plot of mean junction angle of first- and second-order with receiving streams of similar or higher order, in part of the Lexington Plain, Kentucky.

Below: Plot of mean junction angles of first-order streams with first-, second-, and third-order receiving streams in four areas of the east, central and western United States.

stream runoff: see Chapter 9.1), as do the following length parameters (fig. 2.II.7):

1. The length of overland flow (L_g), which is the mean horizontal length of the flow path from the divide to the stream in a first-order basin. This parameter is a measure of stream spacing, or degree of dissection, and is approximately one-half the reciprocal of the drainage density $\left(L_g \approx \dfrac{1}{2D}\right)$.

   As the mean velocity of unconcentrated overland flow is less than $\frac{1}{5}$ that of concentrated channel flow (30,000 cm/hr, versus 160,000 cm/hr), L_g is also an important control over lag time. A less-meaningful measure of stream spacing is the mean stream interval (MI), calculated from sampling stream intersections with a grid.

2. The length of the main stream of the basin (L') (usually designated by following up from the mouth those streams which make the least angle with the next lower segment). This is sometimes continued to the basin margin (then called the 'mesh length').

3. The distance to the 'centre of gravity' of the drainage basin (L_{ca}). This is usually measured up the main stream to a point where one half of the drainage basin area lies headward of it, but for most basins $L_{ca} = 0.5L$ is a good approximation. It is sometimes measured to the centre of gravity of the basin area.

4. The length of the longest basin diameter (L_o), measured from the basin mouth to the most distant point on the perimeter. This is useful in the calculation of basin shape.

The entrance angle (Z_c) between a tributary developed in a valley-side slope (of $\theta°$) and joining a larger stream of lower slope ($\gamma°$) would be approximately expressed by:

$$\cos Z_c = \frac{\tan \gamma}{\tan \theta}$$

As the slopes are degraded through time, one might expect θ to approach γ, and consequently Z_c to decrease. It is characteristic that mean values of junction angle increase as the order of the receiving stream increases (i.e as the difference between γ and θ increases), and it is inversely related to relief (for given orders of junction), probably because high relief imparts especially high gradients to the receiving streams (fig. 2.II.8).

3. Areal aspects of the basin

Basin area (conventionally referred to a horizontal datum plane) is hydrologically important because it directly affects the size of the storm hydrograph and the magnitudes of peak and mean runoff (fig. 2.II.9) (See Chapter 9.1). It is interesting that the maximum flood discharge per unit area is inversely related to size, because the most intense storms are usually of the smallest size (fig. 2.II.9) (More, 1967, p. 166). In a given large drainage basin developed in a homogeneous region

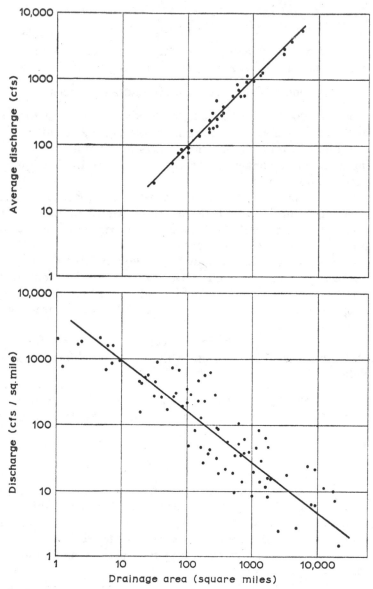

Fig. 2.11.9 Relations between basin area and stream discharge.

Above: Mean discharge (cfs) versus drainage area for all gauging stations on the Potomac River (After Hack: From Strahler, 1964).
Below: Maximum flood discharge (cfs) per square mile versus drainage area for basins in Colorado (From Follansbee, R. and Sawyer, L. R., 1948, *U.S. Geological Survey Water Supply Paper* 997).

basin areas of given order show a logarithmic-normal distribution (fig. 2.11.10), the means of which approximate a direct geometric series (fig. 2.11.11). By relating characteristic discharge to drainage area (the relationship $Q_{2\cdot33} = 12A^{0\cdot79}$ was obtained for basins in central New Mexico, where $Q_{2\cdot33}$ is the flood discharge equalled or exceeded on average once every 2·33 years, or the mean annual

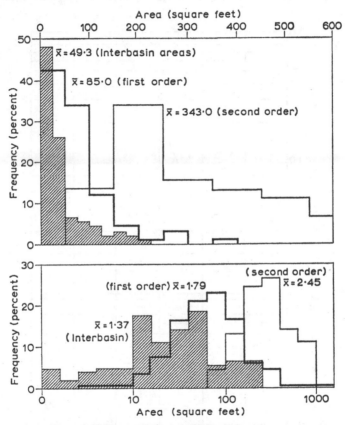

Fig. 2.11.10 Frequency-distribution histograms of first- and second-order basin areas and interbasin areas for the Perth Amboy Badlands, New Jersey (From Schumm, 1956).

Above: Actual basin areas.
Below: Logarithms of basin areas.

flood: see Chapter 7.11), and then area to order, it is possible to show an exponential relationship between order and discharge (fig. 2.11.12).

The relationship of stream length to basin area is important because plots of basin area draining into various locations along the main stream (i.e. area–distance curves) give an idea of the pattern of runoff (fig. 2.11.13), and also because the relationship of total stream lengths of all orders to basin area is one of the most sensitive and variable morphometric parameters, and one which controls the texture of landscape dissection and the spacing of streams. Thus drainage density (*D*), for example, is defined as the total stream length per unit area of basin (e.g. in miles per square mile). Drainage density exhibits a very wide range of values in nature and is commonly believed to reflect the operation of the complex factors controlling surface runoff. Common values of *D* are, for example, 3–4 in the sandstones of Exmoor and the Appalachian plateaus,

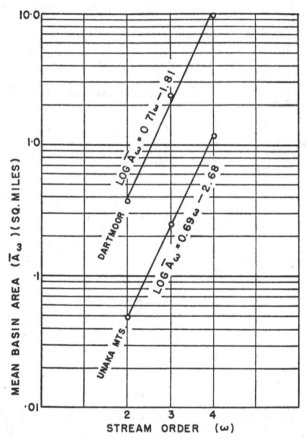

Fig. 2.11.11 Regressions of mean area versus order for drainage basins in the Unaka Mountains, Tenn. and N. Car., and Dartmoor, England (From Chorley, R. J. and Morgan, M. A., 1962, *Bulletin of the Geological Society of America*).

20–30 for the scrub-covered Coast Ranges of California, 200–400 for the shales of the Dakota badlands, and up to 1,300 for unvegetated clays (fig. 2.11.14). An allied measure is the constant of channel maintenance which is the area (in square feet) necessary to maintain 1 ft of drainage channel. Drainage density affects the runoff pattern, in that a high drainage density removes surface runoff rapidly, decreasing the lag time and increasing the peak of the hydrograph (see Chapter 9.1). Of course, basin shape (itself largely controlled by geological structure) is an important control over the geometry of the stream network (fig 2.II.15). Examples of simple shape measures are:

1. The circularity ratio – (see Chapter 9.1).
2. The elongation ratio – the diameter of a circle having the same area as the basin, as a ratio of the maximum basin length (L_0).
3. The measure $(L' . L_{ca})^{0.3}$, found by experience to be a good predictor of basin lag.

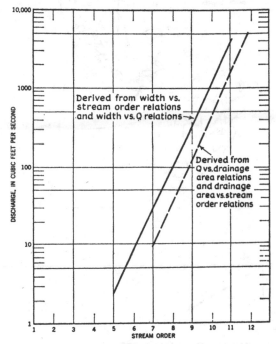

Fig. 2.II.12 Relation of discharge to stream order for ephemeral streams in New Mexico derived by two separate types of analysis (From Leopold, L. B. and Miller, J. P., 1956, *U.S. Geological Survey Professional Paper* 282–A).

It is interesting that, unless pronounced structural control is present, drainage basins differ relatively little in shape, although basins tend to become more elongate with strong relief and steep slopes.

4. Relief aspects of the basin

Longitudinal profiles of stream channels are characteristically concave-up, and it has been suggested that this is due to the increase of discharge downstream not being balanced by any commensurate increase in frictional losses – the increase in the wetted perimeter being more than compensated for by the increase in cross-sectional area and by the decrease of bed material grain size (due to sorting and abrasion during transportation). Many attempts have been made to fit

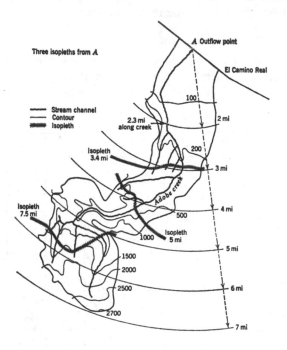

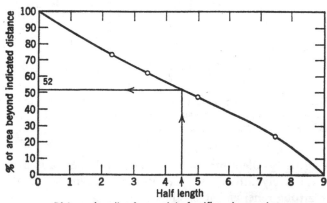

Fig. 2.11.13 Area as a function of channel distance from the basin mouth for Adobe Creek, near Palo Alto, California (Area 11 square miles; drainage density 2·18 miles/square mile) (From De Wiest, 1965).

Above: Adobe Creek, showing channel distance isopleths.
Below: Distribution of area as a function of distance from the basin mouth.

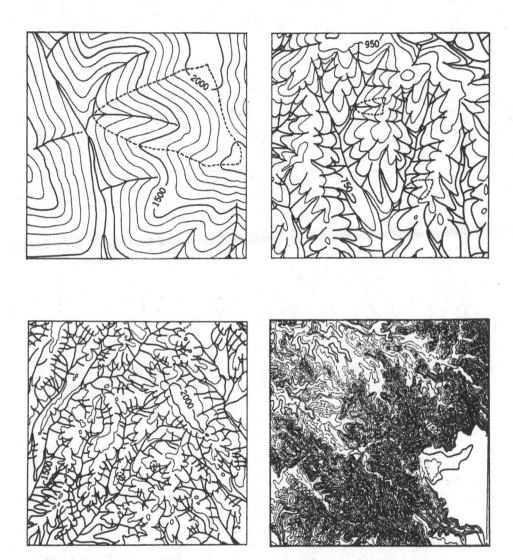

Fig. 2.11.14 Four areas, each of 1 square mile, illustrating the natural range of drainage density (From Strahler, 1964).

Top left: Low drainage density: Driftwood Quad., Penn.
Top right: Medium drainage density: Nashville Quad., Ind.
Bottom left: High drainage density: Little Tujungo Quad., Cal.
Bottom right: Very high drainage density: Cuny Table West Quad., S. Dak.

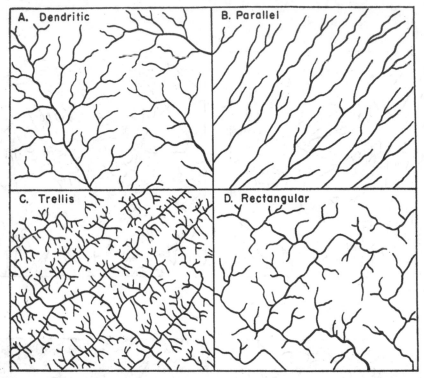

Fig. 2.11.15 Four basic drainage patterns, each occurring at a wide range of scales (From Howard, A. D., 1967, *Bulletin of the American Association of Petroleum Geologists*).

mathematical curves, chiefly logarithmic and exponential, to complete stream profiles, largely to extrapolate them in order to identify supposed ancient base levels. Even the exponential curve, which is most theoretically attractive because it also applies to the rate of attrition of transported debris, usually provides an imperfect fit, and long stream sections seem to be divided into segments by discontinuities which are due to changing discharge, calibre of bedload, or both, or to changing channel characteristics. The influence of discharge is shown by the segmentation on the basis of stream orders in fig. 2.11.16. It has been suggested that in a given basin mean channel gradient bears an inverse geometric relationship to stream order (fig. 2.11.17), although the imperfection of this relationship has led to other more complex ones being suggested. For practical purposes other stream-gradient measures are employed, including:

1. Some measure of the average slope of the main channel in a basin. This is commonly the arithmetic mean slope (S) of the whole channel, or the slope of the 'equivalent' stream (i.e. one having the same length and flood peak travel time) (S_{st}).

2. The simple gross slope of the channel, obtained by dividing the elevation difference between head and mouth by the length of the main stream.

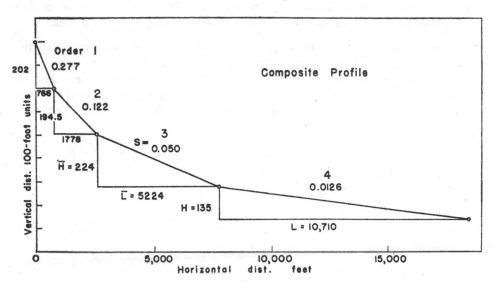

Fig. 2.11.16 Plot of the longitudinal profile of Salt Run, Penn., showing the difference in mean slope of each of the four segments of differing order (From Broscoe, 1959, *Office of Naval Research Technical Report 18, Project NR 389–042, Contract N6 ONR 271–30*).

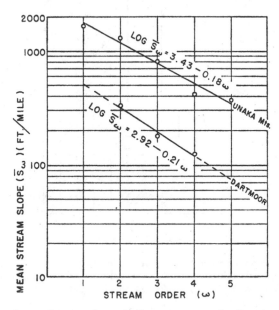

Fig. 2.11.17 Regressions of mean channel slope versus order for streams in the Unaka Mountains, Tenn. and N. Car., and Dartmoor, England (From Chorley, R. J. and Morgan, M. A., 1962, *Bulletin of the Geological Society of America*).

3. The mean slope of the whole channel system, computed by averaging the gradients of all channels draining at least 10% of the total basin area.

These slopes are important because channel slope exercises an important influence over the magnitude of the runoff peak (see Chapter 9.1).

Orthogonal ground-slope angles (S_g) in a basin are commonly measured at the maximum gradient of given valley-side profiles, or expressed as the mean angle of a complete valley-side profile, or sampled over the whole basin. Both the

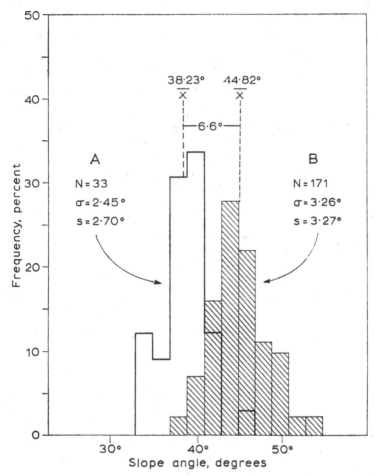

Fig. 2.11.18 Comparison of maximum valley-side slope angles in the Verdugo Hills, California, (A) protected at the base by talus and slope wash, and (B) actively corraded at the base (From Strahler, A. N., 1954, *Journal of Geology*).

maximum and mean valley-side slope angles within a basin are commonly normally distributed (fig. 2.11.18), and, although they are not individually related to the gradient of the basal stream, a mean relationship seems to exist when different regions are compared. The character of the distribution of slope angles sampled over the whole basin depends on the height distribution within it; for a 'just mature' basin with limited flat summit or floodplain areas the distribution is normal, but other basins give skewed distributions, the direction of skew depending on whether the small angles are concentrated on the summits ('youth') or on the floodplains ('late maturity'). Again, simple measures of average ground

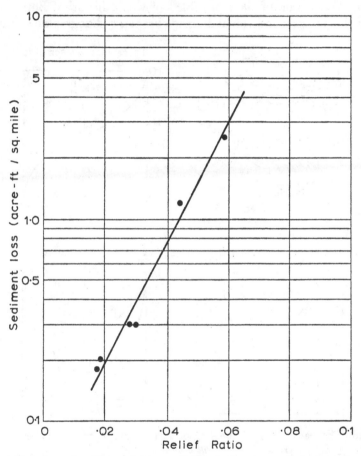

Fig. 2.11.19 Relation of sediment loss to relief ratio for six small drainage basins in the Colorado Plateaus (From Schumm, 1956).

slope within a basin are employed in hydrological analysis, such as the mean basin slope $\left(\dfrac{\text{Total length of contours} \times \text{Contour interval}}{\text{Basin area}}\right)$, and the relief ratio ($R_h$), which is the ratio of the maximum basin relief to the horizontal distance along the longest basin dimension parallel to the main drainage line. Even such crude measures as these can be used to rationalize basin dynamics, it being found that mean basin slope influences the form of the hydrograph and that the relief ratio exercises an important control over rates of sediment loss from some basins (fig. 2.11.19).

The distributional characteristics of land elevations have long been of geomorphic interest, because concentrations of area with elevation (i.e. surfaces of low slope) were believed to be indicative of ancient base levels. Plots of mean land slope versus elevation (clinographic curves) and of amount of surface area versus elevation (hypsometric curves) were prepared for large upland areas to assist in the evaluation of their possible polycyclic histories. Both techniques have been re-applied to individual drainage basins; a clinographic curve giving a

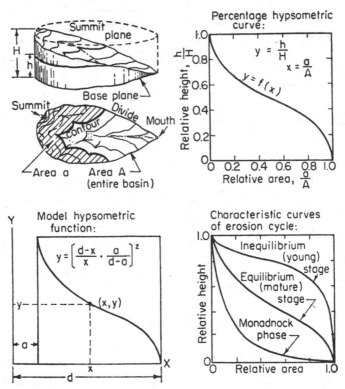

Fig. 2.11.20 The calculation of the hypsometric curve (From Strahler, A.N., 1957, *Transactions of the American Geophysical Union*).

more accurate estimate of land slope than average figures when evaluating the form of the hydrograph, and the dimensionless hypsometric curve representing in some instances the relative stage of basin degradation through time with reference to an assumed original uneroded block (fig. 2.11.20). An especially important hydrological property of the basin related to the distribution of elevations is the amount of floodplain storage available, the effect of which is to make the rising limb of the hydrograph less steep, increase the lag time, and make the peak lower and less pronounced. A knowledge of the distribution of elevations also enables better estimates of rainfall, snowfall, and evaporation in the basin to be made.

In the past, morphometric analysis from maps has been a rather tedious and time-consuming task, but recently techniques have been devised to give it greater facility. The most important of these is the digitizer of the 'pencil-follower' type, which both records the rectangular coordinates of points on a map and also gives a continuous read-out of point coordinates sufficiently closely spaced (maximum recording rate is 20 points per second) to define lines – i.e. contours, stream channels, basin perimeters, etc. The card or tape output can be edited, elaborated, and then fed directly into a computer, together with programmes which will automatically calculate areas, shapes, drainage densities, mean aximuths, maximum slope angles, and the like. Such programming is in its infancy, but already it promises to release the masses of data locked up in topographic maps and will obviously allow much more extensive sampling and generalization of morphometric properties. Before too long these methods will be applied directly to the output from aircraft and satellite scanning equipment, obviating the necessity for the actual compilation of many maps.

5. Some considerations of scale

Considerations of changing scale, both linear and temporal, introduce a certain element of sophistication into studies concerned with morphometric relations.

Similarities such as appear to exist generally between the bifurcation, length, and area ratios of basins of differing size developed on bedrock lacking pronounced structural control have prompted speculation that erosional drainage basins in differing hydrological environments may show a close approximation to geometrical similarity when mean values are considered. If complete geometrical similarity existed one would expect to find all length measurements between corresponding mean points to bear a fixed linear scale ratio, and all corresponding angles to be equal. Although this does not seem to be the case for all morphometric properties (i.e. there appears to be a changing relationship between main-stream length and basin area as basins increase in size within a region, which seems due to both an increasing elongation of larger basins and to the increasing sinuosity of the stream), the similar geometrical relationships between some linear properties suggest a more general significance to the 'laws of morphometry'.

Although it has not been the purpose of this contribution to discuss the relationships between morphometry and erosional processes, it has doubtless become apparent that two quite distinct approaches to this are possible, dependent upon the time-scale with which one is concerned. In the long term it is clear that the hydrological events which compose 'process' must be instrumental in determining the morphometry of the landscape, but, when one is concerned with explaining in the short term the factors which control the character of such individual processes, morphometric features (such as gradient) are commonly invoked. Whether morphometric parameters are viewed as mathematically dependent or independent variables is very much a matter of the time scale employed. Figure 2.11.21 gives two plots involving drainage density, one showing it to be strongly controlled by the precipitation effectiveness and the other

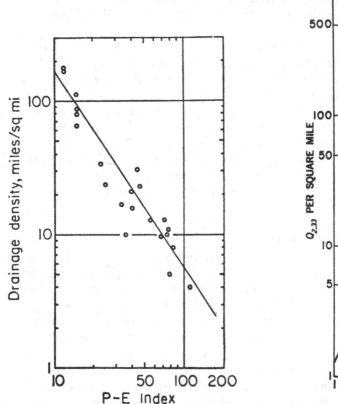

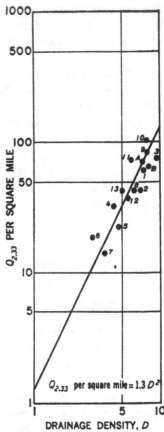

Fig. 2.11.21 The control of drainage density exercised by Thornthwaite's precipitation effectiveness (P–E) index (*Left*) (From Melton, M. A., 1957, *Office of Naval Research Technical Report 16, Project NR 389–042, Contract N6 ONR 271–30*); and (*Right*) the control exercised by drainage density over the mean annual flood ($Q_{2.33}$) for 13 basins in the central and eastern United States (From Carlston, C. A., 1963, *U.S. Geological Survey Professional Paper 422–C*).

indicating how drainage density differences within a region control the peak mean annual flood. A number of the following chapters examine the effects of hydrological processes on aspects of basin morphometry, and Chapter 9.1 uses basin morphometry, among other factors, to analyse the form of the flood hydrograph.

REFERENCES

BECKETT, P. H. T. and WEBSTER, R. [1962], The storage and collection of information on terrain (An interim report); *Military Engineering Experimental Establishment, Christchurch, Hampshire*, 39 pp. (Mimeo).

CHORLEY, R. J., DUNN, A. J., and BECKINSALE, R. P. [1964], *The History of the Study of Landforms*, Volume I (Methuen, London), 678 pp.

CLARKE, J. I. [1966], Morphometry from maps; In Dury, G. H., Editor, *Essays in Geomorphology* (Heinemann, London), pp. 235–74.

DAVIS, W. M. [1899], The geographical cycle; *Geographical Journal*, 14, 481–504.

DE WIEST, R. J. M. [1965], *Geohydrology* (Wiley, New York), 366 pp.

FENNEMAN, N. M. [1914], Physiographic boundaries within the United States; *Annals of the Association of American Geographers*, 4, 84–134.

GOLDING, B. L. and LOW, D. E. [1960], Physical characteristics of drainage basins; *Proceedings of the American Society of Civil Engineers, Journal of the Hydraulics Division*, 86, No. HY 3, 1–11.

GRAY, D. M. [1961], Interrelationships of watershed characteristics; *Journal of Geophysical Research*, 66, 1215–23.

GREGORY, K. J. and BROWN, E. H. [1966], Data processing and the study of land form; *Zeitschrift für Geomorphologie*, Band 10, 237–63.

HORTON, R. E. [1945], Erosional development of streams and their drainage basins: Hydrophysical approach to quantitative morphology; *Bulletin of the Geological Society of America*, 56, 275–370.

KIRKBY, M. J. and CHORLEY, R. J. [1967], Throughflow, overland flow and erosion; *Bulletin of the International Association of Scientific Hydrology*, Year 12(3), 5–21.

LANGBEIN, W. B. *et al.* [1947], Topographic characteristics of drainage basins; *U.S. Geological Survey Water Supply Paper* 968-C, 125–157.

LEE, R. [1964], Potential insolation as a topoclimatic characteristic of drainage basins; *Bulletin of the International Association of Scientific Hydrology*, Year 9, 27–41.

LEOPOLD, L. B., WOLMAN, M. G., and MILLER, J. P. [1964], *Fluvial Processes in Geomorphology* (Freeman, San Francisco), pp. 131–50.

MORE, R. J. [1967], Hydrological models and geography; In Chorley, R. J. and Haggett, P., Editors, *Models in Geography* (Methuen, London), pp. 145–85.

SAVIGEAR, R. A. G. [1965], A technique for morphological mapping; *Annals of the Association of American Geographers*, 55, 514–38.

SCHUMM, S. A. [1956], The evolution of drainage systems and slopes in badlands at Perth Amboy, New Jersey; *Bulletin of the Geological Society of America*, 67, 597–646.

SHREVE, R. L. [1966], Statistical law of stream numbers; *Journal of Geology*, 74, 17–37.

STRAHLER, A. N. [1964], Quantitative geomorphology of drainage basins and channel networks; In Chow, V. T., Editor, *Handbook of Applied Hydrology* (McGraw-Hill, New York), Section 4–11.

VAN LOPIK, J. R. and KOLB, C. R. [1959], A technique for preparing desert terrain analogs; *U.S. Army Engineer Waterways Experiment Station, Vicksburg, Mississippi, Technical Report* 3–506, 70 pp.

WISLER, C. O. and BRATER, E. F. [1959], *Hydrology*; 2nd edn. (Wiley, New York), 408 pp.

WOOLDRIDGE, S. W. [1932], The cycle of erosion and the representation of relief; *Scottish Geographical Magazine*, 48, 30–6.

III. The Drainage Basin as an Historical Basis for Human Activity

C. T. SMITH

Department of Geography, Liverpool University

1. The drainage basin as a unit of historical development

The idea of the drainage basin as a suitable framework for the study and organization of the facts of physical and human geography has a long tradition in the history of the subject. In 1752 Philippe Buache presented a memoir to the French Academy of Sciences in which he outlined the concept of the general topographical unity of the drainage basin. To the cartographers of the late eighteenth and early nineteenth centuries this concept was often caricatured in the mistaken idea that river basins had necessarily to be divided one from another by watersheds so obvious that they could properly be designated by symbols suggesting, in fact, the existence of veritable mountain chains (fig. 2.III.1). Well into the nineteenth century maps sometimes continued to be drawn in this way, even of areas so undistinguished in relief as the plains of European Russia. To some of the academic geographers of the period the identity of the drainage basin, grossly exaggerated as it was by some contemporary cartographers, seemed to offer a concrete and 'natural' unit which could profitably replace political units as the areal context for geographical study (especially in areas of chaotic political fragmentation, such as Germany). It is true that the concept of the watershed as natural boundary and of river systems as unifying networks came into conflict with the idea, also derived to some extent from the study of maps rather than of actual terrain, of the rivers themselves as 'natural' political boundaries, but both concepts disintegrated as knowledge of the earth increased and the work of mapping revealed a much more bewildering complexity (Hartshorne, 1939, p. 45). Ritter, for example, sought refuge in a regional framework which separated upper, middle, and lower portions of stream basins into distinctive types of region (Ritter, 1862), thus reaching a compromise which allowed him to treat of mountain regions in their own right instead of simply as watershed divisions.

Yet in some quarters the idea of the drainage basin as a 'natural' division for regional study survived until quite recently. In his study of the human geography of France, Jean Brunhes based his major divisions of the country on the drainage basins of the Garonne, Loire, Seine, and Rhône–Saône and their major towns. His argument for using this method is based partly on convenience and partly

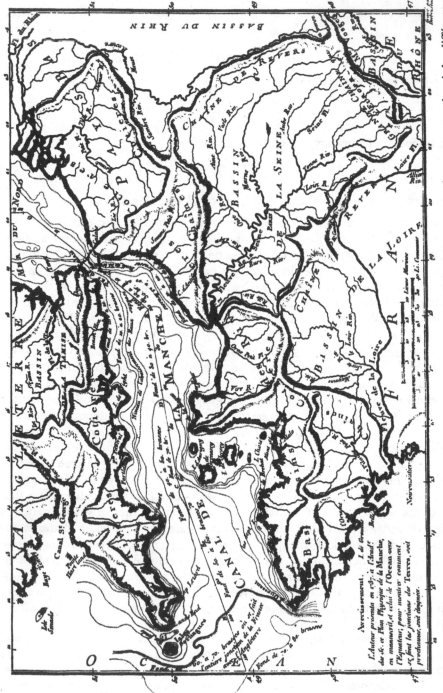

Fig. 2.III.1 Parts of northern France and southern Britain 'dont les eaux s'écoulent directement dans ces mers, depuis les différentes chaînes de montagnes' (From Buache, 1756, Plate XIV).

on a recognition of the importance of water as a link between the earth and man's activities. 'Water is the sovereign wealth of a state and its people. It is nourishment; it is fertilizer; it is power; it is transport' (Brunhes, 1920, p. 93). He goes on to argue that watercourses 'are certainly not the only geographical connection of earth with man, but they do present some principle or possibility of a linkage' (Brunhes, 1920, p. 102). But a more archaic theme is evident in his argument, too, recalling earlier attempts to use the drainage basin as a precisely identifiable areal unit within which to group geographical facts: 'Pays, provinces and regions are not simple natural facts, and are necessarily as variable in their vital and historical expression as they are in their boundaries. Water-courses alone are the precise realities.'

A few years before, in a much more explicit and forward-looking statement of regional divisions of England, C. B. Fawcett also placed considerable emphasis on the drainage basin as a means of territorial division (Fawcett, 1917). His 'provinces' of England were to be the basis for a reorganization of regional government in England, and were arrived at by the application of several criteria. Thus, the size of his regions was to be such that they contained not less than a million people and not more than the total population of the next two smaller provinces; they should contain a regional capital; they should not cut across county loyalties more than was necessary, nor should their boundaries interfere with normal movements of population, particularly between work and home. But their boundaries were also to be based on drainage basins. His argument for this view is worth quoting.

> Since the vital functions of local government include such matters of public health as water-supply and drainage, the making and maintenance of roads and the supply and control of trams, gas and electricity, and since the lines of these are most naturally and easily laid out along the valleys, it will be ordinarily desirable that the boundaries should be drawn near the watersheds, (though) the watershed would only mark out the general trend of a boundary and not govern its details.

In drawing up his map of the proposed provinces, he certainly followed this principle in a number of regions, though not in all. Northern England was to be essentially the basins of the Tyne, Tees, and Eden, bisecting the Lake District and the North York Moors. The Severn region centred on Birmingham consisted more or less of the Severn and Avon basins; the Exe basin was carved from Somerset to be included in the province of Cornwall and Devon. His system was clearly most applicable where watersheds or crest-lines were distinct and where small drainage basins could be grouped together to form a province of suitable size. It was less happy in the Fenland or the Midlands, where divisions had necessarily to be more arbitrary. And although water supply and drainage were frequently valley oriented, one may question whether, in 1917, the provision of trams, gas, and electricity had more than local significance except in the case of London, or whether the cost of road-making was significantly different within drainage basins rather than between them. One is tempted to conclude that Fawcett, like Brunhes to some extent, found the drainage basin and its watershed simply as a convenient means for circumscribing areas which might have been better delineated in terms of urban spheres of influence. Whatever unity had formerly attached to the drainage basin as such in the days of river navigation had largely disappeared in the context of railway-building and the predominance of overland movement. Both Brunhes and Fawcett leave us in doubt as to the precise nature of the link they were seeking to establish, and with the suspicion that such links as there were may have been largely of historical importance by the time they were writing.

Indeed, from the period of primary settlement and colonization to the Industrial Revolution and the revolution in transport, the drainage basin, and more particularly stream networks, had been directly linked with human activity in a variety of ways. Such functional connections could be of two kinds: those which are related to stream networks as water bodies, and those which are related to the existence of systematically arranged patterns of resources (soils, vegetation, local climates, for example), which are themselves organized with respect to the relief, slopes, and stream networks within drainage basins.

The major linkages of the first type would consist of: (a) water supply for domestic purposes and the watering of animals; (b) supplies of fish and game, which are in turn dependent on water supply; (c) a greater ease of accessibility and movement either by navigation or by way of easy overland movement along valley floors; (d) direct water power; (e) water supply for the irrigation of crops. Major linkages of the second type, depending on an association of resources, cannot so easily be generalized, though in areas where the drainage basins are small and relief highly articulated, as in the Alps or in the valleys of peninsular Greece, there may be a complementary pattern of resource which helps to integrate a human society with respect to the drainage basin as a whole. Patterns of seasonal movement between agricultural land in the valleys and highland summer pastures are classic examples of this kind of integration.

Colonization and settlement, particularly before the railway age, have frequently been guided by the lines of accessibility afforded by stream networks, by water supply, or by the concentration of resources afforded by sites in close proximity to the water bodies themselves. In the Anglo-Saxon settlement of England, for example, place-names and archaeological evidence reveal the importance of major river systems as lines of entry and as axes from which subsequent settlement spread towards the watersheds. In the Cambridge region, particularly, early Anglian place-names are found fairly close to the Cam and its tributaries, but the distribution of later clearing and secondary names suggests an expansion of settlement from the valleys towards the watershed regions. The settlement of the East Anglian Stour shows a similar type of sequence, with an added contrast between the nucleated settlement of the valley and the large parishes, dispersed settlement, and late place-names of the watershed. The Welland, the Nene, and the Ouse were similarly lines of entry marked by a scatter of early place-names from which settlement spread towards the watershed zones. In cases such as these easily accessible resources (water supplies, natural meadow, and easily cultivated arable land) may have been more significant than navigability, for the unity of settlement within the drainage basin obviously extends far above navigable limits. And, indeed, primary settlements of the main valleys often continued to retain a leadership in size and wealth long after colonization had been completed (e.g. in the Soar, Nene, Welland, and Cam valleys of the English Midlands).

It would, however, be easy to overemphasize the importance of the drainage basin as such in the primary settlement of north-western Europe, for there were alternative zones of easy movement and attractive resources, such as those offered by the *löss* zone, the relatively open scarp lands of south-eastern England, or the uplands of western Britain.

New links of a different kind were forged in the drainage systems of Western Europe, however, with the emergence of trade and the growth of towns from the tenth and eleventh centuries. In a pre-industrial economy water transport was often more efficient than overland carriage, and for many purposes river navigation was used wherever feasible. Expensive luxuries, such as silks, spices, and dyestuffs, may have been able to stand the cost of transport overland, but bulk trade undoubtedly showed a continued preference for water traffic. The grain trade of the Baltic was channelled along the Oder and the Vistula; timber from central Europe and southern Germany used the waterways; and the expansion and location of viticulture in medieval France and Germany was very strongly controlled by accessibility to navigable water in the Seine basin, the Loire and the Garonne, and in Germany along the Rhine and the Moselle.

Most of the major rivers of western Europe became arteries for the flow of bulk commodities, nourishing a series of urban settlements at their estuaries as seaports, at their lowest feasible bridge-points, at their confluences with major tributaries, at junctions with important overland routes, or at the head of navigation. Indeed, there were very few important towns in western Europe before 1800 which were not located near navigable water. The importance of access to water, even for relatively small towns, may be gauged to some extent from the taxes paid in 1334 by fifty English towns, excluding London. Ten were ports, all on estuaries, and paid an average of £73; twenty-six were on rivers probably used for navigation at this period, and paid an average of £36; ten were not on navigable water, but they paid, on average only £19 in tax; and the remaining four, doubtfully navigable or at best at the head of navigation, paid an average of £14.

It would, indeed, be interesting to test the hypothesis that accessibility by river navigation created a link between the stream networks of a drainage basin and the generation of pre-industrial urban fields, but it would certainly be hazardous to stress too much the dependence of urban settlement and trade on river navigation within the context of individual drainage basins. Long-distance transit trade may have used water transport as much as possible, as for example, between Milan and the Rhine across Italian lakes to the St Gotthard to the Vierwaldstättersee, but the drainage basins they crossed were by no means thereby unified; and man-made obstacles in the form of weirs, taxes, rolls, and brigandage often made overland transport a feasible alternative. And some of the major routes of the Middle Ages, such as the Hellweg across the margins of the Hercynian uplands in Germany, the route from Italy to the Low Countries by way of the Champagne Fairs, or the overland route from Bruges to Cologne, ignored water transport even when it was a viable alternative.

The relative advantage of water routes over land transport has varied greatly, indeed, over time and space. In thickly forested regions difficult of access by any other means water transport has at times provided the only key to exploration, settlement, and exploitation. In the settlement of North America the search for beaver took the fur-traders farther and farther afield to explore the interior of the continent as fur supplies nearer to the Atlantic base were exhausted. For the *coureurs du bois*, living off the country and equipped with the Indian canoe and a few trade goods, the drainage basins were the natural sphere of activity, one linked with another by the shortest possible portages (Brebner, 1933). The short-lived French possession of the Mississippi basin and the foundation of New Orleans were a reflection of this precocious movement. In a similar way, the rapidity of Russian expansion across northern Asia to the Pacific in the seventeenth century also depended on a combination of exploration, the fur trade and the ease of movement along river axes linked by the shortest possible overland portages (Mitchell, 1949). In the Amazon basin the availability of fish and game as well as transport on the major streams supported a more numerous and culturally rich population than the remoter streams and inaccessible watershed regions (Denevan, 1966), and it was precisely these groups who were most strongly affected by the penetration of white populations and cultural change on the opening up of the region by the rubber and cinchona booms of the nineteenth century. In the dense forests of the Amazon basin, the stream network still provides the only integrated transport system, and except in some marginal regions, settlement and commercial activity are almost completely dependent on it.

Changing technologies of transport have, of course, severely eroded such unity as ever attached to the smaller drainage basins of western Europe in terms of navigation and commerce. But in other ways, the coming of steam in the nineteenth century acted as a new stimulus to inland water transport in mid-nineteenth century, for the application of steam power to river navigation from 1807 seemed to emancipate it from the tyranny of sail and variable winds, the tedium of man-power or the restrictions of the towpath. In the Mississippi the great age of the steamboat lasted until the coming of the railway and the re-orientation of the trade of the Mid-West from the south and New Orleans to Chicago and the north-east. In other parts of the world steam navigation seemed to offer the possibility of cheap transport in the opening up of new lands with much smaller capital investment than was needed for railway construction, and ushered in a wave of optimism for the integrated development of major, navigable river basins: the Amazon, the Congo, the Mississippi, the Parana–Paraguay, the Russian rivers, and even the Danube.

Navigation was perhaps the major force creating some sort of unity in the lower parts of drainage basins, but it obviously had no relevance for the upper reaches of shallow waters and high gradients. In a few highly specialized regions of localized industry, these were precisely the zones that achieved some sort of unity through the use of direct water-power for industrial purposes. In the iron industry of the charcoal era, for example, water-power was needed for various purposes, and although the location of iron-making and metal-working industries was usually a product of complex situations and associations of resource, the availability of water-power often controlled the detailed siting of industrial plant, effectively concentrating activity in upland drainage basins. The constraints of water-power were less strong in textile industries before the nineteenth century, though they certainly existed, but waterside locations were needed for wool-washing, fulling, dyeing, and bleaching. Such types of industrial association were to be found in many parts of Western Europe before the Industrial Revolution (Smith, 1967): for example, in the southern tributaries of the Meuse above Liège and Namur (iron and textiles), the Sieg, Lahn, and Wuppertal in Westphalia (iron and textiles), in the minor drainage basins debouching into the St Etienne trough in the Central Massif (metals and textiles), and in various parts of England and Wales (the Yorkshire and Lancashire textile regions, minor valleys above Sheffield, the Stroud valley in Gloucestershire, and in earlier years in the basin of the East Anglian Stour or the Parett basin in Somerset. And where there was navigable water immediately below the zone of exploitable water-power, possibilities existed for the economic integration of a drainage basin on a larger scale, as in the Severn basin or in the Meuse basin below Namur.

2. Irrigation and the unity of the drainage basin

The closest and probably the most widespread association of past human activity with the hydrological balance, relief, slopes, and stream networks of the drainage basin has been achieved through the operation of irrigation systems. Primitive irrigation systems are discussed elsewhere (Chapter 4.III(ii)), but in the context of the drainage basin, the most important types of irrigation are those which involve some degree of communal action for their construction, operation, and maintenance. Of these the chief (in terms of historical importance and geographical spread) are those which use water from a stream network and have more or less elaborate systems of canals for the distribution of water to individual settlements. Irrigation of this type has been characteristic of the great hydraulic societies of India and China or of the New World in pre-Columbian Peru. Limits are set to the extension of irrigation systems of this type by the availability of suitable terrain for cultivation or for the distribution of water, usually within the context of a single drainage basin; by the volume and seasonal regime of water supply; by the technology available; and not least by the scale and nature of social and political organization within which the construction, maintenance, and administration of water-control must be carried out (Forbes, 1965, pp. 4–5). Small-scale and piecemeal irrigation is possible within parts of a drainage basin, and early operations in Western Europe and Peru were often of this kind. But competition for water supply and the pressure of population leading to demands for the extension of irrigation to new land frequently seem to have led at an early stage to the integration of piecemeal systems within the drainage basin as a whole. In coastal Peru, for example, early piecemeal irrigation was soon replaced by canal systems incorporating the whole of the irrigable area of a drainage basin below the Andean foothills. And parallel developments, involving the drainage basin (or the lower part of it) as an integrated unit for defence, religious and administrative control, and the supply of urban centres, point also to the extension of a single political authority over the whole of a valley region (Willey, 1953).

Wittfogel has seen the social and political implications of pre-industrial irrigation systems of water control in terms of coercion and the formation of 'oriental despotism' (Wittfogel, 1957). The relevant part of his argument may be briefly summarized, that the construction and maintenance of large-scale irrigation systems require the assembly of a considerable labour force which may be most efficiently created either by the institution of forced labour or the levy of tribute and taxation or both. A centralized administration is also needed for the maintenance of canals and to control water distribution. The administration in control of the distribution of water is, in effect, in complete control of agricultural activity, and is thus in a position to demand complete authority and complete submissiveness, subject only to mass revolt and rebellion in the face of desperate conditions. Society becomes polarized, in fact, into an illiterate, dependent peasantry and an *élite*, as in the traditional bureaucratic governments of China.

The role of the drainage basin as the fundamental territorial unit on which this type of administrative superstructure may be built and with respect to which the irrigation systems are constructed has not been systematically examined, though it is clearly evident in coastal Peru, where the integration of individual basins was followed by the creation of coastal empires built up from adjacent valley systems, unified by dominant urban centres such as that at Chan-Chan (Bushnell, 1956, p. 114). In China an attempt has been made to establish a relationship between the establishment of new irrigation works and the geographical basis of political power in some of the early Chinese dynasties, in which the unit of the drainage basin has considerable importance (Chao-Ting Chi, 1936). He identified the source of early dynastic power (of the Ch'in and Han dynasties) in the third century B.C. in the zone of the Loess Highlands, where the tributary valleys of the Wei, Fen, Lo, and Chin converge at the great bend of the Hwang Ho. These were small valleys, easily cleared of their woodland and with *löss* soils fertile under irrigation. Streams were relatively short and much more easily controlled than the Hwang Ho itself. The scale of the environment was sufficiently small for the control of these minor drainage basins to be achieved by available technology and administration, but the irrigation systems of these valleys, and particularly the Wei Ho, supported a powerful bureaucratic state with its capital in the Wei valley itself at Chang'an (near Sian). New and larger irrigation works later widened the basis of tribute-collection and thus of political power. They were made possible by new technological advance and by the increase of population, which was itself a measure of the success of earlier irrigation projects. And these new works were now located in the middle course of the Hwang Ho in the North China Plain. Significantly, the capital was shifted to Loyang at the margin of the North China Plain itself, which now replaced the Loess Highlands as the 'key economic area' of Chinese government. Subsequently, however, the shift of irrigation projects farther south into the Hwai River basin, the Red basin of Szechuan, and then the Lower Yangtze created a new basis of economic and political power, and by the time of the T'ang dynasty between the seventh and the tenth centuries A.D. the 'key economic area' had shifted towards central China and the Yangtze basin.

In a variety of ways the drainage basin has formed a framework for human activity: in guiding the direction of primary settlement, in river navigation and the growth of trade and towns, in the provision of water-power for industrial concentrations, and in providing a logical context for irrigation works. But in many cases these functions have created a unity, not in the drainage basin as a whole but rather in those parts of a drainage basin which have relevance for a particular activity. Few recognized, even dimly, the interrelation of the whole drainage basin in any conscious way. The Chinese never appear to have realized, for example, the relationship between deforestation in the Loess Highlands and the floods, silting, and droughts of the North China Plain.

Indeed, the idea of the drainage basin as an appropriate areal unit for the organization of human activity and for regional planning has only recently been revived with the recognition of the basin as an interrelated system in which soil and vegetation cover as well as hydrological balance are involved, and with the recognition of the need for integrated plans and policies to deal with problems posed by flood-control, sedimentation, soil erosion, hydroelectric power production, navigation, and even nature conservation and stream pollution. The Tennessee Valley Authority fulfilled such a need for a regional authority transcending the boundaries of traditional units of government, and it was the prototype from which stemmed other proposals and organizations: a Missouri valley authority; an abortive proposal for a Danube Valley authority, mooted just after the Second World War, when the future of central Europe was still in the melting pot (Kish, 1947); and the regional corporations of the Cauca valley (1954) and the Magdalena valley (1960) in Colombia (Banco de la Republica, 1962). Gilbert White has written of the Mekong valley project as a programme transcending political boundaries, and aiming at an integrated approach to the problem of water control and environmental planning in the greater part of a major drainage basin (White, 1963). And in recent months a programme has been announced for a preliminary study of the feasibility of an integrated development scheme for the basins of the Paraná, Paraguay, Uruguay, and River Plate which will involve co-operation by Argentina, Bolivia, Brazil, Paraguay, and Uruguay (*Peruvian Times*, 1967).

E

REFERENCES

BANCO DE LA REPUBLICA (Colombia) [1962], *Atlas de Economía Colombiana*; Volume 4, Aspectos agropecuarios y su fundamento ecologico.

BREBNER, J. B. [1933], *Explorers of North America 1492–1806;* (London), 501 p.

BRUNHES, J. [1920], *Géographie humaine de la France*; Hanotaux, 6; Editor, Histoire de la France, Vol. 1 (Paris), 493 p.

BUACHE, M. [1756], Essai de géographie physique; *Mémoires de Mathematique et de Physique, Académie Royale des Sciences*, 1752, pp. 399–416.

BUSHNELL, G. H. S. [1956], *Peru* (London), 216 p.

CHAO-TING CHI, T. [1936], *Key Economic Areas in Chinese History* (London), 168 p.

DENEVAN, W. M. [1966], A cultural-ecological view of former aboriginal settlement in the Amazon basin; *The Professional Geographer*, **18**, 346–51.

FAWCETT, C. B. [1917], The natural divisions of England; *Geographical Journal*, **49**, 124–41.

FORBES, R. J. [1965], *Irrigation and Power*; Studies in Ancient Technology; Vol. 2 (Leiden), 220 p.

HARTSHORNE, R. [1939], *The Nature of Geography* (Pennsylvania), 482 p.

KISH, G. [1947], TVA on the Danube?; *Geographical Review*, **37**, 274–302.

MITCHELL, M. [1949], *The Maritime History of Russia, 848–1948* (London), 530 p.

Peruvian Times [1967], news article, 14 July.

RITTER, K. [1862], *Comparative Geography*; translated W. L. Gage (Edinburgh), 254 p.

SMITH, C. T. [1967], *An Historical Geography of Western Europe before 1800* (London), 582 p.

TECLAFF, L. A. [1967], *The River Basin in History and Law* (Martinus Nidhoff, the Hague), 228 p.

WHITE, G. F. [1963], Contributions of geographical analysis to river basin development; *Geographical Journal*, **129**, 412–36.

WILLEY, G. R. [1953], Prehistoric settlement patterns in the Virú valley; *Bureau of American Ethnology*, Smithsonian Institute, Bulletin 155, 453 p.

WITTFOGEL, K. A. [1957], *Oriental Despotism* (New Haven and London), 556 p.

CHAPTER 3

Precipitation

I(i). Precipitation

R. G. BARRY

Department of Geography, University of Colorado

1. General categories

Inhabitants of middle latitudes are familiar with four major precipitation categories – drizzle, rain, snow, and hail. For scientific purposes a more detailed classification is necessary (Table 3.1(i).1).

TABLE 3.1(i).1 The major categories of precipitation

Type	Characteristics	Typical amount
Dew	Deposited on surface, particularly a vegetation canopy (frozen form—hoar frost)	0·1–1·0 mm/night
Fog-drip	Deposited on vegetation and other obstacles from fog (frozen form—rime)	Up to 4 mm/hr
Drizzle	Droplets <0·5 mm in diameter (freezing drizzle when surface temperature below 0° C)	0·2–0·5 mm/hr
Rain	Drops >0·5 mm diameter, typically 1–2 mm diameter	Light <2 mm/hr Heavy >7 mm/hr
Sleet (Great Britain)	Partly melted snow or a rain and snow mixture	
Snowflakes	Aggregations of ice crystals up to several cm across	
Snow grains (granular snow)	Very small, flat opaque grains of ice; the solid equivalent of drizzle	
Snow pellets (graupel, or soft hail)	Opaque pellets of ice 2–5 mm diameter, falling in showers	
Ice pellets (small hail)	Clear ice encasing a snowflake or snow pellet	
Ice pellets (sleet in the U.S.)	Frozen rain or drizzle drops	
Hail	Roughly spherical lumps of ice, 5–50 mm or more diameter, showing a layered structure of opaque and clear ice in cross-section	

2. The precipitation mechanism

Condensation occurs when air is cooled to its dew-point temperature and suitable hygroscopic nuclei are present to initiate droplet formation. Dew-point is the temperature at which saturation occurs if air is cooled at constant pressure. For present purposes it is convenient to differentiate between condensation in the atmosphere and that very close to the ground.

When air rises it is cooled by the adiabatic expansion due to lower pressures, and uplift beyond the level at which condensation occurs leads to the formation of clouds. Precipitation rarely begins until at least 30 minutes after cloud has been observed to form overhead, and many clouds are observed to dissipate without precipitating. This suggests that the growth of cloud droplets, 1–100 microns (1 micron = 10^{-4} cm) in diameter, into rain drops with a diameter of 1 mm or more (1,000 microns) is by no means automatic. In shower clouds with tops not reaching above the freezing level, and more generally in clouds in the tropics, the main mechanism of droplet growth is coalescence. Where ice crystals and supercooled droplets, i.e. liquid at below freezing temperatures, are present in a cloud, the crystals grow at the expense of the droplets because the saturation vapour pressure is lower over ice than over water. Eventually the ice crystals fall from the cloud, and if they pass into air warmer than about 2–3° C they usually melt into rain drops. The microphysics of clouds is treated in most meteorological textbooks; the work of Mason [1962] is specifically concerned with this subject.

Important contributions to the total precipitation are sometimes made by dew and fog. Dew forms on the ground and vegetation surfaces as a result of deposition from the atmosphere when the air is cooled to its dew-point temperature by contact with the radiatively cooling surface. However, in very still air most of the dew originates from vapour derived by evaporation from the soil. The rate of deposition is limited by the rate of removal of latent heat of condensation from the surface. Average dew deposition measured on lysimeters at Coshocton, Ohio, during the snow-free period 1944–55 was between 0·38 and 0·75 mm (0·015 and 0·030 in.) per day.

If the wind stirs the air sufficiently fog droplets are held in suspension. These may accumulate on vegetation and other surfaces as 'fog drip' or, if the drops are supercooled, in frozen form as rime. A fog gauge of wire gauze on Table Mountain, Cape Town, collected 1·7 times more moisture than fell in an ordinary rain gauge during a twelve-month period. Vegetation may be a less efficient collector, but it seems probable that such additions to the moisture budget are important in coastal and montane forests.

3. Genetic types

It is usual to recognize three main types of precipitation, according to the mode of uplift of the air.

A. 'Convective type' precipitation

This occurs in the form of showers or heavy downpours with cumulus and cumulonimbus clouds. Rainfall rates may be of the order of 25 mm (1 in.) per hour. Three subcategories, distinguished according to the spatial organization of the precipitation, are as follows.

1. Summer heating over land causes scattered thundery showers of rain or occasionally hail.
2. Cold, moist air moving over a warmer sea or land surface frequently gives rain (or snow) showers. The convective cells tend to travel with the wind, producing a streaky distribution of precipitation parallel to the wind direction (Bergeron, 1960). The cells may otherwise be organized into bands, some hundreds of kilometres long, perpendicular to the airflow, particularly in association with an advancing wedge of cold air (a cold front or a polar trough).
3. In tropical cyclones cumulonimbus cells become organized about the vortex in spiralling bands of cloud mass. The rainfall can be very heavy and prolonged, affecting large areas. This is often classed as a separate category or as a special case of 'cyclonic type' precipitation. However, a key feature of tropical storms is their mode of energy supply. Energy is generated by the release of latent heat of condensation in cumulonimbus towers. In other words, the small-scale convection maintains the large-scale circulation.

B. 'Cyclonic type' precipitation

In this case the horizontal convergence of airstreams within a low-pressure area brings about widespread upward air motion. In the forward sector of a mid-latitude depression warm air overlies colder air (a warm front) and there is usually deep multi-layered cloud of the nimbostratus type. This gives fairly continuous light to moderate precipitation over very extensive areas. The precipitation may last 6–12 hr or more, according to the width of the rain belt and the speed of the depression. In the rear sector, where cold air tends to undercut warmer air (a cold front, often with associated squall lines preceding the air-mass boundary) heavy showers, quite frequently accompanied by thunder, are usual.

The inadequacy of the simple three-fold classification of precipitation types is shown by the fact that, in the equatorial low-pressure zone, airstream convergence in the easterlies gives rise to westward-moving bands of convective precipitation from cumuliform clouds. This 'cyclonic type' of precipitation could equally well be grouped under A. 2.

C. Orographic precipitation

In the strict sense this term implies that precipitation occurs over high ground when none is falling on the surrounding plains. More frequently, it is a *component* of the total precipitation resulting from the effect of orography on the basic convective and cyclonic mechanisms. The effect is dependent on the size of the barrier and its alignment with respect to the wind. Over narrow uplands the horizontal scale may be insufficient for maximum cloud build-up, and precipitation may be carried over the crest-line by the wind, causing a lee-side maximum.

In middle and higher latitudes it seems that where onshore westerlies are forced to rise sharply over coastal mountains precipitation may increase with height up to 2,000 m (6,500 ft) or more. Walker [1961] considers that maximum precipitation occurs at the level of the cloud base and estimates that in the western Cordillera of British Columbia the maximum precipitation zone occurs as follows (in thousands of feet):

	Coast		Interior	
	South	*North*	*South*	*North*
Summer	6	4	7	6
Winter		4		5

Farther inland the maximum may occur well below the summit levels. For example, in Norway it is located about half the horizontal distance along the windward slope due to the considerable width of the mountain belt. In Java there is a marked decrease above about 2,000 m, and in Hawaii the maximum of more than 800 cm (320 in.) occurs on the eastern slopes of the mountains at only 1,000 m. Yet on some of the Hawaiian Islands peaks rising to 2,000 m receive their maximum on the summit. The reason for these variations appears to de-

TABLE 3.1(i).2 The occurrence of precipitation types in England and Wales, 1956–60 (after E. M. Shaw and R. P. Mathews)

Station	*Warm front*	*Warm sector*	*Cold front*	*Occlu-sion*	*Polar low*	mP	cP	*Arctic*	*Thunder-storms*
Cwm Dyli, Snowdonia (324 ft)	18	30	13	10	5	22	0·1	0·8	0·8
Squires Gate Blackpool (33 ft)	23	16	14	15	7	22	0·2	0·7	3
Rotherham, Yorkshire (70 ft)	26	9	11	20	14	15	1·5	1·1	3
Cranwell, Lincolnshire (208 ft)	27	10·5	14	19	9	11	2·0	1·9	5

mP = maritime polar air
cP = continental polar air

pend not only on the vertical distribution of water content, and consequently on the type of cloud system, but also on the precise effects of the mountains on the airflow. The influence of a given mountain range is, of course, markedly affected by the movement of weather systems from different directions. The simple climatological concepts of rainfall increase with height and lee-side rain shadow need to be replaced by more realistic models for a variety of synoptic situations in each mountain area.

Convergence and uplift occur over coastal areas when air moving inland is slowed down by friction. This special type of orographic effect is evident in the patterns of average seasonal precipitation over south-east Sweden, for example, and also on leeward shores of the Great Lakes and Hudson Bay in early winter.

Table 3.1(i).2 provides a more detailed breakdown of annual rainfall at stations in England and Wales during 1956–60. The orographic component is particularly evident for warm-sector precipitation at Cwm Dyli. Air mass showers (mP) diminish in importance eastward across the country.

4. Precipitation characteristics

Basic information about daily precipitation amounts is supplied by rain-gauge and climatological stations. From these data are compiled statistics of average monthly and average annual precipitation, annual variability, and the number of rain-days ($\geqslant$0·01 in. or >0·2 mm in Britain). Invaluable as such records are, it is essential in hydrological studies to know more about the characteristics of individual rainstorms. Three important parameters of storm rainfall are intensity, frequency, and areal extent.

A. Intensity

Rainfall intensity (= amount/duration) is of vital interest to hydrologists concerned with flood prevention and conservationists dealing with soil erosion.

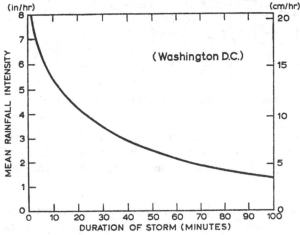

Fig. 3.1(i).1 Generalized relationship between precipitation intensity and duration for Washington, D.C. (After Yarnell, 1935).

Intensity has to be determined from chart records ('hyetograms') of rate-of-rainfall recorders. The results can be presented in the form of an intensity-duration graph as illustrated in fig. 3.1(i).1 for Washington, D.C. Analysis of record precipitation rates from different parts of the world shows that the expected *Global extreme* intensity (in./hr) $\approx \dfrac{14\cdot3}{\sqrt{\text{Duration}}}$. However, this particular assessment overlooked an occurrence of 73·6 in. (187 cm) in 24 hr on the island of Réunion, off Madagascar during March 1952 (Paulhus, 1965). Further discussion is given in Chapter 3.1(i).7.c.

B. Frequency

In many design studies, especially for systems of flood control, it is essential to know the *average* time-period within which a rainfall of specified amount or

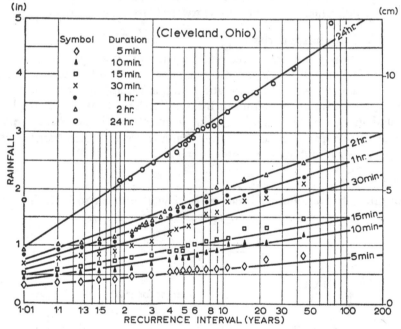

Fig. 3.1(i).2 The return period for annual rainfall totals at Cleveland, Ohio, 1902–47 (After Linsley and Franzini, 1955).

intensity can be expected to occur once. This is known as the 'return period' or 'recurrence interval'. The method of calculation is dealt with in Chapter 11.1.2. Figure 3.1(i).2 illustrates a graph of this type for Cleveland, Ohio.[1]

C. Areal Extent

Storm totals obviously depend on the type and scale of system – local thunderstorm, tropical disturbance or extra-tropical depression – and its rate of move-

[1] A valuable analysis of recent British data is given in *British Rainfall*, 1961 (H.M.S.O. 1967).

ment. In general, the effect of thunderstorm downpours is limited to areas on the scale of individual catchments, whereas steady depression rainfall may affect extensive drainage basins causing large-scale flooding if there is a slow-moving depression. For the United States, maximum 24-hr totals over different areas have been estimated by Gilman [1964] as follows:

ml²	in.	cm
10 (25·9 km²)	38·7	98·3
10²	35·2	89·4
10³	30·2	76·7
10⁴	12·1	30·7
10⁵	4·3	10·9

5. Precipitation statistics

The basic statistical measures of precipitation are concerned with average amounts for a specified time interval and the dispersion of the individual values about the average.

A. The average

1. The *mean*, or arithmetic average, for a specified time interval

$$\bar{p} = \frac{1}{n} \sum_{i=1}^{n} p_i$$

where p_i = precipitation amount for the ith term;

$\sum_{i=1}^{n}$ = summation of the terms from $i = 1$ to n.

2. The *median* is the term which occurs exactly at the midpoint in the series when the terms are ranked. It is a more useful indicator of 'average' precipitation than the mean in arid areas, where perhaps 75% of years, or months, in a series may have a value less than the mean. For example, if in a 35-year series of monthly totals 18 or more values are zero, then the median is zero, whereas the mean must exceed zero unless there is no rain in all the years.

B. Dispersion about the average

1. The simplest indicator of dispersion in a series is the *range* between the extremes. In the British Isles the extreme range of annual precipitation at any locality is about 40–180% of the mean, whereas in the arid, south-west United States it is approximately 25–270%. The range increases enormously when shorter time intervals are considered. This measure is not very satisfactory, because it fails to indicate the frequency of a deviation of specified magnitude.

2. Two measures of dispersion are associated with the mean. The simplest to calculate is the *mean deviation*:

$$\text{M.D.} = \sum_{i=1}^{n} |p_i - \bar{p}|$$

where $|p|$ denotes the absolute value of p, without regard to sign. This statistic of dispersion has been widely used in rainfall studies, but it lacks the versatility of the *standard deviation*, σ, in further statistical application, especially the assessment of probabilities.

$$\sigma = \sqrt{\frac{\sum\limits_{i=1}^{n}(p_i - \bar{p})^2}{n}} \quad \text{or} \quad \sqrt{\left[\frac{\sum\limits_{i=1}^{n}p_i^2}{n} - (\bar{p})^2\right]}$$

3. In connection with the median it is usual to specify the upper and lower *quartile* values, i.e. at the 75 and 25% positions in the ranked series. The quartiles delimit the central 50% of the frequency distribution. The uppermost and lowermost *deciles* (90 and 10%, respectively) in the series may also be of interest.

C. Relative variability

In order to compare the deviations for places with different average values it is necessary to express them as a percentage of the mean. The simple measure of relative variability (R.V.) is

$$\text{R.V. (\%)} = \frac{\text{M.D.}}{\bar{p}} \times 100$$

This index shows a marked tendency to increase sharply with low annual precipitation totals.

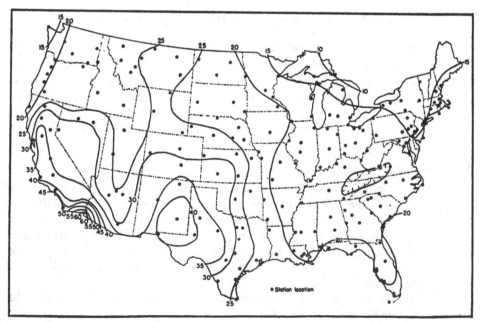

Fig. 3.1(i).3 The coefficient of variation of annual rainfall (%) over the United States (From Hershfield, 1962).

A preferable measure is the *coefficient of variation* (C.V.)

$$\text{C.V. (\%)} = \frac{\sigma}{\bar{p}} \times 100$$

In the United States the C.V. for annual precipitation ranges from about 15% in the north-east and 20% in Florida to more than 35–40% in the arid south-west (fig. 3.1(i).3). Both R.V. and C.V. take no account of the sequence of the data, which may be of considerable importance. Wallén uses the following expression for inter-annual variation (I.A.V.):

$$\text{I.A.V. (\%)} = 100 \frac{\sum_{i=1}^{n} |p_{i-1} - p_i|}{n-1}$$

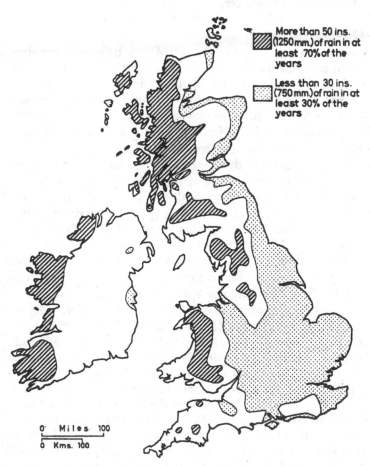

More than 50 ins. (1250 mm.) of rain in at least 70% of the years

Less than 30 ins. (750 mm.) of rain in at least 30% of the years

0 Miles 100
0 Kms. 100

Fig. 3.1(i).4 Areas of reliably high and occasionally low annual rainfall in the British Isles (From Gregory, 1964).

The unshaded areas receive between 30 and 50 in. in at least 70% of years.

D. Probability

If the frequency distribution for a series of annual precipitation totals is symmetrically distributed about the mean and the latter is more or less coincident with the median (i.e. a Normal or Gaussian distribution), probabilities of the occurrence of a specified amount can be determined. The method is set out in most statistical texts; see especially Gregory [1962].

The method has been applied to annual precipitation in several areas of the world; for example, East Africa (Glover *et al.*, 1954) and Great Britain (Gregory, 1957) as illustrated in fig. 3.1(i).4. This shows that large areas of eastern England have at least a 30% probability of receiving less than 75 cm (30 in.), while most upland areas and western Ireland can expect more than 125 cm (50 in.) in at least 70% of years. Such maps are of major significance to agriculturalists and in the assessment of water resources.

6. Dryness and wetness

Abnormal amounts for one part of the world may be quite normal in another. The official British definition of 'absolute drought' is a period of 15 or more consecutive days with each day receiving less than 0·01 in. of rain; a 'wet day' is one with ⩾0·04 in. The former would be quite inappropriate in areas with a long dry season, while the latter would be equally unsatisfactory in the humid tropics. More useful drought definitions involve assessment of effective precipitation and the moisture balance (see Chapter 4.1).

In many parts of the world there seems to be a tendency for dry weather to occur in spells, while the occurrence of a wet day is often independent of the previous conditions. Lawrence [1957] finds that in southern England a dry spell has an increasing probability of continuing (positive persistence) up to about 10 days, whereas after about 30 dry days there is definite likelihood of change (negative persistence). At Tel Aviv an interesting pattern is observed with wet days, ⩾0·1 mm precipitation, in winter. A wet day has a 66% probability of succeeding a wet day, but there is no significant change in the probability if the preceding two or three days were also wet. The occurrence of rain at Tel Aviv is virtually independent of conditions two or more days earlier. This pattern can be described, though not explained, by the statistical model known as a Markov chain. Weiss [1964] shows that the Markov chain model may have wide application to sequences of both wet and dry days in such diverse climatic locations as San Francisco, Moncton (New Brunswick), and Harpenden (England).

The spatial extent of wet and dry extremes on the annual time-scale is also of considerable interest. Glasspoole [1926] analysed the records at 250 stations in the British Isles for 1868–1924 and showed that while 1872 was the wettest for 49% and 1887 the driest for 40% of the whole region, 46 of the 57 years were the wettest or driest of the series *somewhere* in the British Isles. This illustrates the considerable spatial variability of precipitation, even in a zone of predominantly depression control. Spatial variability in areas of low precipitation is commonly underestimated because of the sparse network of rainfall stations.

7. Precipitation characteristics in different macroclimates

Only the briefest sketch of the global variability of precipitation characteristics is possible. We may identify the following properties as being of interest:

annual totals (see Chapter 1.1.4) and their variability;
annual regime;
diurnal regime;
frequency and intensity characteristics.

A. Annual totals and the annual regime

These are the primary climatological characteristics of precipitation. Figure 3.1(i).5 illustrates the theoretical distribution of annual precipitation and its seasonal concentration on a hypothetical continent of low, uniform relief according to Thornthwaite. The actual distribution of summer and winter precipitation is shown in fig. 1.1.4. Six types of regimes are commonly distinguished, although departures from these patterns are numerous. Moreover, the existence of similar regimes does not imply that the mechanisms causing the precipitation are necessarily the same. The six types, illustrated for representative stations in fig. 3.1(i).6, are:

1. *Equatorial.* Rain throughout the year, with two maxima at the equinoxes; generally large annual totals, 250–300 cm or more, and small variability from year to year. Rainfall is associated with the equatorial low-pressure trough, although rainy periods are generally the result of some form of perturbation. In East Africa, for example, outbreaks of rain occur in irregular spells lasting several days (Johnson, 1962). The daily rainfall distribution is determined by large-scale patterns of airflow in the troposphere. The equatorial regime is absent over much of the Indonesian region and in South America other than the Pacific coast.

   Poleward the two equinoctial maxima come closer together, creating a winter dry season. In some areas annual totals are again large, although this regime also occurs on the equatorward margins of tropical deserts.

2. *Tropical* (including monsoon areas). Pronounced summer maximum and winter dry season; annual totals range between about 25 and 100 cm (10 and 40 in.) in the savanna areas where the dry season may last for more than six months, to 200 cm (80 in.) or more in the humid tropics. The regime also extends into the subtropics in eastern Asia. The rainfall in southern Asia and West Africa occurs mainly with disturbances in the monsoon flow south of the equatorial low-pressure trough. There are both convectional downpours and periods of steady rain. In many tropical and subtropical areas late-summer hurricanes make significant contributions to the rainfall.

3. *'Mediterranean.'* Pronounced maximum in the winter half of the year and a dry summer; moderate annual totals of the order of 60–75 cm. It occurs in west-coast subtropical areas. Most of the rainfall is associated with depressions in the westerlies.

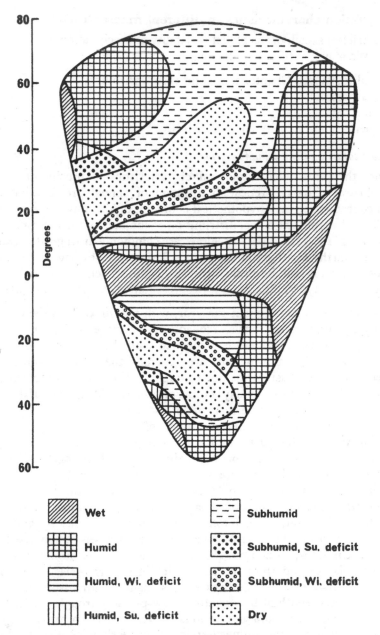

Fig. 3.1(i).5 The hypothetical distribution of precipitation regimes and annual totals on a continent of low, uniform elevation (After Thornthwaite).

4. *Temperate continental interior.* Annual totals of 35–50 cm, mainly occurring as convective rain showers in spring and summer; light winter snowfall; considerable year-to-year variability.

5. *Temperate oceanic (west coast).* Precipitation all year with a maximum in winter or autumn. Moderately high totals (75–100 cm) increasing markedly over coastal mountain areas to over 200 cm; about 200 rain days per year

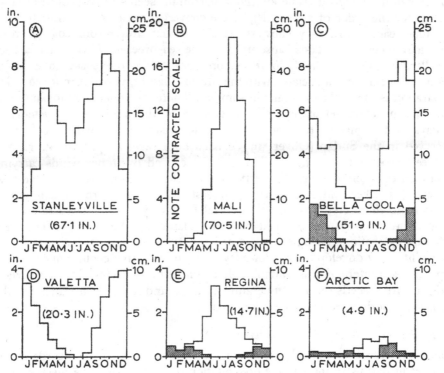

Fig. 3.1(i).6 Examples of the major types of precipitation regimes. Stippled portions indicate Snowfall. A-Equatorial; B-Tropical; C-Temperate Oceanic; D-Mediterranean; E-Temperate Continental; F-Arctic.

and low variability. The predominant control is frontal depressions. In mountain areas and higher latitudes a considerable proportion of the total may fall as snow.

6. *Arctic.* Low annual totals, generally 12–40 cm (5–15 in.), mainly occurring as rain in summer; late summer or autumn maximum; only light winter snowfall due to the very cold, dry air. Convectional activity is at a minimum in these regions, and most of the precipitation occurs with depressions.

B. Diurnal regime

Important diurnal rhythms are often superimposed on the basic seasonal pattern. For example, diurnal heating may lead to an afternoon maximum of precipitation as a result of convective downpours. This regime is popularly regarded as characteristic of tropical climates, but many instances of markedly different diurnal variations are known. Over the tropical oceans a nocturnal maximum is common, while some stations have different regimes in different seasons, so that the diurnal pattern for the year as a whole is indeterminate. At Guam, a small island in the west Pacific ($13°$ N, $145°$ E), the light showers of the 'dry season' occur mainly between 2200 and 1000 hours, whereas in the 'wet season' there is no clear diurnal rhythm. The afternoon maximum seems to be typical of large islands (of the order of 2,500 ml^2), where convergent sea-breezes initiate deep cumulus leading to heavy showers. It also occurs away from coastal areas, especially where mountains cause uplift of the sea-breezes. This is the case on the Pacific slopes of the Colombian cordillera, while the coast zone has a nocturnal maximum, associated with the effects of land breezes. A more complex illustration is the Malacca Straits in summer. There, the nocturnal maximum is due to convection set-off by the convergence of the land-breeze systems of Malaya and Sumatra (Ramage, 1964). A surprising nocturnal maximum is observed in the Sudan (Oliver, 1965). Khartoum receives 77% of its rainfall between 1800 and 0600 hours. This may be caused by upper winds carrying afternoon thunderstorms south-westward from the hills bordering the Red Sea. However, Bleeker and Andre [1951] suggest that the nocturnal maximum of rainfall and thunderstorms over the central United States is related to a large-scale circulation system induced over the plains by the Rocky Mountains. Several explanations have been suggested for the oceanic nocturnal maximum. Instability may develop in the cloud layer due to radiative cooling of the cloud tops. Air–sea interaction is undoubtedly a contributory factor. The sea surface temperature exceeds the air temperature at night, and so the heat transfer to the air is greatest then, reaching a peak near dawn.

C. Frequency and intensity characteristics

The most generally available frequency statistics are limited to records of the number of days with measurable precipitation or 'rain days'. The mean rainfall per rain day is a rough indicator of rainfall intensity in different climates. For example, this increases from about 4 mm (0·15 in.) per rain day in south Australia to 18 mm (0·70 in.) in tropical north Australia. In the equatorial Kenya Highlands the figure is about 8·4 mm (0·33 in.), whereas in arid northern Kenya it reaches 12·2 mm (0·44 in.) per rain day. Representative values for the United States range from about 5 mm (0·20 in.) per rain day over the Prairies and Great Plains to 14–15 mm (0·55–0·60 in.) in Oklahoma–Arkansas. These values are very low compared with 107 mm (4·2 in.) per rain day in June at Cherrapunji, Assam.

The relative raininess of different types of airflow can be assessed in a similar manner by calculating the 'specific precipitation density'. This is the mean rainfall per rain day for a particular type of airflow as a percentage of the mean for all rain days. At Southampton, England, the mean rainfall intensity in January and July is approximately 5 mm (0·2 in.) per rain day. For different airflow types the specific precipitation–density index is given in Table 3.1(i).3:

The light, showery nature of precipitation with Northerly and North-westerly types is quite apparent. There is an unexpected seasonal change in raininess of Westerly and Cyclonic types.

TABLE 3.1(i).3

Airflow type	Specific precipitation density (%)		
(H. H. Lamb's classification)	January	(1921–50)	July
Northerly	41		54
Easterly	105		121
Southerly	123		130
Westerly	129		80
North-westerly	47		81
Cyclonic	108		163
Anticyclonic	34		14

(from Barry, 1967)

A widespread, perhaps even global, characteristic of rainfall is the occurrence of most of the annual total on a few days. Half of the annual precipitation is accounted for by 13% of the rain days in the Kenya Rift Valley, 16% in the basin of the upper Colorado river, and 10–15% in Argentina. At Concord, New Hampshire, 6% of the rain days gave 23% of the total precipitation during 1885–1935. This characteristic seems, therefore, to be independent of the precipitation regime, annual total, and geographical location.

There are many case studies of short-term precipitation intensity, and although it would be premature to attempt a generalized picture for the various climatic regimes, some pointers in this direction can be indicated. Intensities are generally greater in areas of summer rainfall than winter rainfall. For example, daily intensities in the Middle East average only 5 mm (0·2 in.), with a peak of perhaps 12–13 mm once a year in the mountainous areas, whereas at Tucson, Arizona, summer convective storms with an intensity of 25 mm or more per day account for some 25% of the annual precipitation. For the United States, Paulhus and Miller [1964] have mapped the percentage contribution of daily amounts of 0·5 in. (12·7 mm) or more to the average annual precipitation. The figure ranges from 20% in the Great Basin to 90% on the Gulf Coast. The occurrence of rainfalls exceeding 0·5 in. is considered to be a factor in gully erosion potential. At Namulonge, Uganda, storms giving more than 25 mm per day contributed 30% of the annual total during 1950–55, and peak rates in three rainstorms were between 250 and 350 mm (10 and 13·8 in.) per hour. In tropical northern Australia the daily maximum intensity exceeds 100 mm once a year, and on the Queensland coast it may exceed 150 mm. Daily totals of 150 mm or more are expected only once in a century in Britain, and then principally in upland districts of the west. By contrast, investigations in Idaho and coastal British Columbia indicate no relationship between elevation and intensity.

Rather, the increase with elevation of total amount seems to reflect a longer duration of precipitation.

REFERENCES

BARRY, R. G. [1967], The prospect for synoptic climatology: a case study; In Steel, R. W., and Lawton, R., Editors, *Liverpool Essays in Geography* (Longmans, London), pp. 85–106.

BECKINSALE, R. P. [1957], The nature of tropical rainfall; *Tropical Agriculture*, 34, 76–98.

BERGERON, T. [1960], Problems and methods of rainfall investigation; In *Physics of Precipitation, Geophysical Monograph No. 5* (Washington), pp. 5–30.

BLEEKER, W. and ANDRE, M. J. [1951], On the diurnal variation of precipitation, particularly over the central U.S.A., and its relation to large-scale orographic circulation systems; *Quarterly Journal of the Royal Meteorological Society*, 77, 260–71.

CHATFIELD, C. [1966], Wet and dry spells; *Weather*, 21, 308–10.

COOPER, C. F. [1967], Rainfall intensity and elevation in southwestern Idaho; *Water Resources Research*, 3, 131–7.

FOSTER, E. F. [1949], *Rainfall and Runoff* (Macmillan, New York), 487 p.

GABRIEL, K. R., and NEUMANN, J. [1962], A Markov chain model for daily rainfall occurrence at Tel Aviv; *Quarterly Journal of the Royal Meteorological Society*, 88, 90–5.

GILMAN, C. S. [1964], Rainfall; In Ven te Chow, Editor, *Handbook of Applied Hydrology* (New York), Section 9.

GLASSPOOLE, J. [1926], The driest and wettest years at individual stations in British Isles; *Quarterly Journal of the Royal Meteorological Society*, 52, 237–48.

GREGORY, S. [1957], Annual rainfall probability maps of the British Isles; *Quarterly Journal of the Royal Meteorological Society*, 83, 543–9.

GREGORY, S. [1962], Statistical Methods and the Geographer (Longmans, London), 240 p.

HARROLD, L. L., and DREIBELBIS, F. R. [1958], *Evaluation of Agricultural Hydrology by Monolith Lysimeters, 1944–55*; Technical Bulletin No. 79, United States Department of Agriculture (Washington), 166 p.

HASTENRATH, S. L. [1967], Rainfall distribution and regime in central America; *Archiv für Meteorologie, Geophysik und Bioklimatologie*, Ser. B, 15(3), 201–41.

HERSHFIELD, D. M. [1962], A note on the variability of annual precipitation; *Journal of Applied Meteorology*, 1, 575–8.

JENNINGS, J. N. [1967], Two maps of rainfall intensity in Australia; *Australian Geographer*, 10, 252–62.

JOHNSON, D. H. [1962], Rain in East Africa; *Quarterly Journal of the Royal Meteorological Society*, 88, 1–19.

LAWRENCE, E. N. [1957], Estimation of the frequency of 'runs of dry days'; *Meteorological Magazine*, 86, 257–69 and 301–4.

MASON, B. J. [1962], *Clouds, Rain and Rainmaking* (Cambridge), 145 p.

NAGEL, J. F. [1956], Fog precipitation on Table Mountain; *Quarterly Journal of the Royal Meteorological Society*, 82, 452–60.

NYBERG, A. and MODEN, H. [1966], The seasonal distribution of precipitation in the area east of Stockholm and the daily distribution in a few selected cases; *Tellus*, **18**, 745–50.

OLASCOAGA, M. J. [1950], Some aspects of Argentine rainfall; *Tellus*, **2**, 312–18.

OLIVER, J. [1965], Evaporation losses and rainfall regime in central and north Sudan; *Weather*, **20**, 58–64.

PAULHUS, J. L. H. [1965], Indian Ocean and Taiwan rainfall set new records; *Monthly Weather Review*, **93**, 331–5.

PAULHUS, J. L. H. and MILLER, J. F. [1964], Average annual precipitation from daily amounts of 0·50 inch or greater; *Monthly Weather Review*, **92**, 181–6.

RAMAGE, C. S. [1964], Diurnal variation of summer rainfall in Malaya; *Journal of Tropical Geography*, **19**, 62–8.

RODDA, J. C. [1967], A country-wide study of intense rainfall for the United Kingdom; *Journal of Hydrology*, **5**, 58–69.

SAWYER, J. S. [1956], The physical and dynamical problems of orographic rainfall; *Weather*, **11**, 375–81.

SHAW, E. M. [1962], An analysis of the origins of precipitation in northern England, 1956–60; *Quarterly Journal of the Royal Meteorological Society*, **88**, 539–47.

SUZUKI, E. [1967], A statistical and climatological study on the rainfall of Japan; *Papers in Meteorology and Geophysics*, **18**, 103–82.

WALKER, E. R. [1961], *A synoptic climatology of parts of the western Cordillera*; Publications in Meteorology No. 35, Arctic Meteorology Research Group, McGill University (Montreal), 218 p.

WALLÉN, C. C. [1955], Some characteristics of precipitation in Mexico, *Geografiska Annaler*, **37**, 51–85.

WEISS, L. L. [1964], Sequences of wet or dry days described by a Markov chain probability model; *Monthly Weather Review*, **92**, 149–76.

YARNELL, D. L. [1935], Rainfall intensity frequency data; *U.S. Department of Agriculture, Miscellaneous Publication 204* (Washington, D.C.).

I(ii). The Assessment of Precipitation

JOHN C. RODDA

Institute of Hydrology, Wallingford

Rain, snow, hail, and sleet, together with dew, rime, and similar phenomena, make up the various forms of precipitation. Most water reaches the surface of the earth as rain, but, of course, snow and dew are important in certain regions. However, both snowfall and dew are difficult to determine, and only rainfall is gauged extensively and with any degree of certainty. Rain gauges are extremely varied in design, and their usage differs considerably. Some are little more than

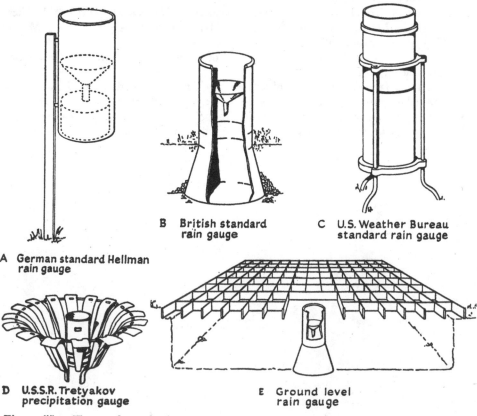

A German standard Hellman rain gauge

B British standard rain gauge

C U.S. Weather Bureau standard rain gauge

D U.S.S.R. Tretyakov precipitation gauge

E Ground level rain gauge

Fig. 3.1(ii).1 Types of standard rain gauges.

a bucket, but others are highly complicated devices that can be interrogated from a distant base for flood-warning purposes. Radar is employed to determine the areal distribution of rainfall, but for quantitative results comparisons are necessary with records obtained from rain gauges.

The history of the rain gauge is lengthy and devious. Some authorities maintain that it commenced in India well over 2,000 years ago. Today there must be hundreds of different types in use, some giving a continuous record of rainfall, but the majority requiring inspection by an observer at a fixed time. However, there is a considerable degree of uniformity in gauge type and observation practice within most national rain-gauge networks – a uniformity which has usually been achieved during the last 100 years. On the other hand, there are appreciable differences from country to country (fig. 3.1(ii).1). At one extreme is the standard gauge of the Soviet Union, standing 2 m high and surrounded by a shield; while at the other is the British standard gauge, a brass cylinder only 1 ft high and 5 in. in diameter. Other national gauges fall between these limits in terms of design and method of installation, but this variety raises the question of how comparable are the results produced by the different gauges. Extensive tests made at the same site show that there are differences from gauge to gauge, so what is recorded as 1 in. of rain on one side of a frontier could be registered as something different on the other. Hence the existing rainfall maps on global and continental scales are not as meaningful as they might be if the same type of gauge were used all the world over. The W.M.O. Interim Reference Precipitation Gauge was introduced in an attempt to provide a basis for comparison, but like most other gauges, its performance varies from site to site, largely due to the effect of wind. Of course there are a number of other sources of error (fig. 3.1(ii).2), but wind is by far the most important. Together they cause the standard gauge to under-register. Wind interacts with features of the site and with the gauge itself to produce turbulence and eddies. These in turn act on the raindrops, particularly in the region immediately over the gauge, where the smaller drops are diverted past the funnel. The higher the gauge, the greater is the effect of wind. On the other hand, the lower the gauge, the greater is the risk of splash from the ground surface.

Shields, walls, and fences can reduce the effect of wind, but the most satisfactory way of overcoming it is to install the gauge so that its rim is flush with the ground surface. Splash can be avoided by surrounding the gauge with a matting surface or by placing it in a shallow pit covered by a grid made of narrow strips of metal or rigid plastic. Such a gauge is considered to give a measure of rain nearer the true value than any other type. However, the true rainfall at a point is not known, because there is no absolute standard of rainfall measurement, as for example, there is in the case of discharge. Hence all rainfall measurements made in the conventional manner are relative, and in spite of the numerous experiments with different gauges, there is still no method of measuring the quantity of rain falling at a particular point on the earth's surface to a known degree of accuracy. This is a fact that is rarely taken into account by hydrologists and meteorologists.

Comparisons of standard gauge observations with those made in nearby

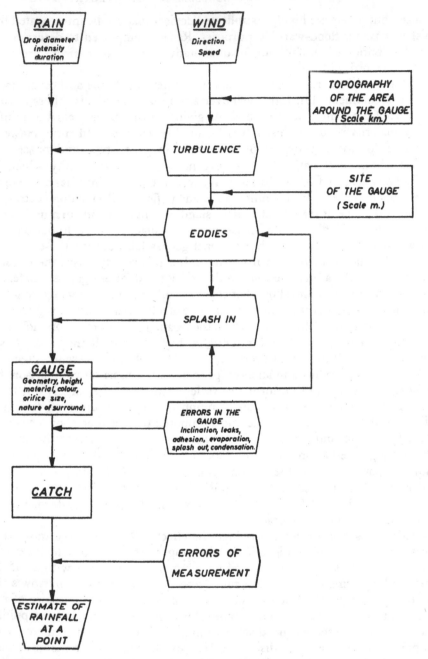

Fig. 3.1(ii).2 Conceptual model of the processes involved in determining rainfall with a conventional rain gauge.

ground-level gauges show that the catch at ground level can be appreciably higher than the catch obtained in the conventional way. Other comparisons have been made using accurate lake-levels measurements and weighing lysimeter records, with similar results. Obviously the difference between the catch at ground level and that obtained at the standard height varies not only with gauge type but also from site to site and with climate. For example, in Britain differences could range from 3 to 10% for annual totals, but for single storms the ground-level catch has been known to be 30 or 40% greater. However, in tropical areas, such as East Africa, the differences are likely to be smaller, because of larger drop sizes and lower wind speeds. Errors in gauging snowfall are much greater than for rain, because wind has more pronounced effects on the falling flakes. By way of compensation, it is relatively simple to measure snow depth and take samples to assess water equivalent and density. Nevertheless, in countries where some of the precipitation occurs in the form of snow the *systematic* error in measurement is likely to be considerable.

Questions arise about the significance of this *systematic* error, first from the point of view of the water balance and then in the application of rainfall data. For even though more rainfall reaches the ground than is measured in the conventional gauge, it is obvious that no extra water is available at the ground surface, because rainfall is balanced by runoff, evaporation, and storage changes in the soil and rock. In fact, errors in the assessment of these other factors obscure that occurring in the rainfall term – the measurement of evaporation being particularly suspect. Hence in terms of water resources the significance of a systematic error need not be large, particularly where climate and site are favourable. On the other hand, over short periods and where conditions of climate and site are unfavourable, the measurement of rainfall is likely to be seriously in error. This error must have important consequences, especially where standard rain-gauge records are employed for practical purposes.

The problems of instrumentation are but one aspect of assessing the mean depth of precipitation over an area. There are also difficulties concerned with the design of the instrument network and in determining the mean from a series of point measurements. Few rain-gauge networks have a rational basis for their present form, most having developed where observers are available rather than on the grounds that a record of rainfall was required at a particular point. Nevertheless, there are a number of methods of network design that are objective and do not rely on arbitrary rules of siting. The distribution of gauges at random over an area has the advantage that statistically valid estimates of the mean rainfall can be obtained. However, there are practical difficulties associated with this method and with a variation – the use of gauges which are moved at random within a specified area. An alternative is the systematic method of design, where gauges are installed at fixed distances over an area, their spacing being controlled in both horizontal and vertical planes. Stratified random-sampling methods combine some of the advantages of both systematic and random design techniques, but there is the difficulty of defining strata in a meaningful way. Of course, the design of any network should be compatible with the use made of the information and with the nature of the topography of the area. For example, a network of gauges installed as a basis for a flood-warning system will be different from one set up to evaluate the water resources of an area, although one set of records could assist in the other objective.

Recording and transmitting gauges, put in a pattern that anticipates the distribution of storm tracks, would produce signals of excessive rates of rainfall and others indicating that some predetermined amount of rain had fallen in a given time. Such a flood-warning system would suffer from the problem of delimiting the area of a storm and the difficulties of estimating intensities at places between the gauges. One solution might be to use a secondary gauge network and build up a history of space-intensity relationships from past storm records. Another would be to provide radar coverage, but this is not often possible because of cost. Non-recording gauges would be employed in the case of a water-resources survey. A wide coverage of the area would be aimed at to provide information on spatial variations, while variations in time could be accounted for by incorporating long-established gauges in the network. It could be argued that networks designed for the same purpose would need to be more dense in mountainous areas than in flat country, because of the wider variations in rainfall that are to be found where differences in relief are most marked. However, this matter is not clear, because similar patterns of rainfall occur over mountainous areas, patterns which appear to be highly correlated with one another; as opposed to the more random distribution of rain that takes place in flat areas.

The transformation of point measurements of rainfall into an estimate of the mean for an area can be carried out in several ways. Where gauges are evenly distributed over an area and relief is subdued use can be made of the arithmetic mean. An advance on this is the construction of 'Thiessen' polygons around each gauge, as then each gauge record is weighted according to the area of the polygon around it. This method is objective by contrast with the isohyetal and isopercentile techniques, which are largely subjective. One other method is the use of regression analysis, but this is only successful in regions where topography controls the distribution of rain.

It is highly probable that the error in determining the mean rainfall for an area will be appreciable, even when the most satisfactory instruments are combined with the best techniques of network design and computation of the mean. Where snow is important this error will be even greater.

REFERENCES

PONCELET, L. [1959], *Sur le comportement des Pluviometres*; Publications, Series A, No. 10 Institut Royal Meteorologique de Belgique.

KURTYKA, J. C. [1953], *Precipitation Measurements Study*; Report of Investigation No. 20 State Water Survey Division, Illinois.

RODDA, J. C. [1967], The Rainfall Measurement Problem; *Proceedings of the Bern Assembly*, International Association of Scientific Hydrology.

WORLD METEOROLOGICAL ORGANIZATION [1965], *Guide to Hydrometeorological Practices*; Bulletin No. 168 t.p. 82.

II. The Role of Water in Rock Disintegration

R. J. CHORLEY

Department of Geography, Cambridge University

1. Some properties of water

Water molecules are formed by the bonding of two hydrogen atoms to one oxygen atom as a result of the former sharing their single negatively charged electrons with the oxygen, giving the latter the optimum eight in its outermost shell (fig. 3.11.1(a)). This bonding is termed *covalent* and produces a molecule in which the hydrogen atoms have a net positive charge and the oxygen a double negative one. The asymmetrical bonding of the hydrogen atoms makes the molecule dipolar, in that one end is charged negatively and the other positively (fig. 3.11.1(b)). This has a number of consequences:

1. Under some circumstances the molecule can separate into two oppositely charged ions (H^+ and OH^-) which make it more available for some chemical reactions. OH^- is the hydroxyl which is important in many weathering reactions, and H^+ is the cation whose effective concentration imparts to an aqueous solution its acidity. The range of the latter is so large that acidity (pH) is expressed as the negative logarithm of the free H^+ concentration in grams per litre (ph $= 7$ is thus 0·0000001 g of free H^+ ions per litre of water; this is the neutral value, higher pH values indicate alkalinity, lower values acidity). The solution of atmospheric carbon dioxide by rain-water produces free H^+ ions in the resulting carbonic acid:

$$CO_2 + HOH \rightleftharpoons H^+ + HCO_3^-$$

2. By orienting themselves in an electrical field water molecules can weaken it. This explains the very effective solvent action of water when it is in contact with other molecules, the atoms of which are simply held together because the component ions are of opposite electrical charge (i.e. ionic bonding). The best example of this is the solution of common salt by the negative and positive ends of the water molecules attaching themselves respectively to the Na^+ and Cl^- ions, neutralizing their charges so that the least mechanical agitation will float the sodium and chlorine ions apart.

3. The positioning of the two hydrogen atoms means that the water molecule is basically four-cornered, with the negative charges concentrated at two points on the surface of the oxygen atom. When water molecules come into association, therefore, they tend to join together by ionic bonding (hydrogen

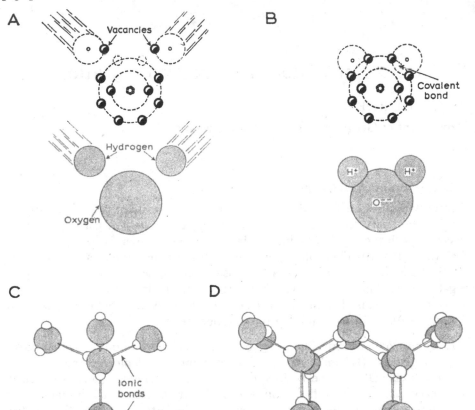

Fig. 3.11.1 The chemical structure of water (After Davis and Day, 1964).

A. The joining of two hydrogen with one oxygen atom.
B. The covalent bond in the water molecule.
C. The ionic bonding of water molecules.
D. The structure of ice.

bonding) into tetrahedral groups of four (fig. 3.11.1(*c*)). The positive charge of the hydrogen nucleus is attracted to the oxygen ion with a force which is only about 6% of that which binds the hydrogen to the oxygen atoms within the molecule. This phenomenon of hydrogen bonding gives rise to such water properties as surface tension and capillarity (i.e. cohesion), and to adhesion to some surfaces – particularly those having surface oxygen atoms (e.g. glass, quartz, clay minerals, etc.). It also explains the high theoretical tensile strength of water, which is about the same as some steels.

4. As a liquid water is atypical, in that it seems to be composed of clusters of molecules connected by hydrogen bonds, separated from one another by

unbound water molecules which can rotate freely, forming lubricating layers and allowing flowage (fig. 3.11.2). The structure of water is not static, and molecules exchange rapidly between the clusters and the flow layers and, on average, each intermolecular hydrogen bond breaks and reforms 10^{12} times a second. As the temperature of water decreases the thermal agitation of the molecules also decreases until the maximum density is reached at about 4° C, this increase of density, together with an increase in the number of hydrogen bonds (and therefore of cluster size), means that viscosity (or internal resistance to deformation) is also inversely related to

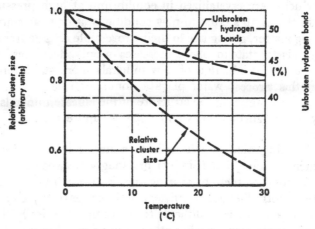

Fig. 3.11.2 The effect of temperature on the relative number of unbroken hydrogen bonds and the relative cluster size in pure water (Data from Nemethy and Scheraga. From Gross, M. G., *Oceanography*, Merrill Physical Science Series, 1967).

temperature (fig. 3.11.2). At 20° C the absolute viscosity of water is little more than half that at 0° C.

5. When the temperature of water falls to about 4° C the expansion due to the widespread formation of the open hydrogen bonds exceeds the contraction due to the decreasing molecular thermal agitation, and the water begins to expand in volume and assume a lower density as more and more of the tetrahedral clusters are taken up into a hexagonal structure (fig. 3.11.1(d)). This expansion continues until at a temperature of −22° C ice achieves its minimum density and maximum expansive pressure under confining conditions. At lower temperatures ice some 3% more dense than water begins to form, and at −70° C the crystal habit changes from hexagonal to cubic. In its usual hexagonal form cleavage is parallel to the basal plane, and the maximum growth rate is normal to this. The former property is of great importance in the flow properties of glacial ice, and the latter in its mechanical action in weathering. Except beneath a more or less flat water-table near enough to the surface to freeze, water in natural circumstances seldom freezes in completely confined conditions, but under conditions

which allow an outside water supply to the growing crystals, which tend to develop in clusters of parallel needles at right angles to the freezing surface (i.e. the air or a rock surface). Under these conditions the stresses developed by ice growth may be ten times those associated with the simple expansion of water during freezing, for they are limited only by the tensile strength of water, which is drawing the water molecules through the capillary films to the ends of the growing ice crystals.

2. Weathering of igneous minerals and rocks

The weathering of igneous rocks may be defined as the response of mineral assemblages which were crystallized in equilibrium at high pressure and temperature within the earth's crust to new conditions at or near contact with air, water, and living matter, giving rise to their irreversible change from the massive to the clastic or plastic state involving increases in bulk, decreases in density and particle size, and the production of new minerals more stable under the new conditions. In this process water plays a dominant role, which can only be understood by first examining the structure of the silicate minerals which compose virtually all of igneous rocks and make up more than 90% of all rock-forming minerals.

The basic building block of the silicate minerals is the silica tetrahedron (SiO_4), in which a silicon ion (Si^{++++}) fits snugly between four oxygen ions (O^{--}) covalently bonding them by sharing one electron with each (fig. 3.III.3(a)). This structure is both chemically efficient and geometrically compact, making the tetrahedra strong and very difficult to break up chemically. However, the silicate minerals are also composed of other positive metal ions (cations), such as aluminium (Al^{+++}), iron (Fe^{+++} or Fe^{++}), magnesium (Mg^{++}), calcium (Ca^{++}), sodium (Na^+), and potassium (K^+), and these form generally the weaker links in the crystalline structures. In particular, the smaller electrical charges of the latter (especially of Na^+ and K^+) make them very susceptible to being neutralized by the dipolar activity of water if it can enter the crystal lattice. It is clear, then, that the most resistant minerals are those formed exclusively of interlocking silica tetrahedra, and the less the interlocking and the greater the interpolation of the other metal cations, the more readily the mineral will break down under the action of water. Apart from orthoclase (potash feldspar), muscovite mica and quartz, the main igneous rock-forming minerals can be divided into two groups. These are, firstly, the group of individually discrete ferromagnesian minerals, where the silica tetrahedra are joined by Fe^{++} and Mg^{++} ions (and in some instances certain others), and, secondly, the continuous series of plagioclase feldspars in which varying proportions of the Si^{++++} ions of the silica tetrahedra have been replaced by Al^{+++} ions, plus additional cations to compensate for the resulting loss of positive charge. The weakest of the common ferromagnesian minerals is olivine (($MgFe)_2SiO_4$), composed of isolated silica tetrahedra linked on all sides by Mg^{++} and Fe^{++} ions, giving the silica : oxygen proportion of 1 : 4. Augite ($Ca(Mg, Fe, Al)S_{12}O_6$) is rather more resistant, in that the silica tetrahedra form single chains by sharing one oxygen atom (silica : oxygen ratio

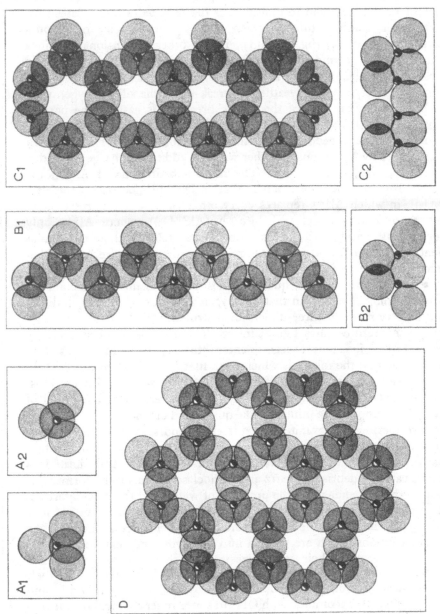

Fig. 3.11.3 The structure of the silicate minerals (After Leet and Judson, 1965).

A. The silicon–oxygen tetrahedron viewed from the side (1) and from above (2).
B. A single chain of tetrahedra viewed from above (1) and from the side (2).
C. A double chain of tetrahedra viewed from above (1) and from the side (2).
D. A tetrahedral sheet viewed from above.

$= 1 : 3$) (fig. 3.11.3(b)), but are weakly linked on four sides by other cations. Hornblende (of complex formula, of the general form $(OH)CaMgFeAlSiO_n$) has double silica tetrahedron chains (fig. 3.11.3(c)) linked by other ions (silica : oxygen ratio $4 : 11$); and in biotite mica the tetrehedra form plates with each tetrahedron sharing three oxygen atoms with their neighbours (silica : oxygen ratio $= 2 : 5$) (fig. 3.11.3(d)), in the other dimension two silica sheets being joined in a sandwich by Al^{+++}, Fe^{++}, and Mg^{++} ions and these sandwiches very weakly linked by K^+ ions. So it can be seen that there is a general increase of resistance to weathering break-up as the silica : oxygen ratio increases, but even in the case of the biotite break-up is easy in the parallel planes containing the K^+ ions. With the plagioclase feldspars there is a similar sequence of resistance, depending upon the proportion of Si^{++++} ions, which have been replaced by Al^{+++} ions, together with the additional charges provided by Ca^{++}, Mg^{++}, Na^+, and K^+ ions. The more substitutions of Al^{+++}, the weaker the structure becomes, such that calcic plagioclase (anorthite; $Ca(Al_2Si_2O_8)$), in which Al^{+++} replaces every other Si^{++++}, is less resistant to weathering than sodic plagioclase (albite; $Na(AlSi_3O_8)$), where Al^{+++} only replaces every fourth Si^{++++}. In the case of orthoclase feldspar there is a similar replacement ($K(AlSi_3O_8)$), and muscovite mica is similar to biotite, except that there are more Al^{+++} ions in place of some of the Fe^{++} and Mg^{++} ions giving a more resistant structure. Finally, quartz is the most resistant of the common igneous rock-forming minerals, in that all oxygen atoms are shared in all directions, giving a very resistant three-dimensional structure.

At higher temperatures silica tetrahedra are linked together more easily by other ions, and Al^{+++} ions find it easier to enter into the silica tetrahedra, so that it is a general rule that the minerals which form first at highest temperatures in a cooling melt (e.g. olivine, augite, and anorthite) form rocks which are most susceptible to surface weathering (e.g. peridotite), whereas rocks composed of the relatively low-temperature minerals, like quartz and orthoclase, are relatively resistant to weathering (e.g. granite) (fig. 3.11.4). Of course, where the minerals composing the rock have very different susceptibilities granular disintegration occurs, as, for example, when the weathering of hornblende and biotite in a quartz diorite causes a debris of quartz and orthoclase crystals to be formed.

Most weathering depends upon the presence of water, and the decomposition of silicate minerals is mainly accomplished by hydrolysis, in which the H^+ ions displace the metal cations in the silicates and the OH^- ions combine with the latter to form solutions which are washed into the rivers and seas. H^+ ions are chemically active because their small size enables them to penetrate many crystal structures, because they carry a large electrical charge relative to their size, because they form other compounds by providing hydrogen bonds, and because they can readily recombine with OH^- in some minerals (e.g. hornblende) to form water. At high concentrations of H^+ ions silica (SiO_2) and alumina (Al_2O_3) are linked by the H^+ ions which have displaced the metal cations to form complex clay minerals. Thus the type, as well as the degree, of weathering depends very much upon the amount of water available.

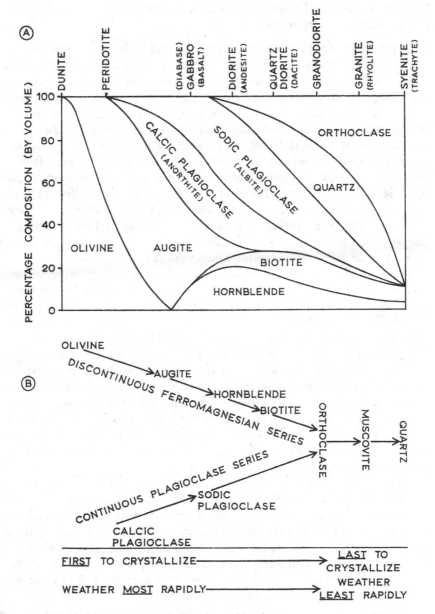

Fig. 3.11.4 The main igneous rock-forming minerals.

A. The average percentage composition of the most common igneous rocks.

B. The sequence of mineral crystallization and weathering.

There are a number of sources responsible for the production of H^+ ions in the water which weathers rocks:

1. The primary source is through the dissolving of CO_2 from the atmosphere, giving an average equilibrium pH of 5·7 at 25° C for rainfall and making it acid. The pH of rainfall, however, ranges from about 4 to more than 9 (for example, Hawaii 4·8–6·3, with a mean of 5·2; Uganda 5·7–9·8, mean 7·8; west coast of Ireland 5·9–7·6, mean 6·5), but it is difficult to generalize about variations on a global scale. Even in an area as small as Sweden, the rainfall is acid along the west coast and more neutral in the north. Apart from the natural decrease of dissolved chloride with distance from the coast, it has been suggested that areas of heavy rainfall of high intensity composed of large drops have lower acidity values because the rain has less opportunity for dissolving atmospheric CO_2. It seems that some tropical regions experience rainfall of much higher pH than that, for example, measured on the average in Northern Europe (5·47), due to some extent to the higher tropical temperatures which inhibit CO_2 solution.

2. The acidity of rain-water is largely irrelevant to many weathering processes, because it is usually changed drastically as soon as the water enters the soil.
   (a) CO_2 is dissolved much more readily from the soil atmosphere than from the free atmosphere.
   (b) Some clay minerals (e.g. bentonite) yield H^+ ions.
   (c) Acids are produced by humus and soil organisms. Controlled experiments with bacterial acids lasting only a few hours have shown that, although resisted by alumino-silicate minerals (indeed silica seems less soluble in humic acids than in pure water), muscovite and some basic minerals may lose $\frac{1}{5}$–$\frac{1}{2}$ of their weight by solution. However, the real effect of organic acids as a weathering process has not been fully investigated, although one suspects that it may be very great, especially in the humid tropics.
   (d) The roots of plants provide H^+ ions by osmosis, which they exchange for the nutrients provided by Ca^{++}, Mg^{++}, and, especially, K^+ cations. Some plants also accumulate significant quantities of dissolved silica, and 3·5% of the dry weight of some tropical hardwoods and up to 10% of bamboos is composed of silica. Assuming a conservative mean value of 2·5%, this would imply the removal of 0·4 tons of silica/acre/year in some tropical areas, which alone would account for 1 ft of denudation every 5,000 years on a basalt having 49% silica.

As weathering processes develop, however, by reactions involving the free H^+ ions, the acidity of water in soil and rock will decrease due to chemical reactions. In British Guiana, for example, rainfall of pH 7 quickly yields soil water of ph 8 on hornblende-rich rocks. Prolonged solution of quartz (pure silica) has little effect, maintaining a neutral pH of 6–7, but feldspars yield a solution of pH 8–10, augite 8–11, and hornblende of 10–11, again supporting the standard scale of chemical susceptibilities. Of course, different minerals are differently susceptible to chemical attack by soil-water solutions, as are different rocks lying in close juxtaposition, so that chemical weathering by altering the character of these fluids sets up chain reactions of weathering of a very complex kind.

1. Where the weathering fluid has a pH of 10 or more, Al_2O_3 is very soluble and SiO_2 relatively so (fig. 3.11.5), and both are carried away in solution, providing enough rainfall is available. Where rainfall is scanty, or evaporation is very high (or both), the decay products of Al_2O_3 and SiO_2 are not removed, but combine to form clay minerals such as montmorillonite and illite.

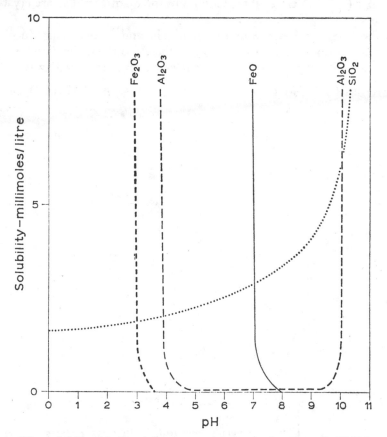

Fig. 3.11.5 The solubility of silica, alumina, and iron as a function of pH (After Keller, 1957, and Loughnan, F. C., *Journal of Sedimentary Petrology*, Vol. 32, 1962).

2. Where the pH is more neutral (e.g. 7–8), as for example where heavy tropical rains are falling on a surface rock containing hornblende, Al_2O_3 is almost insoluble, whereas SiO_2 is still partly soluble (below this pH its solubility is very small, although it increases with temperature). Under these conditions hydrated Al_2O_3 remains as a residue, usually to form gibbsite ($Al_2O_3.3H_2O$), a valuable aluminium-ore mineral.

3. If soil water is acid, as for example in temperate latitudes, where rain falls on quartz-rich rocks covered in vegetation, the solubility of both Al_2O_3 and SiO_2 is very low, and both remain (providing the rainfall is not so unduly heavy as to flush away the II^+ ions) and combine to form clay minerals like kaolinite ($Al_2(OH)_2Si_4O_{10}$) together with quartz debris.

The weathering of a granodiorite under humid temperature conditions illustrates both the relative susceptibilities of its varied mineral constituents to weathering and the main types of products of igneous rock weathering:

1. Quartz (SiO_2). Very slight solution, but the main products are crystal fragments.
2. Orthoclase, combines with carbonic acid and water to produce soluble potassium carbonate (used by plants), the clay mineral kaolinite, plus some soluble silica and fine colloidal particles which are washed away.

$$2KAlSi_3O_8 + H_2CO_3 + nH_2O \longrightarrow K_2CO_3 + Al_2(OH)_2Si_4O_{10}.nH_2O + 2SiO_2$$

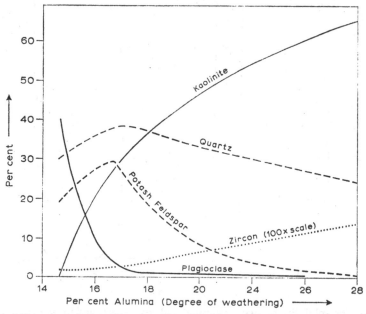

Fig. 3.11.6 Mineral-variation diagram of weathering of a granite gneiss under humid temperate conditions (After Goldich, S. S., *Journal of Geology*, Vol. 46, 1938).

3. Plagioclase. Anorthite and albite similarly combine with carbonic acid and water to produce soluble calcium and sodium bicarbonates, together with kaolinite.

$$CaAl_2Si_2O_8.2NaAlSi_3O_8 + 4H_2CO_3 + 2(nH_2O) \longrightarrow$$
$$Ca(HCO_3) + 2NaHCO_3 + 2Al_2(OH)_2Si_4O_{10}.nH_2O$$

4. Biotite combines with oxygen, carbonic acid, and water to produce soluble potassium and magnesium bicarbonates, limonitic iron, kaolinite, soluble silica, and water.

$$2KMg_2Fe(OH)_2AlSi_3O_{10} + O + 10\,H_2CO_3 + nH_2O \longrightarrow$$
$$2KHCO_3 + 4Mg(HCO_3)_2 + FeO_3.H_2O + Al_2(OH)_2Si_4O_{10}.nH_2O +$$
$$2SiO_2 + 5H_2O$$

5. Hornblende weathers similarly to biotite, but more rapidly.

Figure 3.11.6 shows the changes in relative mineral composition during the progressive weathering of a granite gneiss under humid conditions indicating the stability of quartz, as distinct from the feldspars, which decompose to produce the increasing clay content, with orthoclase being more resistant than plagioclase. The influence of climate on the mechanisms and products of igneous rock weathering is shown by recent work on a weathered quartz diorite in Antarctica, which, although apparently considerably decomposed, was found to be little changed chemically. No clay had been produced, the only chemical change being the oxidation of some iron, and the disintegration seemed mainly due to physical weathering by the growth of ice and sea-salt crystals. Where chemical weathering is dominant, however, igneous rocks break down to produce metal cations in solution, some silica in solution (much of colloidal size), a varying amount of silica wreckage (chiefly in the form of quartz particles), and clay minerals.

3. Weathering of sedimentary rocks

The breakdown of minerals forming igneous rocks thus leads to the production and isolation of more stable minerals, like clay and silica debris, as well as cations, largely of calcium. After the clastic debris is washed away by rivers the clay fraction of it is commonly separated from the coarser material, usually in a marine environment, to form beds which may be lithified to form *shale*. The composition of the shale is largely dependent on the types of clay minerals involved and partly on the environment of deposition. Their future weathering history is very much governed by the presence of metal cations in the open crystal lattice which controls their oxidation and hydration. The coarser clastic quartz material tends to be separated from the clay by the fluvial and marine agents of transportation and forms quartz sandstone (*orthoquartzite*). Obviously the quartz is very resistent to further weathering, and the weathering of sandstone is largely controlled by the material cementing the quartz grains. This is commonly iron, silica introduced subsequently in solution, or calcite and gypsum. The first is the most widespread; where the second occurs the resulting sandstone is extremely resistant to sub-aerial processes (as in the sandstone ridges of the Appalachians); and the third is usually the result of sub-aerial deposition in an arid environment. Formations like the Wingate and Entrada Sandstones of the Colorado Plateau are aeolian sands poorly cemented with calcite. A third type of clastic rock is *greywacke*, a poorly sorted sandstone containing at least 20% shale and characteristic of rapid sedimentation in geosynclines.

TABLE 3.11.1 Composition (% by weight)

Formula	Earth's crust	Average igneous rock	Average granite	Average basalt	Average sandstone	Average greywacke	Average arkose	Average shale	Average limestone
SiO_2	59·08	59·12	70·8	49·9	79·7	65·8	76·1	62·2	5·2
Al_2O_3	15·23	15·34	14·6	16·0	4·8	14·4	11·5	16·5	0·8
CaO	5·10	5·08	2·0	9·1	5·6	3·6	1·6	3·3	43·0
Na_2O	3·71	3·84	3·5	3·2	0·5	3·5	2·0	1·4	0·1
FeO	3·72	3·80	1·8	6·5	0·3	4·3	—	2·6	—
MgO	3·45	3·49	0·9	6·3	1·2	3·0	0·1	2·6	8·0
K_2O	3·11	3·13	4·1	1·5	1·3	2·1	5·7	3·5	0·3
Fe_2O_3	3·10	3·08	1·6	5·4	1·1	1·0	2·4	4·3	0·5
CO_2	—	—	—	—	5·1	1·6	0·4	2·7	41·9

The weathering of greywacke is primarily a function of the breakdown of the shale. The last important clastic rock is the least common – *arkose*. This is a sandstone containing less than 25% quartz and less than 20% shale, the rest being composed of feldspar. Although arkose is formed in a wide variety of environments from the breakdown of granitic rocks, it is most characteristic of rapid burial of partly weathered granite residue under arid conditions. Fossil alluvial fans flanking granite fault blocks have formed much of the present arkose, which is naturally quite susceptible to the further weathering of the feldspar if exposed at the surface. Of all the rocks exposed at the surface of the present continents, fully 52% are clays and shales and 15% the various types of sandstones. The metal cations, particularly Ca^{++}, which forms about $\frac{1}{2}$ of the total dissolved load of the rivers of the world, are washed into the oceans. Although the oceans are not saturated with these cations, the calcium is combined with CO_3 (dissolved from the atmosphere) and precipitated as reef and pelagic limestone by marine organisms. Limestone covers about 7% of the present continental surfaces, being mostly coral limestone, coral breccia, or pelagic limestone. Table 3.11.1 shows the major chemical compounds in some common rocks and indicates the possibility that most of the sedimentary rocks have been formed from the breakdown of igneous rocks.

Solution is, of course, an important weathering process for all rocks, but it is especially destructive to the carbonate sedimentary rocks. The breakdown of the two important carbonate minerals of calcite and dolomite under the action of carbonation by carbonic acid is the best example, and one which is most active at low temperatures:

$$CaCO_3 + H_2O + CO_2 \rightleftharpoons Ca(HCO_3)_2$$

(Calcite)   (Soluble calcium bicarbonate)

$$CaMg(CO_3)_2 + 2H_2O + 2CO_2 \rightleftharpoons Ca(HCO_3)_2 + Mg(HCO_3)_2$$

(Dolomite)   (Soluble calcium and magnesium bicarbonate)

The susceptibility of limestone to weathering partly depends upon its purity, but even more on the absolute amount of water available. Thus one finds that in desert areas limestone outcrops form residual hills and escarpments, whereas other rocks are more rapidly denuded. An example of the relative susceptibility of an arkosic sandstone, a granite, and a quartzite is given in fig. 3.11.7 relating to part of the Sangre de Cristo mountains in New Mexico.

Oxidation, the combination of oxygen with another atom causing the latter to lose an electron and take on a positive charge, is also a weathering process to which a wide variety of rock-forming minerals are susceptible. The simplest example is the production of the iron oxide hematite:

$$2Fe + 3O_2 \longrightarrow Fe_2O_3$$

Oxidation is probably almost entirely effected by the intermediary action of water, particularly in the soil, where there is a plentiful supply of CO_2 to the

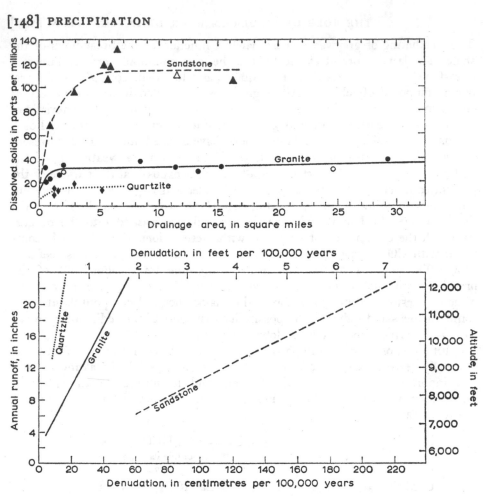

Fig. 3.11.7 Solution in the Sangre de Cristo Range, New Mexico (After Miller, J. P., *U.S. Geological Survey Water Supply Paper* 1535–F, 1961).

Above: Relation of dissolved solids in streams to drainage area (open symbols represent weighted mean values below tributary junctions). Each curve intersects the *Y*-axis at the average value of dissolved solids in snow (i.e. approx. 5 p.p.m.).

Below: Denudation as a function of altitude, calculated from the annual runoff, showing the relative resistance of quartzite, granite, and sandstone.

soil atmosphere by organic decomposition. In terms of sedimentary rock weathering, oxidation is particularly important in the weathering of the clays with especially open crystal lattices (e.g. montmorillonite), where oxygen combines with the Mg^{++} and Fe^{+++} ions.

One of the chief means of breaking down some clays, however, is by hydration – the simple adsorption of water into the crystal lattice with no fundamental chemical change, but accompanied by considerable swelling and the setting up of physical stresses. Where continuous cycles of wetting and drying occur, some clays become very fragmented. Clays are hydrous silicates containing metal cations. Their fundamental building blocks are sheets of silica tetrahedra sharing oxygen atoms with an associated octahedral sheet composed of O^{--} and OH^- ions, grouped around the metal cations (fig. 3.11.8). Kaolin possesses pairs of these sheets (fig. 3.11.8) closely bonded together with the small H^+ ions. This small space allows little further reaction and little ionic exchange to take place with the metal cations. Kaolin is produced by the alteration of feldspar-rich rocks under acid and humid conditions where most of the cations (except

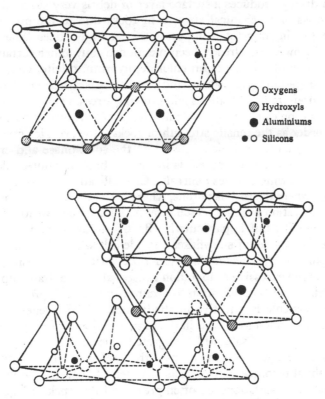

Fig. 3.11.8 The atomic structure of (*above*) kaolinite and (*below*) montmorillonite (After Grim).

Al^{+++}) are leached away and H^+ ions introduced. Illite is composed of octahedral sheets sandwiched between two tetrahedral sheets and the sandwiches strongly held together with K^+ ions. Although some internal chemical reactions occur, there is little ionic exchange or hydration. Illite derives mainly from the weathering of feldspars and micas under alkaline conditions with abundant Al^{+++} and K^+ ions. The most complex of the three common clay minerals is montmorillonite $((OH)_2(SiAl)_4(Al,Fe,Mg)_{2-3}(Na,Ca)_{1-3}O_{10})$ (fig. 3.11.8). Its structure is similar to illite, but the sandwiches are more widely spaced, allowing the entry of water and great expansion of the lattice (i.e. swelling). This allows much ionic exchange within the octahedral sheets, where Fe^{+++} and Mg^{++} are substituted for Al^{+++}, but the bonding remains very loose because the electrical charges between the ions never balance. Montmorillonite forms from basic rocks under alkaline conditions in the presence of Ca^{++} and Mg^{++} ions and a deficiency of K^+ ions. Besides influencing weathering, hydration affects the rates of erosion of shale and clay outcrops, for example, the sodium-rich parts of the Mancos Shale in the Colorado Plateau may, when subjected to free swelling, increase in volume by almost 60% under hydration, and repeated wetting and drying produces a surface layer of debris very susceptible to creep.

Another class of mechanical stresses in rocks due to the presence of water involves the growing of crystals in pore spaces and interstices. The effectiveness of ice-crystal growth has already been mentioned, but under certain conditions the crystallization of salts (sodium chloride, gypsum, calcite, etc.) is important in rock disintegration, granulation, and cavernous weathering. Salt crystallization exhibits preferred orientations and growing stresses similar to ice, and with 1% supersaturation calcite may crystallize against a pressure of 10 atmospheres – of the same order as the tensile strength of rocks. However, ice crystallization is different, in that all the solution enters into the solid phase and crystallization begins from the outside and proceeds inwards – both attributes allowing ice to develop higher growing stresses than salts. Crystallization occurs near the surface of an outcrop, where the rock is porous, where both salts and water are abundant, and where evaporation permits crystal growth. This type of weathering occurs in many environments, but is particularly effective in arctic and desert areas. In high latitudes the nuclei of snowflakes provide salt which, because melting and runoff is small, tends to accumulate near rock surfaces and disintegrate them. In arid regions the excessive evaporation causes salts to be drawn up from depth in capillary films and crystallization to occur at the surface, producing weathering which is particularly effective in shady locations. In general, the present desert areas have tended to be arid in the past, and therefore many of the underlying rocks are sources of salts which migrate to the surface.

It has often been assumed in the past that much desert weathering can be explained without recourse to the effects of water. However, despite the fact that diurnal temperature variations are often great, and that rocks are generally poor thermal conductors and composed usually of minerals having different coefficients of thermal expansion, observation and experiment have shown that thermal dilation is only important in breaking up rock surfaces in the presence of water. The existence of pronounced chemical rotting, particularly in shady sites where more moisture is available, shows that chemical weathering is dominant even in arid areas, as does the expansion of exfoliation shells on boulders, which is obviously due to expansion accompanying chemical alteration. Occasional rainstorms and, particularly, nocturnal dew form a significant supply of desert moisture, the presence of which can be detected several feet below the surface of boulders. Observations on a quartz monzonite boulder in the Mojave Desert have shown that a diurnal temperature range of 24° C would cause a significant linear expansion of 0·0084%, but that the temperature gradients within the rock are sufficiently uniform to allow the whole mass to expand and contract with little differential stress within it. The products of desert weathering, however, differ from those of more humid areas, being on the average rather coarser and not possessing such a high proportion of clay or organic material. These characteristics partly explain differences between arid and humid geomorphic processes and forms, in that creep of the organic- and clay-rich humid soils often contrasts with sheet erosion in desert areas, and that repose slopes in humid areas tend to be of lower angle.

Although in this discussion of the weathering of sedimentary rocks processes and environments have been mentioned which also relate to other rock types, the dominant primary weathering process of igneous rocks is hydrolysis, whereas solution, oxidation, and hydration are more important in most sedimentary rock decomposition. Metamorphic rocks are generally very susceptible to chemical weathering in the presence of water because the usual effects of metamorphism are to produce secondary high-temperature minerals of low stability, together with banding and schistosity, which encourages the entry of surface water. One major exception is recrystallized orthoquartzite, which is probably the most resistant of all rocks to weathering and universally forms strong relief.

4. The weathered mantle

Under suitable conditions, particularly in the humid tropics, weathering can proceed to considerable depths. Shales in Brazil have been altered down to 400 ft, and in Georgia granite has weathered *in situ* to 100 ft. Depth of weathering is controlled by the bedrock, climate, biological action, topography, and time, but even within a single rock body it can be extremely varied, especially where closely spaced joints expose a huge internal weathering surface (fig. 3.11.9), allowing weathering to proceed rapidly to considerable depths. Granite in Hong Kong is weathered to a maximum of 300 ft, but the removal of this mantle would expose a surface of considerable relief. The formation of tors and even tropical inselbergs has been ascribed to deep weathering around more-resistant and less-jointed parts of the rock body, accompanied or succeeded by the evacuation of debris either associated with rejuvenated stream downcutting or climatic change.

The mantle/rock contact exhibits considerable variety. Some rocks, notably basic rocks in the humid tropics, have a sharp transition a few mm wide between the weathered mantle and the bedrock, along what has been termed the 'weathering front' by analogy with the advance of some metamorphic processes. (Indeed, some authors refer to weathering as 'katamorphism'.) Usually, however, the contact is gradational, and granites in the tropics commonly have a transitional layer several metres thick. It appears to be sharpest where the rock is least permeable, where a water-table lies close to the surface, where the minerals weather rapidly, so that the rock weathers uniformly layer by layer from the top down, or where there are few minerals resistant to weathering. Much of the confusion of the contact in granite is due to the high proportion and irregular distribution of stable quartz. It should be stressed that erosion can produce a sharp topographic surface even with a transitional weathering contact, because relatively little weathering is necessary to loosen bedrock sufficiently for erosion to take place.

Rate and character of weathering is thus affected by the texture, fabric, and structure of rock bodies, as well as by their mineral composition. Even where solution is dominant in limestones, thin black rendzina soils remain under humid conditions, supporting short grass. The biological influence is important, not

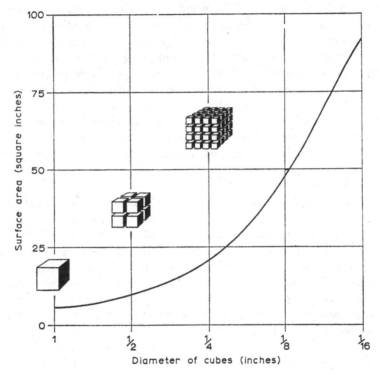

Fig. 3.11.9 The relation of the exposed surface area of a cube to the diameter of smaller cubes into which it is decomposed (Adapted from Leet and Judson, 1965).

only because humic and bacterial acids assist in mineral decomposition and because roots remove K^+ ions but because the accumulation of humus in the upper (A_0) layer of the soil (fig. 3.11.10(a) and (d)) both improves aeration and increases infiltration and water-holding properties. Indeed, much of the climatic influence over soil production is effected by the *soil climate*, in particular by the amount and movement of infiltrating water. Under humid conditions (e.g. pedalfer soils) the amount of clay is greater (fig. 3.11.10(c)) and the leaching of the silica-enriched A horizon is accompanied by deposition of clays, colloids, and cations in the Al^{+++} and Fe^{++} enriched B horizon. Under dryer conditions (e.g. pedocal soils) the discontinuous downward transport by water means that horizons are less well developed, and the periodic upward movement of alkaline ground water usually causes calcium carbonate (fig. 3.11.10(b)) and salts to accu-

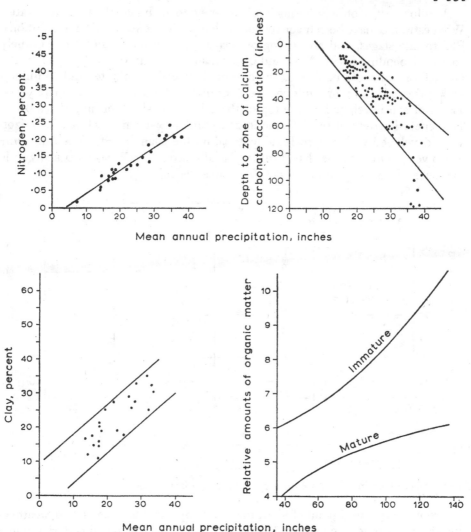

Fig. 3.11.10 The relation of mean annual precipitation to:
Top left. Nitrogen content.
Top right. Depth to zone of calcium carbonate accumulation (Data from Jenny and Leonard, 1934, and Russell and Engle, 1925).
Bottom left. Percentage of clay (Data from Jenny, 1941).
Bottom right. Relative amounts of organic matter
(Mostly from Leopold, Wolman, and Miller, 1964).

mulate below the surface which precipitation is insufficient to leach away. Topography represents another class of factors influencing weathering and soil formation, partly because of its control over micro-climate, vegetation, and drainage but also because downslope movements of soil influence soil thickness and structure.

Absolute rates of weathering are less easy to estimate than relative rates. What estimates have been made rely on such datable events as volcanic eruptions, Pleistocene stages, and archaeological structures. Absolute rates are extremely varied, depending on rock type and climate, and Kellogg was of the opinion that an inch of soil could be formed in any time between 10 minutes and 10 million years! Indeed, 14 in. of crude soil were formed on Krakatoa pumice in 45 years, 12 in. of soil developed on the top of the calcareous slabs forming the walls of Kamenetz Fortress in the Ukraine in 230 years, and 1·8 m of clayey B horizon were developed within pyroclastic material on the island of St Vincent in some 4,000 years. On the other hand, many glacially-scoured rock surfaces show little evidence of 10,000 years of post-Glacial weathering.

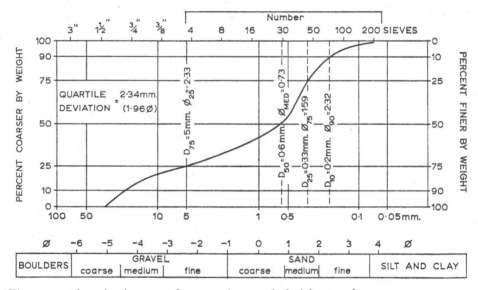

Fig. 3.II.11 A grain-size curve for a poorly-sorted glacial outwash.

From the practical point of view, the most important aspects of rock weathering are the chemical and physical characteristics of the weathered material. Some features of the chemical composition have already been mentioned and will be referred to again elsewhere in this volume. The simple parameters of the grain-size characteristics are shown in fig. 3.II.11. The two most important classes of size parameters are measures of absolute size and the range of sizes present, not least because they strongly influence permeability, shearing resistance, stability, and frost-heaving characteristics of the weathered material. These parameters are expressed in terms of diameter of particle (in mm) on a logarithmic scale, because of the huge range of sizes present in many soils which contain both gravel sizes and minute clay particles. Because of this range it has become the practice to use the ϕ (phi) scale, where $\phi = -\log_2$ diameter (mm), such that 1 mm $= 0 \phi$, 2 mm $= -1 \phi$, 4 mm $= -2 \phi$, $\frac{1}{2}$ mm $= 1 \phi$, $\frac{1}{4}$ mm $= 2 \phi$, etc. Because the engineering properties of soils are so much influenced by the proportions of finer material present, another statement of size is the D scale; D being the diameter (in mm) for which certain percentages by weight are finer. Thus the median size (ϕ_{med}) is D_{50}, and the very diagnostic size for which 10% is finer (ϕ_{90}) is D_{10}. One of the simplest measures of range of sizes present in a soil is the quartile deviation equal to $\dfrac{\phi_{75} - \phi_{25}}{2}$.

REFERENCES

BECKINSALE, R. P. [1966], Soils: Their formation and distribution; Chapter 24 in *Land, Air and Ocean*, 4th edn. (Duckworth, London), pp. 361–99.

BURMISTER, D. W. [1951], *Soil Mechanics*; Vol. 1 (Columbia University Press, New York), 155 p.

DAVIS, K. S. and DAY, J. A. [1964], *Water: The Mirror of Science*; The Science Study Series Number 21 (Heinemann, London), 195 p.

GILLULY, J., WATERS, A. C. and WOODFORD, A. O. [1960], *Principles of Geology*; 2nd ed. (Freeman, San Francisco), 534 p. (especially pages 43–57).

HENDRICKS, S. B. [1955], Necessary, Convenient, Commonplace; in *Water*, U.S. Department of Agriculture Yearbook (Washington, D.C.), pp. 9–14.

KELLER, W. D. [1957], *The Principles of Chemical Weathering* (Lucas Bros., Columbia, Missouri), 111 p.

KELLER, W. D. [1957], *Chemistry in Introductory Geology* (Lucas Bros., Columbia, Missouri), 84 p.

KRUMBEIN, W. C. and PETTIJOHN, F. J. [1938], *Manual of Sedimentary Petrography* (Appleton-Century-Crofts Inc., New York), 549 p.

LEET, L. D. and JUDSON, S. [1965], *Physical Geology*; 3rd edn. (Prentice-Hall, New Jersey), 406 p. (especially pp. 76–90).

LEOPOLD, L. B., WOLMAN, M. G. and MILLER, J. P. [1964], *Fluvial Processes in Geomorphology* (Freeman, San Francisco), 522 p. (especially pp. 97–130).

MOHR, E. C. J. and VAN BAREN, F. A. [1954], *Tropical Soils*; (Interscience, London and New York), 496 p. (especially pp. 133–78).

REICHE, P. [1950], A Survey of Weathering Processes and Products; Revised Edn., *University of New Mexico Publications in Geology*, No. 3, 95 p.

III. The Interaction of Precipitation and Man

R. J. CHORLEY and ROSEMARY J. MORE
Department of Geography, Cambridge University and formerly of Department of Civil Engineering, Imperial College, London University

1. Man's intervention in the hydrological cycle

Every part of the hydrological cycle has been tampered with; runoff is stored behind dams, evaporation is reduced by coating water surfaces with suitable monolayers, transpiration losses are reduced by removing phreatophytes, and ground water is recharged by water spreading and pumping (fig. 3.III.1). It is now clear that, under suitable conditions, natural precipitation can be artificially modified. Even where the actual amount of water cannot be tampered with, man's response to its occurrence is far from passive, in that he changes its circulation by the use of irrigation and, as a last resort, gambles on its occurrence by crop insurance.

Since precipitation is the principal input into the hydrological cycle, man's attempts to modify it will have consequences throughout the working of the cycle. For example, precipitation modification would influence magnitude and time of runoff, soil moisture reserves, and ground-water storage. It is because of these all-pervading climatic, hydrologic, and social repercussions and possible side-effects throughout the hydrological system (some known and many unknown) that man has hesitated to embark upon unbridled precipitation changes, although technologically he is increasingly in a position to be able to make them. However, man is only at the beginning of weather modification, being able to achieve only local effects of small magnitude, so that the wider-scale consequences which surround his endeavours have not yet been encountered.

There are two main approaches to conscious precipitation modification by man, quite apart from the locally important inadvertent effects associated with the construction of large urban and industrial complexes, the creation of large lakes, and the modification of the surface vegetational cover. The first is to attempt to achieve increases of precipitation of a known magnitude (5 or 10%) by available techniques, generally with a view to increasing soil moisture for plants or increased water yield from catchments for public water supply and other uses. An alternative, and broader, viewpoint is to try to assess the ideal water requirement for any location. Although this is difficult to evaluate, involving as it does the determination of the changing uses of the water and reconciliation of competing interests, it is probably a sounder aim than the more limited one of simply increasing precipitation. Thus attention should ideally be

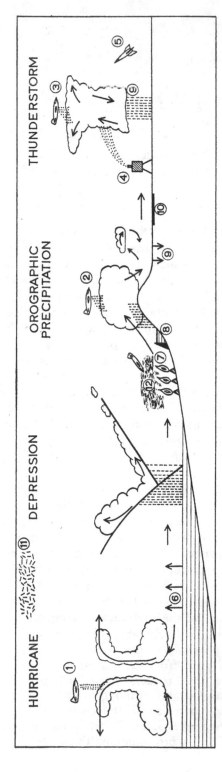

Fig. 3.III.1 Some points of human intervention in the world hydrological cycle.

1. Seeding of hurricane eye-wall.
2. Seeding of orographic cloud.
3. Seeding of thunderstorm.
4. Ground-based silver iodide seeding.
5. Dispelling hail by rocket.
6. Interference with sea-surface evaporation.
7. Irrigation below dam.
8. Artificial reservoir.
9. Water spreading and ground-water recharge.
10. A 'thermal mountain'.
11. Needles in orbit.
12. Local fog dispersion.

given not only to increasing precipitation in areas where there is too little but also to decreasing it where an excess results in poor harvests, flooding and other disasters. Figure 3.III.1 shows some principal points where human intervention in the hydrological cycle is possible. One is at the ocean surface, where moisture rises by evaporation to form moist air masses. A correlation has been shown to exist, for example, between Pacific Ocean surface temperatures 800 miles west of the Sierra Nevada, California, and precipitation falling on this range. Similarly, high sea temperatures and evaporation rates were recorded in the Western Mediterranean before the disastrous floods on the Arno at Florence in 1966. It would seem logical that if atmospheric conditions favouring evaporation from the oceans could be optimized (by chemical methods, heating the surface water, increasing wind speeds, etc.) more moisture would be drawn up into the atmosphere to be available for precipitation on land. Little work has been done on this approach, largely because of the lack of data on ocean temperatures and their correlation with evaporation and precipitation, but satellite sensing of meteorological conditions over oceans may remedy this deficiency in the near future and make experimental work possible.

The saturated air masses over oceans are transported inland largely by depressions following storm tracks, and a second approach to weather modification might be possible by influencing the preferred paths of these storm tracks. Such a large-scale intrusion into the general atmospheric system is not yet possible, but may be more so as storm generation and movement is better understood.

2. Rain-making

The most fruitful point of intervention has been by *cloud seeding*. Rain-making experiments of this type are based on three main assumptions:

1. Either the presence of ice crystals in a super-cooled cloud is necessary to release snow and rain (according to the Wegener–Bergeron theory); or the presence of comparatively large water droplets is necessary to initiate the coalescence process.
2. Some clouds precipitate inefficiently or not at all, because these components are naturally deficient.
3. The deficiency can be remedied by seeding the clouds artificially with either solid carbon dioxide (dry ice) or silver iodide to produce ice crystals, or by introducing water droplets or large hygroscopic nuclei.

Such seeding is thus only productive under limited conditions of orographic lift and in thunderstorm cells, when nuclei are insufficient to generate rain by natural means. Natural precipitation occurs preferentially within certain upper-air temperature ranges – for example, some 80% of winter precipitation in the state of Washington falls when the 700-mb (10,000-ft) temperature is not lower than $-10°$ C and is especially prevalent at $-4°$ to $-8°$ C. Artificial precipitation stimulation must exploit these preferences, and seeding is thus effective within a limited temperature range. Below $-20°$ C natural nuclei, such as dust, become active to form snowflakes, usually in sufficient numbers so that

A. Recognition: radar, balloon, and observing aircraft.

B. Treatment: ground-based silver iodide generator and aircraft dispensing silver iodide or dry ice.

C. Treatment: cloud developing intensively and precipitation falling.

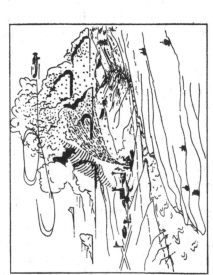

D. Evaluation: precipitation, depth, and cloud information transmitted to headquarters for evaluation.

Fig. 3.III.2 The operation of cloud seeding (Atmospheric Water Resources Program, 1968).

additional silver iodide particles are not needed, and under some conditions are actually detrimental. Cloud seeding may be effected by burning silver-iodide-impregnated fuels or solutions at ground level to produce a smoke which is carried upwards by wind into the effective zone, by firing rockets containing nuclei into the effective zone, or, more usually, by dropping the nuclei from aircraft (fig. 3.III.2).

Cloud seeding by these means has been attempted in many parts of the world, notably in Australia and the western United States. The need for fresh sources of water now and in the future is so acute in the United States that the Bureau of Reclamation has initiated the nation-wide Project Skywater to investigate the possibilities of water management through artificial rain-making. The purpose of one such scheme is to increase winter precipitation over the mountains of the Upper Colorado River, thus augmenting spring runoff, which would be stored and regulated to meet demand by the existing reservoirs. An increase of the November–April precipitation by 15% over 14,200 square miles of target areas is expected to yield an average additional runoff of 1,870,000 acre-feet annually. An important advantage of water provision by cloud seeding methods is that a 10% increase in rainfall can result in a 17–20% increase in runoff, because evaporation does not increase in proportion to the greater precipitation, so that there is more water available for runoff. Preliminary investigations have shown that the optimum conditions for seeding are when there is a thin (less than 5,000 ft thick) saturated air-mass layer, the temperature at the top of which is not less than −20° C, and the temperature over the target area is less than −10° C (fig. 3.III.3). Eight major areas, lying generally above 9,500 ft where annual natural runoff is over 10 in., contribute 75% of the total Upper Colorado basin runoff, although they form only 13% of the basin land area. Most precipitation comes in a few big storms, and since these storms are the main precipitation generators, it is important to take advantage of the limited opportunities they offer. Increasing the total precipitation, however, also increases the variability of its occurrence, since the fall from large storms is increased, the smaller rainfalls remaining the same. The cost of new water in the Upper Colorado, provided by cloud seeding, has been estimated to be approximately $1.00–$1.50 per acre-foot. (This is the operating and running cost, exclusive of the research necessary to make the work feasible.)

An associated method of artificially increasing precipitation is a corollary of cloud seeding and involves attempts to reduce the water loss by the evaporation of precipitation which takes place between the cloud base and the ground. In the Sonora Desert, Mexico, it has been estimated that, whereas 40 in. of rain is annually available at the cloud base, only 9 in. reaches the ground. The problem here is to find an agent which will increase the drop size and keep the drops large, and so far such an agent has not been found.

In summary, it is clear that there is a limited range of natural conditions in which significant artificial interventions can be made to produce, increase, or conserve precipitation. The seeding of some cumulus clouds at temperatures of about −10° to −15° C probably produces a mean increase of precipitation of

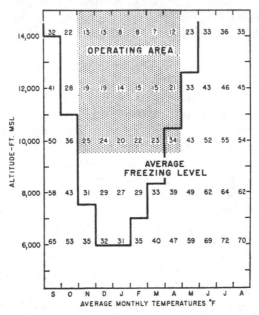

Fig. 3.III.3 Average temperature–altitude chart for Grand Junction, Colorado, showing the average freezing level and the optimum operating area for seeding (i.e. between November and April above 9,500 ft) (Hurley, P. A., 1967, *Augmenting Upper Colorado river basin water supply by weather modification;* Annual Meeting American Society of Civil Engineers, New York, October 18, 1967).

some 10–20% from clouds which are already precipitating or 'are about to precipitate', with comparable increases up to 250 km downwind, and increases of up to 10–15% have resulted from the seeding of winter orographic storms. On the other hand, the seeding of depressions has produced no apparent increases, and it appears that clouds with an abundance of natural nuclei, or with above-freezing temperatures throughout, are not susceptible to rain-making. At present it is often a difficult statistical matter to determine whether many of man's attempts have produced significant increases in precipitation; for example, six experiments in Washington and Oregon produced the following probabilities that rainfall had been increased: 95, 67, 50, 50, 50, and 41%! Another instance serves also to highlight the possible legal problems which attempts at rain-making will provoke. In Quebec a recent 25% increase in rainfall coincided with rain-making attempts, causing extensive floods, crop damage, and disruption of the tourist industry. Following a large public outcry, the Federal Government announced that the effect of the seeding had been to *decrease* the possible rainfall receipt by 5%!

Besides rain-making, other human interventions in precipitation involve the successful local dissipation of freezing fogs over airports by spraying with propane gas, brine, or dry ice, causing snow to fall and clear the air. The Russians have also claimed success in dissipating damaging hailstorms by the use of radar-directed artillery shells and rockets to inject silver iodide into high-liquid-content portions of clouds, which freezes the available super-cooled water and prevents it from accreting as shells on growing ice crystals.

It is probable, however, that man is on the brink of much larger interventions into the hydrological cycle on a scale of hundreds of square miles or more. Hurricanes cause, on average, some $300 million worth of damage annually in the United States and Canada. At present the $9 million spent on forecasting, warning and protection is estimated to save some $25 million of property, with only about 20% of the affected population being involved in protective action. It has been estimated that improvements in forecasting and warning systems might increase the saving to some $100 million. More ambitious projects make it likely, however, that hurricanes can be suppressed or 'damped down' by the seeding of the rising air in the cumulus eye-wall, widening the ring of condensation and updraught, decreasing the angular momentum of the storm and thus the maximum speed of its winds. The spreading of the sea ahead of the storm with oily materials might be used to cut off surface evaporation and thus the hurricane energy supply. Even such apparently beneficial attempts may represent potentially dangerous tampering with the natural global moisture economy, especially so in this instance when it is remembered that 30% of the August rainfall of the Texas coast, 30% of the September rainfall of the Louisiana and Connecticut coasts, and fully 40% of the September rainfall at Atlantic City, New Jersey, are derived from hurricane circulations. Even more speculative schemes involve putting huge quantities of dust or metallic needles into stationary orbit to locally reduce sea temperatures and decrease evaporation; as well as creating 'thermal mountains' by painting desert surfaces black to increase their conservation of solar heat, stimulate convection, and thereby increase cloudiness and precipitation downwind. The unknown dangers attendant upon such large-scale tampering with the delicately balanced world hydrological cycle must postpone such schemes until theoretical mathematical models simulating the behaviour of the earth–atmosphere system have been developed so that all the possible effects can be predicted in advance.

3. Spray irrigation

An alternative method of precipitation modification is to supply the rain artificially at ground level by means of sprinkler equipment. This method is used most commonly for crop irrigation in areas of supplemental irrigation (e.g. North-West Europe, Eastern United States) or in arid and semi-arid areas with a highly capitalized agriculture (e.g. California and Israel). The amount of irrigation water needed is calculated by estimating the potential evapotranspiration of the cropped area by the most appropriate of the available formulae (e.g. by Penman, Lowry and Johnson, Blaney-Criddle; see Chapter 4.1). In many areas a supply of soil moisture is built up in the wet season, and this reserve can be deducted from the calculated potential evapotranspiration, as it is stored water available for the plant's use in the dry season. Calculations of additional water needs for cropping are easier in arid countries than in areas of supplemental irrigation (i.e. in arid areas no rain may be expected), whereas in semi-humid areas calculations have to have a probability factor depending on the assessment of likelihood of growing-season rainfall.

The rain produced by sprinkler irrigation may be called *artificial rain* (in contrast to the natural rain from atmospheric intervention methods), in that it can be controlled, in its time of occurrence, duration, intensity, uniformity, and drop size, by the type of sprinkler equipment chosen and the spacing of the equipment. Precise specifications for artificial rainfall are most closely achieved by sprinklers used in laboratory catchments (see Chapter 2.1). In addition to fixed sprayers and spraylines, more sophisticated methods have been used in the Western United States and in Hawaii, where sprays are mounted on wheels and can be moved by electronic means across the crop, thus simulating the passage of a shower of rain. With this type of equipment evaporative losses can be kept to a minimum, especially if the sprinkling is done at night, at dawn or in the evening (when the sunlight is least), or when the wind speed is low. Winds can seriously interfere with the efficiency of the distribution, particularly from rainguns, although special sprinklers can be used if the prevailing winds are high. Sprinkler irrigation is the most mechanized and probably the most efficient method of applying water to a great range of crops, and the costs are correspondingly higher than those for other methods of irrigation. An interesting elaboration is that similar equipment, with a freezer added, has been used for snow generation in the ski-ing resorts of the Swiss, Austrian, and Italian Alps and in the Cairngorms (Scotland).

4. Gambling with water

While man has made a certain amount of progress in increasing, conserving, and redistributing rainfall, he has done very little to protect himself from the effects of too much. Future architectural advances in wide-arc roofing, already fore-shadowed by the Houston *Astrodome*, may make it possible for urban precincts, and even whole cities, to have controlled climates. Most rural activities, with the limited exception of glasshouse protection for certain high-value crops, must proceed under the threat of the occurrence of droughts, floods, hailstorms, blizzards, and the like, although their probabilities of occurrence vary very much from place to place. Although certain industries, notably building construction, transport, and those geared to a minimum level of streamflow, are victims of these meteorological events, it is the farmer who must gamble most heavily on hydrological probabilities. One of the most obvious ways of doing this is to diversify cropping so that, whatever the weather, some minimum level of production is reasonably assured. Such diversification is particularly apparent in regions of subsistence agriculture, where rainfall is highly variable. In East Africa, for example, more than one crop failure in ten can be disastrous where native farmers have no capital reserves, in contrast to which local commercial agriculture finds one rainfall failure (i.e. less than 30 in. of rain per annum) in three 'acceptable'. Game theory has been used to show how an unconscious perception of climatic probabilities has been employed by primitive farmers to maximize their possibility of continual survival against nature by always planting a balanced range of crops with differing responses to deficiency or abundance of rainfall.

Although diversification is also an essential feature of much advanced farming practice, the changing commercial economic base means that in many regions the narrowing profit margins require ever-increasing production per acre, which, in turn, forces increasing specialization and mechanization, both of which imply economic pressures towards monoculture. At the same time the attendant rising capital costs and narrowing profit margins mean that a single year's crop may represent an investment of four or more years' profits. The narrowing of the economic base of regional agriculture makes climatically-controlled variations in agricultural production of great significance, because all over the world agriculture is still largely at the mercy of the weather, despite advances in weather-resistant crops and the use of weather forecasts (Sewell, Kates and Phillips, 1968, p. 267). Many studies have been made regarding the economic effects of weather variations on single regional crops, for example the effect of drought on wheat failure in the central Great Plains (Hewes, 1965), of July rainfall and mean temperatures on corn production in Indiana (Visher, 1940), of rainfall on rice yields in West Bengal (Hore, 1964), and of January precipitation on milk butterfat contents and South Auckland, New Zealand (Maunder, 1968). The last-named study, for example, showing that the occurrence of a wet January (on average 1 in 6) is 'worth' about $2 million (New Zealand) to the dairy farmers of South Auckland when all the benefits have been evaluated. Even under the reasonably-predictable British climatic conditions economically-significant moisture variations are of great importance. For example, the generally wet and cloudy summers of the 1950's retarded plant growth, encouraged diseases and made cereal harvesting difficult or impossible; and the severe snowfalls of 1947 and 1963 caused severe stock losses. Even grass, as in New Zealand, is very sensitive to moisture variations and, under high stocking conditions in south-east England without irrigation, the milk output may fall from 600 gallons per acre in a wet year to 200 gallons in a dry one, turning a profit of £24 per acre into a loss of £16. More sophisticated studies have been concerned with the relationships between climatic studies and farm organization and management programmes (Curry, 1952 and 1962) and with the multivariate effect of many climatic parameters not only on agricultural production but on the whole economy of a region (Maunder, 1966).

One partial answer to such climatically-controlled variations in agricultural returns is crop insurance, and in the United States more than $2·7 billion worth is held by farmers. This covers twenty-three crops, the most widespread being for wheat, maize, soyabeans, cotton, tobacco, oats, and barley (in that order). For some crops (e.g. tobacco) insurance commonly guarantees a stated income per acre, whereas for others (e.g. maize) it guarantees a given production per acre, at a stated quality. Premiums and conditions naturally vary greatly, depending on fertility, as well as climatic and other risks, and in a single United States' county the guaranteed levels of maize production may range from 30 to 50 bushels per acre. Crop insurance claims number more than 50,000 per year and, of those paid, 39% are for damage by drought, 14% for 'excess moisture', and 10% for hail damage, as against 11% for insect damage, 10% frost, 6% wind, and 5% disease.

REFERENCES

ATMOSPHERIC WATER RESOURCES PROGRAM [1968], *Project Skywater: 1967 Annual Report: Vol. I, Summary*; U.S. Department of the Interior, Bureau of Reclamation (Washington, D.C.), 79 p.

BAILEY, F. JR. [1965], Can you afford to lose your crops?; *The Farm Quarterly*, **20** (1), 66–7, 136, 138, 140, and 142.

BARRY, R. G. and CHORLEY, R. J. [1968], *Atmosphere, Weather and Climate* (Methuen, London), 319 p.

BATTAN, L. J. [1965] *Cloud Physics and Cloud Seeding*; The Science Study Series Number 27 (Heinemann, London), 144 p.

CURRY, L. [1952], Climate and economic life: A new approach with examples from the United States; *Geographical Review*, **42**, 367–83.

CURRY, L. [1962], The climatic resources of intensive grassland farming: The Waikato, New Zealand; *Geographical Review*, **52**, 174–94.

GOULD, P. R. [1963], Man against his environment: A game theoretic framework; *Annals of the Association of American Geographers*, **53**, 200–17.

HENDERSON, H. J. R. [1963], Climatic factors and agricultural productivity; *University College of Wales, Aberystwyth, Memorandum No. 6*.

HEWES, L. [1965], Causes of wheat failure in the dry farming region, central Great Plains, 1939–1957; *Economic Geography*, **41**, 313–30.

HORE, P. N. [1964], Rainfall, rice yields and irrigation in West Bengal; *Geography*, **49**, 114–21.

MASON, B. J. [1962], *Clouds, Rain and Rainmaking*; (Cambridge), 145 p.

MAUNDER, W. J. [1966], Climatic variations and agricultural production in New Zealand; *New Zealand Geographer*, **22**, 55–69.

MAUNDER, W. J. [1968], Effect of significant climatic factors on agricultural production and income: A New Zealand example; *Monthly Weather Review*, **96**, 39–46.

SEWELL, W. R. D., Editor [1966], Human dimensions of weather modification; *University of Chicago, Department of Geography, Research Paper 105*, 423 p.

SEWELL, W. R. D., KATES, R. W., and PHILLIPS, L. E. [1968], Human response to weather and climate; *Geographical Review*, **58**, 262–80.

SUGG, A. L. [1967], Economic aspects of hurricanes; *Monthly Weather Review*, **95**, 143–6.

TAYLOR, J. A. [1965], Weather hazards in agriculture; *The Royal Welsh, Annual Edition*.

VISHER, S. S. [1940], Weather influences on crop yields; *Economic Geography*, **16**, 437–43.

CHAPTER 4

Evapotranspiration

I. Evaporation and Transpiration

R. G. BARRY

Department of Geography, University of Colorado

Investigation of the transfer of moisture from the surface of the earth to the atmosphere concerns workers from a number of disciplines. On the practical side there are agriculturalists, foresters, and hydrologists, and on the theoretical side meteorological physicists and plant physiologists. The physical controls on evaporation have been recognized since 1802, when John Dalton first stated the basic principles, but it is only during the last twenty or so years that the active exchange of practical and theoretical findings between research workers has begun to provide a coherent body of knowledge. Inevitably there are innumerable specialized papers on evaporation reflecting these different approaches, and the treatment here is necessarily restricted to a statement of the basic concepts and an outline of some applications and results.

More detailed accounts of the theory are provided by King [1961], Sellers [1965, pp. 141–80], and Thornthwaite and Hare [1965].

1. Basic mechanisms of evaporation

Net transfer of water molecules into the air occurs only if there is a vapour-pressure gradient between the evaporating surface and the air, i.e. evaporation is nil when the relative humidity of the air is 100%. Evaporation from a moist surface involves a change of state from liquid to vapour, and therefore necessitates a source of latent heat. To evaporate 1 g of water requires 540 cal of heat at 100° C and 600 cal at 0° C. An external heat source must therefore be available. This may be solar radiation, sensible heat from the atmosphere, or from the ground. Alternatively, it may be drawn from the kinetic energy of the water molecules, thus cooling the water until equilibrium with the atmosphere is established and evaporation ceases. In general, solar radiation is the principal energy source for evaporation.

In addition to the two primary controls, the evaporation rate is affected by wind speed, since air movement carries fresh unsaturated air to the evaporating surface. Within approximately 1 mm of the surface the upward movement of vapour is by individual molecules ('molecular diffusion'), but above this surface boundary layer turbulent air motion ('eddy diffusion') is responsible. The temperature of the evaporating surface also affects evaporation. At higher temperatures more water molecules can leave the surface due to their greater kinetic energy. Salinity depresses the evaporation rate in proportion to the solution concentration. For sea-water the rate is about 2–3% lower than for fresh water.

2. Plant factors

Water loss from plants – *transpiration* – takes place when the vapour pressure in the air is less than that in the leaf cells. About 95% of the diurnal water loss occurs during the daytime, because water vapour is transpired through small pores, or *stomata*, in the leaves, which open in response to stimulation by light. Transfer of water vapour to the atmosphere is the initiating process in the movement of water from the soil via the plant. It is a passive process so far as the plant is concerned, but it performs a vital function in effecting the internal transport of nutrients and in cooling leaf surfaces. Transpiration considerably exceeds the direct water needs of the plant. Nevertheless, the transfer of water to the air is unavoidable. In the absence of a plant cover evaporation would still occur from the soil.

Interaction between soil-moisture content and root development is a complicating factor. If soil water is not replenished over a period of weeks vegetation with deeper roots, especially trees, will transpire more than shallow-rooted plants, other things being equal (see Chapter 4.1.8). Some support for this idea is provided by catchment studies. Run-off from catchments under grass generally exceeds that from catchments under woodland. However, this problem remains a subject of considerable controversy.

Resistances to water movement, both in the soil and in plant tissues, must be considered. These include soil-water tension, the resistance of cell walls in the roots and leaves to water transport, and the resistance of stomata to vapour transfer. The *internal* (stomatal) *resistance* of a single leaf to diffusion is an important control on transpiration, and it is dependent on the size and distribution of the stomata. For a crop or vegetation cover with several leaf layers the effective stomatal resistance (r_s) is reduced to approximately 30% of that of an individual leaf, owing to the decreased ventilation within the cover. Seasonal variations associated with changes in the leaf area affect r_s, as do diurnal variations. The latter result partly from the opening and closing of the stomata with light intensity and partly from the effects of transpiration stress on the stomata when water uptake lags behind transpiration. A separate *external resistance* of the air to molecular diffusion (r_a) arises through frictional drag of air over the leaf (larger leaves have lower transpiration rates) and the interference between diffusing molecules of water vapour. A decrease in r_a may be due to higher wind speeds or greater 'roughness' of the vegetation surface, which causes increased turbulence in the air flow. Generally the stomatal resistance r_s is larger than r_a, although, as discussed below (Chapter 4.1.8), the interaction of r_s and r_a is an important determinant of evaporation rates.

A further effect of a vegetation cover is that it intercepts precipitation before it can reach the surface. A forest canopy may retain up to 30% of total precipitation (more for conifers than deciduous species), and the proportion is larger for light, showery precipitation.

The amount reaching the ground via stem flow varies according to tree-type, but the bulk is evaporated without entering into the soil–plant part of the cycle. This might be regarded as an excessive loss of moisture compared with a grass cover, especially in the winter. However, the radiant energy used in evaporating intercepted water is unavailable for other evapotranspiration, and hence interception is not as serious a problem as it might appear to be.

3. Potential evapotranspiration

Moisture transfer from a vegetated surface is often referred to as evapotranspiration,[1] and when the moisture supply in the soil is unlimited the term potential

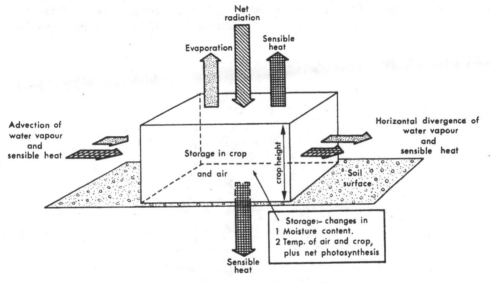

Fig. 4.1.1 The energy balance at the surface (After King, 1961).

evapotranspiration (PE) is used. It has been suggested that PE can be defined more specifically as the evaporation equivalent of the available net radiation, i.e. $PE = R_N/L$, where L is the latent heat of vaporization (59 cal cm^{-2} ≈ 1 mm evaporation). In some cases this equivalence may be invalid. For example, if an irrigated area is surrounded by dry fields evaporation rates can exceed R_N/L by 25–30%. Air heated by passing over the dry areas upwind maintains the high rates through the downward transfer of sensible heat to the irrigated section – the so-called 'oasis effect'. Horizontal transport (advection) of sensible heat *through* the vegetation cover (fig. 4.1.1) can also cause anomalous evaporation rates – termed the 'clothesline effect'. This occurs when a study plot is not

[1] In agricultural studies the term 'consumptive use' (CU) of water by crops is commonly used. However, in irrigation engineering, where the term is applied to irrigated crops, CU = PE. At certain times of year CU is less than PE. Consequently, there is a risk of confusion and misinterpretation.

surrounded by a zone with identical vegetation cover and environmental conditions. The 'buffer zone' necessary to eliminate these effects varies in size, but may exceed 300 m radius. Nevertheless, for all short crops of approximately the same colour and completely covering the ground the *PE* rate is essentially determined by the total available energy as long as there is unlimited soil water. Plant physiology is important in the case of specialized crops, such as rice and sugar cane (high water use rates) and pineapple (low usage).

4. Actual evaporation

It is known that when the moisture supply in the soil is limited plants have difficulty in extracting water, and the evaporation rate (*E*) falls short of its maximum value (*PE*). The precise nature of this relationship is controversial. One view is that the potential rate is maintained until soil-moisture content

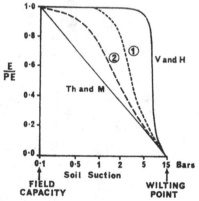

Fig. 4.1.2 The relationship between the ratio of actual to potential evapotranspiration $\left(\frac{E}{PE}\right)$ and soil moisture (After Holmes, 1961, and Chang, 1965).

V and H = Veihmeyer and Hendrickson.
Th and M = Thornthwaite and Mather.
1 and 2 = Schematic curves for a vegetation-covered clay-loam under low evaporation stress and a vegetation-covered sandy soil under high evaporation stress, respectively.
1 Bar = 1,000 millibars (10^6 dynes/cm²).

drops below some critical value, after which there is a sharp decrease in evaporation, while another is that the rate decreases progressively with diminishing soil moisture. At field capacity (maximum soil moisture content under free drainage) $E/PE = 1$, i.e. evaporation proceeds at the maximum potential rate. Veihmeyer and Hendrickson consider that no change takes place in this ratio until the plant is near wilting point (fig. 4.1.2). Thornthwaite and Mather assume the decrease below field capacity to be a logarithmic function of soil suction, but recent work suggests that $E/PE \approx 1$ as long as the moisture content is at least 75% of field capacity. Undoubtedly the soil type and climatic conditions are important; field capacity ranges from 25 mm in a shallow sandy soil to 550 mm in deep clay-loams. Chang [1965] indicates that Veihmeyer and Hendrickson's results may apply to a heavy soil with vegetation cover in a humid, cloudy region, whereas in sandy soils with a vegetation cover under arid conditions a rapid decline in E/PE is likely. Experimental work by Holmes [1961] supports this view; see lines (1) and (2) on fig. 4.1.2.

5. Meteorological formulae

There are two principal lines of approach to estimating evaporation through physical relationships; one is the aerodynamic (or mass transfer) method, the other is the energy budget method.

A. Aerodynamic method

This method considers factors controlling the removal of vapour from the evaporating surface. These are the vertical gradient of humidity and the turbulence of the air flow. The mathematical expression relates evaporation from (large) water bodies to the mean wind speed at height z (u_z), and the mean vapour pressure difference between the water surface and the air at level z $(e_w - e_z)$,

$$E = K u_z (e_w - e_z) \tag{1}$$

where K is an empirical constant. e_w is calculated for mean water surface temperature. The method has been applied to ocean areas in particular, but only for monthly averages, since it assumes that the temperature lapse rate is adiabatic, and this does not apply on many individual occasions. More elaborate forms of equation (1) incorporating complex wind functions have been developed for land surfaces and other lapse-rate conditions, but their value is limited mainly to the provision of independent estimates of evaporation for research purposes.

B. Energy budget method

From fundamental principles of the conservation of energy it follows that the net total of long- and short-wave radiation received at the surface (R_N) is available for three processes (fig. 4.1.1): the transfer of sensible heat (H) and of latent heat (LE) to the atmosphere and of sensible heat into the ground (G). That is,

$$R_N = H + LE + G \tag{2}$$

The fraction of R_N used in plant photosynthesis is generally negligible. Accordingly, evaporation can be determined by measurement of the other terms

$$E = \frac{R_N - H - G}{L} \tag{3}$$

R_N can be measured by the use of a net radiometer, and G is calculated from data on the soil-temperature profile or by direct measurement of soil heat flux, but H cannot readily be estimated. An indirect method is to employ Bowen's ratio $\beta = H/LE$. This is calculated from the ratio of the vertical gradients of

G

temperature and vapour pressure. However, the determination is unreliable when the surface is dry and H is large. On substitution of β in (3)

$$E = \frac{R_N - G}{L(1 + \beta)} \tag{4}$$

The use of Bowen's ratio assumes that the vertical transfer of heat and water vapour by turbulence takes place with equal efficiency. Recent work in Australia (Dyer, 1967) shows that this assumption is universally valid. Given the requisite observational data, the energy budget approach is a practicable one for determining evaporation over periods as short as an hour.

C. Combination methods

A number of methods have been developed to combine the aerodynamic and energy budget approaches, thereby eliminating certain measurement difficulties which each presents. The most widely used combination method was derived by Penman [1963, p. 40]. He expresses PE as a function of available radiant energy (R_N) and a term (E_a) combining saturation deficit and wind speed.

$$R_N = 0.75S - L_N \tag{5}$$

where $0.75S$ = solar radiation absorbed by a grass surface;

L_N = net long-wave (terrestrial) radiation from the surface.

$$E_a = f(u)(e_s - e) \tag{6}$$

where $f(u) = 0.35 (1 + 0.01u)$ for short grass;

u = wind speed at 2 metres (miles/day);
e_s = saturation vapour pressure (mm mercury) at mean air temperature;
e = actual vapour pressure at mean air temperature and humidity.

The expression for PE from short grass[1] is

$$PE \text{ (mm/day)} = \frac{\left(\dfrac{\Delta}{\gamma}\dfrac{R_N}{L} + E_a\right)}{\dfrac{\Delta}{\gamma} + 1} \tag{7}$$

where $\dfrac{\Delta}{\gamma}$ = Bowen's ratio;

$\gamma = 0.27$ (mm mercury/° F), the psychrometric constant;

$\Delta = \dfrac{de_s}{dt}$, the change of saturation vapour pressure with mean air temperature (mm mercury/° F).

It is worth while examining certain of these terms further. In equation (5) the 0.75 weighting of the incoming solar radiation is due to the 25% albedo (re-

[1] In Penman's original formulation evaporation was determined first for an open-water surface (PE_0), then a weighting factor (f) of 0.6–0.8 was used according to season and type of surface. $PE = f . PE_0$.

flection coefficient) of short grass. Most green crops have a similar reflectivity, but values for coniferous forest and heath are approximately 15%, while an average value for water is 5%. This factor should augment evaporation from a water surface compared with grass or crops, but at least a partial compensation is provided by the greater aerodynamic roughness of these surfaces (see Chapter 4.1.8). Rutter, for example, finds an annual evaporation of 679 mm from Scots Pine (*Pinus sylvestris*) in Berkshire for 1957–63, compared with a calculated open-water evaporation of 597 mm. Observational evidence is by no means unanimous, however, on this point. The meaning of the saturation deficit term

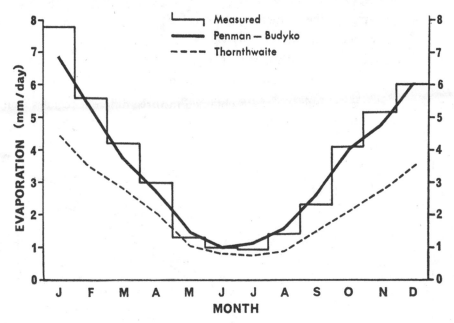

Fig. 4.1.3 Lysimetric observations at Aspendale, Australia (38° S), compared with estimates from the Penman–Budyko and Thornthwaite methods (Based on data from McIlroy and Angus, and Sellers, 1965).

$(e_s - e)$ in equation (6) is a common source of misunderstanding. It represents the 'drying power' of the air, but this need not be directly related to evaporation. In fact, $(e_s - e)$ is likely to be greatest when the surface is very dry and no moisture is available for evaporation. Certain aspects of this approach require further comment. First, the basis of the formulation involves the assumptions of Bowen's ratio, discussed in the previous section. Second, the transfer of sensible heat into the ground is neglected. Third, the use of mean temperatures and humidities makes it unsuitable for short-period estimates (<24 hours) of evaporation rates.

Budyko independently derived a similar expression for *PE*, and fig. 4.1.3 illustrates the accuracy of the Penman–Budyko approach compared with lysimetric observations in Australia.

A much simplified approach incorporating saturation deficit and radiation has been developed by Olivier. The equation for PE (mm/day) is

$$PE = (T - T_w) \frac{L}{\bar{L}^2} \qquad (8)$$

where $L = \dfrac{S}{S_v}$, $\bar{L} = \dfrac{\bar{S}}{\bar{S}_v}$;

$S =$ total solar radiation under clear skies for the latitude of the station for a particular month;

$S_v =$ vertical component of S;

$\bar{S} =$ average of the 12 monthly values of S;

$\bar{S}_v =$ average of the 12 monthly values of S_v;

$T =$ mean monthly temperature (° C);

$T_w =$ mean monthly wet-bulb temperature.

Figure 4.1.4 shows the estimate of PE obtained in three different regimes using Penman's and Olivier's formulae.

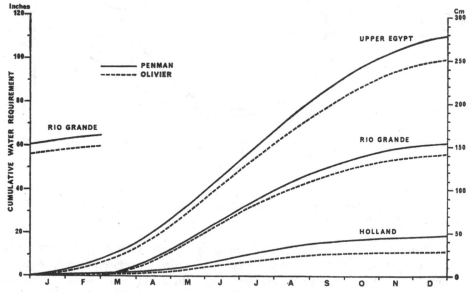

Fig. 4.1.4 Estimates of potential evapotranspiration in three different climatic regimes using Penman's and Olivier's methods (After Olivier, 1961).

D. Temperature formulae

One of the best-known methods of estimating PE was developed by Thornthwaite. He related observations of consumptive use of water in irrigated areas in the western United States to air temperatures, with adjustments for daylength.

$$PE \text{ (mm/month)} = 16\left(\frac{10T}{I}\right)^a \qquad (9)$$

where T = mean monthly temperature (° C);

a = an empirical function of I;

$$I = \sum_{1}^{12} \left(\frac{T}{5}\right)^{1.514}$$

The values can be readily calculated from published tables or nomograms. The method has been widely applied, although in some climatic regimes it gives unreliable results, as fig. 4.1.3 indicates for south-east Australia.

A more soundly based relationship has been illustrated by Budyko. He shows that if heat storage and sensible heat transfer are effectively zero annual PE is given approximately by

$$PE \text{ (mm/year)} \approx \frac{R_{N0}}{L} \approx 0.18 \sum T \tag{10}$$

where R_{N0} = the net radiation budget of a wet ground surface;

$\sum T$ = the sum of daily mean temperature which exceed 10° C.

Another temperature formula has been developed by Turc for *actual* evapotranspiration. Annual amounts (mm) for catchments are given by

$$E = \frac{P}{\sqrt{\left\{0.9 + \left(\frac{P}{I}\right)^2\right\}}} \tag{11}$$

where P = annual precipitation (mm);

$I = 300 + 25T + 0.05T^3$;

T = mean air temperature (° C).

This method has not been tested sufficiently to make any sound assessment possible.

6. Evaporation measurement

There are four main types of measuring device, although each has its limitations and disadvantages.

A. Atmometers

These are water-filled glass tubes having an open end through which water evaporates from a filter-paper (Piche type) or porous plate (Bellani type). The tube supplying water is graduated to read evaporation in mm, but the evaporation (termed 'latent evaporation') can only be compared with readings from another such instrument in a similar exposure, and may bear little or no relation to evaporation from land or water surfaces, since it only reflects the saturation deficit of the air (see p. 175). The instrument is apparently more responsive to wind speed than radiant energy.

B. Evaporation pans

The 'Class A' pan of the United States Weather Bureau, which is approved by the World Meteorological Organization, is 122 cm (48 in.) in diameter and 25 cm (10 in.) deep. Problems can arise through splashing, heating of the pan walls, and interference by birds or animals, while the installation position (sunken or mounted on or above the surface) is particularly critical. Unfortunately, pan evaporation (E_p) is not related to lake evaporation (E_L) in any simple or constant manner. Kohler found that E_L/E_p is generally within the range 0·6–0·8 for the United States.

In general, evaporation decreases as the size of the water body increases. In part, this arises from the 'oasis' effect. Air travelling over a large water surface picks up sufficient moisture to reduce the evaporation rate towards the leeward shore. Water depth is another cause of pan–lake differences. Much energy in spring goes into heating a deep lake, thereby suppressing evaporation rates. Morton calculates the average annual evaporation from Lake Ontario as 813 mm (32·0 in.), whereas that from Lake Superior is only 546 mm (21·5 in.). In the case of pan–lake comparisons these seasonal effects are even more serious, and it is only safe to use pan data for annual estimates of lake evaporation.

Even allowing for the regional and other factors which affect the ratio E_L/E_p (the 'pan coefficient'), the measurements provide no indication of evaporation from a land surface. The other two approaches now to be described are directed to this end.

C. Lysimeters

A lysimeter is an enclosed block of soil with a vegetation cover (usually short grass) similar to that of its surroundings. Figure 4.1.5 illustrates the installation at Hancock, Wisconsin. At regular intervals the weight change (ΔS), precipitation (P), and percolation (r) are measured. By means of the moisture-balance equation, evapotranspiration is determined as a residual.

$$E = P + \Delta S - r \tag{12}$$

The use of a large block allows an accuracy of 0·01 in. of water-depth.

A simpler version for PE determination is the 'Thornthwaite' type of evaporimeter. Here the moisture supply is maintained by 'irrigating' the block when necessary, so that the soil-moisture storage can be regarded as constant. The percolation (r), precipitation (P), and added water (W) are measured.

$$PE = P + W - r \tag{13}$$

With both types of device the presence of the tank's base may interfere with the soil moisture profile compared with that in a natural soil unless special precautions are taken.

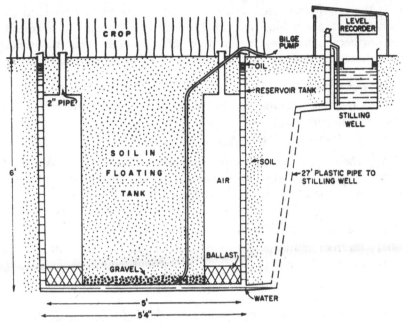

Fig. 4.1.5 Lysimeter installation at Hancock, Wisconsin (From King, 1961). Here the soil block floats in a tank of water. Changes of water level are recorded instead of weighing the block.

D. The 'evapotron'

Attempts have recently been made to measure the vertical transfer of moisture directly. Instruments developed by C.S.I.R.O. in Australia measure the magnitude and direction of vertical eddies which transfer water vapour upwards. There are many difficulties with this approach. In particular, there is a need for instruments that measure instantaneous changes of both the vapour content and vertical velocity of the air. The subsequent determination of average evaporation rates requires a computer to integrate the results. Moreover, effects of advection and storage below the measuring level (see fig. 4.1.1) may present serious difficulties unless measurements are made near the surface. This technique seems likely to be limited to research applications.

7. Budget estimates

The moisture-budget equation already referred to can be used in two very different ways to estimate evapotranspiration from large areas over a time period of the order of a month.

A. Catchment estimates

If suitable allowance can be made for storage in the catchment system, or if it is assumed constant, over a sufficiently long time interval,

$$E = r - P \qquad (14)$$

where r is the runoff measured by river gauging. Checks of this kind showed that estimates of evaporation from the Thornthwaite formula in northern Finland and northern Labrador–Ungava were 80% too high. This may reflect low snowfall estimates as well as low water use by moss and lichen surfaces.

B. Aerological estimates

In analogous manner evaporation can be estimated from data on atmospheric moisture (see Chapter 1.1.5).

$$E = \Delta D - P \pm \Delta S \qquad (15)$$

where ΔS is the storage change in the overlying air column and ΔD is the net divergence (or convergence) of water vapour out of (or into) the column. This method requires very complete aerological records.

8. Evaporation rates in different macroclimates

The global pattern of annual evaporation has been outlined in Chapter 1.1 and we can now look in more detail at variations in seasonal regime.

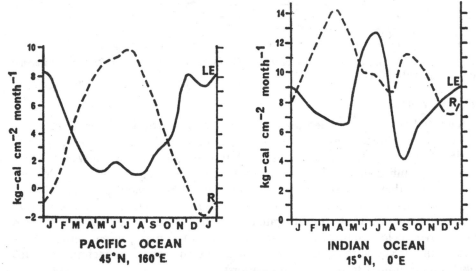

Fig. 4.1.6 The seasonal march of net radiation and evapotranspiration over ocean areas (After Budyko, 1956).

For the ocean areas of the world an average of 90% of annual R_N is used for evaporation, whereas for the continents the figure is only just over 50% and the remainder represents sensible heat transferred to the soil and the atmosphere. The inverse correlation of the seasonal march of R_N and LE over the oceans is therefore at first sight unexpected (fig. 4.1.6). This is a result of the complex interaction of heat storage and heat transfer by ocean currents. Much of the required energy for ocean evaporation is derived from the water itself, and the rate is mainly determined by wind speed and the vapour-pressure gradient. Over the Indian Ocean the summer maximum is caused by the higher wind speeds, while cloudiness diminishes the radiation receipt. The secondary winter maximum is due to the advection of dry trade-wind air. Off the eastern shores of Asia and North America there are large evaporation losses in winter as cold, dry continental polar air flows across the warm Gulf Stream and Kuro Shio Currents. In summer, however, reduced wind speeds and low air–sea vapour-pressure differences suppress evaporation rates. It may be recalled that this is the

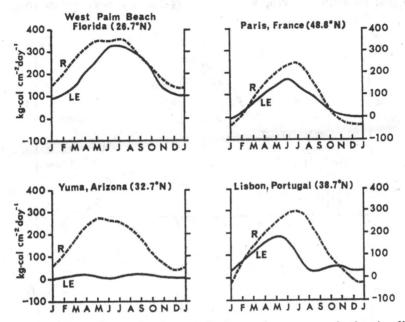

Fig. 4.1.7 The seasonal march of net radiation and evapotranspiration in different climatic regimes (After Budyko, 1956 and Sellers, 1965).
West Palm Beach, 27° N, 80° W; Paris, 49° N, 2° E; Yuma, 33° N, 115° W; Lisbon, 39° N, 9° W.

season when large horizontal moisture fluxes are directed from the continent (Chapter I.1).

Over land, the seasonal regime generally reflects the occurrence of maximum net radiation receipts and maximum surface-air vapour pressure difference. Where precipitation occurs mainly in summer, or has an even distribution throughout the year, there is a simple summer maximum and winter minimum of evaporation. Figure 4.1.7 illustrates typical profiles for West Palm Beach, Florida, and Paris. In areas of summer drought and winter rains there is a spring evaporation maximum, such as at Lisbon, while in districts with rains in autumn and winter there may be a double maximum in spring and autumn as at Yuma, Arizona.

Regional and local differences in evaporation rate arise not only from variations in the meteorological controls but also from largely independent soil and vegetation factors. For example, observations in the Canadian Subarctic indicate that $LE/R_N \approx \frac{1}{3}$ over lichen (*Cladonia* spp.) surfaces. The low evaporation rate is apparently due to the negligible extraction of moisture from the soil by non-vascular vegetation. Local differences may also reflect the varying external (r_a) and internal (r_s) resistances of vegetation surfaces to vapour diffusion (see 4.1.2). Theoretical computations by Monteith indicate a potential transpiration in the Thames valley, England, of 47 cm/year for short grass, compared with 58 cm/year for a tall farm crop with smaller r_a due to greater surface roughness.

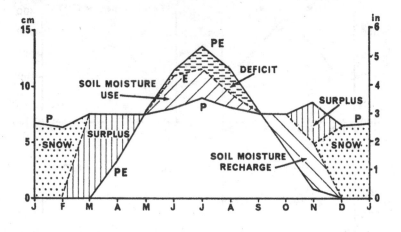

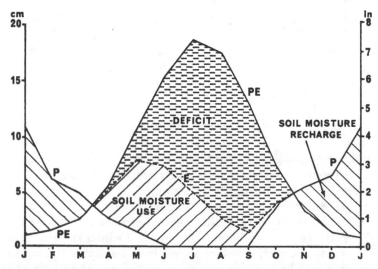

Fig. 4.1.8 Moisture budget diagrams for Concord, New Hampshire (*above*), and Aleppo, Syria (*below*) (Based on data in Thornthwaite and Hare, 1965, and Mather, 1963).

The method assumes that 50% of the soil water surplus runs off in the first month, 50% of the remainder in the next, and so on, unless additional surplus forms.

The loss from a pine forest (48 cm/year) is almost the same as from grass because increased stomatal resistance offsets the influence of lower albedo (15% compared with 25% for grass and green crops) and greater surface roughness of the forest which otherwise tend to promote transpiration.

9. The moisture balance and some applications

The principal difficulty encountered in computing moisture budgets for individual localities is the problem of assessing soil-moisture storage and actual evaporation. Only the simplest of the models outlined in 4.1.4 has been extensively applied in practice – namely that of Thornthwaite and Mather (Mather, 1963). Details of Budyko's more complex method are summarized by Sellers [1965, p. 175]. Figure 4.1.8 illustrates typical moisture budget diagrams for mid-latitude stations in humid (Concord, New Hampshire) and semi-arid (Aleppo, Syria) regimes. The relative amounts of annual moisture surplus (S) and deficit (D) provide one of the bases for Thornthwaite's 1948 classification of climates. In the revised 1955 version of this classification the moisture index (Im) is

$$Im = \frac{100\,(S - D)}{PE} \tag{16}$$

In North America forest predominates in regions where the humidity index (100 S/PE) > 35 and the aridity index (100 D/PE) < 10, so that there is ample soil moisture in nearly all months. Where S and D are small and approximately equal, the vegetation is typically tall grass prairie, but where the aridity index is of the order of 30 this gives way to short grass. Sagebrush and other desert vegetation occurs with aridity indices >40. For regions where data are inadequate to calculate a complete budget, estimates of ($P - PE$) may provide a useful climate parameter (Wallén, 1966; Davies, 1966). Davies, for example, discusses the relationships between vegetation and ($P - PE$) isopleths in Nigeria. In the Near East, however, Perrin de Brichambaut and Wallén [1963] find that the limit of dry-land farming is determined more by the amount and reliability of rainfall and soil-moisture storage than by potential evapotranspiration, which does not vary much over short distances.

REFERENCES

CHANG, J-H. [1965], On the study of evapotranspiration and the water balance; *Erdkunde*, **19**, 141–50.

CURRY, L. [1965], Thornthwaite's potential evapotranspiration term; *Canadian Geographer*, **9**, 13–18.

DAVIES, J. A. [1966], The assessment of evapotranspiration for Nigeria; *Geografiska Annaler*, **48**, Ser. A., 139–56.

DYER, A. J. [1967], The turbulent transport of heat and water vapour in an unstable atmosphere; *Quarterly Journal of the Royal Meterological Society*, **93**, 501–8.

HOLMES, R. M. [1961], Estimation of soil moisture content using evaporation data;

In Proceedings of Hydrology Symposium No. 2. Evaporation. Department of Northern Affairs and National Resources (Ottawa), pp. 184–96.

KING, K. M. [1961], Evaporation from land surfaces; In *Proceedings of Hydrology Symposium No. 2. Evaporation,* Department of Northern Affairs and National Resources (Ottawa), pp. 55–80.

MATHER, J. R. [1963], Average Climatic Water Balance Data of the Continents, No. 2 Asia (excluding U.S.S.R.); *Publications in Climatology XVI. No. 2* (Centerton, New Jersey), 262 p.

MCCULLOCH, J. S. G. [1965], Tables for the rapid computation of the Penman estimate of evaporation; *East African Agricultural and Forestry Journal,* 30, 286–95.

MONTEITH, J. L. [1965], Evaporation and Environment; *In The State and Movement of Water in Living Organisms,* Society for Experimental Biology, 19th Symposium (Cambridge), pp. 205–34.

MORTON, F. I. [1967], Evaporation from large deep lakes; *Water Resources Research,* 3, 181–200.

OLIVIER, H. [1961], *Irrigation and Climate* (London), 250 p.

PELTON, W. L. [1961], The use of lysimetric methods to measure evapotranspiration; In *Proceedings of Hydrology Symposium No. 2. Evaporation,* Department of Northern Affairs and National Resources (Ottawa), pp. 106–22.

PENMAN, H. L. [1963], *Vegetation and Hydrology;* Technical Communication No. 53, Commonwealth Bureau of Soils (Farnham Royal), 124 p.

PERRIN DE BRICHAMBAUT, G. and WALLÉN, C. C. [1963], *A study of Agroclimatology in Semi-Arid Zones of the Near East;* World Meteorological Organization, Technical Note No. 56 (Geneva), 64 p.

RUTTER, A. J. [1967], Evaporation in Forests; *Endeavour,* 26, 39–43.

SELLERS, W. D. [1956], *Physical Climatology* (Chicago), 272 p.

THORNTHWAITE, C. W. [1948], An approach towards a rational classification of climate; *Geographical Review,* 38, 55–94.

THORNTHWAITE, C. W. and HARE, F. K. [1965], The loss of water to the air; In Waggoner, P. E., Editor, *Agricultural Meteorology, Meteorological Monographs,* 6, No. 28 (Boston, Mass.), pp. 163–80.

WALLÉN, C. C. [1966], Global solar radiation and potential evapotranspiration in Sweden; *Tellus,* 18, 786–800.

WALLÉN, C. C. [1967], Aridity definitions and their applicability; *Geografiska Annaler,* Ser. A, 49, 367–84.

WORLD METEOROLOGICAL ORGANIZATION [1966], Measurement and estimation of evaporation and evapotransportation; *Technical Notes No. 83.*

II. Soil Moisture

M. A. CARSON

Department of Geography, McGill University

A large literature exists on the physics of soil moisture and upon the way in which soil water influences the nature of the soil in which it exists. This essay is not intended as a summary of these topics. Such standard works as *Soil Physics*, by Baver [1956], and *Soil*, by Jacks [1954], already serve this purpose admirably. The purpose of this essay is to outline the part played by soil moisture in some aspects of the denudation of the landscape.

1. The nature of soil moisture

A number of forces are capable of attracting water into dry soil. One is the simple affinity of the soil particles for water vapour in the soil atmosphere, although such hygroscopic water forms only a very small percentage of the water existing in most soils. A more important mechanism is the capillary suction which exists on the menisci of water films in contact with soil particles. This suction is usually explained by analogy with the rise of water in a capillary tube. Insertion of a thin

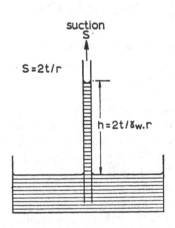

$S = 2t/r$

$h = 2t/\delta_w.r$

t : surface tension of water

δ_w: density of water

r : radius of tube and meniscus

Fig. 4.II.1 Suction on water in a capillary tube.

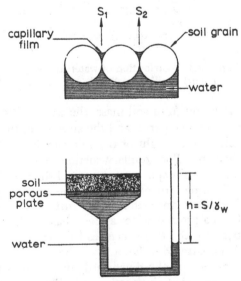

$h = S/\delta_w$

Fig. 4.II.2 Suction on water in soil.

tube into a tank of standing water (fig. 4.11.1) produces a rise in the level of the water in the tube relative to the level outside it. The extra height of the water in the tube is attributable to capillary suction acting against the force of gravity, which, alone, would maintain the same level inside and outside the tube. The magnitude of the suction is determined by the surface tension of the water and the radius of the meniscus. The height of capillary rise in the tube is determined by the ratio of the suction and the weight of a unit volume of water. The menisci in the pores of soil possess similar suction (fig. 4.11.2), and this is also measured by noting the amount of water displaced against gravity. As capillary water is

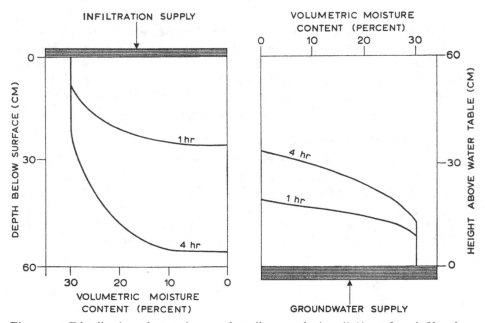

Fig. 4.11.3 Distribution of water in a sandy soil mass during (*left*) surface infiltration, and (*right*) capillary rise from a water table (After Liakopoulus, 1965a).

drawn out of a soil mass, the water which remains in the soil occupies smaller and smaller pores, and the suction on this water increases in precisely the same way as the height of rise in a capillary tube increases with decreasing radius of the tube. The capillary suction in a soil, together with the gravity force, determines the major movement and distribution of water in soil.

The pattern of change in the distribution of water in a soil mass during infiltration from water on the surface, and also during the entry of water upward from a ground-water system, has been treated by many workers. The similarity between the two processes has been emphasized by Liakopoulos [1965a] and is demonstrated in fig. 4.11.3. The entry of water into a soil mass from the surface proceeds by the downward advance of a wetting front where there is a sudden change from wet to dry soil. The amount of moisture in the soil shows little change with depth at a distance behind the wetting front, although it decreases rapidly in the area immediately behind it. A similar pattern of change occurs with the upward advance of a wetting front from a ground-water supply. In the soil just beneath the wetting line there is an increase in water content with depth until a constant moisture content is attained. This special value is identical in the two cases, and in the sandy soil used by Liakopoulos was about 30% by volume, which represents about three-quarters saturation of the pore space. These results agree with those of Bodman and Colman [1943] and other early workers.

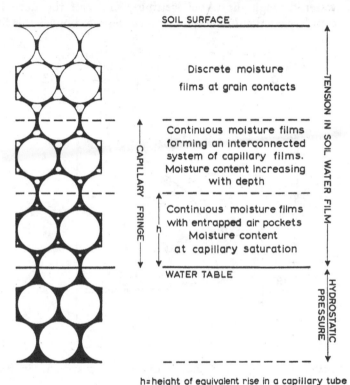

Fig. 4.11.4 The zones of capillary moisture in soil.

The ultimate equilibrium position to which the system tends differs in the two cases. In the case of capillary rise (fig. 4.11.4) a number of distinct zones exist. Immediately above the water-table is a belt of soil with constant moisture content at capillary saturation. Above this there is a systematic decrease in the amount of moisture in the soil. The area of continuous capillary water above the water-table is termed the capillary fringe, and above it the moisture films separate into discrete menisci. The thickness of the capillary fringe depends upon the size of the pores in the soil mass in much the same way as the height of water in a capillary tube depends upon the radius of the meniscus. The theoretical extent of the capillary fringe is very great in the case of a clay with very small pores, whereas it is negligible in a sandy soil. The rate of capillary flow in a soil with very small pores is usually so slow, however, that the state of ultimate equilibrium is never attained, although, even then, the extent of this zone is appreciably greater than in sandy soils.

The ultimate moisture distribution in the infiltration case must differ from the capillary rise case, since the water supply will eventually exhaust itself. The pattern of moisture redistribution after the stage when all the water has entered the soil mass is unfortunately not clearly known. All the water above the wetting front will not continue to drain down through the soil. The gravity force will draw some water through the initial wetting front, but the development of menisci and capillary suction in the soil mass will tend to oppose this force. This

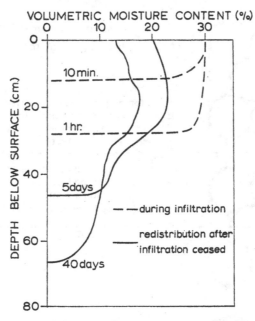

Fig. 4.11.5 Distribution of water in a sandy soil mass during and after infiltration from the surface (After Liakopoulos, 1965b).

was noted by early workers who distinguished between 'movable' and 'immovable' water. A soil which has just shed all its drainable water and still retains its maximum capacity of water which is immovable under the attraction of gravity alone was stated to exist at its *field capacity*. In this state the soil water would be held by a suction of about 20–40 in. of water. The amount of moisture in the soil at field capacity would depend upon the number of small pores and was thought to range from about 30% by volume for a clay soil to less than 10% for sandy soils. The practical value of the concept of field capacity has been criticized by a number of workers. The tests by Liakopoulos [1965b] show that even in a fine sand (fig. 4.11.5) drainage past the initial wetting front proceeds a long time after infiltration has ended.

The water which drains past the initial line of the wetting-front moves towards an underlying ground-water system. During this movement the water is still subject to lateral capillary suction and, as pointed out by Sherman [1944], it may never reach the water-table unless most of the soil pores are large. This depends also on the height of the water-table. The subsequent history of movable water beneath the initial wetting front depends, in addition, on the nature of the solid rock underlying the soil mantle. A shattered mass of rock with large gaps will facilitate downward percolation to the water-table, whereas a highly impermeable stratum may lead to a temporary ground-water system above the main one at depth.

The picture presented above is a simplified account of the movement and distribution of water in a soil mass during infiltration from the surface on level ground. A very different situation must exist on a steeply sloping land surface. The tests by Whipkey [1965] suggest that a wetting front advances into the soil mantle in much the same way as on a level surface. These tests do suggest, however, that the soil may attain complete saturation. Another feature is that during and after infiltration there is a marked downslope flow within the soil mass, and this is especially marked where an impermeable soil layer retards vertical entry of water. Such water will by-pass the underlying ground-water system and return to the stream through the soil mantle. It is important to realize therefore that the infiltration of storm water into a soil may differ radically on slopes from a flat land surface. These tests indicate that temporary water systems perched above the main water-table may occur often during prolonged rainstorms and offer a mode of subsurface flow entirely different from the pattern which occurs within a soil mass beneath flat ground.

2. Soil moisture and denudation

The most obvious part played by soil moisture in the denudation of the land-scape is the assistance given in the weathering of solid rock masses into loose debris and the direct transport of material as solutes and colloids out of the waste mantle. This direct loss of material from the waste mantle by moving water has attracted little interest in the past. Attention has been focused upon the conditions which induce subsequent redeposition at another level within the mantle and thus create a soil profile.

There is never uniformity with depth in a soil mantle. The upper parts of the mantle are more humic, and the lower parts tend to be more moist. Soil minerals which are unstable in the upper levels and taken into solution by the soil water may attain greater stability at depth in the mantle and redeposition occurs. This eluviation and illuviation inevitably accentuate the differences between the upper and lower parts of the mantle, and a soil profile ensues. A well-leached soil commonly shows a pallid layer of silt and sand grains which overlies a horizon of illuviated clay and other minerals. The emergence of a soil profile is complicated in dry areas by the upward movement of soil water and the deposition of salts in the upper crust with evaporation.

The classic issue of the development of soil profiles usually assumes a flat surface. The loose material which mantles most hillslopes often shows, in contrast, little sign of illuviation: the only noticeable change with depth is an increase in the amount of unweathered debris. The absence of a soil profile on slopes may be due to many reasons. The soil mantle on a hillslope is subject to continual erosion of different types, and it is maintained only by the compensating supply of new soil through the weathering of the underlying solid rock. Although a soil mantle may exist permanently on a slope, it exists in a dynamic state (Nikiforoff, 1949), and a particular mass of soil may not stay on the slope for a sufficiently great length of time to develop a profile. Another explanation may be the tendency of soil water to seep downslope rather than vertically, as noted by Whipkey [1965], and under these circumstances a downslope change in soil type rather than a vertical profile might materialize. Such a pattern would

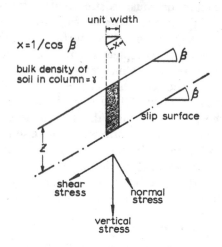

vertical stress $= \gamma . Z / x = \gamma . Z . \cos \beta$

shear stress $= \gamma . Z . \sin \beta / x = \gamma . Z . \sin \beta . \cos \beta$

normal stress $= \gamma . Z . \cos \beta / x = \gamma . Z . \cos^2 \beta$

Fig. 4.11.6 The stresses imposed by gravity at a point in a soil mass.

be comparable to the catena sequence, except that it would be the product of subsurface soil water.

The transport of material out of the soil mantle by moving water within the soil mass is only a minor process, at least directly, in the denudation of the landscape. A residual soil mass will always survive this process, and much of the material lost from one part of the mantle may be expected to be deposited in another part. The major landforms are shaped by the processes of soil wash and mass-movement of different types which act upon this residual mantle. Soil moisture has a distinctive role in each of these. The influence of antecedent soil moisture in determining the amount of surface runoff and the extent of soil wash is discussed in Chapter 5.1. The less obvious role of soil moisture in the mass movements of shallow landslides and seasonal soil creep is discussed here.

Soil moisture plays a vital role in determining the stability of a soil mantle on a hillside. The shear stresses which exist within a soil mass and the shear strength to withstand these stresses both depend partly upon the amount of moisture in the soil. The shear stress on the plane of failure in fig. 4.11.6 is dependent upon the angle of the slope, the depth of the plane, and the density of the soil mass above it; the last of these will vary with the amount of moisture in the soil. The shear strength of a soil mass is derived from a cohesive element and a frictional component. The amount of internal friction developed along a plane of failure depends upon the effective stress normal to that plane as well as the angle of shearing resistance of the soil. The effective normal stress itself depends not only upon the component of the weight of the soil column at right angles to the

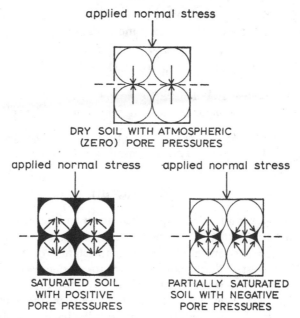

Fig. 4.11.7 The influence of moisture on pore pressures in soil.

plane (fig. 4.11.6) but also upon the pressure in the pores of the soil. This is shown in fig. 4.11.7. In a soil mass which is completely dry the normal stress applied by the overlying material is neither supplemented nor alleviated by the air pressure in the pores, since the pressure is atmospheric. When the soil pores are partly filled with water the pressure in the water films under the menisci is less than atmospheric (Skempton, 1960), so that the overall pore pressure is negative and a suction force augments the applied normal force in drawing the soil grains together. In a soil mass which is saturated with free-draining water the pore pressures are positive relative to the atmospheric datum, and this acts against the applied normal stress. The moisture content of the soil, through its influence on the effective normal stress and thus on the amount of internal friction which may be developed upon a potential plane of failure, is a vital consideration in the stability of a hillside soil mantle. It has indeed been suggested by Vargas and Pichler [1957] that the majority of natural landslides owe their origin to the development of positive pore pressures in a soil mass during prolonged rainstorms.

An implication of this is that, assuming that a soil mass is typified by particular pore-pressure values, the angle of limiting stability in any area will congregate around particular values determined by the pore pressure and the shear strength of the soil. The maximum angle of a stable slope occurs when the shear stress and the shear strength of the soil are just balanced. In the situation depicted in fig. 4.11.6 this is given by:

$$\gamma \cdot z \cdot \sin \beta \cdot \cos \beta = c' + (\gamma \cdot z \cdot \cos^2 \beta - u) \tan \phi'$$

where γ is the density of the soil mass;

z is the depth of the plane of failure;

β is the angle of slope;

c' is the cohesion of the soil;

u is the pore pressure;

ϕ' is the angle of shearing resistance of the soil.

The effect of pore pressure is most conveniently illustrated by dealing with soils which have negligible cohesion. The maximum stable slope in the situation in fig. 4.11.5 is then given by:

$$\tan \beta = \tan \phi' \left(1 - u/(\gamma \cdot z \cdot \cos^2 \beta)\right)$$

Soil mantles which never attain complete saturation and never fully dry out will always possess negative values of u, the pore pressure, and may thus stand at angles which exceed the angle of shearing resistance of the soil material. Schumm [1956] suggested that this occurs on 40–45-degree badland slopes in South Dakota, where there is sufficient silty material to provide lasting capillary suction. In contrast, soils which are essentially loose rock fragments are unlikely to attain complete saturation when they mantle hillsides due to the large pores, and for the same reason they are unlikely to maintain capillary water films permanently. They are thus characterized by pore pressures which are essentially atmospheric ($u = 0$), and the maximum stable slope is in this case the same as the angle of shearing resistance of the material. This is very probably the reason why so many scree slopes exist at angles near to 35 degrees, since this value approximates the angle of shearing resistance of small loose rocky matter. The majority of hillslopes, however, at least in humid areas, stand at angles which are less than the angle of shearing resistance of the soil mantle, and it seems very likely that this is due to the development of positive pore water pressures at times of prolonged rainstorms which give rise to perched water-tables. The pore-water pressures in free-draining water depend upon the pattern of flow, but in the case of ground-water flow parallel to the surface (fig. 4.11.8) the pressure at any point is given by:

$$u = \gamma_w \cdot z \cdot \cos^2 \beta$$

where γ_w is the density of water.

Substitution of this value in the previous equation gives:

$$\tan \beta = \tan \phi' \left(1 - \gamma_w/\gamma\right)$$

and since the bulk density of most surface soils is about twice the density of water, this indicates that the maximum stable slope under these circumstances should approximate to half the angle of shearing resistance in tangent form. The work of Skempton and DeLory [1957] suggests that, in the London Clay at least, this hypothesis is supported by the field evidence: angles of limiting slope approximate 8–9 degrees, and this agrees with a residual angle of shearing

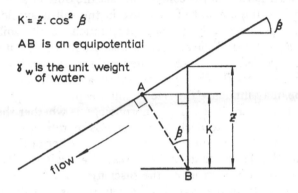

$$K = Z . \cos^2 \beta$$

AB is an equipotential

γ_w is the unit weight of water

pore-water pressure = O at A and u_1 at B

positional potential = $\gamma_w Z \cos^2 \beta$ at A and O at B

total head at B = total head at A

$$\therefore \ u_1 = \gamma_w Z \cos^2 \beta$$

Fig. 4.11.8 The relation between pore-water pressure and depth in a soil mass with ground-water flow parallel to the surface.

resistance which is near to 16 degrees. In the case of an area which experiences widespread artesian pore-water pressures the limiting slope angle will be less than the value predicted on the basis of the previous model, although such pressures are possibly rather rare.

Soil moisture thus plays a conspicuous part in the denudation of the landscape under landslides and, through the attainment of particular values of pore pressures, may lead to the emergence of special angles of limiting slope in any one area. A hillside which is stable against rapid mass-wasting is not immune to still further transport of debris downslope, and in humid areas subsequent denudation is mostly through moisture-induced soil creep.

There are two types of soil creep which act upon hillslopes. One is the unidirectional movement downslope under the impetus directly of gravity and designated by Terzaghi [1950] as shear creep. The rate of this type of movement is probably very slow. Superimposed upon this continuous creep is seasonal soil creep. This is produced by random and cyclic disturbances which operate on flat land as well as on slopes. In the latter case, however, they are given a systematic bias downslope by the component of the gravity force. The major mechanism underlying movement in seasonal soil creep in humid temperate areas is the expansion and shrinkage of soils with changes in the moisture content of the soil. Some pioneer tests by Young [1960] indicated that under this mechanism alone soil creep might be expected to transport 0·5–1·0 cm³ of soil past a line 1 cm wide across the slope in an average year. Subsequent work (Young, 1963) suggests that this approximates the actual amount of soil creep on slopes in the British Isles, and the conclusion that the bulk of seasonal soil creep is due to changes in the amount of moisture in the soil is supported by other workers. The creep rate clearly depends very much on the soil type on the slope. Sandy soils swell very little and clays considerably when water is absorbed, and it might be expected that the creep rate would increase with the amount of colloidal material in the soil. This is a possibility, although other features, such as angle of slope, may influence the rate of soil creep.

One of the major controversies in geomorphology is whether the processes of denudation cause hillslopes to decline in steepness or retreat at an unchanging angle. It is doubtful whether soil creep will produce either of these two changes on a straight hillside. The decline or the retreat of a straight hillslope demands that there is a systematic increase in the discharge of soil moved at different points downslope. This, in the case of soil creep, means that there must be one or both of an increase in the rate of soil creep and an increase in the thickness of the moving mantle downslope. The meagre evidence available suggests that neither of these occurs and, by implication, a straight slope acts essentially as a plane of transport for material from upslope with no effective erosion on the straight slope. The product of this is the replacement of the straight slope by the encroachment downslope of it of a convex hilltop, and there is neither retreat nor decline in the straight slope during this sequence.

The evolution of a landscape which is stable against rapid mass-wasting is, at least when it is free from the action of moving ice and the particular effects of karst limestone, determined primarily by the strength of soil wash as against soil creep. The balance between these two processes may depend a great deal on the soil type. Schumm [1956] showed this in the badlands of South Dakota and Nebraska: some clays absorbed all surface water and were subject only to creep while more impermeable clays induced surface run-off and soil wash. The effect of micro-climate on the balance between the two processes was discussed by Hack and Goodlett [1960]: moist slopes in the Appalachians appeared to be subject mostly to creep, while soil wash dominated on dry slopes. The major distinction between the two processes undoubtedly occurs with the differences in climate on the world scale. Soil creep in semi-arid areas is usually dwarfed in importance by the vast amount of soil wash which occurs in torrential storms on bare hillslopes. The results of work in humid areas suggests, in contrast, that soil creep is far more important and that it is only in the most intense and pro-longed rainstorms that run-off and soil wash may take place. It is perhaps not mere coincidence, therefore, that the slopes of arid areas are dominated by straight profiles and sharp crests, while the upper hillslope convexity assumes its most distinctive development in the humid temperate areas of the earth.

The existence of soil moisture is clearly fundamental to the major processes of denudation. It is not surprising that differences in soil type and climate, affecting the amount and movement of soil moisture, are translated into differences in the efficacy of these processes and thus mirrored in the earth's landforms.

REFERENCES

BAVER, L. D. [1956], *Soil Physics*; 3rd Edn. (New York).

BODMAN, G. B. and COLMAN, E. A. [1943]. Moisture and energy conditions during downward entry of water into soils; *Proceedings of the Soil Science Society of America*, **8**, 116–22.

HACK, J. T. and GOODLETT, J. C. [1960], Geomorphology and forest ecology of a mountain region in the central Appalachians; *U.S. Geological Survey Professional Paper 347*.

JACKS, G. V. [1954], *Soil* (Edinburgh), 221 p.

LIAKOPOULOS, A. C. [1965a], Theoretical solution of the unsteady unsaturated flow problem in soils; *Bulletin of the International Association of Scientific Hydrology*, **10**, 5–39.

LIAKOPOULOS, A. C. [1965b], Retention and distribution of moisture in soils after infiltration has ceased; *Bulletin of the International Association of Scientific Hydrology*, **10**, 58–69.

NIKIFOROFF, C. C. [1949], Weathering and soil evolution; *Soil Science*, **67**, 219–30.

SCHUMM, S. A. [1956], The role of creep and rainwash on the retreat of badland slopes; *American Journal of Science*, **254**, 693–706.

SHERMAN, L. K. [1944], Infiltration and the physics of soil moisture; *Trans. American Geophysical Union*, **25**, 57–71.

SKEMPTON, A. W. [1960], Effective stress in soils, concrete and rocks; *Pore Pressure and Suction in Soils*, 4–16.

SKEMPTON, A. W. and DELORY, F. A. [1957] Stability of natural slopes in London Clay; *Proceedings of the 4th International Conference of Soil Mechanics*, **2**, 378–81.

TERZAGHI, K. [1950], Mechanism of landslides; *Bulletin of the Geological Society of America, Berkey volume*, 83–122.

WHIPKEY, R. Z. [1965], Subsurface storm flow from forested slopes; *Bulletin of the International Association of Scientific Hydrology*, **10**, 74–85.

VARGAS, M. and PICHLER, E. [1957], Residual soil and rock slides in Santos (Brazil); *Proceedings of the 4th International Conference of Soil Mechanics*, **2**, 394–8.

YOUNG, A. [1960], Soil movement by denudational processes on slopes; *Nature*, **188**, 120–2.

YOUNG, A. [1963], Soil movement on slopes; *Nature*, **200**, 129–30.

III(i). Water and Crops

ROSEMARY J. MORE

Formerly of the Department of Civil Engineering, Imperial College, London University

1. Plants and Water

Water is needed by plants to supply carbon dioxide and oxygen in solution to the cells for photosynthesis and respiration; to transport raw materials, manufactured and waste products within the plant; and to maintain the rigidity (*turgidity*) of the plant structure. Plant cells consist of a permeable *cell wall*, enclosing a layer of *cytoplasm* (a viscous fluid containing fine particles and

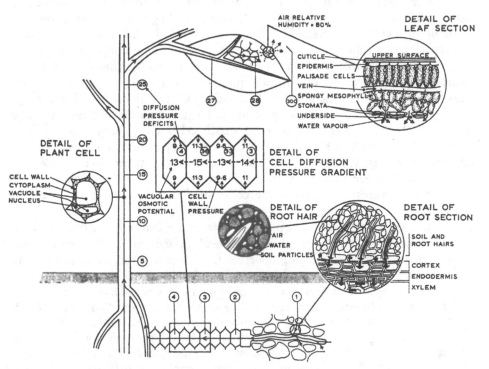

Fig. 4.III(i).1 Diagram illustrating the water circulation between soil, plant, and atmosphere. The possible diffusion pressure deficits at various points of this hydrostatic system are shown, assuming a 400-ft tree, readily available soil water, and an atmospheric relative humidity of 80% (Partly from Knight, 1965).

globules), which has the property of being *semi-permeable* (i.e. allowing the passage of some molecules but prohibiting others). The cytoplasm surrounds the *vacuolar sap*, an aqueous solution of salts, sugars, and organic acids (fig. 4.III(i).1). The more saline the sap, the lower the kinetic energy of the random movements of the water molecules, such that the solution is said to have a *diffusion pressure deficit* (D.P.D.) with respect to pure water. Therefore, where a permeable membrane separates two solutions of different salinity there is a net tendency for water molecules to pass from the less towards the more saline solution. The kinetic energy of the dissolved salt molecules, however, operates in the reverse sense, being greatest in the more saline solution. In flow between plant cells the semi-permeable property of the cytoplasm prevents the reverse flow of salt molecules, and only the net flow of water occurs, by a process called *osmosis*. The maximum force with which water can enter a cell is equal to the D.P.D. of the vacuolar sap, and were it not for the pressure of the stretched cell wall (W.P.), this would be identical to the *osmotic potential* (O.P.) of the sap. In reality:

$$D.P.D. = O.P. - W.P. \tag{1}$$

Water moves in the plant as a continuous stream from the root hairs to the leaf surfaces in response to the cell *diffusion pressure-deficit gradient*. This gradient is made up of a chain of increasing D.P.D.s, and the D.P.D. is thus a measure of the capacity of a saline cell to absorb water, either from an adjacent (generally less saline) cell or from soil water. As equation (1) shows, the D.P.D. is not entirely a function of the cell osmotic potential, and water may pass between cells having equal osmotic potential (i.e. salinity), providing the cell-wall pressures are different. The cell-wall pressure is largely a function of the extent to which the cytoplasm is forced against the elastic cell wall by changes in the water content of the vacuolar sap. Cells can lose turgidity either by *plasmolysis* (see later) or by *wilting*. Wilting occurs whenever evaporation takes place from leaf or stem surfaces faster than water can be supplied to the cells, such that the vacuolar sap decreases, reducing the cell-wall pressure, causing the plant to droop.

The movement of water up from the roots to the leaves is controlled by its rate of evaporation (*transpiration*) from the leaf surfaces, and is thus termed the *transpiration stream*. One of the important functions of plant water is that the moist leaf surfaces absorb atmospheric carbon dioxide, but this maintenance of moisture implies a continuous water supply to the leaf surface to replenish that which is inevitably evaporated. Generally speaking, the upper surfaces of plant leaves are covered with a rather impermeable waxy *cuticle*, through which only about 10% of the total leaf transpiration occurs. The majority takes place through the cells of the small pores (*stomata*) in the *epidermis* on the leaf undersides. The rate of transpiration (assuming complete saturation of the leaf *mesophyll* spaces) depends, firstly, on the D.P.D. between the stomatal cell moisture and that of the air adjacent to the leaf (this being a function of the relative humidity of the air, the temperature of the air and leaf, and the rate of air movement), and secondly, on the size of the stomatal aperture (which differs between plants and is also controlled by the light intensity, such that stomata are closed and transpiration ceases at night). As the stomatal cell walls lose transpired moisture cell salinity increases, together with the osmotic potential of the cell, and shrinkage decreases the cell-wall pressure, both having the combined effect of increasing the D.P.D. and allowing available water to move by osmosis from adjacent cells. This process continues, in a highly complex manner, down the plant, and the water molecules from the weakly-saline soil water diffuse

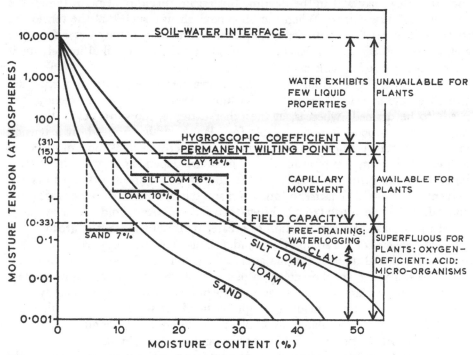

Fig. 4.III(i).2 The relationship between moisture content and moisture tension in four different types of soil, indicating the range of moisture available for plants (Partly from Buckman and Brady, 1960).

through the root cells into the *xylem*, the principal water-carrying cells of the plant.

Figure 4.III(i).1 shows this transpiration stream for a tall tree, the diffusion pressure-deficit gradient in a block of four root cells, and a series of cell diffusion pressure deficits at a number of points, indicating its gradual increase upwards, with the highest value occurring at the critical stomatal plant/air interface.

Returning to the link between the plant roots and the soil moisture, it is important to note that not all soil water is equally available for plant use. Figure 4.III(i).2 shows the moisture content (as a percentage of the total volume of the soil plus water) of four main types of soils, together with the soil-moisture tensions associated with each, indicating that as the interstitial soil-moisture films diminish, their tension increases, and it is less easy for the plant roots to extract moisture from the soil. After a rainstorm some of the water filling the soil interstices is free to drain away under gravity, leaving films held by a tension of about ⅓ atmosphere. In this condition the soil is at *field capacity*, and plants are free to absorb both water and air from the soil. Unless the soil water is replenished, plant water use and direct evaporation from the soil combine to reduce the moisture content by shrinking the films. The accompanying increase of soil-moisture tension progressively denies the plant a freely-available water supply such that they may wilt in the daytime but regain turgidity at night (when the transpiration loss ceases). When the films have shrunk such that the tension of their surfaces reaches 15 atmospheres plants can no longer draw water from the soil, *permanent wilting point* has been reached, and the plant will ultimately die if water is not made available soon.

It is obvious that soil moisture exists in several degrees of utility for plants; that in excess of field capacity is superfluous, leads to oxygen deficiency, and needs to be drained; whereas, at the other extreme, moisture below the permanent wilting point is not available for plant growth – indeed, at a tension greater than the *hydroscopic coefficient* (31 atmospheres) it loses most of its liquid properties. The amount of available soil moisture depends partly on the characteristics of the plant and partly on the soil type. Figure 4.III(i).2 shows the percentage of *available water* in four main soil types, sand having the least (7%) and silt loam the most (16%). Finer-textured soils can hold more available moisture because of a combination of greater total pore space, the greater wettable surface area of the soil particles and, usually, a greater proportion of water-retaining colloidal matter. Factors affecting the ability of plants to avail themselves of soil water are the extension of the root system, the drought-resistant properties of the plant, and the stage and rate of its growth. Plant roots only draw soil water from their immediate neighbourhood. It is not clear whether soil moisture is equally available to a plant throughout the whole range from permanent wilting point to field capacity, and it has been suggested that the *optimum moisture zone* occurs at a moisture content considerably above the permanent wilting point.

2. Plants as crops

Since the Neolithic Revolution man has cultivated many plants as crops. Of the four factors significant in cropping, the plant, the atmosphere, the soil, and its water, man cannot modify the atmosphere (except in the most limited sense), and he can do only a small amount to modify the physical and chemical composition of the soil over large areas (e.g. by the application of fertilizers). However, he can exploit and utilize the genetic variations within plant species and selectively breed those varieties of plants which are resistant to moisture and temperature extremes, deficiencies in soil chemistry, the activity of pests, etc. Some of the most striking changes in cropping are made, however, by the degree to which man can control the amount and quality of water available to plants, either by irrigation, drainage, or improvement of water quality. His aim is to keep soil moisture as close as possible to the optimum for the soil type and the crop, within the limits of engineering and economic feasibility. However, in extensive semi-arid areas of the world man has found it uneconomical to make appreciable modifications to the soil-water environment and has developed *dry farming* techniques in association with drought-resistant plant strains. Cultivation loosens the soil and allows it to absorb and retain more water than if it remained in a dense untilled state. In the Great Plains of the United States, for example, dry farming involves in fallow year deep early-spring tilling which allows greater absorption of the succeeding rainfall and limits surface evaporation losses. The accompanying increase in weed growth is counteracted by periodic cultivation or the use of weed-killers. This exploitation of the cultivated soil as a moisture-storage reservoir during a fallow year conserves a variable amount of the annual rainfall, usually about a quarter, and after such a fallow year wheat yields in Kansas and North Dakota, for example, may be almost doubled. Dry farming, however, is still at the mercy of annual rainfall variations, and the amount and quality of yields inevitably strongly reflect this variability.

3. Irrigation

The purpose of irrigation is the control of soil moisture between a lower limit that will not restrict plant growth and an upper limit that avoids the disadvantages of water-logging. One calculation of irrigation need is based on the concept of *potential transpiration*. Penman has shown that during the May-to-September growing season in Western Europe the energy used in transpiration is the largest single term in the net exchange of energy between the sky, atmosphere, and ground. The following seven terms are involved in the energy balance (H) (the percentages giving the relative orders of magnitude recorded in a series of British observations during 1949):

Total incoming solar radiation (R_c) (100%)
Transpiration (E) (39%)
Back radiation from the surface (R_B) (34%)
Reflection from the surface ($R_c \cdot r$: where r = the albedo) (20%)
Heating the air (K) (4%)
Heating the soil (2%)
Plant growth (1%)

Of these, the last two are small enough to be omitted, leaving the following relationships:

$$H = E + K \simeq R_c - R_c \cdot r - R_B \qquad (2)$$

in which R_c, r, and R_B can be determined from meteorological observations (duration of bright sunshine, air temperature, humidity, and cloudiness) and K can be eliminated by expressing it as a ratio of E. The ratio of E_0/K (where E_0 is the estimated evaporation from a theoretical open water surface) is obtained by inserting measured meteorological values (wind-speed variations; water temperature and saturation vapour pressure; and air temperature and vapour pressure) into other equations. E_0 can be transferred to the *potential transpiration* (E_T, which is considered equal to E) by recognizing that the rate of transpiration from a short green crop (e.g. grass) with an adequate water supply is the same as the evaporation rate from a water surface, and by merely supplying a weighting factor (for Britain during May to September $E_T/E_0 = 0.8$) related to the length of daylight during which the stomata are open and transpiration takes place. Substitution into equal (2) thus gives the amount of potential transpiration which would occur with an adequate water supply, but when this is not available any deficiency of precipitation with respect to potential transpiration forms a *moisture deficit* which, for adequate cropping, must be made up from pre-existing soil-moisture supplies or from applied irrigation water.

There are two restrictions to the wide application of this ingenious approach to the calculation of plant water needs, firstly, the lack of detailed meteorological measurements in many localities and, secondly, that it is less well adapted to regions of high temperature, low humidity, and sparse vegetation and crop covers. In the western United States a simplified calculation for crop water use (shown as *consumptive use*) has been devised by Blaney and Criddle, such that:

$$u = kf \tag{3}$$

where u = monthly consumptive use (in inches);

k = consumptive use coefficient dependent upon mean monthly temperature and crop stage of growth;

f = monthly consumptive-use factor $\dfrac{t \times p}{100}$;

t = mean temperature in ° F;

p = percentage of daytime hours of the year, occurring during the period (from tables).

The following Table 4.III(i).1 gives the values used in calculating monthly consumptive use by alfalfa in the Upper Salinas Valley of California:

TABLE 4.III(i).1

Month	t	p	f	k	u
April	57·9	8·85	5·12	0·60	3·07
May	62·5	9·82	6·14	0·70	4·30
June	65·7	9·84	6·46	0·80	5·17
July	68·4	10·00	6·84	0·85	5·81
August	67·8	9·41	6·38	0·85	5·42
September	66·6	8·36	5·57	0·85	4·73
October	62·2	7·84	4·88	0·70	3·42
Total consumptive use for the irrigation season					31·92

Irrigation is used to supply any deficit existing between potential transpiration and precipitation. The engineering techniques by which this is accomplished vary according to the type of crop and the physical environment, but are mostly controlled by the economic level of the society concerned. Primitive irrigation usually involves local, temporary storage or diversion of surface runoff, whereas in advanced societies irrigation is economically linked to multiple-purpose

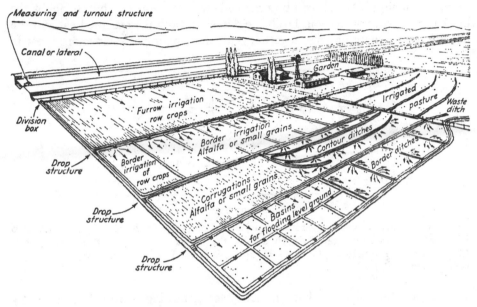

Fig. 4.III(i).3 Irrigation techniques. Various methods of applying irrigation water to field crops (From Israelson and Hansen, 1962).

projects concerned with large dams, canals, and a variety of sophisticated water-spreading techniques (fig. 4.III(i).3).

4. Drainage

The roots of most plants and cultivated crops develop above the perched soil water-table, where both air and water are available. However, in some fine-grained soils, particularly clays, the available air in the interstices below the capillary fringe is so meagre that plant roots cannot penetrate much below this fringe, and there is evidence that most crop roots do not extend deeper than some 12 in. above the average height of the water-table in the soil. Saturation results in a reduction in the absorption of soil oxygen and plant nutrients, leading to inefficient photosynthesis and consequent reduction in transpiration. Artificial drainage is thus aimed at lowering the soil water-table to allow a sufficient depth of the aeration zone wherein a healthy root system can develop well supplied by water, air, and soil nutriments. The optimum water-table and rooting depths differ between crops, the former being about 20 in. for celery and 36 in. for cereals and sugar beet in fen peat soils of Eastern England, but fig. 4.III(i).4 provides some general relationship between crop yield and depth of water-table for four soil types in Holland, showing that as the soil texture becomes heavier maximum yields are obtained with deeper water-tables. Some crops are very water tolerant, for example, rice, which has a mechanism for obtaining oxygen direct from the air, but others, like tobacco, are very sensitive to excess water. Soil drainage also has other generally beneficial effects on crops by increasing the nitrogen supply which can be obtained from the soil and by changing the thermal properties of the soil. Drainage decreases both the specific heat and the thermal conductivity of soils, together with the amount of solar energy lost in

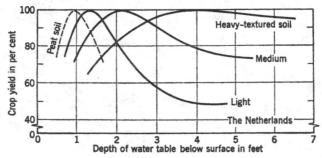

Fig. 4.III(i).4 Depth of water versus crop yield. The general relationship between crop yield and constant water-table depth during the growing season in the Netherlands (After Visser: From Schwab, G. O. *et al.*, 1966, *Soil and Water Conservation Engineering* (Wiley, New York)).

evaporation from soils, and has the net effect of making more soil heat energy available for plant growth.

From the foregoing it is apparent that it is usually more difficult to calculate optimum drainage requirements than optimum soil moisture or water-quality requirements.

5. Improvements in soil water quality

Quality of soil water is as important as quantity in achieving optimum plant growth, for certain amounts of dissolved salts derived from mineral and organic soil constituents (and fertilizers) are vital for plants. In humid soils the ready availability of water generally causes the soil water to be dilute, but in arid regions short intense storms do not permit infiltration into more than 1–2 ft of the soil, and the rapid direct evaporation of the rainfall leads to higher salt concentrations in the upper part of the soil profile which exert injuriously high osmotic potentials. The accompanying shrinkage of water films near the surface causes saline water to be drawn up from depth (as long as the continuous film stage exists), the evaporation of which leaves the soil and soil water more saline. Salinity can also be inadvertently increased artificially by applying saline irrigation water. Salinity is controlled by the amount of soluble salts (chiefly sodium, calcium, magnesium, and potassium) which are present, and can be measured

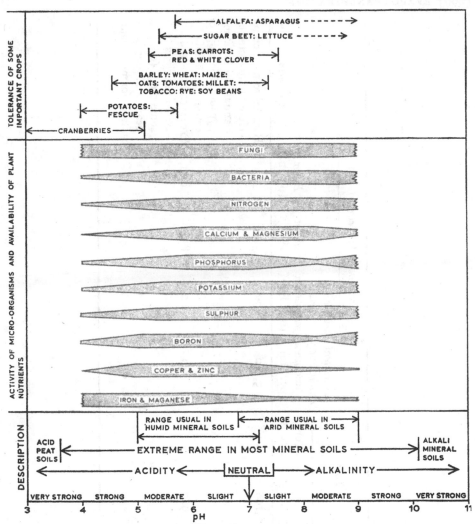

Fig. 4.III(i).5 The relationship between the activity of micro-organisms and the availability of plant nutrients, on the one hand, and of mineral soil pH. The width of the band
represents the pH ranges of greatest activity and availability. The pH tolerances of some
important crops are also shown (Adapted from Buckman and Brady, 1960).

in the field by simple electrical instruments because salinity exerts a direct
control over the electrical conductivity of water.

If plant roots are in contact with soil water of higher salinity than that of their
vacuolar sap osmotic withdrawal of the sap causes the cytoplasm to shrink away
from the cell wall and the space between to be invaded by saline soil water
(which passes into the cell through the permeable cell wall). This process of
plasmolysis can be reversed and turgidity restored if soil water of sufficiently low
salinity to allow the cell to absorb water molecules by osmosis becomes available,

H

TABLE 4.III(i).2

Source	Water class (U.S. Salinity Lab.)	Total Salt content (p.p.m.)	Total Salt content (Tons/acre-ft)	Electrical conductivity (micromhos per cm)	Calcium and magnesium (m.e. litre)	Sodium (%)	Boron (p.p.m.)	Comments
					Some important constituents			
Pecos River, Texas	3	6,198	8·4	9,150	47·8	52	—	Harmful to most crops. Here counteracted by natural soil lime and gypsum, fair crops of cotton and alfalfa
Coachella Valley, California, ground water	2 Plus excess sodium and boron	910	1·2	1,740	2·2	85	0·71	Excess sodium and boron injurious to some crops (e.g. beans and grapes)
Rio Grande, New Mexico	1	641	0·8	870	5·1	4·0	—	Suitable for most plants

The availability to the plant of chemical and biological nutrients from the soil, besides being obviously related to soil mineral composition, temperature, etc., is largely influenced by the pH values of the soil water (fig. 4.III(i).5). Just as there is an optimum range of soil water, so the pH range of 6–7 is optimum for plant growth, below which important nutrients (notably phosphorus) are unavailable in the acid soils, and above which occur the alkaline soils of the arid and semi-arid regions.

Such arid soils, which present special water-quality problems, can be divided into:

(*a*) Saline soils: with excess soluble salts, producing plasmolysis and impairing productivity.
(*b*) Saline/alkali soils: with excess soluble salts plus an injurious excess of sodium.
(*c*) Alkali soils: with an excess of sodium and pH of 8·5–10.

Treatment of such soils is largely effected through a control of soil water by:

(*a*) Lowering the soil water-table so that soluble salts can be leached away by rainfall or applied irrigation water. This is the standard treatment for saline soils, but with saline/alkali and alkali soils a soluble calcium compound (e.g. gypsum) must first be added to leach away the sodium salts to improve the permeability so that leaching can proceed.
(*b*) Not applying irrigation water to excess, thus keeping the water-table low and avoiding drawing up salts to the capillary fringe near the surface.
(*c*) Cultivation designed to promote free percolation of surface water into the soil and preventing waterlogging.
(*d*) Control over the quality of applied irrigation water.

The quality of irrigation water in arid regions is commonly alkali, and it is necessary to apply it in controlled amounts to crops with appropriate levels of salt tolerance (fig. 4.III(i).5). The table 4.III(i).2 exemplifies three classes of irrigation water. Class 2 and 3 water must be applied with care to crops, because the salt concentration of soil water is commonly 2–100 times that of the applied irrigation water (depending on the amount of direct evaporation from the soil), and the accumulation of salts may build up, ultimately making the soil unsuitable for cropping, as in large areas of West Pakistan.

6. Interactions of irrigation, drainage, and water quality

It is obvious that these three aspects of water management are interdependent, and that a balance must be achieved to provide an optimum moisture environment for crops, both as regards quantity and quality. A classic example of integrated water management is the recent reclamation of an area in West Pakistan, made saline by previous uncontrolled irrigation. This was effected by a balance of water-table reduction by pumping, combined with the application of just enough irrigation water to allow leaching without waterlogging.

Water management is only one aspect of total crop management, however, and must be combined with fertilizer application, pest control, good soil management, selection of appropriate plant strains, and other sound agricultural practices in an effort to achieve maximum productivity within the limits of the plant's total environment.

REFERENCES

BUCKMAN, H. O. and BRADY, N. C. [1960], *The Nature and Properties of Soils* (The Macmillan Co., New York), 567 p.

FOGG, G. E. [1963], A digression on water economy; Chapter 3 in *The Growth of Plants* (Penguin Books, London), pp. 61–87.

KNIGHT, R. O. [1965], *The Plant in Relation to Water* (Heinemann, London), 147 p.

ISRAELSON, O. W. and HANSEN, V. E. [1962], *Irrigation Principles and Practices* (John Wiley & Sons, New York), 447 p.

LUTHIN, J. N., Editor [1957], *Drainage of Agricultural Lands* (Madison, Wisconsin), 620 p. (especially Chapter 5).

MORE, R. J. [1965], Hydrological models and geography; In Chorley, R. J. and Haggett, P., Editors, *Models in Geography* (Methuen & Co., London), pp. 145–85.

PENMAN, H. L. [1963], *Vegetation and Hydrology* (Commonwealth Bureau of Soils, Harpenden), Technical Communication No. 53, 124 p.

III(ii). Primitive Irrigation

ANNE V. KIRKBY

Department of Geography, Bristol University

Primitive irrigation is characterized by its great variety; in many different physical environments it is practised by peoples with varying cultures and levels of technology and on scales ranging from individual farmers to the great hydraulic societies of India and China. It includes the flooded ricefields of the Tonkin delta, the small maize patches of the Hopi Indians in Arizona, the water meadows of medieval England, and the wheatfields of ancient Persia. Despite the variety of physical settings and methods used, most primitive irrigation schemes show a sophisticated awareness of local hydrological conditions and are economic in that they produce an increased return over rainfall farming for the increased input of labour and equipment. Once a community has decided to irrigate, the irrigation system employed is chosen within two sets of constraints – those of the physical environment and those of their own culture, especially the level of technology and the economic framework. Conversely, once the method of irrigation is decided in accordance with these constraints, the irrigation system itself can set constraints on the economy and society of the irrigators and can locally modify the physical environment so that an important feedback between water and man takes place through irrigation. In its ultimate form, absolute state power may be achieved through control of water resources and their distribution.

Man may move water in two ways: in units, such as bucketfuls, or as a continuous flow, as in a canal. The movement of water for irrigation may be divided into three phases: (1) abstraction from the water source; (2) distribution throughout the irrigation system; and (3) application on to the crops. In each of these three phases the water may be moved as units or as a continuous flow, and many primitive irrigators combine both types of movement within one system. The sources of water available for irrigation are limited by the physical environment, the most common natural sources being rivers, lakes, springs, and ground water, but where these are not available primitive irrigators may depend on more unusual sources; for example, water stored in limestone sink holes or in sand dunes. In primitive systems water from surface sources is generally abstracted by means of continuous flow, in canals, and water from ground-water sources, which requires lifting to the field surface, is most easily taken out in units. However, within the great range of primitive irrigation schemes known, examples may be found of lifting surface water in units, as is done from rivers in India, by means of a lever device called the *shaduf*; and ground water can be brought to the surface as a continuous flow without the use of modern pumps, by constructing a complex system of wells and tunnels through which the ground water flows to the surface under gravity, as in the *ghanats* of Iran and the *galerias* of Tehuacán, Mexico (fig. 4.III(ii).1).

The second, or distributional, phase of the irrigation scheme is usually done by means of canals, which vary in construction and efficiency from unlined,

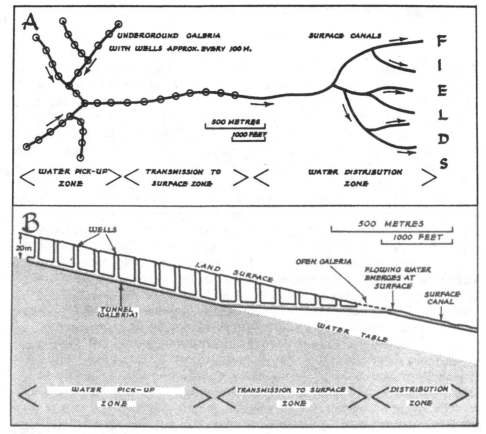

Fig. 4.III(ii).1 Typical *galeria* irrigation system from the Tehuacán Valley, Mexico; (A) plan, and (B) cross-section.

meandering earth channels with high evaporation and seepage losses, to tiled or cemented canals and aqueducts which minimize water losses. Water can be distributed without using canals, and in some schemes it is moved underground through tunnels (*ghanats* and *galerias*), and in pot irrigation in Mexico it is distributed in pots or buckets which the irrigator carries from the nearest of several wells in his small plot to each plant individually.

The third phase of the system, that of applying the water to each plant, employs one of the following methods:

1. individual canals (including furrows);
2. standing sheets of water (e.g. swamp rice cultivation);
3. flowing sheets of water (e.g. simple flood water farming);
4. unit application by pots or buckets.

Each of these methods have different efficiencies in terms of crop production per unit of irrigation water, per man hour, per plant, and per unit area, and it is the balance between the availability and cost of these various factors that determines the real efficiency of water application. Where water is scarce and labour is abundant and cheap, as is common in primitive irrigating communities, the most efficient method of application is by unit delivery, as in pot irrigation. Plants neither receive too little water, with resulting decline in production per plant, nor too much, with resulting decline in efficiency per unit of water. In contrast, individual canal or furrow irrigation is more efficient in terms of labour, but does not distribute water evenly, so that plants nearest the entry canal in the field receive more water than they need, while plants lower down the furrows lack water. Furrow irrigation is therefore less efficient per unit of water and per plant, but efficiency per unit area can be maintained by closer plant spacing.

Primitive irrigators must also be aware of the importance of adequate drainage to remove the build-up of salts in the soil. If land has been made useless to crops through salt accumulation, as in parts of Iran, the irrigation area may be abandoned because remedial measures, such as lowering a high water-table by pumping, or increasing percolation by adding massive amounts of water to leach away salts, often require capital outlay which is beyond the primitive near-subsistence irrigator. Primitive irrigation therefore tends towards preventive medicine by good irrigation management rather than drastic measures to save the dying patient.

The feed-back that *can* occur between irrigation methods and the social and economic development of the irrigating community may be illustrated by two villages in the semi-arid Valley of Oaxaca in Mexico. The first village is situated on the crest of a ridge in the piedmont zone, where the water-table is far below the surface but where one of the few perennial streams in the area can be led on to the ridge crest by means of a take-off canal beginning in the mountains high above the village. The fields lie below the village on both flanks of the spur and are irrigated by a distributary system of earth canals and finally by means of individual furrows. The take-off and main canals are communally owned and maintained by the farmers of two villages, and because the water is distributed from a single source which cannot irrigate all the fields at the same time it is allocated to farmers on a rota basis of once every fifteen days. Crops are therefore selected from those which grow best with fifteen days between waterings so that there is a strong tendency for all farmers to grow the same irrigated crop, alfalfa, and the resulting land-use pattern is very uniform. The system of water distribution not only establishes the necessity for, and the idea of, co-operation between farmers but also establishes the machinery, in the form of water overseer and committee, for organizing community action in other spheres and for enforcing the majority opinion on the minority by the threat of withdrawal of water rights. Community projects, such as road building and the installation of electricity, therefore play an important role in the life of this village.

The second village, although only eighteen miles away, is in strong contrast to the first. It is located on the flat valley floor away from the mountains, where there is no perennial surface water, but a high water-table at between 5 and 12 ft deep provides a widespread source of irrigation water. Instead of receiving water from a single source via a communal distributory system, farmers can pot irrigate from many shallow wells. Thus abstraction, distribution, and application are all within each farmer's individual control, and he requires little co-operation with his neighbours for the success of his irrigation scheme. The amount and frequency of watering may be adjusted to the requirements of each crop and even to each plant, and is controlled by each farmer so that the range of crops grown is diverse and the resulting land-use pattern is very varied, in contrast to the monotony of alfalfa fields in the piedmont village. There is little co-operative enterprise, and no machinery exists for enforcing co-operation, so that community action plays a small role, and even schemes for the direct benefit of farmers, such as flood-water control and the co-operative ownership of irrigation pumps, fail in the face of the aggressive independence of family groups.

In these two Mexican villages therefore two very different irrigation schemes have resulted from, on the one hand, a single water source from which water must be distributed to the fields; and on the other hand, a widespread water source which is readily obtainable within each field. In such primitive irrigation communities as these there is therefore a real two-directional interrelationship between water and man, between the physical environment and the economic and social life of the communities which exploit it.

Acknowledgement. To Professor Aubrey Williams, Dept. of Sociology and Anthropology, University of Maryland, U.S.A., for ethnographic information about the Valley of Oaxaca, Mexico.

REFERENCES

CLARK, C. [1967], *The Economics of Irrigation* (Oxford), 116 p.

ISRAELSEN, O. W. and HANSEN, V. E. [1962], *Irrigation Principles and Practices;* 3rd Edition (New York), 447 p.

KIRKBY, A. V. [IN PREPARATION], *Land and Water Use in Present and Ancient Oaxaca, Mexico;* unpubl. Ph.D. thesis, The Johns Hopkins Univ., U.S.A.

WITTFOGEL, K. A. [1957], *Oriental Despotism* (New Haven, U.S.A.), 556 p.

CHAPTER 5

Surface Runoff

I. Infiltration, Throughflow, and Overland Flow

M. J. KIRKBY

Department of Geography, Bristol University

When the rainfall that has not been intercepted by vegetation reaches the ground surface part of it fills small surface depressions (depression storage), part percolates into the soil, and the remainder, if any, flows over the surface as overland flow. Each component of this equation is highly variable, and depends not only on the intensity of the rainfall but also on soil, vegetation, and surface gradient. The amount of water intercepted by vegetation depends on the type of plants and their stage of growth, but it is usually close to a value given by all of the first 1·0 mm of rainfall and 20% of the subsequent rainfall in any one storm. Rainfall reaching the soil surface has to fill the small depressions on the surface before any overland flow can occur, even on a totally impermeable surface. This depression storage does not vary with the amount of rainfall but with the nature of the surface, especially with slope gradient, vegetation cover, and land-use practices. Under natural conditions depression storage absorbs about 2–5 mm of rainfall in any one storm. Contour ploughing is particularly effective in increasing depression storage by as much as ten times.

1. Infiltration

Infiltration rate is defined as the maximum rate at which water can penetrate into the soil. For initially moister soils the infiltration rate is lower throughout storms, and for all soils it decreases during the course of a storm. The rate at which water can travel through the soil depends on the number and size of pore spaces in the soil and the distribution of water within them. In effect, the infiltrating water has two components, a transmission component, which is constant and represents a steady flow through the soil; and a diffusion component, which is an initially rapid, and then an increasingly slow, filling-up of air-filled pore spaces, from the surface downwards. These components can be expressed in the infiltration equation (Philip, 1957):

$$f = A + B \cdot t^{-\frac{1}{2}} \tag{1}$$

where f is the instantaneous rate of infiltration;
t is time elapsed since the beginning of rainfall;
A is the 'transmission constant' of the soil; and
B is the 'diffusion constant' of the soil.

In this equation the transmission and diffusion terms can be identified with the two components of an idealized model. The transmission term represents unimpeded laminar flow through a continuous network of large pores. The diffusion term represents flow in very small discrete steps from one small pore space

TABLE 5.1.1 Variation of minimum infiltration rates with soil grain size, initial moisture content, and vegetation cover

(a) The effect of grain size in initially wet soils without vegetation cover

Grain size class	Infiltration rates (mm/hr)
Clays	0–4
Silts	2–8
Sands	3–12

(b) The influence of moisture content for Illinois clay-pan soils (after Musgrave and Holtan, 1964)

Initial moisture content (%)	Infiltration rates (mm/hr)		
	Good grass cover		Poor weed cover
	Topsoil > 13 in. thick	Topsoil < 13 in. thick	Topsoil < 13 in. thick
0–14	17	19	6
14–24	7	7	4
24 +	4	4	3

(c) The influence of ground cover for Cecil, Madison, and Durham soils (after Musgrave and Holtan, 1964)

Ground cover	Infiltration rate (mm/hr)
Old permanent pasture	57
Permanent pasture; moderately grazed	19
Permanent pasture; heavily grazed	13
Strip-cropped	10
Weeds or grain	9
Clean tilled	7
Bare ground crusted	6

to the next, in a random fashion. The only reason that a net diffusion flow results is that more pores are dry lower down, so that there is greater opportunity for downward movement than for upward. In an actual soil the two phases of this model cannot be separated, as all pores show a combination of the two types of behaviour, but equation (1) remains a good approximation for measured infiltration rates.

Table 5.1.1 shows some of the range of variation which can be expected in infiltration rates under a range of vegetation and moisture conditions, and in fig. 5.1.1 a comparison is made between typical infiltration rates and expected storm rainfall rates.

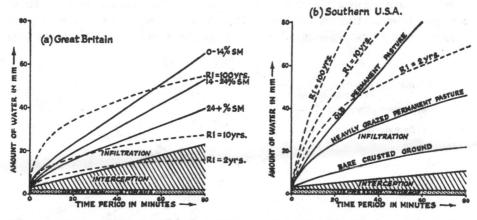

Fig. 5.1.1 Comparison of expected rainfall intensities with infiltration rate, interception, and depression storage, with representative values to demonstrate the relative frequency of overland flow in Great Britain and southern United States.

(a) Great Britain: rainfall (Data from Bilham, 1936), interception values for 100% vegetation cover, infiltration rates for Illinois clay-pan soils with good grass cover (Data from Musgrave and Holtan, 1964).

(b) Southern United States: rainfall (Data from Yarnell, 1935), interception values for 50% vegetation cover, infiltration rates for Cecil, Madison, and Durham soils (Data from Musgrave and Holtan, 1964).

2. Overland flow and throughflow

'Horton overland flow' is defined as overland flow which occurs when rainfall intensity is so great that not all the water can infiltrate, and is described by Horton [1945]. This type of overland flow is a fairly common phenomenon in semi-arid climatic conditions, but is relatively rare in humid and humid–temperate conditions. The role of vegetation is thought to be a critical cause of this distinction. Vegetation increases the infiltration rate by promoting a thicker soil cover, a better soil texture, and by breaking the impact of raindrops on the surface. Its effect on soil structure is mainly to build up an organic-rich A horizon with a relatively open pore structure and high permeability. If raindrops strike the surface without being impeded by vegetation fine material is thrown into suspension by the impact and is redeposited as an almost impermeable surface skin which can lower infiltration by as much as ten times. Vegetation therefore has a controlling influence on Horton overland flow by increasing both the initial depression storage and the infiltration rate, so that where a dense vegetation cover is established Horton overland flow is very unusual. Soil compaction by animals and vehicles reduces the infiltration rate while increasing depression storage, so that its net influence is problematic.

Within a small drainage basin, where soils are more or less homogeneous, it may be expected that interception, depression storage, and infiltration rate will not vary greatly, so that the conditions for Horton overland flow will be satisfied by comparable intensities and durations of rainfall, and overland flow will occur simultaneously all over the basin. Typical velocities for overland flow are about 200–300 m/hr, so that in a rainfall of 1 hour water from all points of a basin (200–300 m is a typical distance from divide to stream in Britain) will be reaching

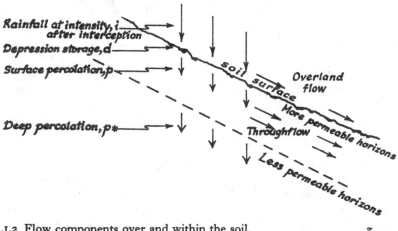

Fig. 5.1.2 Flow components over and within the soil.

a stream channel, and flow over the slope surface will have reached a steady state, represented by the equation:

$$q_0 = (i - f) \cdot a \tag{2}$$

where q_0 is the overland flow discharge per unit contour length;
i is the rainfall intensity after interception;
f is the infiltration rate; and
a is the area drained per unit contour length (equal to the distance from the divide if all the contours are straight lines).

Thus, provided that the rainfall intensity is high enough (or the infiltration rate low enough) for this type of overland flow the actual magnitude of the flow will be strongly dependent on the area or distance of overland flow, and will be almost independent of the storm duration, provided that it exceeds a reasonable minimum value. This is Horton's [1945] classic overland flow model.

Some of the water which infiltrates into the soil passes downward to recharge the water-table, and some, usually the greater part, flows down the hillside within the soil layers as 'throughflow' and ultimately contributes to streamflow (fig. 5.1.2). Within the soil, permeability varies and is generally highest in the open-textured organic A_0 horizon and the eluviated A_1 horizon. B horizons tend to be less permeable because clays are washed down into them, and in some cases because of the development of a hardpan. Conditions in the parent bedrock vary widely from limestones, with open solution fissures, to totally impermeable consolidated clays and shales. Wherever permeability is decreasing downward within the soil, and this occurs most commonly at the base of the A horizon, part of the water which is percolating downwards cannot penetrate into the lower layers fast enough, and is deflected laterally within the upper layer, as through-flow. This is similar to the production of Horton overland flow at the surface, except that within the soil the reduction of permeability is usually gradual, leading to a progressive deflection of throughflow. Table 5.1.2 shows an example of the progressive decline of permeability through the soil profile for a hillslope in Ohio, U.S.A.

TABLE 5.1.2 Variations of permeability and soil type with depth in an Ohio forest soil at a gradient of 15° (after Whipkey, 1965)

Soil depth (cm)	Textural class	Bulk density (gm/cc)	Saturated permeability (mm/hr)
0–56	Sandy loam	1·33	—
56–90	Sandy loam	1·41	286
90–120	Loam	1·78	17
120–150	Clay loam	1·80	2

If rainfall continues for a long time soil layers become saturated and through-flow is deflected closer and closer to the surface, so that the upper, more permeable soil layers are filling up from their bases because the throughflow is unable to carry away the water fast enough. In time the soil will become saturated right up to the surface and 'saturation overland flow' will occur. Under steady rainfall this condition will ultimately be attained under rainfall intensities much lower than are required to produce Horton overland flow. Since soil thickness and the velocity of throughflow vary much more over a small area than does permeability, and since the base of a slope tends to become saturated sooner than the divide, certain parts of a hillside are likely to produce saturation overland flow preferentially, in contrast to the rather widespread production of Horton overland flow, when it occurs at all.

Throughflow, travelling through soil pore spaces rather than over the ground surface, moves at very much lower velocities than overland flow. Rates of 20–30 cm/hr for throughflow are of the order of a thousand times lower than overland flow rates, so that periods of about 1,000 hours rainfall are needed for a steady state of flow to be achieved throughout an average basin. In practice, such a steady state is never attained for throughflow, and equation (2) for overland flow must be replaced with equation (3) for throughflow:

$$q_T = (p - f_*) \cdot v \cdot t \qquad (3)$$

where q_T is the throughflow discharge per unit contour length;

$\quad$ p is the rate of surface percolation, equal to i or f, whichever is smaller;

$\quad$ f_* is the rate of infiltration at the base of the more permeable soil;

$\quad$ v is the velocity of throughflow; and

$\quad$ t is the time elapsed (strictly $v \cdot t$ should be replaced by the area/unit contour length within a distance $v \cdot t$ upslope).

In this equation the *time elapsed* is the most important control over the flow, in place of the *distance from the divide* for Horton overland flow. In reality, the flows are delayed by the time of transmission from the surface to a zone of decreasing permeability, a period which may be a matter of minutes or a few hours.

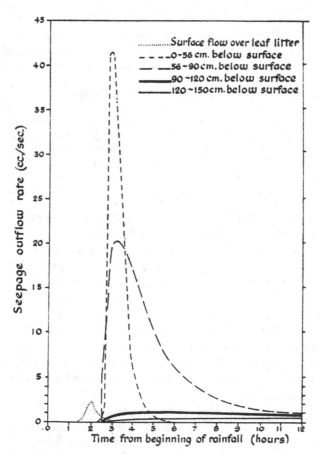

Fig. 5.1.3 Discharge hydrographs of flow within the soil resulting from a simulated storm of 5·1 cm/hr, lasting 2 hr, on a 16° slope which had previously drained for more than four days. The rapid, although small, Horton overland flow results from the initially low permeability of the dry surface soil, which rapidly increases with wetting. The lag before throughflow begins is the time taken for rain to infiltrate vertically to the 90-cm-deep, less-permeable interface (After Whipkey, 1965, p. 81).

If outflow from various depths in the soil is measured throughout a storm of uniform intensity there will be an initial transmission period of little or no flow, followed by a period in which the flow is increasing rapidly (though not linearly, because v in equation (3) is itself varying) until the moment when rainfall ceases, after which the flow will decrease more slowly than it increased. In other words, the flow should broadly resemble a flood hydrograph of a stream, despite the absence of surface run-off. In fig. 5.1.3 (Whipkey, 1965) actual values are shown for a rainfall intensity of 5·1 cm/hr falling for 2 hours on the soil whose properties are described in Table 5.1.2. It is apparent from fig. 5.1.3 that even at this high intensity Horton overland flow is negligible, and that throughflow from shallower soil layers is later than from deeper layers due to the time lag for transmission of water down through the soil, followed by saturation from the base up.

Equation (3) contains an unknown quantity, namely the velocity of through-flow, v. In order to evaluate the throughflow more exactly, Darcy's law, which states that flow through a permeable medium is proportional to the pressure gradient, must be combined with the continuity equation, which states that differences between inflow and outflow must be accommodated by changes in moisture content. For a soil layer of uniform permeability and moisture content (in depth) Darcy's law for soil on a slope is

$$Q = z \cdot \cos \alpha \left\{ K \cdot m \cdot \sin \alpha - D \cdot \frac{\partial m}{\partial x} \right\} \qquad (4)$$

and the continuity equation is

$$\frac{\partial m}{\partial t} + \frac{\partial Q}{\partial x} = i \qquad (5)$$

where Q is the downslope discharge measured in a horizontal direction;
    x is the distance downslope measured in a horizontal direction;
    K is the soil permeability (a function of soil moisture);
    D is the soil diffusivity (a function of soil moisture);
    α is the surface slope angle;
    m is the soil moisture content;
    z is the thickness of the soil layer;
    i is the rainfall intensity after interception; and
    t is the time elapsed.

Solution of these equations is necessarily numerical, but in a simple actual example of a uniform soil on a uniform gradient (Hewlett and Hibbert, 1966) estimates of permeability and diffusivity were as shown in Table 5.1.3; and cal-culated soil-moisture patterns during (a) uniform rainfall and (b) uniform drain-age were as shown in fig. 5.1.4. It can be deduced from the data of Table 5.1.3 that the permeability (K) term in equation (4) is the more important for inter-mediate values of moisture, while the diffusivity (D) term becomes dominant at extreme low and high values of moisture. The low-moisture case is of little interest, but the high-moisture case, when the moisture content is approaching

TABLE 5.1.3 Values of permeability (K) and diffusivity (D) for varying soil moisture, computed from the results of an experiment carried out during soil drainage of a trough of soil inclined at 22° (Hewlett and Hibbert, 1963)

Moisture content (expressed as % of saturated moisture content)	Permeability (K) (cm/hr)	Diffusivity (D) (cm/hr)
68	0·000	0·0
73	0·004	3·2
78	0·081	12·5
83	0·46	19·3
88	1·58	40·8
93	4·00	100·4
98	8·67	304·0
99	10·64	517·0
100 (saturated)	12·08	Infinity

saturation, is of great hydrologic significance. In this important case soil moisture is constant at saturation, so that equation (5) becomes extremely simple, namely

$$\frac{\partial Q}{\partial x} = i \qquad (6)$$

What this means in practice is that the near-saturated soil zones respond very rapidly to changes in rainfall intensity, even before overland flow begins. Most

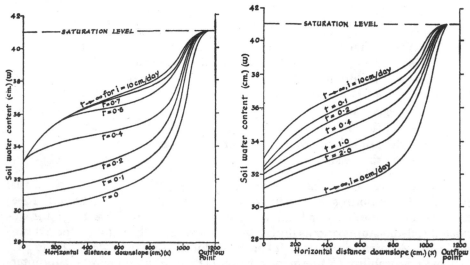

Fig. 5.1.4 Calculated moisture distributions in a 12-m-long soil trough, inclined at 22° during: (left) rainfall at a constant rate of 100 mm./day, and (right) drainage after indefinite rainfall of this intensity.

The data are calculated from the results of an experiment by Hewlett and Hibbert (1963).

important is the saturated zone which is often at the side of flowing streams, for the outflow from this zone is given by:

$$q_T = i . x_s + q_T{}^*$$ (7)

where x_S is the width of the saturated zone;

   q_T is the throughflow discharge per unit contour length;

   $q_T{}^*$ is the throughflow contribution from farther upslope; and

   i is the rainfall after interception.

Equations (4) and (5) also provide a basis for assessing the influence of changing gradient or (with slight modification) of contour curvature on the throughflow discharge within the soil. The following four regions of a slope are the most likely to become saturated, and hence provide more rapid response to rainfall and more frequent saturation overland flow:

   1. areas adjacent to perennial streams;
   2. areas of concave upwards slope profile;
   3. hollows (areas of concave outward contours);
   4. areas with thin or impermeable soils.

3. Areas contributing to stream flow

The relative infrequency of overland flow in humid regions, together with the very low velocities of throughflow, suggest that most of the rainfall which falls on hillslopes is unable to reach a channel until long after the rainfall has stopped and the stream flood peak has passed. In other words, only water from a relatively small 'contributing area' is able to reach a channel in time to contribute to the flood hydrograph of the stream. In its simplest form, for a rainstorm of constant intensity, the contributing area, A_c, is defined as

$$A_c = \frac{\text{Stream discharge}}{\text{Rainfall intensity}}$$ (8)

Where the rainfall intensity varies during the storm its value in the equation is necessarily an average one, weighted towards the most recent intensities. The contributing area is continuously changing during the storm (fig. 5.1.5), and is generally at its greatest at about the same time as the peak discharge in the stream. For basins measured in North Carolina the maximum contributing area varied relatively little from storm to storm, but from basin to basin it ranged from 5 to 85% of the total drainage area. These measurements of contributing area can be compared with models derived from the two types of hillside flow: 1. 100% overland flow, and 2. 100% throughflow. During 100% overland flow water from all parts of the basin commonly reaches a channel within about an hour, so that for a storm lasting longer than an hour the contributing area, expressed as a percentage of the total drainage area, is

$$\frac{\text{Rainfall intensity} - (\text{Infiltration and surface losses})}{\text{Rainfall intensity}} \times 100\%$$

Under conditions where vegetation and soil are thin or absent, this contributing area may be large, and the value of 85% contributing area refers to an abandoned copper strip-mining area with less than 36% vegetation cover.

During 100% throughflow the contributing area consists of:

1. the area of the stream channels themselves, which is usually 1–5% of the drainage area;
2. the areas of saturated or near-saturated soil, mainly adjoining channels, which respond rapidly to changes in rainfall intensity as is shown in equations (6) and (7);
3. a narrow strip of hillside around the saturated areas, the width of which is determined by the slow rates of throughflow in unsaturated soils.

In the North Carolina basin with a 5% contributing area, the actual channel area and a swampy area backed up behind the stream-gauging installation

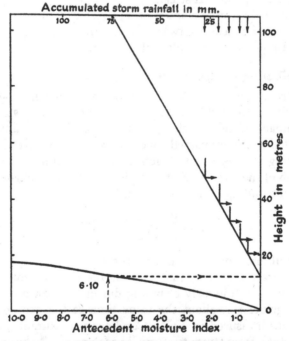

Fig. 5.1.5 An early concept of the variation of contributing area (considered as area below elevations shown), with initial moisture conditions and accumulated storm rainfall, for Bradshaw Creek, Tennessee (From T.V.A., 1964, Fig. 12).

accounted for almost the whole of the contributing area. Actual values of contributing area for humid drainage basins usually lie between these extremes, commonly in the 10–30% range, depending on soils, hillslope gradients, land use, and drainage texture.

Soils influence contributing area through their infiltration rate and the thickness of their permeable horizons, and attempts have been made to prove that contributing areas coincide with thin soil areas in a small drainage basin. Slope gradients influence the rate of throughflow (equation (4)), and hence the distribution of saturated soil and saturation overland flow. Vegetation cover and cultivation practices strongly affect the permeability of surface soil layers, mainly through the effect of vegetation in reducing rainsplash impact and through the effect of cultivation on depression storage and soil structure. In a basin with a high drainage density a relatively large area of hillslope is close to a channel, so that under throughflow the contributing area will also be relatively large; and under Horton overland flow there will be a relatively short time lag between the start of rainfall and a condition of maximum contributing area. Clearly these factors are not independent of one another, but each has a separate influence on contributing area.

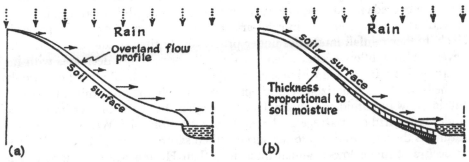

Fig. 5.1.6 Patterns of hillslope flow during Horton overland flow and throughflow. Arrow lengths show relative discharges over or through the soil.

(a). Horton overland flow (After Horton, 1945, p. 316). Thickness of water layer on surface is drawn proportional to actual thickness.
(b). Throughflow. Thickness of water layer below surface is drawn proportional to soil moisture content. Soil moisture from progressively earlier rainfalls is shown by progressively darker shading. The subsurface layer does not indicate the depth of infiltration into the soil.

Part of stream baseflow is derived from the water-table, which is itself supplied by deep percolation of water from the soil, but this contribution to baseflow is probably large only where well-defined aquifers are present. A large part of baseflow also comes from throughflow in the soil, which will take months to reach a channel from interfluve areas, and produces sufficient water to supply the measured baseflows in many areas. Since the same rain-water, much of it via throughflow, is responsible for both high and low flows in streams, all of the factors described above, which tend to produce high contributing areas during rainstorms, and hence a large proportion of total runoff during storms, also lead to reduced storage of rain-water after the flood flows have subsided, and hence to lower baseflows.

A final important contrast between the overland flow and throughflow models is that, whereas in overland flow it is the rain-water which is actually falling that flows into the stream during a rainstorm (fig. 5.1.6(a)), in throughflow, much of the water flowing into the channels is not physically the same as the rain-water which is currently falling (due to the time lag involved). It has been shown in infiltration experiments that almost all water flowing through the soil flows out in the order in which it flows in. This means that infiltrating water has to displace all of the soil water downslope before it can itself flow into the stream (fig. 5.1.6(*b*)), so that most water flowing into the stream, even at high flows, has been stored in the soil for a matter of weeks or months, and so has been able to come to chemical equilibrium with the soil. This soil water storage has obvious implications for interpreting the dissolved load of streams.

4. Summary

There are two extreme models of hillside water flow; the Horton overland flow model and the throughflow model. Horton overland flow occurs when rainfall intensity exceeds infiltration rate, and when it occurs at all in a basin, it is widespread. It is most common in semi-arid climates, and only occurs at progressively higher rainfall intensities under progressively thicker soil and vegetation covers. Throughflow occurs whenever the soil permeability decreases with increasing depth in the soil within some portion of the soil profile, most commonly at the base of the A horizon. Throughflow is probably the predominant mode of hillside flow in humid and humid–temperate areas, but it is of lesser importance under more arid or less-vegetated conditions with thin soils. When throughflow saturates the soil profile up to the surface, then saturation overland flow occurs. It occurs at much lower rainfall intensities than Horton overland flow, and is usually much more localized in its distribution, being commonest near streams. Under suitable conditions, both overland flow and throughflow may occur at any point although their relative frequencies will vary greatly from point to point. The separation of hillside flow into its two components, overland flow and throughflow, and a recognition of the distinct properties of each, allows a clearer understanding of the mechanisms of both streamflow and hillside erosion.

REFERENCES

BETSON, R. P. [1964], What is watershed runoff?; *Journal of Geophysical Research*, **69**, 1541–52.

BILHAM, E. G. [1936], Classification of heavy falls in short periods; *British Rainfall*, **75**, 262.

HEWLETT, J. D. and HIBBERT, A. R. [1963], Moisture and energy conditions within a sloping mass during drainage; *Journal of Geophysical Research*, **68**, 1081–7.

HORTON, R. E. [1945], Erosional development of streams and their drainage basins: hydrological approach to quantitative morphology; *Bulletin of the Geological Society of America*, **56**, 275–370.

HUDSON, N. W. and JACKSON, D. C. [1959], Results achieved in the measurement of erosion and runoff in Southern Rhodesia; *3rd Inter-African Soil Conference, Dalaba*, Paper No. 63.

JENS, S. W. and MCPHERSON, M. B. [1964], Hydrology of Urban Areas; In Chow, V. T., Editor, *Handbook of Applied Hydrology* (New York), Section 20, 45 p.

KIRKBY, M. J. and CHORLEY, R. J. [1967], Throughflow, overland flow and erosion; *Bulletin of the International Association of Scientific Hydrology*, **12**, 5–21.

LINSLEY, R. K., KOHLER, M. A., and PAULHUS, J. L. H. [1949], *Applied Hydrology* (New York), 689 p.

MUSGRAVE, G. W. and HOLTAN, H. N. [1964], Infiltration; In Chow, V. T., Editor, *Handbook of Applied Hydrology* (New York), Section 12, 30 p.

PHILIP, J. R. [1957–8], The theory of Infiltration; *Soil Science*, **83**, 345–57 and 435–48; **84**, 163–77, 257–64, and 329–39; **85**, 278–86 and 333–7.

TENNESSEE VALLEY AUTHORITY [1964], Bradshaw Creek – Elk River: A pilot study in area-stream factor correlation; *Office of Tributary Area Development, Knoxville, Tennessee*, Research Paper No. 4, 64 p. and 6 appendices.

TENNESSEE VALLEY AUTHORITY [1966], Cooperative Research Project in North Carolina: Annual Report for Water Year 1964–1965; *Division of Water Control Planning, Hydraulic Data Branch*, Project Authorisation No. 445.1, 31 p.

WHIPKEY, R. Z. [1965], Subsurface stormflow from forested slopes; *Bulletin of the International Association of Scientific Hydrology*, **10**, 74–85.

YARNELL, D. L. [1935], Rainfall intensity-frequency data; *U.S. Department of Agriculture, Misc. Publ. No. 204.*

II. Erosion by Water on Hillslopes

M. J. KIRKBY

Department of Geography, Bristol University

Water detaches and transports hillside material through the effect of raindrop impact; by unchannelled flow over the surface and within the soil; and by the formation and enlargement of a network of rills, gullies, and channels. Rain-splash detaches soil particles, and may on its own produce a net downhill transport of debris on a hillslope. Throughflow carries material in solution and in suspension within the soil mass, and in some cases carries material selectively along certain lines, which may be simply lines of greater permeability or may form small tunnels. Overland flow transports soil particles detached by rain-splash and may erode distinct channels, some of which may become a permanent part of the drainage channel network, whereas others are obliterated between storms by mass-wasting processes. Lateral and vertical cutting by pre-existing streams may erode and steepen the lower parts of slope profiles. These processes are the direct contribution of flowing water to hillslope erosion, but through the part played by soil moisture, water also enters indirectly into almost all other slope processes (Chapter 4.II). This section is concerned with a closer examination of the erosional processes caused directly by flowing water.

1. Surface and subsurface wash

On a level surface the splash-back following raindrop impact has been observed to move 4-mm stones distances up to 20 cm; 2-mm stones up to 40 cm; and smaller stones up to 150 cm. The stones are not transported in any one direction, but are part of a random exchange. On a slope the movement of individual stones is preferentially downhill, so that there is a net transport of material downslope. At any given rainfall intensity the *distance* of movement decreases with increasing particle size, but the *mass transport*, defined as the product of mean distance travelled and particle mass, *increases* with particle size for particles of up to 50 mm diameter (for rainfall intensities of 25–50 mm/hr). It is probable that there is not an indefinite increase, but a peak mass transport in the size range 100–400 mm diameter (which includes movement initiated by splash undermining). In areas with similar rainfall intensities the total rate of mass transport on a vegetation-free surface increases with mean annual rainfall. However, because vegetation cover also increases with mean annual rainfall, and vegetation shields the ground surface from rainsplash impact, mass transport does not increase indefinitely with mean annual rainfall but reaches a maximum and then decreases with further increase in rainfall (fig. 5.11.1). The maximum rate of 2·6 cm³/cm/yr occurs at a rainfall of 375 mm per year in the Western U.S.A. This rate may be compared with measurements of 0·09 cm³/cm/yr under a continuous vegetation cover in Britain, showing that rainsplash erosion is almost completely suppressed in humid areas except where there are local breaks in the vegetation cover.

Clays are certainly transported downwards through the soil from the A horizon to the B horizon by percolating water, and this is an essential part of soil development. As subsurface flow is predominantly lateral (throughflow), it is natural to assume that water also moves clays downslope, although there have

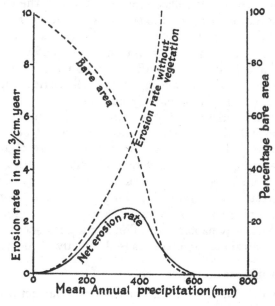

Fig. 5.11.1 Vegetation cover, erosion on bare ground, and erosion under natural vegetation, as each varies with mean annual rainfall in the south-western United States.

been few, if any, direct measurements of this process. In humid areas throughflow is more common than overland flow, but throughflow velocities are much less, so that it is not known which process is the more important. Dendritic networks of lines along which the permeability is greater and the A horizon is deeper are found, sometimes leading into the heads of stream channels. These lines carry greater throughflow and exhibit deeper weathering, and they may therefore be lines of more subsurface wash than in the surrounding soils, but these 'percolines' may also merely be the result of infilling of former extensions of the stream network. 'Piping', consisting of a subsurface network of small tunnels in soil and poorly cemented sediments, is widespread in semi-arid areas. Appreciable areas, especially on mesa tops, drain entirely through these piping systems, which supply springs lower down the hillside and carry all the material eroded from the areas they drain. Limestone cavern systems should perhaps be quoted as an extreme example of subsurface sediment transport.

2. Rill and gully erosion

As water flows over a hillside its depth varies in relation to the irregularities of the ground surface, being deepest in the depressions. It appears that most small depressions, even on an unvegetated hillslope, do not develop into linear channels, but any newly eroded channels that do form must begin with the concentration of flow in some of these small depressions. Under conditions of widespread overland flow, a steady state will soon be set up in which overland flow discharge, on average, increases linearly downslope with increasing distance from the divide. Horton has shown how a critical distance from the divide may be calculated, at which the hydraulic power of the flow is great enough to overcome the strength of the soil and vegetation mat, and begin to erode a channel. There should therefore be a line of channel heads across a hillside, each channel head being at the critical distance from the divide. Falling from this line there should be a set of many sub-parallel rills, or shoe-string gullies, covering the hillside. This is Horton's classic runoff model, in which the critical distance from the divide is controlled by the overland flow intensity, the gradient, the hydrodynamic roughness of the surface, and the strength of the vegetation mat. A rill system of this kind does not generally last for very long, but is rapidly converted into a normal dendritic network through the dominance of a few rills which are enlarged to form gullies, each in its own small, eroded valley. Horton has described the process of cross-grading by which water overtops the small divides between adjacent rills and leads to progressive diversion of water into the rill which is at the lower level. This process can lead to the development of an overall gradient at an angle to the direction of flow of the rills, and the rills will gradually turn in direction so that they again flow down the line of steepest slope (Leopold, Wolman, and Miller [1964], Chapter 10).

Where rills exist cross-grading may be an important process of drainage integration, but rill systems are not, in fact, common except on hillslopes with little or no vegetation cover. Schumm has shown that where they do exist they are often a seasonal phenomenon; the pattern he studied being destroyed each winter by frost action. Rill patterns may also be seasonally destroyed by field cultivation, and both of these processes of rill destruction average the erosion of the rills across the width of the hillslope. Figure 5.11.2(a) shows how rill erosion, alternating with obliteration by tillage and surface wash, is tending to smooth out the irregular profile of a ploughed field into a smooth, convexo-concave form. Measurements show that the increasing total amount of debris transported with increasing distance downslope (fig. 5.11.2(b)) is achieved almost entirely by a rapid increase of transport in the gullies, and almost no increase between them. This is probably caused by a correspondingly rapid increase of discharge in the gullies and a progressive lateral diversion of overland flow into them from the interfluve areas.

Under rainfall conditions where Horton overland flow is being uniformly

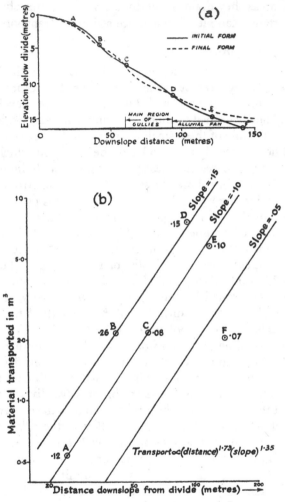

Fig. 5.11.2 Erosion and deposition by surface wash and small gullies in a ploughed field in Maryland, U.S.A., during the course of one year.

(a) Generalized longitudinal profile of field showing net erosion or deposition on same vertical scale.

(b) Correlation of transport rate (T) with distance from the divide (x) and tangent slope (s). Best-fit regression equation is $T \propto x^{1.73} s^{1.35}$, which is 'probably significant' at 95% level.

produced over a hillslope, the erosive force of the flowing water increases with both distance from the divide and with gradient. On a hillside of uniform gradient the erosive force is therefore greatest at the base of the slope; but on a convexo-concave slope the erosive force is greatest at a short distance downslope from the region of steepest gradient, so that it is here that Horton overland flow may be expected to initiate gullies. Saturation overland flow is preferentially produced in certain areas so that resultant gullying is also localized in these areas, namely; areas adjacent to perennial streams, areas of concave-upward slope profile, areas of thin or impermeable soils, and hollows (Chapter 5.1). These areas are strongly influenced by local factors of soil and three-dimensional valley shape, because the throughflow which allows saturation overland flow to develop travels so slowly that the water flows only a short distance while the rain is still falling. Distance from the divide is therefore not a control on the location of gullies which arise from saturation overland flow. Where Horton overland flow is involved, the factors are reversed in importance: with high flow velocities, distance from the divide is of paramount importance and, in comparison, local factors are much less important. A second distinction between the distribution of the two types of gullies is that Horton overland flow can initiate gullies which may be entirely separate from the main channel network, whereas saturation overland flow generally extends existing channel systems.

3. Slope profiles and drainage texture

The way in which overland flow and hydraulic erosion vary down a hillslope profile has important consequences for the form of the profile. Traditionally, convexities have been ascribed to soil creep and similar processes, and concavities have been ascribed to hydraulic processes, on the grounds that the erosive force of overland flow increases with increasing distance downslope. However, there is an effective increase of transporting capacity with distance only along water-eroded channels, and it is this that gives gullies their characteristic eroded, concave long profiles. On hillsides between channels the influence of distance from the divide is much less than has been formerly supposed for three reasons: 1. much of what has formerly been attributed to sheetwash is actually due to rainsplash, which is not influenced by distance from the divide; 2. even when overland flow is widespread, interfluve areas do not show a systematic increase of transporting capacity with distance downslope because a part of the flow is continually being diverted laterally into channels; and 3. where throughflow and saturation overland flow occur, distance from the divide is of little importance, since a steady state of flow is not achieved. For these three reasons, much of the force of the argument relating slope concavities to surface erosion by water is based on a premise which is, at best, partially true. Unchannelled hydraulic surface erosion is, therefore, like creep, likely to produce only convex hillslope forms.

Concavities at the bases of slopes must therefore be explained by mechanisms which do not rely on sheet erosion. There seem to be three possibilities: 1. Where a rill pattern is seasonally formed and destroyed, the movement of the rills from season to season allows their influence to be averaged across the slope so that the whole hillside is lowered uniformly at an average rate. Instead of the stream profiles becoming concave and the interfluves remaining convex, the changing positions of the rills lead to a general concavity. 2. A concentration of flow caused by convergence in a hollow will lead to rapidly increasing surface and subsurface flow, and a consequent progressive increase downslope in soil transport by all slope processes. The result will be a decrease in profile convexity, which may even become concave. 3. Any slope process which is in-filling a valley bottom will produce a concavity. In-filling may follow gullying, headward extension of existing streams, or lateral migration of streams. Slopes examined in the field show both depositional and erosional concavities.

The distribution of overland flow, through its effect on the exact positions of gully and stream heads, influences not only the form of the slope profile but also the texture of the whole drainage net. Since approximately half of the total length of streams in a basin is in first-order streams, a small change in the position of stream heads has a major effect on drainage density. Gullying appears to be the beginning of a vicious circle, since gullying produces a small valley which diverts more water into the gully, which in turn increases the rate of erosion. However, drainage densities do not increase indefinitely, because between major storms mass movement is able to partially refill the gullies, which are then able to carry more subsurface flow than before without saturation. Even under uniform conditions of agriculture and climate, the exact position of each stream head depends on the magnitude of the last major storm, and the period of time that has elapsed since. Thus, within a homogeneous area, neighbouring valleys may show different drainage densities reflecting a differing incidence of major storms, but indicating no difference in their basic equilibria. This variability of stream-head positions raises problems for the operational definition of stream networks, as the existence of eroded channels in the field is generally considered to be the basic criterion for comparison with maps and air-photography data.

4. Factors influencing rates of erosion

In determining the net erosional effect of hydraulic erosion, little work has been done on the relative contribution of each process. The bulk of the measurements have been of sediment collected either from artificially sprinkled plots in cultivated fields, or from natural streams. Table 5.II.1 summarizes some representative data from soil-erosion experiments, and shows that average rates of soil stripping may be as high as 30 mm/yr. The values are perhaps less important than the factors which control the rate of soil erosion. These factors fall within four main groups: 1. the depth, permeability, and other properties of the soil; 2. the land gradients; 3. the frequency distribution of rainstorms which are able to produce overland flow; and 4. land use, especially the amount of vegetation cover during the rainstorm periods of the year. In these plot experiments land use and soil properties are to some extent within the control of the experimenter, whereas stream measurements of erosion are usually done for larger areas and under more natural conditions, in which vegetation and soil are strongly correlated with climate. The measured rates of stream erosion are therefore expressed in terms of climatic and topographic variables. Langbein and Schumm [1958] have correlated erosion in basins of 50–5,000 km² with rainfall, across the United States, and have shown that there is a maximum erosion at a rainfall of about 250–350 mm p.a. (fig. 5.II.3). Evidence from both humid and arid regions suggests that most stream sediment is derived directly from its banks (as much by mass movement as by hydraulic action), and this is in accordance with the

TABLE 5.11.1 Measured rates of soil erosion from experimental plots

Area	Mean annual rainfall (mm)	Vegetation cover	% runoff	Soil loss (mm/yr)	Source
South-east U.S.A.	2,500–4,000	Oak forest	0·8	0·008	Meginnis [1935]
		Bermuda grass pasture	3·8	0·030	
		Scrub-oak woodland	7·9	0·10	
		Barren abandoned land	48·7	24·4	
		Cultivated: rows on contour	47·0	10·6	
		Cultivated: rows downslope	58·2	29·8	
Rhodesia	1,000	Dense grass	2·7	0·018	Hudson and Jackson [1959]
		Bare ground	38·0	2·3	

Erosion $\propto$ (Slope)^a × (Distance)^b × (Rainfall energy factor) × (Soil erodibility factor)

Slope exponent, $a = 0·6 - 2·0$

Distance exponent $\begin{cases} b = 0·0 - 0·7 \text{ without gullying} \\ b = 1·0 - 2·0 \text{ with gullying} \end{cases}$

decrease of drainage density with increased vegetation cover and infiltration rate which has been observed in the United States in regions of more than 250–350 mm of rainfall p.a. (fig. 5.11.4).

It seems that the principal cause of high erosion rates in semi-arid regions is the relative lack of vegetation cover, which leads to low infiltration rates. The result is increased overland flow, which in turn leads to high drainage densities and to an increase of runoff and erosion. Within humid areas the completeness of the vegetation cover varies relatively little with areal differences in rainfall. Therefore, as one moves from dryer to wetter areas (starting from 250–350 mm p.a. for the United States) the rates of erosion initially decrease rapidly to the point (about 600 mm p.a. for the United States) at which a total vegetation cover is established, and thereafter change very little (fig. 5.11.3). Despite the close correlation with annual rainfall, it is the distribution of rainfall within the year, much more than the annual total, which controls the vegetation and amount of erosion. More complex causal chains have been proposed, in which low-intensity rainfalls are said to have most influence on vegetation cover, whereas high-intensity rainfalls have most influence on peak stream runoffs. Thus a climatic change from low-intensity winter rains to high-intensity summer rains (as has occurred in the American South-west) will significantly increase rates of erosion, even though the change in mean annual rainfall is slight.

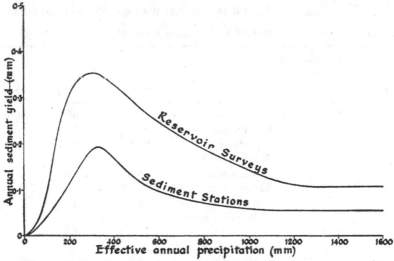

Fig. 5.11.3 The variation of total basin erosion rate with effective mean annual precipitation in the United States, for river sediment stations (of average area of 4,000 km²) and
reservoir surveys (of 80 km² average area). The difference illustrates the normal variation
of sediment yield ∝ (basin area) $^{-0.15}$ (After Langbein and Schumm, 1958).

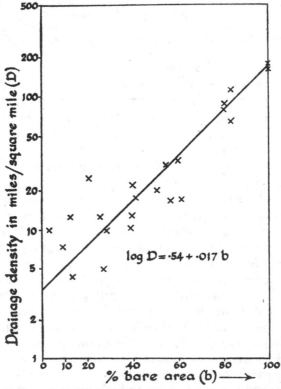

$$\log D = \cdot54 + \cdot017\, b$$

Fig. 5.11.4 The variation of drainage density measured in the field with the percentage
of bare, unvegetated area, for third- and fourth-order basins in the United States (After
Melton, 1957).

Factors which influence the areal pattern of erosion through climate and vegetation also influence the temporal pattern of erosion as conditions change at a point. There is evidence of alternate periods of gullying and valley deposition in many different climatic regions. These alternations are most striking in semi-arid regions like the American South-west, but they are also present in temperate regions like the eastern United States and Britain. Work in semi-arid areas shows fairly good correlation between dry periods and periods of valley cutting, although the most recent period of valley cutting in the south-western United States, around 1900, has been ascribed to either a minor climatic change or, alternatively, to destruction of vegetation through over-grazing.

In more humid regions fluctuations in climate are not usually of sufficient magnitude to destroy vegetation through aridity. The role of man, through clearing the vegetation for cultivation, has therefore been relatively greater. Experiments have shown that the greater the rainfall and the thicker the natural vegetation cover (up to a point), the greater will be the acceleration of erosion consequent on stripping the vegetation. However, this effect is offset by the more rapid regeneration of vegetation in humid climates, and by the generally greater thickness of soil which must be stripped before unweathered parent material is exposed. Perhaps for these reasons the soils most sensitive to erosion appear to be in areas which are semi-arid, and in these areas it is most common for irreversible soil erosion to occur, leading to bedrock slopes on which vegetation cannot effectively regenerate.

5. Summary

To summarize the current state of knowledge of hillslope fluvial erosion processes; it appears that subsurface wash is of unknown, but probably minimal, importance; surface wash between channels is produced mainly by raindrop impact, is very dependent for its efficiency on local or general absence of vegetation, and is also of minor importance. Only the erosive power of water flowing in clearly-defined channels is a truly effective transporting and eroding agent. The positions of channel heads, which control the drainage texture, vary areally between basins and fluctuate through time within a basin. At any moment their positions depend on the incidence of major storms which are extremely local in their effect; but the average positions of channel heads over a period of time are controlled mainly by soil, slope gradient, precipitation, and vegetation cover. Man can interrupt the natural balance by his control of vegetation and, to a lesser extent, by tillage, and most agriculture would lead to gullying if no preventive measures were taken. However, the extent to which conservation measures are necessary varies widely, according to the extent to which overland flow is increased by cultivation. Vegetation is seen to play a major role in the rate at which hillslopes are eroded by running water, and the influence of both man and climatic change is chiefly through their effect on changing the vegetation cover.

I

REFERENCES

BRYAN, K. [1928], Historic evidence of changes in the channel of the Rio Puerco, a tributary of the Rio Grande in New Mexico; *Journal of Geology*, **36**, 265–82.

BUNTING, B. T. [1964], Slope development and soil formation on some British sandstones; *Geographical Journal*, **130**, 73–9.

CARLSTON, C. W. [1966], The effect of climate on drainage density and streamflow; *Bulletin of the International Association of Scientific Hydrology*, **11**, 62–9.

ELLISON, W. D. [1945], Some effects of raindrops and surface flow on soil erosion and infiltration; *Transactions of the American Geophysical Union*, **26**, 415–29.

FLANNERY, K. V., KIRKBY, A. V. T., KIRKBY, M. J., and WILLIAMS, A. W. [1967], Farming systems and political growth in Ancient Oaxaca; *Science*, **158**, No. 3800, 445–54.

HACK, J. T. [1942], The changing physical environment of the Hopi Indians of Arizona; *Reports of the Awatovi Expedition, Peabody Museum, Harvard University*, Report No. 1, 85 p.

HACK, J. T. and GOODLETT, J. G. [1960], Geomorphology and forest ecology of a mountain region in the central Appalachians; *U.S. Geological Survey Professional Paper 347*, 66 p.

HORTON, R. E. [1945], Erosional development of streams and their drainage basins: hydrophysical approach to quantitative morphology; *Bulletin of the Geological Society of America*, **56**, 275–370.

HUDSON, N. W. and JACKSON, D. C. [1959], Results achieved in the measurement of erosion and runoff in Southern Rhodesia; *3rd Inter-African Soil Conference, Dalaba, Paper No. 63*.

LANGBEIN, W. B. and SCHUMM, S. A. [1958], Yield of sediment in relation to mean annual precipitation; *Transactions of the American Geophysical Union*, **39**, 1076–84.

LEOPOLD, L. B. [1951], Rainfall frequency, an aspect of climatic variation; *Transactions of the American Geophysical Union*, **32**, 347–57.

LEOPOLD, L. B., WOLMAN, M. G., and MILLER, J. P. [1964], *Fluvial Processes in Geomorphology;* (Freeman & Co., San Francisco), 522 p.

MELTON, M. A. [1957], An analysis of relations between climate and landforms; *Office of Naval Research, Technical Report 11; Project N.R. 389–042*, 102 p.

MELTON, M. A. [1958], Correlation structure of morphometric properties of drainage systems and their controlling agents; *Journal of Geology*, **66**, 442–60.

PARKER, G. G. [1963], Piping, a gemorphic agent in landform development in the drylands; *International Association of Scientific Hydrology*, Publication No. 65, 103–13.

SCHUMM, S. A. [1956], Evolution of drainage systems and slopes in badlands at Perth Amboy, New Jersey; *Bulletin of the Geological Society of America*, **67**, 597–646.

SCHUMM, S. A. [1964], Seasonal variation of erosion rate and processes on hillslopes in western Colorado; *Zeitschrift fur Geomorphologie*, Supplementband **5**, 215–38.

III. Overland Flow and Man

M. A. MORGAN

Department of Geography, Bristol University

It is obviously impracticable to draw too fine a distinction between the movement of water over the surface and immediately beneath the surface, a distinction, that is, between overland flow and throughflow, since in practice the one shades imperceptibly into the other. The infiltration capacity of a soil, and hence the relationship between overland flow and throughflow, depends on the texture of the soil, on its thickness and its degree of compaction. Surface runoff can only occur when the intensity of the rainfall exceeds the infiltration capacity of the soil. A clay soil composed of many fine particles and containing much interstitial water has a low infiltration capacity, while a dry sandy soil will at least for a time absorb and pass water fairly easily (Table 5.III.1).

TABLE 5.III.1 Infiltration capacities of field soils

Soil texture	Infiltration capacity (in./hr)
Clay loam	0.1–0.2
Silt loam	0.3–0.6
Loam	0.5–1.0
Loamy sand	1–2

(after Kohnke and Bertrand)

Hard driving rain will cause compaction of the surface of a normally permeable soil, reducing the pore spaces and thereby lowering the infiltration capacity. A good vegetation cover will prevent compaction by rain, and thus increase the infiltration capacity. Obviously climate, geology, soil, and vegetation all affect the character of runoff.

The ideal situation from the point of view of man is where ample gentle rain falls at the right time for his crops, penetrates the soil without violence, maintaining just sufficient moisture around the soil particles without filling the interstices, and passes down to nourish a stable water-table, which in turn feeds a system of streams and rivers whose regime is accommodating and predictable. Unfortunately even in the most favoured areas these conditions are rarely completely satisfied. The runoff in a sense may be regarded as a measure of the extent to which any particular hydrological system departs from the ideal from a human point of view. Too great, irregular, or unpredictable runoff shows that the system is not working at its best for the farmer. A great deal of the history of farming is the story of the attempts to find ways of improving the plant–soil–water relationships, and most of the world's productive agricultural areas show the imprint of generations of effort directed to redressing some of the imbalances caused by undesirable runoff characteristics.

1. Soil erosion

Certainly in human terms the most far-reaching and devastating consequence of excessive uncontrolled runoff is soil erosion, and there are many areas in which constant vigilance against its effects are essential. In the humid and sub-humid regions natural vegetation usually binds the soil particles in place and slows down the flow of surface water. Although erosion takes place, natural processes normally create at least as much soil as they remove. Interference by man with the natural vegetation, which is usually finely and uniquely adapted to the totality of natural conditions, can start the process of erosion. Once bare soil is exposed on sloping land, whether by ploughing, overgrazing, or by too enthusiastic felling of timber, it is vulnerable. Wherever rain falls faster than it can soak into the bare earth the raindrops loosen the soil particles. The effectiveness of raindrops in detaching soil particles depends on the intensity of the rainstorms and at what season they occur. Generally summer thunderstorms are the most violent. High rainfall intensities produce relatively large raindrops, which have high velocities, and these in turn can detach more and bigger fragments of soil. Such particles are borne away by the surface water as it flows downslope, carrying with it mineral salts in solution as well as organic material. The wholesale removal of layers of soil from slopes is known as sheet erosion. Where the process is maintained for any length of time the water is gradually concentrated into definite channels, giving rise to rill erosion. All too frequently in the past, in the absence of remedial measures the rills in turn have grown into gullies, which can extend at rates of a hundred feet a year, carving deep into farmland and often making it impossible to use machinery. Streams and rivers become choked with sediment, which they spread in a suffocating blanket over cropland in the valley floors. The loss of texture and porosity in the denuded soils that are left on the slopes makes them even less capable of absorbing the rain, so the runoff increases with time, and the regimes of the streams and rivers become even more erratic. The nature of the soil, the severity of the slopes, the geological structure, the rainfall characteristics, and the type of farm enterprise all play their part in affecting the nature and extent of soil erosion in any one area. Table 5.III.2 represents an attempt to indicate in very simple terms the relationships between the more important variables, and is formulated in terms of the conditions in the United States. While it is easy to overdramatize the effects, it is none the less the case that in the United States alone in the late 1950s erosion was estimated to be the main or dominant conservation problem on over 700 million acres, more than half the agricultural land in the country, and 234 million acres of cropland needed constant attention against erosion.

TABLE 5.III.2 Classification of runoff-producing characteristics

Designation of watershed characteristics	Runoff-producing characteristics			
	100 *Extreme*	75 *High*	50 *Normal*	25 *Low*
Relief	(40) Steep, rugged terrain; average slopes generally above 30%	(30) Hilly; average slopes of 10-30%	(20) Rolling; average slopes of 5-10%	(10) Relatively flat land; average slopes of 5%
Soil infiltration	(20) No effective soil cover; either rock or thin soil mantle of negligible infiltration capacity	(15) Slow to take up water; clay or other soil of low infiltration capacity, such as heavy gumbo	(10) Normal, deep loam; infiltration about equal to that of typical prairie soil	(5) High, deep sand or other soil that takes up water readily and rapidly
Vegetal cover	(20) No effective plant cover; bare or very sparse cover	(15) Poor to fair; clean-cultivated crops or poor natural cover; less than 10% of drainage area under good cover	(10) Fair to good; about 50% of drainage area in good grassland, woodland, or equivalent cover; not more than 50% of area in clean-cultivated crops	(5) Good to excellent; about 90% of drainage area in good grassland, woodland, or equivalent cover
Surface storage	(20) Negligible; surface depressions few and shallow; drainage ways steep and small; no ponds or marshes	(15) Low, well-defined system of small drainage ways; no ponds or marshes	(10) Normal; considerable surface-depression storage; drainage system similar to that of typical prairie lands; lakes, ponds, and marshes less than 2% of drainage area	(5) High; surface-depression storage high; drainage system not sharply defined; large flood-plain storage or a large number of lakes, ponds, or marshes

Each column shows the contribution of the different watershed characteristics, relief, soil infiltration, vegetation, and surface storage to a particular amount of runoff. For example, under extreme conditions of 100%-runoff, relief accounts for 40% of the runoff, while the other three elements account for 20% each. The right-hand column shows that a low proportion of runoff is considered to be 25%, to which relief contributes twice as much as either soil infiltration, vegetation, or surface storage.

From *Farm Planners' Engineering Handbook for the Upper Mississippi Watershed*, U.S. Soil Conservation Service, Milwaukee, Wis., 1953.

Nor is erosion by running water the only hazard. The exposure of bare earth to strong drying winds can result in the soil particles being detached and carried along by the wind. Very small particles less than 0·1 mm in diameter can be knocked off the surface by larger ones in motion, and carried off in suspension distances of several hundreds, even thousands, of miles. Sand-sized particles varying in diameter between 0·1 and 0·5 mm are rolled along by the wind. Since there is no wind at all at the actual surface of the ground, only the tops of such particles are directly affected by the wind, and they start to spin, reaching several hundreds of revolutions a second. As a particle rolls a partial vacuum is created above it, lifting it into the air rather like a plane when it reaches take-off speed. As it rises the particle loses lift and falls back to the ground. In fact, it moves with the wind in a series of bouncing hops, a process known as saltation. Soil fragments slightly larger, between 0·5 and 3·0 mm in diameter, are too heavy to be lifted, but they can be rolled along the ground, causing surface creep. Only exceptional winds can move particles bigger than 3 mm in diameter. Sandy soils, whose particles are mostly the best size for saltation to be effective, are most liable to wind erosion when conditions are right, but since wind erosion is selective of particle size, the finest and often the most fertile parts of good soils, the smallest fragments, can be progressively winnowed out by the wind, leaving behind an impoverished and coarse soil whose fertility is much reduced. The Dust Bowl is the classic area in the United States, and was created initially by too regular cropping for wheat and by the practice of leaving the soil exposed during the period of summer fallow.

Normal conservation practices, largely developed in the United States, are based on watersheds as the management units. They are devoted to stopping active erosion and to reducing runoff by increasing the infiltration capacity of the soil. Conservation methods include soil treatment, introduction of improved crop rotations, contouring and terracing, gully damming, drainage diversion, grassing waterways, cover cropping, pasture improvement, tree planting, and woodland conservation. Conservation schemes are costly, but most governments have systems of subsidies and grants to encourage farmers to make improvements. The economic and social benefits of such conservation programmes are often direct and immediately apparent, and this tends to decrease the hostility with which rural communities might normally be expected to regard innovations. Many schemes pay for themselves in a few years. For example, in many parts of the United States it is normal practice to dam streams in the upper reaches of drainage basins. The lakes thus formed are used to store runoff water. But they also serve other purposes as well. They can be used for recreation, and by furnishing emergency water supplies for livestock and for fighting farm fires they can directly benefit the farmer by lowering his costs. They also reduce the cost to the Government of flood compensation. Even so, it must be admitted that social and institutional resistance to improvement schemes of this kind are still a force to be reckoned with in the United States, let alone in other countries, whose rural communities are often far less sensitive to economic and financial incentives.

While attention has properly been focused on the problems of soil erosion in rural areas, it is in some ways almost as great a problem in some of the rapidly expanding urban and suburban areas. Construction sites for motorways, freeways, factories, and housing estates may have acres of raw earth stripped and exposed by bulky excavators and earth-moving equipment. Such sites can remain open for several years, even the single dwelling creates its own small problem for several months at least. Some cities in the United States have regulations designed to reduce erosion rates under such conditions, but they are usually neither comprehensive nor easy to enforce. Official practice in conservation-conscious areas encourages temporary cover of exposed places, with annual sown grasses or even tarpaulin, and drainage ways to take storm water gradually off the site by way of sediment traps, so that as little top-soil as possible is lost. Reliable information on rates of loss is not easy to find, but a recent study by Wolman and Schick of conditions on construction sites in Maryland, where soils are deep, slopes generally less than 10%, and annual rainfall about 42 in., suggests that rates are between twice and several hundred times as great per unit area as from the surrounding rural areas, and 'the equivalent of many decades of natural or even agricultural erosion may take place in a single year from areas cleared for construction'. Comparative figures suggest a typical sediment yield from undisturbed areas of between 200 and 500 tons per square mile per year, and on construction sites figures range from several thousand up to 140,000 tons per square mile per year. The very high yields are recorded on very small sites, and in general the bigger the site, the smaller the sediment yield, since the big site allows internal readjustment of material, much of which never gets carried beyond the site to contribute to the load of external rivers and streams. The increased stream loads give rise to many problems. Domestic and industrial water supplies are often drawn from this source, and where the water is polluted by increased turbidity additional costs are incurred in removing contamination. Reservoirs silt up more rapidly. The ecology of rivers and ponds can be changed very quickly as the channels accommodate themselves to the increased burden of sediment and fishing and sporting interests are naturally affected. Recreational land use possibly suffers most of all, and though it is almost impossible to put a value on amenity, it is beyond dispute that recreational land near the margins of the large cities grows more precious each year, and anything that contributes to despoiling such amenities is not only to be deplored but to be resisted.

2. Rural drainage

Successful farming in much of the temperate area in the world depends to an extent much greater than commonly realized upon the success of schemes for getting rid of excess soil moisture. Plants only use the water that forms a thin film around the soil particles. The interstitial water is a nuisance, its presence makes the soil heavy, and the object of any drainage scheme is to get rid of it, to lower the water-table, thereby raising the soil temperature through reducing evaporation, and at the same time aerating the soil.

Many of the oldest drainage schemes in Britain were undertaken by the early shepherds on the mountain and hillslopes. Over the course of many centuries they created a network of drainage channels designed to remove surplus water from the grazing areas in such a way that it did the least possible damage. On flattish land, however, too intricate a network of drainage ditches impedes cultivation, the ditches need frequent attention to clear them of aquatic plants, and by themselves they will not usually deal effectively with really wet soils. One of the greatest but often underestimated drawbacks of the medieval open-field system of farming was that it was not easy to drain the great fields effectively. The open ditches and deep furrows that the farmers made were all too often hopelessly inadequate, and after heavy rain water could stand on the land week after week, gnawing away at its fertility.

Another and usually superior way of draining superfluous moisture from the soil is by the use of underdrains. The Enclosure movement in Britain created compact individual land holdings in much of the area previously dominated by the open-field system, and released from the restrictions of a communal system, individuals were free within limits to experiment with new methods to drain the heavy lands. The first textbook in Britain devoted to the study of farming techniques was probably *The Improver Improved*, by Capt. Walter Bligh, published in 1650. It is apparent from this and later books that underdraining was already being practised in some areas. One popular method of the time consisted of cutting parallel trenches from 18 to 48 in. deep so arranged as to run down the slope. The bottoms of the trenches were filled with a layer of blackthorn, other brushwood, or straw, covered with a turf, which in turn was covered with stones and then earth until all traces of the trench had been removed. In time the replaced soil would consolidate above the brushwood or straw, so that when that eventually decayed a tunnel would be left below the surface through which surplus water would drain to a suitable outfall on the edge of the field. Stones were used instead of brushwood where they occurred locally in sufficient quantity. Arthur Young, in his travels in England in the latter half of the eighteenth century, noted with approval the extensive use made of drains of this type. Agricultural techniques at the time were still composed of a blend of traditional practices, on the one hand, and the enlightened empiricism of a few men of vision operating without the benefit of a truly scientific background, on the other, so we need not be surprised that many different approaches to the problems and purposes of underdrainage all found the active support of enthusiastic advocates, and nearly all had the merit that at least they worked. In 1831 James Smith of Deanstone in Perthshire published an article explaining the methods of drainage by means of which he succeeded in the space of a few years in transforming his marshy and sour farmland into a rich and fertile property. His work attracted the attention of a Select Committee of the House of Commons charged with an inquiry into the State of Agriculture at a time of agricultural distress in 1836. The report devoted much of its space to an investigation of drainage, and its chairman at one point ventured the (probably excessive) opinion that drainage was 'the only thing likely to promote the general improvement of agriculture'. The report is a valuable guide to the nature and extent of the drainage methods in use at the time.

Smith's system was designed both to drain and to break up the hard sub-soil pan that develops in many wet soils. Parallel field drains were laid 30 in. deep and between 16 and 21 ft apart. Then Smith used his own design of plough, a formidable implement drawn by a large team of horses to cut deep into the soil, crumbling the hard pan, breaking up the sub-soil, aerating it, warming it, and encouraging it to yield up its water more easily to the underlying drains. Josiah Parkes (1793–1871) believed that deeper drains were preferable, and there followed a period of controversy sustained for several years in the fashion of the

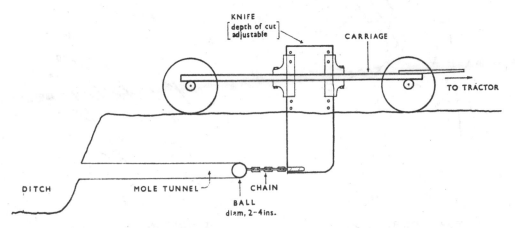

Fig. 5.III.1 Method of forming a mole drain.

times and nourished by a succession of passionately partisan lectures, papers, and articles.

An important advance was marked by the invention of a cheap, seamless clay pipe in 1843. On large estates with deposits of suitable clay it was common practice to build tile factories as near as possible to where the pipes were required. Cartage of pipes and 'tiles' was expensive, and where no local clays were found stone or straw continued to be used in trenches for several decades. Landlords found that their tenants were in many cases not averse to paying increased rents for drained lands. Increases of 5s. per acre with drains at 18-ft intervals and 2s. 6d. per acre with drains at 36-ft intervals were quoted in 1836 for tenants of the Duke of Portland in Ayrshire.

With the advent of steam engines many complicated and ingenious machines were invented to trench, lay the clay pipes, and fill in in a single operation. Where the soil was a tough homogeneous clay and where the slopes were fairly simple, it was easier and cheaper to use the mole plough. This was developed towards the end of the eighteenth century, and is still used (fig. 5.III.1). As it is drawn through the soil it leaves a 'pipe' in its wake through which the water drains away. A

Royal Commission in 1880 reported evidence that between 1846 and 1873 more than 10 million acres of farming land were drained by one means or another, and as one would expect, the greatest periods of activity coincided with the times when farming was generally at its most prosperous.

Modern field drainage practice is not substantially different from that of the last century, although we now understand a great deal more about the processes involved. Any underdrainage system is bound to be expensive, so care must be taken to design it properly. Typical drainage networks are illustrated in fig.

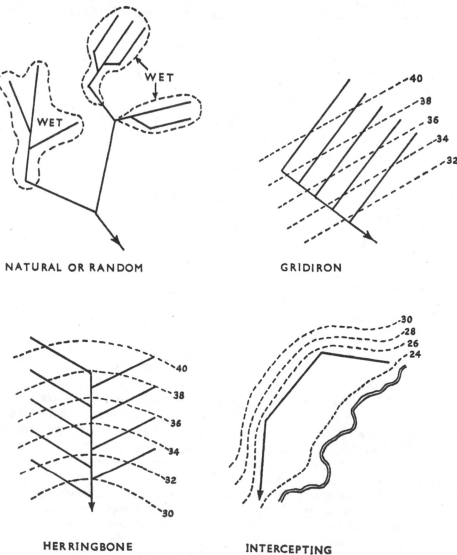

NATURAL OR RANDOM

GRIDIRON

HERRINGBONE

INTERCEPTING

Fig. 5.III.2 Some common types of drainage systems. (Adapted from Linsley and Franzini, 1964).

5.III.2. Standard underdrain pipes used to be about 2 in. in diameter, but it is now normal to use 4-in. ones. The distance between drains is controlled pre-eminently by the permeability of the soil to be drained. Where this is high, then the drains can be widely spaced, where low, then the drains have to be more closely spaced. In really heavy and sticky clay soils the density of drains might well have to be so high as to be too costly, in which case a mole plough might well be the answer. Gradients should not be less than 0·2%, to give a velocity of 1 ft/sec when flowing full, to avoid clogging by sediment. Spacing of drains usually varies from about 50 to 150 ft. The depth of the drains is mainly affected by the type of crop to be grown, the soil type, and the source and salinity of the water to be removed. The drains create a water-table with ridges and troughs aligned in the direction of the drains, the drains themselves cor-responding to the troughs, and the zones between them to the ridges in the

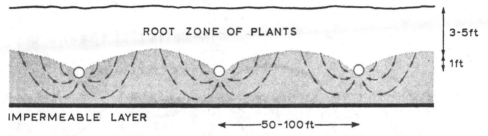

ROOT ZONE OF PLANTS   3-5ft

1ft

IMPERMEABLE LAYER   ←——50-100ft——→

Fig. 5.III.3 The effect of underdrainage on the water table. The arrows indicate very approximately the direction in which water moves to the drains. The stipple represents saturated soil.

water-table (fig. 5.III.3). The spacing must be such that the highest point in the ridge is below the root zone of the plants to be grown. In humid regions a depth of from 2 to 3 ft is suitable for most crops, though orchards, vineyards, and alfalfa need the water-table lower, from 3 to 5 ft. Deep-rooted and most irrigated crops need a depth of around 5 ft.

3. Large-scale drainage schemes

The problems posed by such schemes are different in kind from those we have so far examined, involving as they do spectacular engineering works and for-midable quantities of surplus flood waters at certain critical periods. In Britain the largest drainage scheme is in the Fens. The initial engineering work was carried out by the Dutchman Vermuyden for the Earls of Bedford and their financial associates between 1630 and 1652. The sluggish and meandering River Ouse carried the waters flowing from the hills to the east and south out to the Wash. But the gradients were so gentle that the Fen basin was largely a swampy marshland, a haven for fishermen and fowlers. The first attempts at drainage were initially very successful. Vermuyden cut two straight channels, each 21 miles long, between Earith Sluice in Huntingdon and Denver in Norfolk. The purpose of the sluice at Denver was to prevent the tides flowing back up the course of the Ten Mile River towards Ely, and that at Earith was to allow the engineers to decide in an emergency how much of the waters passing the sluice should be allowed to flow down the straight channels (the Bedford Levels) and how much down the original course of the Old West River (fig. 5.III.4). The two new channels reduced by 10 miles the length of the original course. By mani-

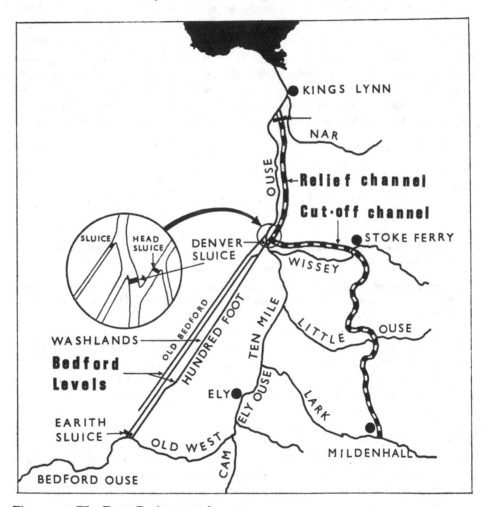

Fig. 5.III.4 The Fens. Drainage works.

pulating the two sluices, and by using the considerable storage capacity of the 'Washlands' between the two channels, the flood dangers could be minimized. The scheme was shortly to reveal unexpected difficulties. The peaty soils in the drained area very soon began to shrink as they dried out, and in addition bacterial action on the oxidized outer layers exposed to the air caused extra wastage. The loss of soil was less worrying than the fact that the level of the drained fields was steadily falling in relation to the rivers into which the waters drawn from the fields were drained. Soon it became necessary to use windmills to operate pumps, and before long, since the lifting power of pumps is limited, 'flights' of windmills linked in series became a common sight in the Fens. The shrinkage has continued. In Holme Fen (Hunts) there is a well-known cast-iron pillar which reveals a shrinkage of nearly 13 ft since 1848. Unfortunately the limit of series-linked windmills was reached before, in some cases, new techniques could be developed. Centrifugal pumps driven first by steam and after 1913 by diesel engines have now given way to electric motors and automatic axial-flow pumps in the struggle to keep the rich farmlands dry and safe from floods. Disastrous floods like those of 1947 serve as a constant reminder of the tenuous nature of man's control and have stimulated the latest engineering enterprise, the £8 million Great Ouse Flood Protection Scheme. This involves first the construction of a wide, deep, and straight Relief Channel from the sluices at Denver to the sea beyond King's Lynn. Second the straightening of the Ten Mile River and the Ely Ouse. Thirdly a 30-mile-long cut-off channel is being built around the eastern rim of the Fens to intercept the flood waters of the Lark, Little Ouse, and Wissey before they can reach the low-lying Fens. Such expensive undertakings are only justified by the great value of the high-quality land whose cultivation they permit.

4. Urban drainage

Towns and cities, with their considerable areas of man-made impermeability alternating with cultivated gardens and open spaces, present special problems of disposal of rain-water (storm water) which historically have been linked with the disposal of domestic sewage and industrial effluent (foul water). Early systems of urban sewerage were often designed so that storm water passed through the same system of pipes as foul water, thus flushing the system periodically. Modern sanitary engineering practice is based on the complete separation of foul and storm water in unconnected but often spatially parallel systems.

Wide variations in permeability, and hence runoff, are found within a typical city, and average figures are given in Table 5.III.3 as a guide to the magnitude of the range in Britain. In densely built-up areas drained almost entirely by a piped sewerage system the flow from the paved and other relatively impermeable surfaces will almost always be greater than the flow from any permeable surfaces, and the latter flow, being out of phase, can be ignored for design purposes. Where a significant part of the catchment area is not yet built upon and is still primarily drained by streams and rivers, the total runoff has a large component from the latter source. This secondary flow is generally assumed to reach its peak at a time equal to twice the estimated time of concentration used for the primary flow.

In designing an urban sewerage system the engineer has to consider not only the total quantity of water to be handled but also the time period over which a particular pattern of rainfall will be associated with the system. A storm, for instance, will fall with varying intensity over a catchment area, but different

TABLE 5.III.3 Impermeability factors, storm-water sewers, urban and sub-urban areas

Density (houses/acre)	Percentage impermeability	
	Primary flow	Secondary flow*
Industrial	50–100	12·5–0
Special areas	40–60	15·0–10
30	68	8
25	64	9
20	55	11
12	38	15·5
9	30	17·5
6	25	18·8
2 Rural	5–10	23·8–22·5
Open spaces	5	23·8

* Runoff coefficients vary according to sub-soil characteristics and cultivation. Typical values are:

60–75% for rocky or clay sub-soil on steep slopes
50–60% undrained clay sub-soil on steep slopes
20–40% drained and cultivated clay sub-soil
10–20% cultivated sandy or loamy sub-soil
5–10% woodland area.

parts of the catchment will discharge into the sewers at different times, and water in the system will take time to pass through it. The storm water will obviously not all have to be handled at one point in time. The design of the system must be such that it will handle all that is presented to it under most, but rarely under all, circumstances. To make a system large enough to cater for even the exceptional 'once in a century' storm would be unnecessarily expensive, so a decision has to be made concerning the severity of the storms the system is going to be designed to handle. In Britain it is the sudden short-lived summer storm that throws the greatest strain on the storm-water sewers. The Road Research Laboratory hydrograph method is now generally regarded as the most suitable for calculating runoff from all but the smallest urban areas (i.e. less than 20 acres of impermeable surfaces). In essence five steps are involved, illustrated in fig. 5.III.5. First, the catchment area is divided into sub-areas of a convenient size. Second, the impermeable area of each sub-area is calculated and an area–time diagram drawn for each. The area–time diagram relates the area contributing to the rate of flow with the time after the start of the rainfall. The graphs for each contributing area are combined to form a cumulative total profile for the whole system (fig. 5.III.5(b)). Third, a rainfall profile is drawn from tables prepared by the Road Research Laboratory. These are based on long-term observations and show the minute-by-minute changes in the amount of rain received during storms of varying severity and frequency. Profiles for storm frequencies

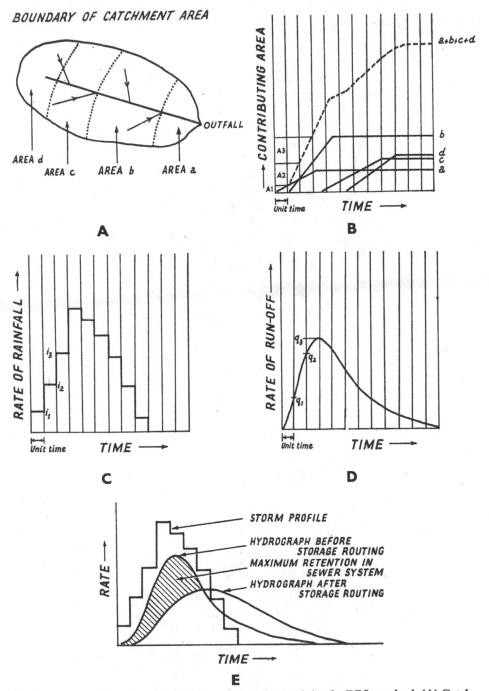

Fig. 5.III.5 Procedure for calculating runoff hydrograph by the RRL method. (A) Catchment area with four sub-areas; (B) Area/time diagram; (C) Rainfall profile; (D) Runoff hydrograph before allowing for storage, $q_1 = (i_1 \times A_1)$, $q_2 = (i_2 \times A_2) + (i_2 \times A_1)$, $q_3 = (i_1 \times A_3) + (i_2 \times A_2) + (i_3 \times A_1)$; (E) The effect on the hydrograph of allowing for retention in the sewer system (After Watkins, 1966. By permission of The Controller of H.M. Stationery Office. Crown copyright reserved).

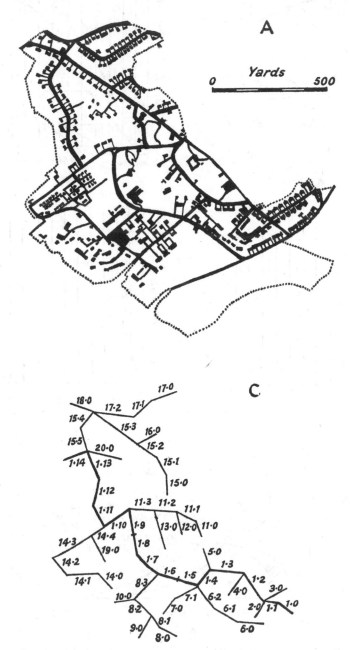

Fig. 5.III.6 Storm-water sewers.

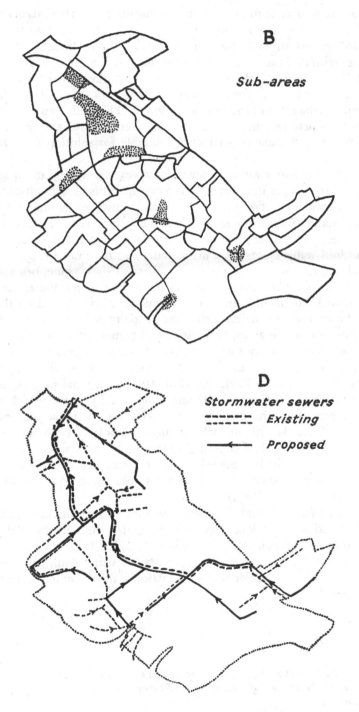

B

Sub-areas

D

Stormwater sewers

------- *Existing*

←— *Proposed*

expected once a year, once in five, once in ten, and once in thirty years are available. A simplified and generalized profile is shown in fig. 5.III.5(c). Fourthly, the two profiles already determined, the cumulative total area–time profile and the appropriate rainfall profile, are combined to give the first hydrograph (fig. 5.III.5(d)). This, however, takes no account of the storage capacity of the system itself. Methods are available for calculating the effects of the additional factor and for determining the form of the final hydrograph. This fifth step is illustrated in fig. 5.III.5(e). In practice, the method gives excellent results, and the number of calculations required make it particularly suitable for solution by digital computer.

Most work on urban sewerage systems involves redesigning existing systems to accommodate changes in input or to remove operational difficulties. A typical problem is illustrated in fig. 5.III.6. The area in question is one of fairly substantial Victorian villas with some more recent development. There are a number of open spaces, and gardens are generally large for an older urban area. The storm- and foul-water systems are inadequate, and both need redesigning. The first map shows the impermeable areas, i.e. those of slate, stone, brick, concrete, stucco, asphalt, etc. The next map shows the way in which the entire area has been subdivided into smaller units, the delimitation being based on the natural direction of movement of surface water (the gentle regional slope is to the outfall in the north-west of the area), and also the alignment of the existing sewerage system. The dotted areas on this map show the areas at present liable to flood. The third map shows the sketch of the sewerage system, defining the main lineaments of the network. Each link is numbered (decimal point 0 always indicates the length of sewer most upstream of the outfall), and the thicker line and the prefix 1 represents the trunk sewer, having its outfall at the end of link number 1 : 14. Once data about each catchment area have been compiled along the lines indicated, a standard computer programme is available very rapidly and therefore cheaply to calculate the expected flows along each link and the size of pipe needed to accommodate them. In the example quoted the final form of the system is shown in fig. 5.III.6(d).

In most large cities alterations to the sewerage systems involve adding extra links to the existing network or selectively increasing the capacity of the system, so the freedom for fundamental redesign is severely limited. In a new town, however, the possibility, at least in theory, exists for using computer methods to optimize not only the operational characteristics of the sewerage system but also to overall design.

REFERENCES

DEPARTMENT OF SCIENTIFIC AND INDUSTRIAL RESEARCH [1963], *A Guide for Engineers to the Design of Storm Sewer Systems;* Road Note No. 35 (H.M.S.O., London).

FUSSELL, G. E. [1952], *The Farmers' Tools;* (Andrew Melrose, London).

HOUSE OF COMMONS, 1836, *Third Report from the Select Committee Appointed to Inquire into the State of Agriculture;* Parliamentary Papers, July.

KOHNKE, H. and BERTRAND, A. [1959], *Soil Conservation* (McGraw-Hill, New York).

LINSLEY, R. K. and FRANZINI, J. B. [1964], *Water-Resources Engineering* (McGraw-Hill, New York), 654 p.

SCHWAB, G. O., FREVERT, R., EDMINSTER, T., and BARNES, K. [1966], *Soil and Water Conservation Engineering* (Wiley, New York).

THORN, R. B., Editor [1966], *Engineering and Water Conservation Works* (Butterworths, London).

WATKINS, L. H. [1962], *The Design of Urban Sewer Systems;* Department of Scientific and Industrial Research, Road Research Technical Paper No. 55 (H.M.S.O., London).

WOLMAN, M. G. and SCHICK, A. P. [1967], Effects of construction on fluvial sediment, Urban and suburban areas of Maryland; *Water Resources Research*, 3 (2), 451–64.

CHAPTER 6

Ground Water

I. Ground Water

J. P. WALTZ

Department of Geology, Colorado State University

1. Definition of ground water

Ground water is water which occurs beneath the surface of the earth within saturated zones where the hydrostatic pressure is equal to or greater than atmospheric pressure. This precise definition is useful in distinguishing between ground water and other types of subsurface water, such as capillary water or soil water. However, more important than fine distinctions between the various modes of subsurface water occurrence is the recognition that all water, whether in the atmosphere, on the surface, or beneath the surface of the earth, is part of a common supply. Man, with increasingly effective means to control precipitation, to create surface water storage, and to utilize the natural ground-water reservoirs for storage and development, is rapidly approaching the point where he merely has to decide where and when to establish water supplies.

Ground water is an intriguing part of the hydrologic cycle. Man cannot see water move through the ground. He can dig a hole and peer into it, but he has disturbed or destroyed part of what he wishes to see. Sophisticated electronic gadgetry may be used to 'look' beneath the surface of the earth, but the information gained by indirect measurements is often inexact and always incomplete. Hence, the movement and occurrence of ground water remain somewhat of a mystery to man.

2. Occurrence of ground water

Generally speaking, ground water can be found by drilling at almost any point on the surface of the earth *if* the hole is drilled deep enough. However, the mere presence of water is not usually what man wishes to determine. More important than the presence of ground water are the volume or *supply* of ground water and the *rate* at which the supply can be removed from the ground. Thus, our discussion of ground-water occurrence will focus first on those physical properties of earth materials which affect the amount of water which can be stored beneath the ground and the ease with which the water can be extracted from the ground.

A. Porosity and permeability

The volume of water which can be held within earth materials is controlled by the *porosity* of the materials. Porosity may be defined as follows:

$$\text{Porosity} = \frac{\text{Volume of voids in a material}}{\text{Bulk volume of the material}}$$

Porosity is usually expressed as a decimal fraction or as a percentage. For example, a rock specimen which contains pores or open spaces equal to one-fourth the total volume of the specimen would have a porosity of 25%. Naturally occurring geologic materials vary widely in porosity. Table 6.1.1 con-

TABLE 6.1.1. List of representative porosities and permeabilities for geologic materials

Geologic material	Representative porosities (% void space)	Approximate range in permeability (gallons/day/ft²; hydraulic gradient = 1)
Unconsolidated		
Clay	50–60	0·00001–0·001
Silt and glacial till	20–40	0·001–10
Alluvial sands	30–40	10–10,000
Alluvial gravels	25–35	10,000–1,000,000
Indurated		
Sedimentary:		
Shale	5–15	0·0000001–0·0001
Siltstone	5–20	0·00001–0·100
Sandstone	5–25	0·001–100
Conglomerate	5–25	0·001–100
Limestone	0·1–10	0·0001–10
Igneous and metamorphic:		
Volcanic (basalt)	0·001–50	0·0001–1
Granite (weathered)	0·001–10	0·00001–0·01
Granite (fresh)	0·0001–1	0·0000001–0·00001
Slate	0·001–1	0·0000001–0·0001
Schist	0·001–1	0·000001–0·001
Gneiss	0·0001–1	0·0000001–0·0001
Tuff	10–80	0·00001–1

tains a list of representative porosities for various geologic materials. In the case of granular sediments, porosity is not directly affected by the *size* of the grains, but is affected by the uniformity of size, the shape, and the packing characteristics of the grains.

The ease with which water can move through earth materials is a function of the *permeability* of the materials. A more exact definition of permeability is given later in this chapter under the topic 'Movement of ground water'. The permeability of granular earth materials is greatly affected by the size of grains as well as by the shape, packing, and uniformity of size of grains. Permeability can be expressed in a number of different ways. For the purpose of this discussion, permeability will be described as a rate of discharge per unit area (e.g. gallons/day/ft^2) under controlled hydraulic conditions. Table 6.1.1. gives approximate ranges of permeability for various geologic materials. It is important to recognize the magnitude of the range of permeabilities for naturally occurring earth materials.

B. Aquifers

A geologic material which yields significant amounts of water to wells is called an *aquifer*. From a practical point of view, an aquifer must yield sufficient quantities of water to make it economically feasible to extract water from it. Obviously, if economics enters into the definition of an aquifer, then the intended use for the water will somewhat determine whether a water-bearing formation can be called an aquifer or not. Hence, a fractured granite may yield enough water for household uses, and would qualify as an aquifer. The same rock would probably not supply sufficient water for agricultural purposes, and in this case would not be called an aquifer.

C. The geologic framework: categories of earth materials

The study of ground water, whether for the purpose of determining its value to man or to see how it interacts with the earth through which it passes, must begin with a study of the geologic framework. Earth materials are usually classified geologically as to origin, i.e., igneous, metamorphic, or sedimentary. This classification, however, is not suited for studies of ground water. In terms of their effect on the occurrence of ground water, two broad categories of earth material are more suitable: a category which includes all unconsolidated materials, and a category which includes all indurated materials.

1. Unconsolidated materials

By definition, unconsolidated deposits do not contain cementing materials in their pore spaces. Thus, these deposits are characterized by relatively high porosities (Table 6.1.1). Included in this category of earth materials are the geologically recent deposits of alluvial or stream-transported sediments, aeolian or wind-transported particles, colluvial or gravity-driven debris, and glacial or ice-transported materials.

An analysis of recent geologic history and the origin of the various types of unconsolidated deposits in a region can be a major contribution to an evaluation of the ground-water resources. The significance of geologic studies in ground-water evaluations lies in the fact that there are predictable relations between geologic processes and the physical properties of the sediments they produce. For example, aeolian deposits are characterized by uniformity of grains in the silt and sand size range. Glacial moraines, in contrast, are poorly sorted and may contain mixtures of particles ranging from clay size to boulder size. Also, knowledge of how a material was deposited can be used to predict the overall geometry of the deposit. An example of this can be seen in the case of wind-blown sand. Because of the formation of dunes, aeolian deposits are extremely variable in thickness. Thus, if two wells, spaced 150 ft apart, were drilled in an area known to contain a buried aeolian deposit of saturated sand the first hole might penetrate 50 ft of the water-bearing sand, while the second hole might miss the deposit entirely.

2. Indurated earth materials

Indurated earth materials (rocks) also play a significant role in the occurrence and movement of ground water. Sedimentary deposits usually become indurated through cementation of the grains by a chemical precipitate of iron oxides, calcium carbonate, or some form of silica. Clay particles which might be present in a deposit also can form a cementing matrix when the deposit begins to consolidate. These cementing agents may fill essentially all of the original void spaces within the deposit. Thus, sedimentary rocks are usually much less permeable and porous than their unconsolidated counterparts. Limestones, however, provide important exceptions to this rule. Fractures in limestone may widen due to solution by ground water, making the rock highly permeable.

Igneous and metamorphic rocks are generally characterized by low permeabilities and extremely small porosities (Table 6.1.1). In most igneous and metamorphic rocks fractures are the primary source of void spaces which can contain and transmit water. Some types of volcanic rocks, however, develop high porosity due to entrapment of gas bubbles within the rock during cooling. Also, lava flows may have relatively high permeabilities because of extensive fracture systems developed during movement and cooling of the lava flow. Examples of fractured volcanic rocks which yield great quantities of ground water may be found in the Hawaiian Islands and areas in the north-western United States. Although fractures may contribute significantly to the permeability of rocks near the surface of the earth, at depths greater than 200 or 300 ft the fractures generally are compressed and will not yield appreciable quantities of water to a well.

Porosity and permeability of indurated earth materials may also be affected by chemical and mechanical weathering processes, but generally not at depths greater than 50–100 ft. The degree and extent of weathering are controlled by climate, topography, time, and the chemical composition of the rock. In general, igneous and metamorphic rocks are more susceptible to chemical decomposition than are sedimentary rocks.

D. Stratigraphy, geologic structure, and topography

Stratigraphy is the branch of geology which deals with the formation, composition, sequence, and correlation of stratified earth materials. In ground-water studies it is important to know the mode of formation of a deposit and the nature of the grains which compose it, because these factors control the porosity and permeability of the deposit. The sequence of stratified deposits in natural environments of deposition is such that sediments of varied physical characteristics are often deposited in distinct layers. For example, sedimentary layers of sand and gravel commonly are found alternating with layers of silt and clay. Since the permeability of the coarse-grained deposits may be many thousand times greater than that of the fine-grained deposits, most of the water which moves through the ground is transmitted through the segregated coarse-grained layers. Finally, and probably the most significant aspect of stratigraphy relative to ground-water studies, is the use of stratigraphic correlation techniques to locate water-bearing strata. Suppose, for example, that the drilling log of a productive water well shows that drilling began in limestone, progressed through 100 ft of shale, and

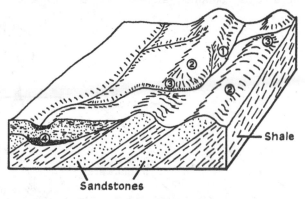

Shale

Sandstones

Fig. 6.1.1 Geologic factors control the occurrence of water.

Alternating strata of sandstone and shale have been tilted and erosion has cut a stream valley (1) into an exposed shale stratum. The sandstone strata, being more resistant to erosion, form the ridges (2) which parallel the valley. Where the sandstone ridges have been breached by the stream (3), the relatively permeable sandstones may receive water from the stream. The buried stream channel (4) contains gravels and other unconsolidated sediments which were deposited by a stream at an earlier time.

then water was obtained in a sandstone layer immediately beneath the shale. Several miles away, another well is to be drilled in the vicinity of a rock outcrop which exposes the contact between the limestone and the shale. The concept of stratigraphic correlation would indicate that the well would have to be drilled through approximately 100 ft of shale before the sandstone aquifer could be tapped. Thus, stratigraphy is important in ground-water studies because it helps to define the nature, location, and extent of aquifers.

Aquifers and other geologic formations may become folded or broken (faulted) because of the stresses which develop within the crust of the earth. These deformations are referred to as the geologic structure. The structure of geologic formations influences ground-water occurrence and movement because the folding and faulting of strata control the localization of areas where ground water may enter and leave each stratum. Figure 6.1.1 illustrates how different ground-water systems could result from various combinations of geologic structure and stratigraphy.

The influence of topography on the occurrence and movement of ground water can also be seen in fig. 6.1.1. Land-surface topography affects the localization of ground-water movement into and out of the ground and also controls the nature and location of surface hydrologic features, such as lakes and streams. Note in fig. 6.1.1. how surface topography is controlled by geologic structure and stratigraphy.

E. Confined (artesian) and unconfined ground water

A discussion of the effects of stratigraphy, geologic structure, and topography on ground water would not be complete without some mention of the two modes of ground-water occurrence: confined (artesian) and unconfined.

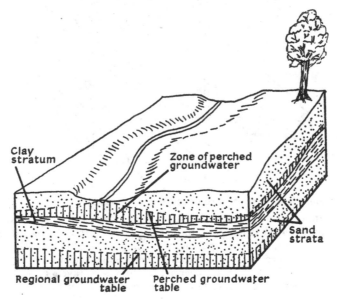

Fig. 6.1.2 Perched ground water.
The illustration shows how a zone of ground water may develop apart from the regional body of ground water. The lateral extent of a body of perched ground water is determined by the extent of the underlying impermeable stratum. The perched water in the illustration is recharged by percolation of stream water from the surface. Recharge may also occur directly from percolation of precipitation. The lower (regional) body of ground water is not recharged by percolating surface waters where it is overlain by an impermeable stratum.

Ground water which is not overlain by relatively impermeable materials is considered to be unconfined. The upper surface of a zone of unconfined ground water is called the *water-table*. By definition, the pressure of the water at every point on the water-table is equal to atmospheric pressure. In many places a zone of unconfined water may exist near the surface of the ground where downward movement of percolating water is impeded by an underlying impermeable stratum. A body of water which is isolated in this way from other ground water is termed *perched* ground water (fig. 6.1.2).

Artesian water is that which is confined beneath a relatively impermeable stratum such that, if a well penetrates the confined zone, water will rise into the well to an elevation above that of the confined zone (fig. 6.1.3). If the artesian pressure in a confined aquifer is great enough water may rise in the well to an elevation above the land surface. This phenomenon is called a flowing artesian well.

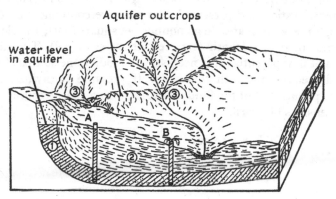

Fig. 6.1.3 An artesian ground-water system.

The basic elements of the artesian system are an aquifer (1), an overlying impermeable stratum (2), and a recharge area (3) for the aquifer. The well at point A taps the artesian aquifer, and water has risen in the well to a point near the ground surface. The well at point B also taps the aquifer, but in this case the water rises within the well to a point above the ground surface. Both wells are artesian, but the well at point B is a *flowing* artesian well. The dashed line which runs through well A and passes above well B represents the height to which water will rise in a well which penetrates the aquifer. The decrease in elevation of the dashed line away from the recharge area indicates that water loses energy as it moves from a recharge area towards a discharge area.

3. Movement of ground water

Ground water is always in motion. Movement is from a recharge area (usually where precipitation from the surface has percolated through the ground) to a discharge area (where ground water emerges from the ground in the form of a spring, seep, or discharge from a well). Because infiltration of precipitated moisture is the principal source of ground water, it is important to recognize that infiltration and percolation vary from point to point on the earth's surface. This variation is produced by the geologic and topographic factors which have been discussed in the preceding sections of this chapter.

The path of ground-water flow and the rate at which ground water moves are determined not only by the geologic conditions already discussed but also by hydraulic conditions. All things considered, water follows the path of least resistance. Movement of water from one point to another is caused by a difference in flow potential or 'head' between the points. In ground water, head usually consists of two components: a *pressure-head* component and an *elevation-head* component. We are familiar with elevation head because it provides most of the flow potential in surface-water streams. Water in stream channels invariably responds to gravity and flows from higher elevations to lower elevations. Pressure head is also no stranger. Everyone has witnessed the flow of a fluid from a point of high pressure to a point of low pressure: when air is released from a balloon or when tooth paste is squeezed from its tube.

Total flow potential or 'head' for ground water at any point in the earth is the sum of the pressure head and the elevation head at that point. How does one go about adding elevation and pressure? The pressure component of total head is measured in force per unit area (e.g., pounds per square foot); the elevation component is measured as a length (e.g., feet above sea-level). To add these two components of flow potential, it is necessary to modify one so that it will become dimensionally equivalent with the other. If pressure is divided by the unit weight (e.g., pounds/ft³) of the fluid which is causing the pressure, the quotient has the dimensions of length.

$$P = \text{Pressure} = \frac{\text{Force}}{\text{Area}}$$

$$W = \text{Unit weight} = \frac{\text{Force (weight)}}{\text{Volume}}$$

$$\frac{P}{W} = \frac{\dfrac{\text{Force}}{\text{Area}}}{\dfrac{\text{Force}}{\text{Volume}}} = \frac{\text{Volume}}{\text{Area}} = \text{Length}$$

Therefore, if we let the elevation head be represented by the letter Z the flow potential or total head (H) at a point is given by

$$H = \frac{P}{W} + Z$$

Thus, total head is expressed as a length. This length has physical significance because it represents the height of a column of water required to produce a pressure at its base equal to the total flow potential.

If water is not moving the total head is a constant at all points in the water. For example, the flow potential at the bottom of a lake is equal to the flow potential at the surface of the lake.

Moving water loses flow potential as it moves. The stored or potential energy is transformed into heat energy because of frictional resistance to flow. Head loss in moving water is a function of the rate of flow and of the resistance to flow. In the case of ground-water flow, resistance to flow is usually represented by the coefficient of permeability, K, as given in

$$Q = KA\frac{H}{L}$$

where Q is discharge rate (gallons per minute), K is the coefficient of permeability, A is the cross-sectional area of flow, and H/L is the unit loss in total head due to flow between two points a distance L apart. This equation, basic to ground-water flow, is known as Darcy's Law. Henri Darcy was a French civil engineer who in 1856 in Dijon, France, reported his experiments on the relationship between head loss and discharge of water as it passed through the sand filters utilized in the water system of the city of Dijon. Darcy's Law states that discharge is directly proportional to head loss.

The discharge as determined by Darcy's Law is related to the actual velocity of ground-water movement. A rearrangement of the terms in Darcy's Law gives

$$\frac{Q}{A} = K\frac{H}{L} = V$$

where V is the discharge per unit area and is called a *volume flux*. Because V has the dimensions of velocity, it is commonly confused with the velocity of ground-water movement. Since V is actually a measure of discharge per unit area, the flow velocity is equal to V only in the case where the cross-sectional area A is completely open to flow. In earth materials this area includes the cross-sectional area of the mineral grains as well as the cross-sectional area of the spaces between grains. Of course, the fluid passes only through the open spaces between grains. The open spaces (porosity) usually constitute from about $\frac{1}{4}$ to $\frac{1}{2}$ of the total volume (or cross-sectional area) of a granular porous material. Thus, for unconsolidated sediments the velocity of flow is approximately two to four times the volume flux. If we represent the porosity of a sediment by the letter p velocity of ground-water movement through the sediment is given by

$$\text{Velocity} = v = \frac{KH}{pL}$$

Ground-water flow velocities in nature may vary in the extreme from several feet per second to less than a foot per year. The normal rate of flow of ground water is probably between 5 ft/yr and 5 ft/day (Todd, 1959). Thus, compared to flow rates in surface streams, motion of ground water is extremely slow.

REFERENCES

DAVIS, S. N. and DEWIEST, R. J. M. [1966], *Hydrogeology;* (John Wiley and Sons, Inc., New York), 463 p.

HEATH, R. C. and TRAINER, F. W. [1968], *Introduction to Groundwater Hydrology;* (John Wiley and Sons, Inc., New York), 284 p.

JOHNSON, A. I. [1967], Groundwater; *Transactions of the American Geophysical Union,* **48** (2), 711–24.

JOHNSON, E. E. [1966], *Groundwater and Wells;* (Edward E. Johnson, Inc., Saint Paul, Minnesota), 440 p.

MEINZER, O. E. [1923], The occurrence of groundwater in the United States, with a discussion of principles; *U.S. Geological Survey Water Supply Paper* 489, 321 p.

TODD, D. K. [1959], *Groundwater Hydrology;* (John Wiley and Sons, Inc., New York), 336 p.

II. The Geomorphic Effects of Ground Water

PAUL W. WILLIAMS

Department of Biogeography and Geomorphology, The Australian National University

Ground water will be considered here as all water of meteoric origin within the soil and underlying rocks. Thus defined its distribution is widespread, but since its geomorphic effects operate primarily through solution, its influence is most striking in carbonate terrains. This brief account of the role of ground water in modifying the relief will therefore focus attention on limestone and dolomite landscapes where solution manifests itself through *karst* landforms.

The karstic process of solution and concomitant rock settling and collapse is largely directed by geological constraints acting on the passage of subsurface water. A brief examination of the principles of ground-water movement through limestones is therefore essential for proper understanding of the resultant topography.

1. Ground water in limestones

The characteristic dryness of limestone areas is a consequence of the highly permeable nature of the rock. This permeability is dependent upon the volume and interconnection of both primary and secondary interstices in the rock, and while the most permeable limestones possess both primary and secondary porosity, the latter is usually more significant for fluid circulation because of the larger voids concerned.

The susceptibility of limestone to solution is fundamental in the enhancement of its permeability, for the stimulation of solution provided by the very act of water circulation is instrumental in the progressive enlargement and modification of the subterranean conduit system. The capacity of carbonates for this progressive development of permeability distinguishes them hydrologically from most other rocks, and in large measure accounts for the distinctive quality of their topography – above and below ground. However, it is clear that water conditions in limestones vary considerably from place to place as lithology, structure, and geomorphic history change. So it is impossible to formulate a neat, all-embracing theory of karst hydrology that may be used to assist in the explanation of surface landforms. Current opinions on the nature of water conditions in limestones waver basically between the ideas of Martel [1894] and Grund [1903], namely between discrete conduit and interlinked network approaches. Anglo-American hydrologists have always favoured the latter, since King's [1899] theory of ground-water motion and Davis's [1930], Swinnerton's

K

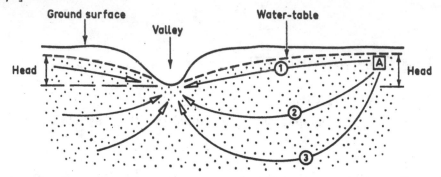

Water at A will divide itself into proportions directly related to the ease of movement. In the three directions illustrated, the largest amount will take the shortest and least resistant route, path 1, and the smallest amount path 3.

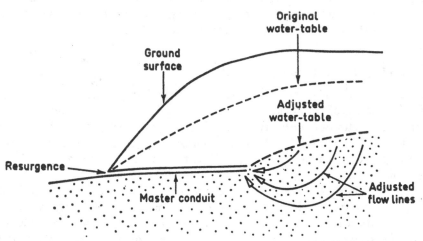

Fig. 6.11.1 *Above:* Water movement beneath the water-table in limestones, after Swinnerton (1932).
Below: Water conduit development and water-table adjustment in limestones, after Rhoades and Sinacori (1941).

[1932], and Rhoades' and Sinacori's [1941] adaptation of it for limestone terrains have emphasized the concept of a unified body of ground water, with the upper limit of the permanently saturated zone being delimited by a hypothetical surface, the water-table (fig. 6.11.1). However, in continental Europe conditions leading to the existence of simple water-table situations have never been fully admitted, for study of the hydrology of the limestone Alps has constantly revealed intricate ground-water systems in which the water-table concept has no apparent viability (see Zötl [1965] for a summary of some valuable modern work).

Although hypotheses on the nature of limestone hydrology and corresponding cave genesis are almost as diverse as the terrains which stimulated them (reviewed in Warwick [1962]), there is still much agreement on the general nature of water movement in the aerated or vadose zone, and this is what concerns us most, for it is there that solution has its greatest effect on surface landforms. In the vadose zone water can be divided into that which percolates through the soil and fissures in the underlying rock, and that which passes rapidly underground via swallow holes and caves. The two categories will be referred to as percolation water and vadose stream water respectively. Drainage is basically vertical towards the zone of phreatic water, the third category, where the rock is permanently saturated and where lateral flow becomes more important.

2. Topographic effects of ground-water solution

Underground water is able to affect topography because of its solvent capacity. The majority of investigators agree that the most important solvent in carbonate terrains is carbonic acid, which is produced by the solution of carbon dioxide from the air, and particularly from the soil atmosphere, where the partial pressure of the gas is usually greater, although numerous other natural solvents are known to exist (Keller, 1957).

Ground water in limestones was subdivided above into percolation water, vadose stream water, and phreatic water. Each has a different erosive role and a distinctive geomorphic effect (fig. 6.11.2). Percolation water infiltrates very slowly into the limestone mass, and hence its corrosive capacity or 'aggressivity' is often spent within a few metres of the surface (Gams, 1966; Pitty, 1968; Williams, 1968). Subterranean stream water, on the other hand, moves rapidly underground, and its solvent capacity may not be satiated until a depth of several hundred metres or more is reached, and in addition to their corrosive role, vadose streams also erode mechanically like their surface counterparts. The activity of phreatic water is difficult to ascertain because of inaccessibility. Swinnerton [1932] and Rhoades and Sinacori [1941] (fig. 6.11.1) postulated maximum solution and cave formation at the level of the water-table where ground-water flow is concentrated in its greatest volume. Their ideas were strongly supported by Kaye [1957], who experimentally showed solution to be proportional to the velocity of solvent motion. Theoretical support for deep subterranean solution has been provided by Bögli [1964], who has indicated how mixing waters with different carbonate concentrations are imparted renewed capacity to dissolve even if the original solutions were saturated.

The morphological effects of all styles of ground-water solution tend in the same direction, namely ultimate overall lowering of the relief, and numerous estimates of this have been made in different countries, the majority falling in the range 0·02–0·1 mm/yr. Nevertheless, the short-term effects of the above three categories of ground water differ. Percolation solution directly lowers the surface because it operates most strongly in the uppermost few metres, but its intensity varies according to geological factors, such as fracturing, which concentrates percolation routes. Zones of high-frequency jointing are therefore especially

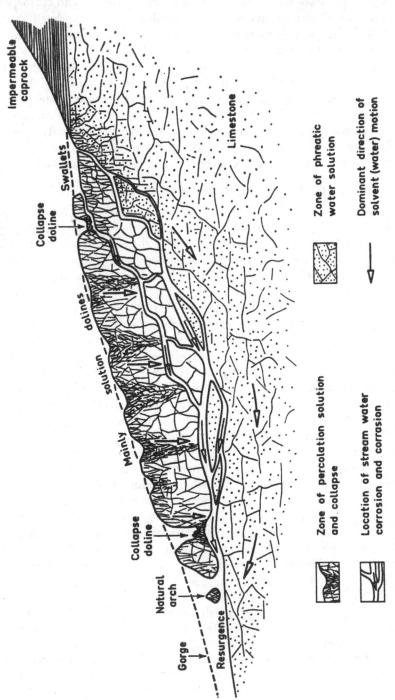

Impermeable caprock

Limestone

Swallets

Collapse doline

dolines

Mainly solution

Collapse doline

Natural arch

Gorge

Resurgence

Zone of phreatic water solution

Dominant direction of solvent (water) motion

Zone of percolation solution and collapse

Location of stream water corrosion and corrasion

Fig. 6.11.2 The distribution of solution through a limestone mass.

prone to attack, and in such locations solution depressions develop (fig. 6.11.3). However, the considerable variation in the density and form of these features in tropical, temperate, and sub-arctic zones suggests that the overriding consideration in their distribution and morphology is climatic. Percolating water passing through thick drift (especially glacial and alluvial) overlying limestone is also frequently responsible for myriads of small enclosed depressions entirely within the deposit: *drift dolines*[1] (fig. 6.11.3). The percolating water erodes the drift partly by selective solution of its contents and partly by downwashing of fines

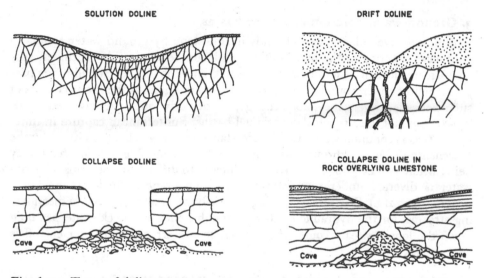

Fig. 6.11.3 Types of doline.

into voids in the underlying rock. This important but frequently overlooked process is known as chemical and mechanical *suffosion*.

It is not yet possible to compare quantitatively the erosive effects of percolation and vadose stream waters, but qualitative differences are immediately apparent. The most obvious contrast is in the distribution of denudation, which in the first case is relatively uniform over the limestone outcrop, but in the second is restricted to the trace of the stream. Vadose stream action, be it corrosion, corrasion, or the pressure effects of flood-water, results in the undermining of limestone outcrops and in the production of steep-sided collapse-induced depressions on the surface that contrast markedly with the rounded forms of percolation water solution depressions (fig. 6.11.3). Collapse depressions frequently have elongate plans (as do solution depressions if there is a particularly dominant joint direction) which may cut across the structure where subterranean streams are followed. But depressions are only produced by collapse above a stream where the cave roof is thin enough to permit the transmission to the surface of roof falls. The critical roof thickness is a function of the span of the

[1] The author now prefers to use the term *subsidence dolines*.

passage, the magnitude of discharge fluctuations in the stream, the mechanical competence of the overlying beds, and the degree to which they are corrosionally weakened. As a rough guide, a cave roof thicker than about 100 m is usually adequate to prevent sporadic collapse effects reaching the surface.

Although phreatic solution is difficult to measure, it has been shown to be important in the early stages of development of many caves (Bretz, 1942). Yet while the morphological effects of phreatic solution are widespread underground, there are few surface manifestations. The main geomorphic effect of deep ground-water solution is to reduce the overall volume of the limestone mass.

3. Ground-water solution and river basins

The progressive enlargement of voids in limestone by ground water results in automatic lowering of the hydrostatic equilibrium level. As this continues, surface streams lose more and more water into their beds until eventually they pass permanently underground. The morphological functions of the lower courses of such beheaded rivers inevitably change, particularly since the engulfed headwaters may not later rejoin their original basins. Subterranean capture in limestone areas is common where rapid water-table lowering is being induced by the entrenching of major through streams; the most favourable position for piracy being where two neighbouring rivers are incised to different levels. In such cases water is diverted underground through the topographic divide to the lower system (fig. 6.11.4). Self-piracy, leaving dry ox-bows, may also occur in meandering streams, for the hydraulic gradient through a meander neck is steeper than down the longer stream course (fig. 6.11.4); so subterranean meander cut-off is encouraged.

4. Surface features associated with subterranean streams

A variety of landforms develop from a combination of surface- and groundwater processes at the points where streams pass underground. The simplest and most widespread feature is the *swallow hole* (stream-sink); this is a small enclosed basin into which a stream flows and passes underground. At the site of ingress the water of a sinking stream is normally aggressive; thus morphological evolution is rapid. Corrosion acts directly and indirectly by widening fissures and inducing collapse. As swallets enlarge, they develop into elongate closed basins with steep downstream ends, more properly known as *blind valleys*. Processes operating in them are similar to those in swallets, and Gams suggested that in northern Yugoslavia their length and width is inversely related to the hardness of inflowing streams. Considerable further enlargement under conditions favourable to flooding, and hence lateral solution plantation, will develop a blind valley into a *karst margin polje* (marginal polje), although such landforms are attributable more to surface corrosion than ground-water action.

The cave roof downstream of a swallow hole is usually relatively thin, and since the zone is subject to intense chemical weathering and violent pressure changes with fluctuating stream discharge, collapse depressions frequently occur even where the cave passes beneath overlying, non-limestone formations (fig.

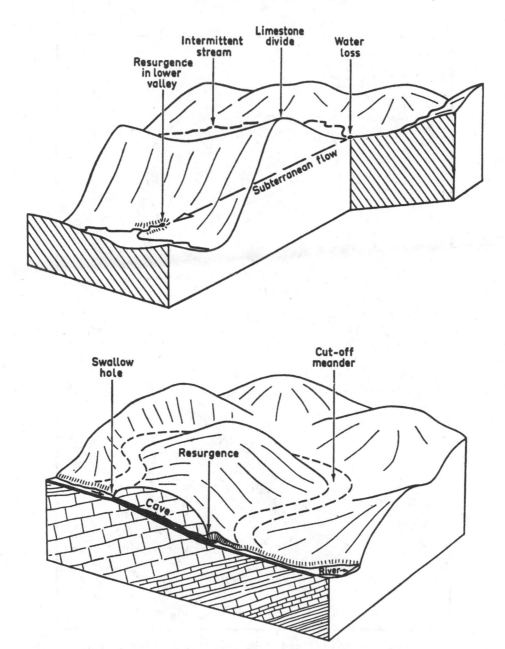

Fig. 6.11.4 Examples of (*above*) subterranean river capture, and (*below*) subterranean meander cut-off.

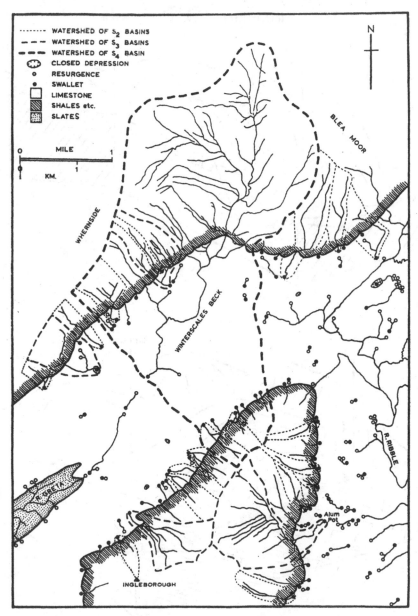

Fig. 6.11.5 Some karst features of the Ingleborough district, Yorkshire, England (From Williams, 1966).

6.11.3). As the stream passes farther and more deeply underground, associated collapse forms on the surface are less frequent, although in shallow systems, the river may be traced for considerable distances. However, subterranean streams do not always maintain discrete courses, especially where they flow into a phreatic network; so surface collapses cannot always be related to simple cavern plans.

At resurgences the morphological situation is in some respects similar to that at swallow holes, but in reverse. An increasing frequency of collapse depressions often heralds the emergence of an underground stream. The point of resurgence may be a *steep-head*, marked by steep or even overhanging slopes which retreat vigorously upstream by undermining and collapse. Downstream of the steep-head, the river commonly flows in a gorge, and cave-roof remnants may form natural arches. Gorges are frequently found where ground water resurges, although in karst areas they are not always produced by cavern collapse.

5. The quantitative analysis of karst landforms

The morphometric approach to the accurate description of individual karst landforms and of their interrelationships (Williams, 1966; LaValle, 1967), although in its infancy, is already casting doubt on the traditional notion that karst landscapes are chaotic and disorganized. Morphometry, as developed from map and air-photograph analysis, is able to describe only the more gross properties of landforms, whereas the more subtle attributes can be observed and measured only in the field. However, even there landform identification is by no means always simple, and the majority of dolines, for example, cannot safely be assigned to a solution or collapse category without examination by excavation. The special nature of karst, where solution and collapse combine in varying proportions to produce a continuum of landforms, renders accurate genetic identification extremely difficult. The morphometric approach must therefore abandon the majority of field-worker's techniques and concentrate on measurable properties of shape, size, and distribution.

In any estimation of the geomorphic effects and morphological role of ground water it is thus advisable first to try to gain some idea of the general nature of the ground-water system itself, as follows (it being assumed that the limestone mass is pure throughout):

1. From a detailed geologic map of the district under examination, delimit the limestone outcrop, measure its area (A_l) (12 square miles)[1] (fig. 6.11.5), and give its thickness (H_l) (750 ft).
2. Locate swallow holes (S) and order them (S_o) (fig. 6.11.6) according to the Strahler order of the stream flowing into them ($\Sigma S_1 = 31$; $\Sigma S_2 = 19$; $\Sigma S_3 = 4$; $\Sigma S_4 = 1$; plus 37 which cannot be ordered). Note their altitudes (H_{so}) ($\bar{H}_s = 1,127$ ft) and distance to their neighbours having the same order (L_{so}). ($L_{s1} = 0.22$ miles; $L_{s2} = 0.43$ miles; $L_{s3} = 1.66$ miles).

[1] The values given for the morphometric parameters relate to 1 : 25,000 maps of the Ingleborough district shown in fig. 6.11.5.

3. Locate every karst spring (rising: K_r) ($\Sigma K_r = 81$) within the limestone outcrop, its altitude (H_r) ($\bar{H}_r = 1,009$ ft) and distance to its nearest neighbour (L_r).

4. Mark the course and calculate the length ($I_{\prime l}$) of every through-flowing stream.

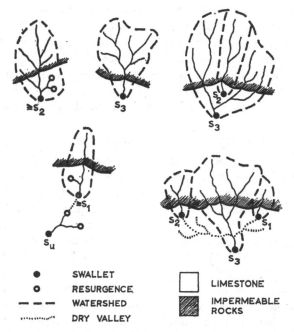

●	SWALLET	▢ LIMESTONE	
○	RESURGENCE	▨ IMPERMEABLE	
– – –	WATERSHED	ROCKS	
··········	DRY VALLEY		

Fig. 6.11.6 Procedure for ordering Swallets and delimiting their catchments (From Williams, 1966).

Given this information, the following attributes of the karst drainage system can be estimated – bearing in mind that the larger the area, the more generalized the information derived:

1. Swallet density $\left(D_s = \dfrac{\Sigma S}{A_l}\right)$ (7·67 per square mile); a measure of the number of streams draining into the limestone per unit area.

2. Rising density $\left(D_r = \dfrac{\Sigma K_r}{A_l}\right)$ (6·57 per square mile); a statement of the number of risings per unit area.

3. The swallet/rising ratio $\left(R_{sr} = \dfrac{\Sigma S}{\Sigma K_r}\right)$ (1·12); indicating the amount and direction of stream branching underground. Purely vadose systems are expected to have R_{sr} values of >1, while areas with important phreatic ground-water bodies will have lower values which could descend to <1.

4. Mean shortest distance of underground flow (L_u) (0·36 miles); the mean straight-line distance between each swallet and its nearest downhill resurgence or river channel.
5. The *vadose index* ($V_i = \bar{H}_s - \bar{H}_r$) (118 ft); a measure of the depth of the zone of aeration (fig. 6.11.7). This is significant only for small areas, for the hydrostatic equilibrium level (which should determine $\bar{H}_r$) varies according to subterranean evolution, structure, and lithology.

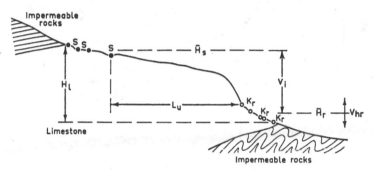

Fig. 6.11.7 The relationships of some karst parameters.

6. The *rising coefficient* $\left(V_{hr} = \dfrac{\sigma H_r}{\bar{H}_r} \times 100 \right)$ (12·5); where σ is the standard deviation estimated from the sample. This defines the variation in altitude of springs; so in areas of pure limestone it is a direct measure of the uniformity of the water-table surface. It is thus an indirect measure of the state of evolution of the ground-water network.
7. *Stream density on limestone* $\left(D_l = \dfrac{\Sigma L_l}{A_l} \right)$ (2·4 miles per square mile); is a statement of the length of permanent streams per unit area, and is thus an indirect measure of permeability in an area of uniform climate.
8. Semi-logarithmic plots of order against mean number of swallets, mean distance to swallets of the same order (L_{so}), and mean area of swallet basins ($\bar{A}_{so}$) summarize other attributes of swallet systems (fig. 6.11.8).

Where additional information is available from hydrologic surveys, bore-hole data, etc., it may be possible to locate phreatic water divides and so delimit ground-water drainage systems. Further characteristics of these systems may be estimated using an index of *relative karst system relief* (H_k) (this for any locality is the difference in height between the basin's lowest resurgence point and the highest place in the area considered) and a *karst relief ratio* (R_k) (this for a given site is the ratio of karst system relief to the distance of the site from the nearest phreatic divide) (LaValle, 1967). The relative karst system relief is related to the vadose index, since it measures the maximum depth of the zone of aeration in any particular area. The karst relief ratio was designed to estimate hydraulic gradient, but its effectiveness is questionable.

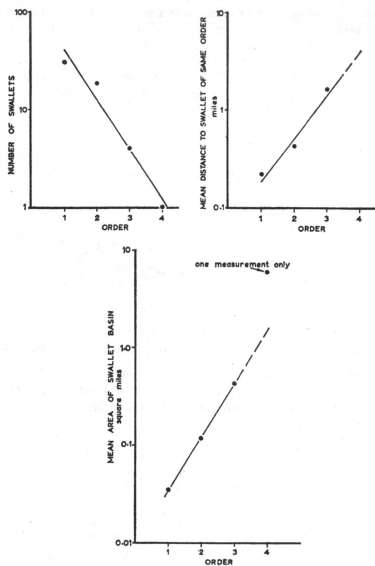

Fig. 6.11.8 Swallet relationships in the Ingleborough districts, Yorkshire, England (From Williams, 1966).

6. Closed depressions

Further analysis should also be made of surface conditions, which in a karst area implies particular attention to closed depressions (swallow holes, solution do-lines, collapse dolines, etc.), including any enclosed hollow from the smallest pit to the largest karst polje. But since ground-water solution plays a subservient role to lateral surface-water solution in the origin of poljes, our concern here is only for the smaller features.

A temperate-zone doline is usually a relatively simple basin set in a gently rolling limestone surface, and, even in alpine areas, it is delimited on a map by a break of slope at its rim, indicated by a scar symbol or closed contour. However, many tropical depressions, with the exception of swallow holes, are different in form and setting. Whereas temperate depressions are often circular or ovoid in plan, the contours of their tropical counterparts are frequently star-shaped. The tropical basins are also interspersed between steep-sided, residual hills from which storm runoff drains into the intervening hollows. The hillsides are thus integral parts of the basin. While temperate depressions are adequately defined by their bounding break of slope, tropical depressions of the cockpit type seem best defined by their topographic divides (fig. 6.11.9); both methods in fact demarcate the immediate catchment areas of the closed basins concerned.

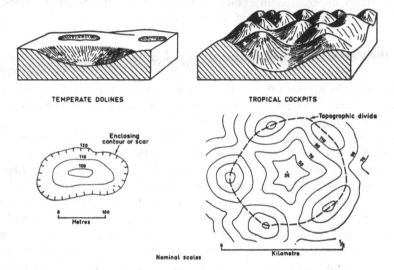

TEMPERATE DOLINES TROPICAL COCKPITS

Fig. 6.11.9 Delimitation of temperate and tropical closed depressions.

Having established the immediate catchment area as the most realistic criterion for delimitation of karst depressions, the following analytical procedure may be followed:

1. Locate and count all closed depressions (C_d) (95); measure their areas (A_{cd}); maximum relief (difference between lowest point in basin and highest point on rim) (H_{cd}); length/width ratios (R_{de}); and long axis orientations in degrees of azimuth relative to the direction of general land slope (θ_s), also where appropriate, with respect to the geological strike direction (θ_l), major joint trend (θ_j), and maximum hydraulic gradient (θ_h).

2. Divide the area under examination into convenient sized units (e.g. 1 km²) and calculate for each unit

   (a) Closed depression density $\left(D_{cd} = \dfrac{\Sigma Cd}{A_l} \right)$ (7·8 per square mile); a simple statement of depressions per unit area.

(b) Mean area of depressions $\left(\bar{A}_{cd} = \dfrac{\Sigma A_{cd}}{N} \right)$ (approximately 0·0005 square miles).

(c) Mean depression relief $\left(\bar{H}_{cd} = \dfrac{\Sigma H_{cd}}{N} \right)$

(d) Index of pitting $\left(R_p = \dfrac{A_l}{\Sigma A_{cd}} \right)$ (2,500); indicating the extent to which the surface is dissected with depressions – unity representing complete pitting.

(e) Mean elongation ratio of depressions $\left(\bar{R}_{de} = \dfrac{\Sigma R_{de}}{N} \right)$.

(f) Mean depression orientation relative to generalized land slope, geological features, and hydraulic gradient.

From the above calculations, spatial variations will become apparent which require explanation with particular attention to ground-water influence. At this stage it should be stressed that here we pass largely into the realm of hypothesis, since as yet there has been little quantitative substantiation of the possible relations that are suggested.

Closed depression density over an area with relatively uniform climate, limestone lithology, structure, and slope should show little variation, because percolation solution should be fairly even. But should above-average depression densities occur, they may be interpretable as due to a greater incidence of collapse, especially if the higher densities are found round the borders of the limestone outcrop near swallets and resurgences. The possible effect of ground-water-induced collapse may therefore be tested by exploring the relation between closed depression density per unit area and mean depth to water-table per corresponding unit area. The level of the water-table (where it exists) should be significant, as it is there that cave passage formation, with concomitant roof collapse, is sometimes thought to be most rapid. A negative correlation should be expected where collapse-induced depressions are important. If depth to the water-table is unknown it can be approximated as varying with distance from the phreatic divide, resurgences, or limestone edge. Where lithology is not uniform, the influence of other variables, such as rock purity and jointing, may be more important determinants of depression density. In order to assess the relative significance of a range of variables on closed depression density, a multiple correlation programme will be required.

Closed depression shape and orientation may also reflect phreatic and vadose stream influence, but such control may only stand out if elongate depressions transgress the structure. Those accordant with structure could result from any combination of solution and collapse, and only field examination is likely to determine their evolution. The proportions of aligned and non-aligned depressions with respect to various aspects of structure should therefore give a minimal estimate of the importance of collapse, but more accurate would be depression orientation in relation to the direction of maximum hydraulic gradient.

Judging from LaValle's work in Kentucky, phreatic water control seems also reflected in the inverse relations of both depression elongation and percentage of structurally aligned depressions with distance from resurgence mouths. An explanation of this apparent causal relation might be that the magnitude of cave passages and corresponding roof falls decrease upstream while roof thickness increases (until, of course, the swallow holes are reached); thus the chances of collapse effects modifying surface landforms diminish upstream until the inflow zone is approached. It has also been suggested elsewhere that the hydrostatic equilibrium level has a direct effect on tropical karst relief, but this relation, like so many others in karst morphometry, is statistically unproven.

7. Conclusions

Karst landscapes are largely the result of solution on the surface and underground, plus collapse induced by solution. The dimensional characteristics and spatial distribution of karst landforms are being quantitatively described and a sensible if complex arrangement of features is being revealed; not the utter chaos that was once ascribed to karst. The techniques involved in this analysis are in the early stages of development, and many possible interrelationships remain to be tested. The problems of isolating and quantifying the landform responses to percolation solution, vadose stream action, and phreatic water solution are very great, and indeed attempts to do so are rather premature at this early stage in karst morphometry, when general patterns and the full range of related landforms have yet to be adequately described.

The writer chose to illustrate the geomorphic effects of ground water by concentrating on karst, not only because of the limits of personal experience but also because in the shaping of karst landscapes ground-water processes are more dominant than any other. Nevertheless, the importance of ground water in the evolution of other landform systems must not be under-estimated, particularly in the tropics, where ground-water weathering (Berry and Ruxton, 1959; Ollier, 1960; Thomas, 1966) and ground-water deposition (Woolnough, 1927; Prescott and Pendleton, 1952; Wopfner and Twidale, 1967) are fundamental processes in the evolution of the landscapes.

Acknowledgement. The author is grateful to Mr J. N. Jennings of the Australian National University for valuable criticism of the manuscript.

REFERENCES

BERRY, L. and RUXTON, B. P. [1959], Notes on weathering zones and soils on granite rocks in two tropical regions; *Journal of Soil Science*, **10**, 54–63.

BÖGLI, A. [1964], Mischungskorrosion – ein Beitrag zum Verkarstungsproblem; *Erdkunde*, **18**, 83–92.

BRETZ, J. H. [1942], Vadose and phreatic features of limestone caverns; *Journal of Geology*, **50**, 675–811.

DAVIS, W. M. [1930], Origin of limestone caverns; *Bulletin of the Geological Society of America*, **41**, 475–628.

GAMS, I. [1962], Blind valleys in Solvenia (summary in English); *Geografski Zbornic*, **7**, 265–306.

GAMS, I. [1966], Factors and dynamics of corrosion of the carbonatic rocks in the Dinaric and alpine Karst of Slovenia (Yugoslavia); *Geografski Vestnik*, **38**, 11–68.

GRUND, A. [1903], Die Karsthydrographie: Studien aus Westbosnien; *Geographische Abhandlungen herausgegeben von A. Penck*, **7**, 103–200.

KAYE, C. A. [1957], The effect of solvent motion on limestone solution; *Journal of Geology*, **65**, 35–46.

KELLER, W. D. [1957], *The Principles of Chemical Weathering* (Columbia), 111 p.

KING, F. H. [1899], Movement of ground water; *Nineteenth Annual Report of the United States Geological Survey*, Part 2, 59–294.

MARTEL, E. A. [1894], *Les Abîmes* (Paris), 580 p.

LAVALLE, P. [1967], Some aspects of linear karst depression development in south central Kentucky; *Annals of the Association of American Geographers*, **57**, 49–71.

OLLIER, C. D. [1960], The inselbergs of Uganda; *Zeitschrift für Geomorphologie*, **4**, 43–52.

PITTY, A. F. [1968], The scale and significance of solutional loss from the limestone tract of the southern Pennines; *Proceedings of the Geologists' Association*, **79**, 153–77.

PRESCOTT, J. A. and PENDLETON, R. L. [1952], *Laterite and Lateritic Soils;* Commonwealth Bureau of Soil Science (Rothamsted), Technical Communication No. 47, 51 p.

RHOADES, R. and SINACORI, M. N. [1941], Pattern of ground-water flow and solution; *Journal of Geology*, **49**, 785–94.

SWINNERTON, A. C. [1932], Origin of limestone caverns; *Bulletin of the Geological Society of America*, **43**, 663–93.

THOMAS, M. F. [1966], Some geomorphological implications of deep weathering patterns in crystalline rocks in Nigeria; *Transactions and Papers of the Institute of British Geographers*, **40**, 173–93.

WARWICK, G. T. [1962], The origin of limestone caverns; in Cullingford, C.H.D., Editor, *British Caving*, (2nd Edition), (London), pp. 55–85.

WILLIAMS, P. W. [1966], Morphometric analysis of temperate karst landforms; *Irish Speleology*, **1**, 23–31.

WILLIAMS, P. W. [1968], An evaluation of the rate and distribution of limestone solution and deposition in the river Fergus basin, western Ireland; *Australian National University, Research School of Pacific Studies, Department of Geography*, Publication G/5, 1–40.

WOOLNOUGH, W. G. [1927], The duricrust of Australia; *Journal and Proceedings of the Royal Society of New South Wales*, **61**, 24–53.

WOPFNER, H. and TWIDALE, C. R. [1967], Geomorphological history of the Lake Eyre basin; in Jennings, J. N. and Mabbutt, J. A., Editors, *Landform Studies from Australia and New Guinea* (Canberra), pp. 118–43.

ZÖTL, J. [1965], Tasks and results of karst hydrology; in Štelcl, O., Editor, *Problems of the Speleological Research*, (Prague), pp. 141–5.

III. Human Use of Ground Water [1]

R. L. NACE
U.S. Geological Survey

1. The fountains of the deep

The plains of the Iranian Plateau are dotted at places with thousands of circular earthen mounds surrounding vertical shafts as much as 200 m deep. These are access shafts to nearly horizontal tunnels, called khanats (fig. 6.III.1). The head segment of each tunnel intercepts and cuts below the water-table in permeable

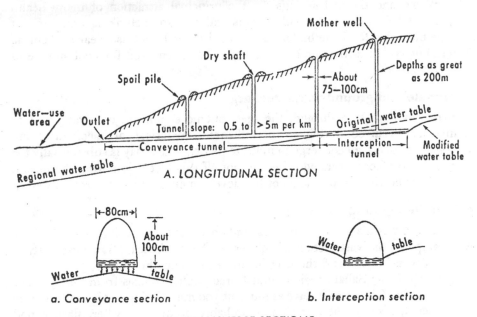

Fig. 6.III.1 Cross-sections of a typical khanat (After Massoumi, 1966).

gravel. Downslope, the grade of the tunnel is less steep than that of either the land surface or the water-table. Where the tunnel reaches the surface, water can be led from it by canal to towns or irrigated fields.

About 25,000 khanats exist in Iran today, the longest being about 70 km. Their origin is lost in antiquity, but as early as 714 B.C., when King Sargon II of

[1] Publication authorized by the Director, U.S. Geological Survey.

Assyria invaded Armenia he found khanats there (De Camp [1963], p. 66). These he destroyed but he took the idea back to Assyria, whence it spread through the East, North Africa, and as far away as China. Much later, Spaniards introduced the khanat to Chilc, and some are still in operation on the Atacama Desert, where they are known as *socavónes* (Dixey [1966], p. 91).

Many other examples are available of effective exploitation of ground water since antiquity in dry areas. The biblical characters who spoke of the fountains of the deep knew whereof they spoke. Wells, ground-water interception and diversion tunnels, enlargement of springs, artificial recharge, all date from antiquity. Only the technology for exploitation has improved.

The uses for ground water are far more varied than merely for domestic supply or irrigation. The spectacular geysers of Yellowstone National Park, which are natural hot-water fountains, have thrilled millions of tourists. Springs that discharge high on the face of a cliff in southern Idaho are led through a penstock to generate hydro-electric power. Icelanders use volcanically warmed water for space heating. In Italy steam from hot ground water drives electrical generators. Mineralized and thermal springs are the principal attraction of many health resorts. Underground pools and rivulets add zest to spelunking and harbour strange biota that fascinate biologists. Some fish hatcheries use clear cool spring water. For centuries the rural spring house has been used for cool storage of perishable food.

2. Climate and ground-water recharge

Climate has fluctuated through the ages, but no significant one-way trend or change has occurred during recent millennia. Historical, archaeological, and geological evidence all converge to indicate that climate today is substantially the same as it was 8,000 years ago. Recognition of this fact is important because it has a direct bearing on problems of the development and use of ground water.

3. Inherited ground water

Some of the world's greatest aquifers contain water that is a legacy of the past. An example is the vast system of aquifers in North Africa, known variously as the Nubian sandstone and the Continental Intercalary Fountain.

In much of the Sahara region annual precipitation ranges from nearly nil to about 250 mm. Few areas receive as much as 500 mm. Owing to high evaporation rates, recharge is extremely small and probably occurs only where flash runoff from intense local storms reaches outcrops of the aquifer. Estimated total storage in the aquifer beneath $6 \cdot 5 \times 10^6$ km^2 of area is of the order of 600×10^3 km^3 (Jones, 1965). Current recharge is relatively negligible, and it is evident that most of the water entered this aquifer during and before the pluvial period at the end of the Pleistocene epoch. Radio-carbon analyses of water samples indicate that some of the water entered the aquifer 30,000–40,000 years ago.

Another example is the Ogallala formation, a sand and gravel aquifer in the High Plains of Texas and adjacent States. South of the Canadian River in Texas and east of the Pecos River in New Mexico, an area of about 90,000 km^2 is isolated from any source of replenishment by underflow. No perennial streams flow through the area and none rise within it. Streams that rise in the eastern area receive but little water by natural discharge from the aquifer. Because of low rainfall, high evaporation, and low permeability of subsoil, estimated recharge from direct infiltration of precipitation is only about 10–15 mm. Some estimates are as low as 4 mm.

The estimated water content of the aquifer in 1938, before the onset of heavy pumping, was about 600 km³, of which about half is theoretically recoverable. Calculations based on underflow velocities and distances through the aquifer indicate that some of the water has been in transit during at least 13,000 years.

Pumpage from the Ogallala formation has reduced storage by nearly 110 km³, and annual withdrawals in recent years have been about 6·2 km³. This is about six to fifteen times the variously estimated natural recharge rates, so the reserve is being depleted and could be exhausted in the foreseeable future. Complete exhaustion, of course, will not actually occur. Declining water levels will lead to lower yields, higher pumping lifts and higher costs, and operations that become uneconomical will end; thus, economics will force a balance between supply and draft. This has already happened in some parts of the High Plains.

Many parts of the world contain inherited ground water in smaller but important aquifers. The water in such aquifers may be considered as non-cyclic because it would not naturally participate in the water cycle within a humanly significant span of time. For example, a sedimentary aquifer in the vicinity of Maracaibo, Venezuela, is about 1,300 km² in extent and has a water-storage volume of 35 km³. Radio-carbon analyses of samples of the waters indicate that they range in age from 4,000 to 35,000 years.

4. Cyclic ground water

The well-known characterization of aquifers as both reservoirs and pipelines is appropriate, but the pipeline analogy is less apt than that of the reservoir. In fissured limestone with openings so large that Darcy-law flow does not occur the speed of ground-water movement may be comparable to that of some rivers at low stage – about 0·03–0·3 m s⁻¹. In cavernous limestone speeds may be as high as 7–10 m s⁻¹ (Pardé [1965], p. 38). The pipeline analogy is apt for such aquifers.

On the other hand, a study of ground-water movement in finely fractured crystalline rock (Marine, 1967) indicated an average speed of about $17·5 \times 10^{-6}$ m s⁻¹ (about 1·5 m da⁻¹). Even slower speeds prevail in some argillaceous sediments. Here the pipeline analogy becomes strained. Well-sorted clean gravel is excellent material, but 'all the water moving through clean, well-sorted gravel in a bed 100 ft thick and 1 mile wide [30 m and 1·6 km] with 1% gradient could be transmitted in a pipe 14 inches [about 35 cm] in diameter at the same gradient' (Thomas and Peterson [1967], p. 71). On the other hand, water stored in 1 square mile (about 4 km²) of the same aquifer could supply the same discharge pipe for about 4 years.

These few data illustrate that though ground water is generally in continuous motion, its detention time in aquifers is long compared to detention of water in river channels. Hence, great apparent age of a ground-water sample is not in itself an indication that recharge is small or that the water is non-cyclic. A study in the Great Artesian Basin of Queensland and New South Wales, Australia, indicated that water in the aquifers 30–80 km from the recharge areas has been underground about 20,000 years (Water Research Foundation, 1964). This implies a travel speed of $1 \cdot 5$–4 m yr^{-1}.

In many aquifers, however, underflow speeds range from a fraction of a metre to several metres per day. Water-detention times range from a few minutes to a few decades. In extensive highly permeable aquifers such as the basalt of the Snake River Plain in Idaho, occupying 30,000 km^2, underflow speeds average 10–15 m da^{-1}, and detention periods range up to several hundred years.

5. Management of ground water

Management of water means controlled use in accord with some plan. Most uses of ground water are simple exploitation as a free good, and the water is generally considered to be self-renewing. Preceding remarks show, however, that recharge of some aquifers is so small or slow that renewal will not occur naturally within spans of time that are relevant to current planning or management. Moreover, many aquifers are so inefficient as pipelines that relatively little of their recoverable water can be extracted where and when it is wanted.

Discharging wells do not lower the water-table or pressure surface uniformly throughout an aquifer. In an aquifer like the Ogallala formation, for example, the water-table is not flat but dimpled, each dimple being a cone of depression around a well. The intervening mounds are in segments of the aquifer that have not been dewatered. Springs and seeps along the natural discharge area of the Texas aquifer will continue to discharge. All pumpage within the plain, therefore, is a depletion draft on reserve storage because pumping of wells will not reduce natural discharge until pumping effects reach the discharge area and lower the hydraulic gradient there. The only measure which will permit pumping from the Ogallala at present rates to continue into the far future is artificial recharge with imported water. The area lacks good surface reservoir sites, but the ground-water reservoir has tremendous unused storage capacity.

Israel is an example of a modern industrial–agricultural economy where water is so scarce that total management of ground water and surface water as a unit resource is practised. This is necessary in order to get the most out of every drop of water and to control sea-water incursion in coastal aquifers. Even humid areas like Western Germany and the Netherlands have had to establish careful control, because increasing pollution of all waters makes it increasingly difficult to produce suitable water for growing demands.

Where water demand is large, surface-water supplies, if available, have generally been exploited first, because the water is visible, the amount available can be measured readily, and run-of-the-river diversions are cheap and require no great engineering works. Ground water generally has been quite independently exploited, as though it were a different or separate resource. In fact, however, ground water and surface water typically are in hydraulic continuity,

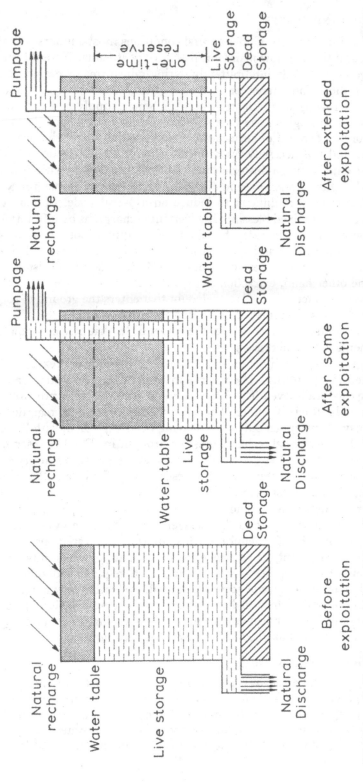

Fig. 6.III.2 Stylized representation of planned exploitation of ground-water storage in order to capture a larger share of current water yield (After Mandel, 1967).

parts of a single system. Under natural conditions most aquifers are full to over-flowing, and the overflow maintains base flow of streams. Were this not so, most rivers would flow only intermittently during rainstorms or periods of snowmelt and for a short time thereafter.

With narrowing of the margin between water supply and demand in many areas, water managers have begun to accept the hydrological facts. They talk about conjunctive use of surface water and ground water, the need to manage aquifers, and the necessity for artifical recharge. The practice of artificial re-charging dates from prehistoric times, but extensive systematic recharging is a twentieth-century phenomenon. Recharging is not a simple process, and in many situations it is difficult to induce on a useful scale. Moreover, it is not necessarily an unmixed blessing. Inadvertent recharge, as by infiltration of excess irrigation water, may increase the salinity of an aquifer, raise the water-table, and waterlog land. This has happened in many irrigated areas, and it is a major problem in agricultural development of the Indus Plain in Pakistan.

On the other hand, extraction of ground water may proceed to the point that the amount extracted exceeds the amount that enters the ground to replenish the supply. The resource may thus be depleted to the extent that the aquifer no longer yields enough water to meet the demand. In that case economic and social adjustments will be necessary.

The ground-water situation in an area where withdrawal exceeds recharge is illustrated in fig. 6.III.2. The reserve in dead storage is water that would not discharge naturally even if recharge stopped entirely. This probably could not be extracted artificially, either. The one-time reserve may be regarded as capital, which may be expended only once. Having been expended, withdrawal must be decreased or additional recharge must be brought in. The live reserve is storage that must be retained as a cushion which may be drawn on in dry years, to be replenished in wetter years. The live reserve may also be a buffer to prevent in-vasion of sea-water, for example.

Water in an aquifer cannot be seen and its amount cannot be measured directly. The first step towards management is accurate evaluation of the re-source: Its amount, its chemical quality, its transient variations in amount and quality, its movement, its sources of recharge, and its availability (feasibility of extraction). Every aquifer is unique, and each requires individual evaluation. Depending on the degree of accuracy needed, evaluation may range from hydro-geological reconnaissance to highly complicated study, including test-drilling, test-pumping, geochemical and geophysical surveying, mapping, use of mathe-matical and analog models, and various other techniques.

6. Artificial recharge

In many irrigated areas deep percolation loss of water from canals and irrigated fields each year is equivalent to a layer of water a few centimetres to a metre in depth over the entire area. In an area of 100,000 ha inadvertent recharge from irrigation thus may amount to 30 to $1,000 \times 10^6$ m^3 of water.

Deep percolation losses from irrigated fields are evidence that in some places recharge is easy to induce. In situations where recharge amounts to a loss of useful water or an aggravation of water-logging, the rate seems high. Where recharge is desired, however, infiltration rates of a few centimetres or metres per year are low. Generally, it is not feasible to inundate tens of thousands of hectares for the sole purpose of recharge. Evaporation loss would be high, and extensive flat areas generally are too valuable for other purposes. Recharging is usually practicable only where the unit-area rate of recharge is high or where water can be injected underground through intake wells. The sands and gravels of Long Island, New York, and of the alluvial fans around the margins of the Central Valley of California are examples of favourable areas for water spreading.

The High Plains of Texas and New Mexico exemplify conditions less favourable for recharge. The plains are dotted by many hundreds of playa basins that fill with water from local runoff during intense storms. The subsoil is permeable, but the basins are largely floored by fine silt, and much of the area is underlain by poorly permeable caliche. Therefore, depending on local conditions, 20–80% of the ponded water evaporates rather than percolating underground (Havens [1966], p. 35). In order to reduce this loss, experiments have been widespread to get the water underground through intake wells. The degree of success has ranged from poor to excellent during short periods. Suspended sediment in the injected water clogs the aquifer, and the wells must be pumped periodically to remove the sediment (Myers, 1964). After a few weeks or months the wells may cease to be effective.

In the Grand Prairie rice-growing region of Arkansas the water supply would be ample if ways could be found to store excess winter streamflow. This could be accomplished by injecting the water underground, but clogging by sediment is a problem, and chemical reactions between the natural ground water and the injected water produce precipitates. Entrained air and micro-organisms also cause clogging (Sniegocki, 1963). In order for recharge to be successful, injected water must be pretreated essentially as though for a municipal water supply. The problem thus is one of economics.

Even at places where recharge is feasible, the inefficiency of aquifers as pipelines again is a drawback. Physical transit of water is not the whole story, because transfer of hydraulic head is more rapid. Accessions by recharge create a so-called recharge wave, and this wave may travel at rates of a fraction of a metre to several hundred metres per day, depending on permeability of the aquifer. So far as water availability is concerned, the effect is the same as physical transfer of water. However, in some aquifers transfer of head is so slow that injected water, for practical purposes, is recoverable only in the vicinity of the injection site.

7. Pollution

All natural water is contaminated in the chemical sense, in that it contains substances other than H_2O. Pollution, on the other hand, is a totally subjective concept, and few definitions are satisfactory. In general, water managers consider that water is polluted when dissolved or entrained substances are present in amounts that make the water unfit or undesirable for specified uses. To the hydrologist, however, *polluted* is an incongruous label for a natural water, which he would call mineralized.

Absence of serious pollution of ground water in some places where sources of pollution are widespread is a consequence of the nature of aquifers. A sand aquifer acts like a sand filter in a water-treatment plant. By sorption processes it can also function as an ion-exchange bed. Conditions within aquifers are largely anaerobic, and lack of oxygen kills many waterborne organisms.

On the other hand, ground water will not necessarily purify itself through any given distance or period of time. This is especially true of aquifers such as cavernous limestone, which carry organic and mineral pollutants rapidly through considerable distances. The generally slow motion of ground water in granular aquifers is highly important in relation to pollution. Once polluted, an aquifer may remain so for years or centuries, because flushing action is very slow. Studies of industrial waste in the ground water of Long Island (Perlmutter, Lieber and Frauenthal, 1963) show that, even if pollution were discontinued, at least ten years would pass before all the polluted water would be discharged from the aquifer.

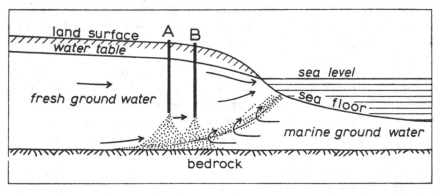

Fig. 6.III.3 Cross-section showing dynamic relation between salt water and fresh in a coastal aquifer. The shaded area represents the zone of transition. The deepest shading represents the zone where the horizontal gradient is nil. (Adapted from Kohout and Kiein, 1967).

Salt-water invasion of aquifers is a growing problem in most areas near the sea or other bodies of salt water. The Ghyben-Herzberg concept of a lens of static fresh water floating on salt water beneath a circular oceanic island is inadequate to describe the situation and processes in coastal areas.

Figure 6.III.3 is a simplified portrayal of a coastal situation. The interface between salty and fresh water is actually a zone of transition from fresh to salt water. Motions of these waters are indicated by curvilinear flow lines. In the transition zone seaward-flowing fresh ground water dilutes and sweeps out the intruding sea-water. The width of the zone is controlled by permeability of the aquifer and velocity of flow, and by the range of oceanic tides. Heavy pumping of the well at A may create an upward tongue of salt water which may reach that well itself and may also migrate to other wells, as at B.

The actual situation in water-table aquifers and in limestone and artesian aquifers along coasts is much more complicated than that in the simplified portrayal. Each aquifer and locality is, in fact, a special case, and each must be studied individually (see Cooper [1959] and Kohout and Klein [1967]).

Acknowledgements. The author gratefully acknowledges constructive review of this chapter and suggestions by his colleagues, J. H. Feth, Howard Klein, B. N. Myers, and A. G. Winslow.

REFERENCES

COOPER, H. H. [1959], A hypothesis concerning the dynamic balance of fresh water and salt water in a coastal aquifer; *Journal of Geophysical Research*, **64**, 461–7.

DE CAMP, L. S. [1963], *The Ancient Engineers* (Garden City, New York), 409 p.

DIXEY, F. [1966], Water supply, use and management; *In* Hills, E. S., Editor, *Arid Lands, a geographical appraisal* (London and Paris), pp. 77–102.

HAVENS, J. S. [1966], Recharge studies on the High Plains in Northern Lea County, New Mexico; *U.S. Geological Survey Water Supply Paper* 1819-F, 52 p.

JONES, J. R. [1965], Written communication to R. L. Nace, 22 October.

KOHOUT, F. A. and KLEIN, HOWARD, [1967], Effect of pulse recharge on the zone of diffusion in the Biscayne aquifer; *International Association of Scientific Hydrology Publication*, **72**, pp. 252–270.

MANDEL, S. [1967], Underground water; *International Science and Technology*, **66**, 35–41.

MARINE, I. W. [1967], The use of a tracer test to verify an estimate of the groundwater velocity in fractured crystalline rock at the Savannah River Plant near Aiken, South Carolina; *American Geophysical Union, Geophysical Monograph Series*, **11**, 171–9.

MASSOUMI, AHMAD, 1966, *Groundwater production in Iran:* Paper presented to United Nations Seminar on Methods and Techniques of Ground-Water Investigation and Development, Tehran, October 16 (duplicated), 12 p, 5 figs.

MYERS, B. N. [1964], Artificial-recharge studies; In Cronin, J. G., editor, A summary of the occurrence and development of ground water in the Southern High Plains of Texas; *U.S. Geological Survey Water Supply Paper* 1693, 56–71.

PERLMUTTER, N. M., LIEBER, M. and FRAUENTHAL, H. L. [1963], Movement of waterborne cadmium and hexavalent chromium wastes in South Farmingdale, Nassau County, Long Island, New York; *U.S. Geological Survey Professional Paper* 475-C, 179–84.

PARDÉ, M. [1965], Influences de la permeabilité sur le régime des riviéres; *Colloquim Geographicum*, **7**, 21–100.

SNIEGOCKI, R. T. [1963], Problems in artificial recharge through wells in the Grand Prairie region, Arkansas; *U.S. Geological Survey Water Supply Paper* 1615-F, 25 p.

THOMAS, H. E., and PETERSON, D. F., JR [1967], Ground water supply and development; In Hagan, R. M. and others, Editors, Irrigation of agricultural lands; *American Society for Agronomy, Agronomy Series*, **11**, 70–91.

WATER RESEARCH FOUNDATION OF AUSTRALIA [1964], Nuclear chemical study of the age and renewal rate of the Great Artesian Basin; *Water Research Foundation 9th Annual Report and Balance Sheet*, 12–13.

CHAPTER 7

Channel Flow

I. Open Channel Flow

D. B. SIMONS

Department of Civil Engineering, Colorado State University

1. Introduction

Flow in open channels has been nature's way of conveying water on the surface of the earth since the beginning of time. Furthermore, these streams have constantly been the subject of study by man since he has been alternately blessed by the life-giving quality of streams under control and plagued by their destructive quality when out of control, such as in time of flood. Hence, the characteristics of rivers are of importance to everyone dealing with water resources, whether from the viewpoint of geomorphology, hydraulics, flood control, navigation, stabilization or water-resources development for municipalities, and industry.

2. Properties of fluids

The following physical properties of fluids influence fluid motion, channel geometry, and help explain sediment transport.

Mass is the amount of substance in matter measured by its resistance to the application of force.

Density is mass per unit volume, and is commonly symbolized by the Greek letter ρ (rho).

Weight is the force that gravity exerts on a mass; $W = gM$, where g is the acceleration of gravity.

Specific Weight is the weight per unit volume and is symbolized by the Greek letter γ (gamma); $\gamma = \rho g$.

Viscosity is the property of fluids that resists deformation and is commonly symbolized by the Greek letter μ (mu).

Shear is a property of fluid motion that is closely related to viscosity. It is the tangential force or stress per unit area that is transmitted through a unit thickness of a fluid. Shear, τ (tau), is related to viscosity by the equation $\tau = \mu \dfrac{dv}{dy}$.

Temperature affects the density of liquids slightly and the viscosity significantly. That is, water is essentially incompressible. The viscosity of liquids decrease with increasing temperature. Water temperature in open channels can vary as much as 40° F within a 24-hour period.

Elasticity and *Surface Tension* have little effect on flow in open channels, including sediment transport.

3. Types of flow

There are several types of flow in open channels. These include laminar flow and turbulent flow; uniform flow and non-uniform (or varied) flow; steady flow and unsteady flow; and tranquil flow, rapid flow, and ultra-rapid flow.

A. Laminar flow

Fluid motion may occur as laminar or turbulent flow. In laminar flow each fluid element moves along a specific path with a uniform velocity. There is no diffusion between the stream tubes, layers, or elements of flow; and accordingly, there is no turbulence. The energy used in maintaining viscous flow is dissipated in the form of heat from the friction within the fluid.

With laminar flow, the shear stress $\tau = \mu \dfrac{dv}{dy}$ being transmitted through each unit of depth varies uniformly from zero at the surface to a maximum at the stream bed, while the velocity curve is parabolic in shape, with its vertex at the surface.

In stream flow disturbances are present in such magnitude that laminar flow is rarely found. As velocity or depth increases, a given condition of laminar flow will reach a critical condition and become turbulent flow.

B. Reynolds' number

Values of *Reynolds' number* (*Re*) can be used to predict the type of flow. This dimensionless number includes the effects of the flow characteristics, velocity, and depth, and the fluid properties density and viscosity.

$$Re = \frac{VR\rho}{\mu} \tag{1}$$

The ratio $\dfrac{\mu}{\rho}$ is a fluid property called the kinematic viscosity, commonly designated ν (nu). Using this property,

$$Re = \frac{VR}{\nu} \tag{2}$$

With the value of Reynolds' number less than 500, laminar flow will prevail; whereas, with values in excess of 750, turbulent flow will prevail for smooth boundary conditions. For natural channels the critical value will be near 500 due to bed roughness.

The Reynolds' number is defined as

$$Re = \frac{\text{Inertia force}}{\text{Viscous force}} \tag{3}$$

That is, *Re* is an index of the relative importance of viscous forces in a hydraulic problem. Using Newton's second law of motion to define the inertial force, the expression $\tau = \mu \dfrac{dv}{dy}$ to define the viscous force and dimensional analysis

$$Re = \frac{\rho \dfrac{L^3 L}{T^2}}{\mu \dfrac{LL^2}{TL}} \qquad (4)$$

substituting

$$V = \frac{L}{T} \quad \text{and} \quad L = D$$

$$Re = \frac{\rho V^2 D^2}{\mu VD} = \frac{VD\rho}{\mu} \qquad (5)$$

Many other dimensionless parameters which have the same form as the foregoing Reynolds' number are utilized in the analysis of open-channel flow problems, such as: $\dfrac{wd}{\nu}$ and $\dfrac{V_* d}{\nu}$,

where $w =$ fall velocity of sediment or bed material;

$\quad d =$ median diameter of the sediment or bed material;

$\quad V_* =$ shear velocity which is equal to $\sqrt{gRS}$.

Problem No. 1. A sheet of water 0·25 ft deep is flowing over a smooth surface at 1·0 ft/sec. Compute the Reynolds' number (*Re*) of the flow. Will the flow likely be laminar or turbulent? (Kinematic viscosity (μ) = 1·21 × 10⁻⁵ ft²/sec).

$$Re = \frac{(1 \cdot 0)\,(0 \cdot 25)}{1 \cdot 21}\,(10^5) = \underline{\underline{20{,}700}} \text{ Turbulent}$$

C. *Froude number*

The Froude number F_r is another dimensionless parameter frequently used to describe flow conditions. It is an index to the influence of gravity in flow situations where there is a liquid–gas interforce – such as in an open channel. The Froude number is usually defined as

$$F_r = \left(\frac{\text{Inertia force}}{\text{Gravity force}}\right)^{\frac{1}{2}} \qquad (6)$$

or

$$F_r = \frac{\text{Velocity of flow}}{\text{Velocity of a small gravity wave in still water}} \qquad (7)$$

Referring to the first definition and dimensional analysis

$$F_r = \left(\frac{\rho \dfrac{L^3 L}{T^2}}{\Delta \gamma L^3}\right)^{\frac{1}{2}} \qquad (8)$$

Substituting

$$V = \frac{L}{T}$$

$$F_r = \left(\frac{\rho V^2 L^2}{\Delta \gamma L^3}\right)^{\frac{1}{2}} = \frac{V}{\sqrt{\Delta \gamma L / \rho}} \tag{9}$$

where L = a length dimension;

$\Delta \gamma$ = difference in specific weight of the fluids – usually air and water;

V = average velocity of flow;

ρ = mass density of the fluid.

In open-channel flow $\Delta \gamma$ is essentially the same as γ for water alone, since the density of air is so small. If D (depth) or hydraulic radius $R = \frac{A}{P}$ is used for the L dimension and $\frac{\gamma}{\rho}$ replaced by its equivalent g, then the Froude number becomes

$$F_r = \frac{V}{\sqrt{Dg}} \tag{10}$$

Note that in wide shallow channels depth of flow and the hydraulic radius are nearly equal.

A Froude number of unity indicates critical flow; less than unity indicates tranquil flow, the common variety of turbulent flow; and greater than unity, 'rapid flow'. In keeping with the second definition of Fr the $\sqrt{gD}$ term is the velocity at which a small wave travels in still water of depth D. For example, one can throw a pebble into a stream and, comparing the velocity of the waves caused by the pebble and the velocity of the flow, determine whether flow is sub-critical, critical, or super-critical.

Problem No. 2. Compute the Froude number of an open-channel flow where the mean velocity is 5 ft/sec and the depth is 1·5 ft. Describe the flow with respect to critical flow.

$$F_r = \frac{V}{\sqrt{gD}}$$

$$F_r = \frac{5}{\sqrt{(32 \cdot 2)(1 \cdot 5)}} = 0 \cdot 73 \underline{\underline{\text{tranquil or subcritical}}}$$

D. Turbulent flow

Turbulence, as a complicated pattern of eddies, produces small velocity fluctuations at random in all directions with an average time value of zero. Energy dissipation is high in turbulent flow due to the continuous interchange of finite masses of fluid between neighbouring zones of flow. The resistance to flow increases with approximately the square of the velocity.

Turbulent flow, as a result of this mixing and exchange of energy, has a more uniform distribution of velocity from top to bottom than laminar flow. The velocity distributions for laminar and turbulent flow are compared qualitatively in fig. 7.1.1.

Several parameters have been developed to describe turbulent flow. The shear stress in turbulent flow is defined as $\tau = \eta \dfrac{dv}{dy}$, in which η (eta) is termed the *eddy viscosity*. A parameter used to describe the magnitude of turbulent velocity fluctuations is the root-mean-square ($\sqrt{V'^2}$) of the deviations from the mean

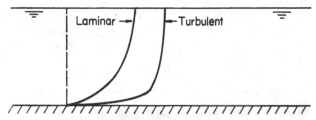

Fig. 7.1.1 Comparison of velocity distribution in laminar and turbulent flow.

velocity. The mean size of the turbulent eddies is measured by the mixing length (l) – the distance through which fluid elements move before diffusing with the surrounding fluid. The diffusion coefficient, $\epsilon = l\sqrt{V'^3}$, is a measure of the mixing process. The general pattern of variation of these parameters in turbulent flow is shown in fig. 7.1.2.

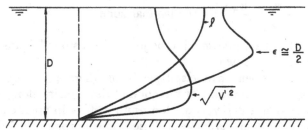

Fig. 7.1.2 Variation in mean turbulence characteristics with depth (After Rouse, 1946).

From a combination of experimental study and theory the distribution of velocity in turbulent flow over smooth boundaries has been determined. A thin layer of laminar flow persists at the boundary surfaces. This layer is called the laminar sub-layer. The theoretical velocity curve is a composite of the logarithmic turbulent flow pattern and the nearly linear laminar pattern joined by a transition curve as shown in fig. 7.1.3.

In fig. 7.1.3 δ' is the thickness of the laminar sub-layer, and it is defined by the equation

$$\delta' = \frac{11 \cdot 6\,\nu}{\sqrt{\dfrac{\tau}{\rho}}} = \frac{11 \cdot 6\,\nu}{V_*} \tag{11}$$

L

If extended towards the bed the logarithmic form of the turbulent velocity distribution will yield zero velocity at a distance y' above the bed. Experiments show that $y' = \delta'/107$.

The term $\sqrt{\dfrac{\tau}{\rho}}$ is a common parameter called the shear velocity (V_*), and is equal to $\sqrt{RSg}$.

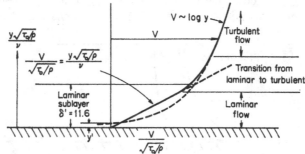

Fig. 7.1.3 Details of flow near the bed of an open channel (After Albertson, Barton and Simons, 1960).

The Karman–Prandtl equation describes the velocity distribution of turbulent flow over a smooth bed as

$$\frac{v}{V_*} = 5\cdot75 \log_{10} \frac{V_* y}{\nu} + 5\cdot5 \text{ for the turbulent zone}$$

and

$$\frac{v}{V_*} = \frac{V_* y}{\nu} \text{ for the laminar zone}$$

The two equations are presumed to be joined by a smooth transition curve at the distance δ' above the bed.

Uniform flow in open channels depends on there being no change with distance in either the magnitude or direction of the velocity along a stream line, i.e. both $\partial v/\partial s = 0$ and $\partial v/\partial n = 0$. Non-uniform flow in open channels occurs when either $\partial v/\partial s \neq 0$ or $\partial v/\partial n \neq 0$. Varied flow in open channels is a type of non-uniform flow which occurs when $\partial v/\partial s \neq 0$. Steady flow occurs when the velocity at a point does not change with time, i.e., $\partial v/\partial t = 0$. When the flow is unsteady $\partial v/\partial t \neq 0$. An example of unsteady flow is a flood wave or a travelling surge.

Unlike laminar and turbulent flow, tranquil flow and rapid flow exist only with a free surface or inner face. The criterion for tranquil and rapid flow is the Froude number $F_r = V/(gD)^{1/2}$. When $F_r = 1\cdot0$ the flow is critical; when $F_r < 1\cdot0$ the flow is tranquil; and when $F_r > 1\cdot0$ the flow is rapid. Ultra-rapid flow involves slugs or waves superposed over the uniform flow pattern, which makes the flow both non-uniform and unsteady.

Uniform flow in an open channel occurs with either a mild, a critical, or a steep slope. With a mild slope the flow is tranquil; with a critical slope the flow is critical; and with a steep slope the flow is rapid.

4. Velocity distribution over rough beds

Stream channels have rough beds. The roughness is expressed in terms of K_s, which is equivalent to the diameter of the sediment grains which compose the bed. The dimension of K_s is larger than that of δ', and therefore the sub-layer ceases to exist for practical purposes. Turbulent flow is assumed to occur throughout the depth. The Karman–Prandtl velocity equation for rough beds is

$$\frac{v}{V_*} = \frac{2 \cdot 303}{\kappa} \log_{10} \frac{y}{K_s} + 8 \cdot 5 \tag{12}$$

where kappa (κ) is the so-called universal velocity coefficient, which is approximately $0 \cdot 4$ for fixed boundary channels. The distribution of velocity in accordance with this equation is a straight line when plotted on semi-log paper with a slope of $\dfrac{\kappa}{2 \cdot 303 \, V_*}$.

5. Velocity and discharge measurements

Velocity is a vector quantity, hence both its direction and magnitude must be measured. The discharge in an open channel or pipe, in the most simple terms, is the product of area and average velocity measured normal to the area.

A. Current meters

In an open channel one of the most common methods of measuring discharge involves integration of the velocity distribution across the flow section using a current meter, pitot tube, or similar device. The current meter consists of an instrument with an impeller mounted on a rod or cable. If a cable is used to suspend the meter in the flow there must be a streamlined weight at its lower end, below the current meter, of sufficient magnitude to overcome the force of the stream, enabling the operator to place the meter at any desired point in the vertical. Having taken sufficient point measurements in a vertical to establish the average velocity, the operator moves to a new vertical, by wading in small streams or by cable car or boat on large streams.

The average velocity in a vertical is located at approximately $0 \cdot 6$ depth below the surface and can be more precisely determined by averaging the point velocities at $0 \cdot 2$ and $0 \cdot 8$ depth. In flows with a depth less than $1 \cdot 5$ ft a single-point measurement is taken in each vertical in the stream cross-section at $0 \cdot 6$ depth. In deeper streams the $0 \cdot 2$ and $0 \cdot 8$ measurements are taken and averaged in each vertical. Discharge determined by the $0 \cdot 2$ and $0 \cdot 8$ measurements is illustrated in the example problem.

Problem No. 3. The computation of stream discharge based on current meter measurements is illustrated in Table 7.1.1.

Similarly, floats can be used to estimate the magnitude and direction of surface velocities. Greater accuracy can be achieved by using a submerged or partly submerged float which measures the velocity at more nearly $0 \cdot 6$ depth. Also, it is more independent of the effect of wind and waves, but may be bothered by debris.

TABLE 7.1.1

| Distance from bank | Depth | Obser- vation depth | Velocity | | | Area | Mean depth | Width | Dis- charge (q) |
			At Point	Mean in vertical	Mean in section				
0	0	0	0	0					
						0·78	1·70	0·85	2	1·33
2	1·70	0·35	1·52	1·56					
		1·35	1·60						
						1·73	4·40	2·20	2	7·61
4	2·70	0·54	1·91	1·89					
		2·16	1·88						
						2·08	13·80	3·45	4	28·7
8	4·20	0·84	2·35	2·27					
		3·36	2·19						
						2·33	8·60	4·30	2	20·1
10	4·40	0·88	2·41	2·38					
		3·52	2·34						
						2·37	8·40	4·20	2	19·94
12	4·00	0·80	2·31	2·36					
		3·20	2·40						
						2·05	13·60	3·40	4	27·9
16	2·80	0·56	1·92	1·74					
		2·24	1·57						
						1·49	4·30	2·15	2	6·41
18	1·50	0·30	1·35	1·24					
		1·20	1·13						
						0·62	1·50	0·75	2	0·93
20	0	0	0	0					

$$Q = \Sigma q = 112 \cdot 9 \text{ cfs.}$$

B. Dye-dilution

Various fluorescent tracer-type dyes can be detected and accurately measured at very low concentrations using a fluorometer. This makes it possible to successfully measure discharge by various dye-dilution techniques. Two methods can be used. One involves

$$Q = q\frac{C_1}{C_2} \qquad (13)$$

where q is the injected discharge, C_1 is concentration of the dye in the injected flow, and C_2 is the concentration of dye in the unknown discharge Q. On small streams injection of a small steady q at known concentration C_1 for about 15 minutes will enable C_2 to stabilize. Only C_2 at its plateau needs to be determined to compute the stream discharge Q since the effect of q is small.

The second method is called the total recovery method. The discharge is evaluated using the relation

$$Q = \frac{(\text{Vol. of dye})(\text{Conc. of dye})}{\displaystyle\int_0^\infty C\,dt} \tag{14}$$

The integral term is the total area under the concentration–time curve, where C is the measured dye concentration at time t at the point of sampling. In both relations any material background fluorescence must be considered in measuring effective dy concentrations.

C. Weirs

The weir is extensively used to measure flow in open channels. It is essentially an overflow structure extending across the channel normal to the direction of flow, see fig. 7.1.4.

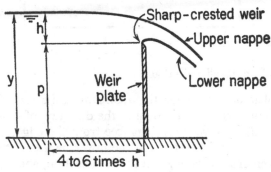

Fig. 7.1.4 Sharp-crested weir.

Weirs are classified according to shape. The most common ones are the standard uncontracted weir, also known as the suppressed weir, the contracted weir, the V-notch weir, the trapezoidal weir, and the broad-crested weir. The first four are sharp crested, as shown in fig. 7.1.4.

Many formulae have been suggested by various experimenters, but only a few are presented.

In 1823 Francis suggested the equation

$$Q = 3 \cdot 33 L h^{3/2} \tag{15}$$

for uncontracted (suppressed) weirs in which L is the length of crest in feet and h is the head on the weir in feet. For the contracted weir

$$Q = 3 \cdot 33 \left(L - \frac{nh}{10} \right) h^{3/2} \tag{16}$$

where n is the number of horizontal end contractors with a simple weir only contracted at the sides of the channel $n = 2$.

For a triangular weir $\qquad dQ = C_d x \sqrt{2gy}\, dy \qquad\qquad\qquad$ (17)

from which $\qquad\qquad Q = \dfrac{8}{15}\, C_d \sqrt{2g}\, \tan\dfrac{\theta}{2}\, h^{5/2} \qquad\qquad$ (18)

see fig. 7.1.5.

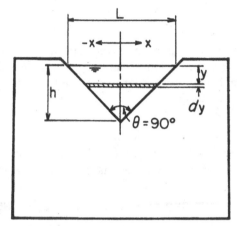

Fig. 7.1.5 Triangular weir.

The weirs should be installed so that:

1. The weir plate is vertical and the upstream face essentially smooth.
2. The crest is horizontal and normal to the direction of flow. The crest must be sharp, so that the water springs free from the edge.
3. The pressure along the upper and the lower nappe is atmospheric.
4. The approach channel is uniform in cross-section, and the water surface is free of surface waves.
5. The sides of the channel are vertical and smooth, and they extend downstream from the crest of the weir.

in order for the equations to apply.

Still greater accuracy can be achieved with all types of measuring devices if they are calibrated after construction.

Many other methods of measuring discharge in open channels and pipes are utilized. The principles utilized are (1) volumetric, (2) use of the Bernoulli momentum and continuity equations, (3) drag on an object in the flow, (4) mass flux measurements, and (5) by the slope area method. The measuring devices include: Venturi meters, Parshall flumes, nozzles, orifices, gates, weirs, spillways, contracted openings, vane meters, rotometers, and wobbling discs. Refer to any fluid mechanics, such as Albertson, Barton, and Simons [1960], for further details.

6. Hydraulic and energy gradients

The hydraulic gradient in open-channel flow is the water surface. The energy gradient is above the hydraulic gradient a distance equal to the velocity head.

The fall of the energy gradient for a given length of channel represents the loss of energy, either from friction or from friction and other influences. The relationship of the energy gradient to the hydraulic gradient reflects not only the loss of energy but also the conversion between potential and kinetic energy. For uniform flow the gradients are parallel and the slope of the water surface represents the friction-loss gradient. In accelerated flow the hydraulic gradient is steeper than the energy gradient, indicating a progressive conversion from potential to kinetic energy. In retarded flow the energy gradient is steeper than the hydraulic gradient, indicating a conversion from kinetic to potential energy. The Bernoulli theorem defines the progressive relationships of these energy gradients.

For a given reach of channel ΔL, the average slope of the energy gradient is $\Delta h_L/\Delta L$, where Δh_L is the cumulative loss through the reach. If these losses are solely from friction, Δh_L will become Δh_f and

$$\Delta h_f = \frac{S_2 + S_1}{2} \Delta L \tag{19}$$

where S_1 and S_2 are the slopes of the energy gradient at the ends of reach ΔL.

7. Energy and head

If streamlines of flow in an open channel are parallel and velocities at all points in a cross-section are equal to the mean velocity V the energy possessed by the water is made up of kinetic energy and potential energy. Referring to fig. 7.1.6,

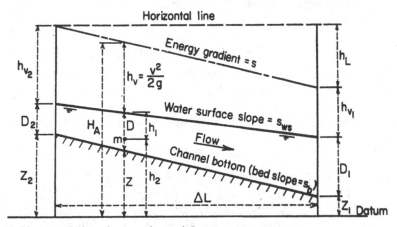

Fig. 7.1.6 Characteristics of open-channel flow.

the potential energy of mass M is $\gamma(h_1 + h_2)$ and the kinetic energy of M is $\gamma \dfrac{V^2}{2g}$. Hence, the total energy of each mass particle is:

$$E_m = \gamma h_1 + h_2 + \frac{V^2}{2g} \tag{20}$$

Applying the above relationship to the total discharge Q in terms of the unit weight of water γ,

$$E = Q\gamma \left(D + Z + \frac{V^2}{2g} \right) \tag{21}$$

where E is total energy per second at the cross-section.

The parentheses term in equation (21) is the absolute head:

$$H_A = D + Z + \frac{V^2}{2g} \tag{22}$$

Equation (22) is the Bernoulli Equation.

The energy in the cross-section referred to the bottom of the channel is termed the specific energy. The corresponding head is referred to as the specific energy head, and is expressed as:

$$H_E = D + \frac{V^2}{2g} \tag{23}$$

Where $Q = AV$, equation (23) can be stated:

$$H_E = D + \frac{Q^2}{2ga^2} \tag{24}$$

8. Flow equations

One of the common open-channel flow equations for estimating average velocity is the Chezy equation

$$V = (8g/f)^{1/2}(RS)^{1/2} = C(RS)^{1/2} = C/g^{1/2}(gRS)^{1/2} \tag{25}$$

Bazin [1897] suggested that

$$C = 157 \cdot 6/[1 + (k_1/R^{1/2})] \tag{26}$$

where k_1 is the roughness coefficient varying from $0 \cdot 11$ for very smooth cement or planed wood to $3 \cdot 17$ for earth channels in rough condition. In this equation the upper limit for C is $157 \cdot 6$.

In 1911 Johnston and Goodrich proposed using an exponential formula of the form,

$$V = CR^p s^q \tag{27}$$

and gave values of C and p, making q uniformly equal to $0 \cdot 5$ for simplicity (Ellis, 1916). This is exactly the same formula as proposed by Chezy, where the numerical value of the Chezy's coefficient C is equal to $0 \cdot 5$. Other open-channel flow equations that are often used are given by N. G. Bhowmik [1965] and Garbrecht [1961].

In an effort to correlate and systematize existing data from natural and artificial channels, Manning [1889] proposed the equation

$$V = (1 \cdot 5/n)R^{2/3} S^{1/2} \tag{28}$$

or

$$Q = AV = A(1 \cdot 5/n)R^{2/3} S^{1/2} \tag{29}$$

in which n is the Manning roughness coefficient which has the dimensions of $L^{1/6}$. By comparing equation (25) with equation (28), the Chezy discharge coefficient C can be expressed as follows:

$$C = 1 \cdot 5 (R^{1/6}/n) \tag{30}$$

and is related to the Manning coefficient n and the hydraulic radius $R = \dfrac{A}{P}$.

The Manning n was developed empirically as a coefficient which remained a constant for a given boundary condition, regardless of slope of channel, size of channel, or depth of flow. As a matter of fact, however, each of these factors causes n to vary to some extent. In other words, the Reynolds' number, the shape of the channel, and the relative roughness have an influence on the magnitude of Manning's n. Furthermore, for a given alluvial bed of an open channel the size, pattern, and spacing of the sand waves vary, so that n varies. Despite the shortcomings of the Manning roughness coefficient, it is used extensively.

The magnitude of Manning roughness is given in Table 7.1.2 for rigid channels

TABLE 7.1.2 Manning roughness coefficients for various boundaries

Boundary	Manning Roughness n in $(ft)^{1/6}$
Very smooth surfaces such as glass, plastic, or brass	0·010
Very smooth concrete and planed timber	0·011
Smooth concrete	0·012
Ordinary concrete lining	0·013
Good wood	0·014
Vitrified clay	0·015
Shot concrete, untrowelled, and earth channels in best condition	0·017
Straight unlined earth canals in good condition	0·020
Rivers and earth canals in fair condition – some growth	0·025
Winding natural streams and canals in poor condition – considerable moss growth	0·035
Mountain streams with rocky beds and rivers with variable sections and some vegetation along banks	0·040–0·050
Alluvial channels, sand bed, no vegetation	
1. Tranquil flow, $F_r < 1$	
Plane bed	0·014–0·02
Ripples	0·018–0·028
Dunes	0·018–0·035
Washed-out dunes or transition	0·014–0·024
Plane bed	0·012–0·015
2. Rapid flow $F_r > 1$	
Standing waves	0·011–0·015
Anti-dunes	0·012–0·020

and alluvial channels. Considering alluvial channels, note that as the form of bed roughness changes from dunes through transition to plane bed or standing waves the magnitude of Manning n decreases by approximately 50%.

9. Natural channels

The natural shape of an open channel may be markedly different from rectangles and trapezoids. However, it is usually possible to break down the complex shape of a natural open channel into simple elementary shapes for analysis. Consider fig. 7.1.7, in which flow is occurring not only in the main channel but also in the

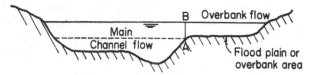

Fig. 7.1.7 Shape of natural channels.

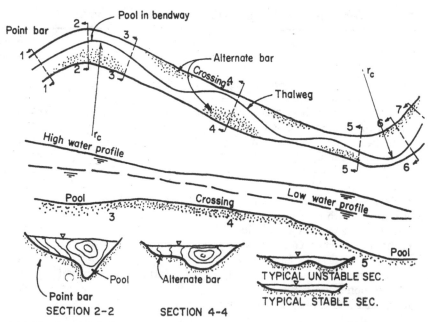

Fig. 7.1.8 Geometry of rivers.

overbank or floodplain area. In this case the hydraulic radius R, which would be obtained by using the area and the wetted perimeter for the entire section, would not be truly representative of the flow. Furthermore, the roughness coefficient in the overbank area is usually different from the coefficient in the main channel. Therefore, such a section should be divided along AB and treated as two separate sections. The plane AB, however, is not considered as a part of the wetted perimenter, since there is no appreciable shear in this plane.

Along a natural channel there are frequently pools and riffles or rapids. At low to moderate discharges the slope of water surface is relatively flat over the pools and steep over the riffles. With a further increase in stage, this condition may be reversed as for the Mississippi River (fig. 7.1.8). Therefore care must be taken in studies of natural streams to consider the correct slope for the particular discharge and reach of stream in question.

10. Forms of bed roughness and resistance to flow in alluvial channels

The primary variables which affect the form of bed roughness and resistance to flow in sand-bed alluvial channels (Simons and Richardson, 1962, 1963) include:

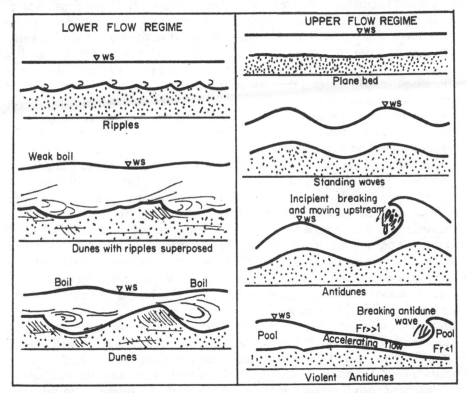

Fig. 7.1.9 Forms of bed roughness in alluvial channels (the term 'flat bed' is now preferred to 'plane bed').

the slope of the energy grade line, depth, physical size of the bed material as related to grain roughness, and fall velocity or effective median fall diameter as related to form resistance. The fall velocity or effective median fall diameter depends on the viscosity and mass density of the water sediment mixture and the mass density, size, and shape of the bed material. It reflects the principal viscous effect on flow in alluvial channels when Re is large. The effective median fall diameter is defined as the diameter of a sphere having a specific gravity of 2·65 and a fall velocity in distilled water of infinite extent at a temperature of

24° C equal to the fall velocity of the particle falling alone in any quiescent stream fluid at stream temperature (Haushild *et al.*, 1961; Simons *et al.*, 1963).

The regimes of flow and various forms of boundary roughness (Simons and Richardson, 1963) which can occur in alluvial channels are illustrated in fig. 7.1.9.

In the lower flow regime flow is tranquil and the water surface undulations are out of phase with the bed undulations. Resistance to flow is large, because separation of the flow from the boundary generates large-scale turbulence that dissipates considerable energy.

With depths of flow ranging from 0·4 to 1·0 ft, ripple heights range from 0·01 to 0·1 ft and length (crest-to-crest) range from 0·5 to 1·5 ft. When depth is small the ripples increase in size with depth, but at greater depths ripple size becomes independent of depth. Therefore, ripples observed in flumes are similar in size and shape to those in natural streams. A decrease in n occurs when depth is increased, indicating a relative roughness effect or when effective fall diameter is increased, which causes a decrease in ripple size. The decrease in n with an increase in effective fall diameter is similar to change reported in Leopold and Maddock [1953]. Apparently, ripples do not form when the median diameter of the bed material is coarser than 0·7 mm.

With depth of flow ranging from 0·4 to 1·0 ft, dune heights range from 0·15 to 1·0 ft and dune length from 4 to 20 ft, based upon flume studies (Simons and Richardson, 1963). In deep rivers dunes 30–60 ft in height and with lengths of several hundred feet have been observed. In the flume studies n increased with depth because size of the dunes and, hence, scale and intensity of turbulence increased. This may not be the case as larger depths are studied. With an increase in slope, n decreased for the fine sand but increased for the coarse sand because dune length increased appreciably with the fine sand but did not with the coarse sand. An increase in effective fall diameter increased n because dune length decreased and dune angularity increased. The long dunes formed by the finer sands exhibited smaller n values than ripples.

In the transition zone n varied from the largest value for the lower flow regime to the smallest value for the upper flow regime. In this zone a well-defined relation between n and boundary shear does not exist. The bed form in the transition zone depended, in addition to the other factors, on antecedent conditions. Starting with dunes, slope and/or depth could be increased to relatively large values before plane bed or standing waves occurred. Conversely, with a plane bed and/or standing waves, the slope and/or depth could be decreased to relatively small values before dunes developed.

In the upper flow regime n values are small because surface or grain resistance predominates. However, the energy dissipated by the wave formation with the standing waves and the formation and breaking of the waves with antidunes increases n. Standing waves and antidunes, from the standpoint of wave mechanics, are rapid flow phenomena.

Standing waves are sinusoidal, in-phase sand and water waves (fig. 7.1.9) that build up in amplitude from a plane bed and water surface and gradually fade away. In the flume studies with depth of flow ranging from 0·2 to 0·6 ft the water wave height (trough-to-crest) ranged from 0·01 to 0·6 ft and was 1·5–2 times the height of corresponding sand waves. Spacing of the waves was from 2 to 5 ft. Both height and spacing of the waves increased with depth. Resistance to flow for standing waves was larger than for a plane bed and, as with a plane bed, increased with an increase in sand size. Standing waves did not occur using the two finer sands, because the mobility of the particles (effective fall diameter) allowed the development of antidunes whenever the Froude number equalled one.

Antidunes are similar to standing waves, except they increase in amplitude until they break. Breaking antidunes are similar to the hydraulic jump. The breaking wave dissipates a large amount of energy that is reflected by increased n. The increase of n is in direct proportion to the amount of antidune activity and the portion of the flume or channel occupied by the antidunes. Antidune activity increases with a decrease in effective fall diameter or with an increase in slope.

11. Comparison of flume and field phenomena

The preceding comments were based on observations of bed configurations and flow phenomena that occurred in the flume experiments and in natural rivers (field conditions) and are equally true for both situations. However, there are major differences between flume and field conditions. In the usual flumes only a limited range of depth and discharge can be investigated, but slope and velocity can be varied within a wide range. In the field a larger range of depth and discharge is common, but slope of a particular channel reach is virtually constant. Larger Froude numbers ($V/\sqrt{gD}$) can be achieved in flume studies than will occur in most natural alluvial channels because natural banks cannot withstand prolonged high-velocity flow without eroding. This erosion increases the cross-sectional area, and this causes a reduction in the average velocity and the Froude number. Rarely does a Froude number, based on average velocity and depth, exceed unity for any extended time period in a natural stream with erodible banks. In fact, rarely are natural channels truly stable when $F_r > 0·25$. In the field, where the slope of the energy grade line is constant, the Froude number is also constant unless there is a change in the resistance to flow,

$$\left(F_r = \frac{C}{\sqrt{g}} \sqrt{S} \right)$$

Flow is more nearly two-dimensional in flumes than in natural streams. However, the main current meanders from side to side in a large flume, as it does in the field, and bars of small amplitude but large area develop in an alternating pattern adjacent to the walls of the flume. (It is on these bars that the bed forms shown in fig. 7.1.9 superpose themselves.) If very large width–depth ratios are maintained by keeping depth of flow shallow these bars may grow to the water surface.

In the field it is even more obvious that the flow meanders between the parallel banks of a straight channel, and the alternate bars which form opposite the apex of the meanders are easier to distinguish. As in a flume, if the depth is decreased the alternate bars increase in amplitude until they are close to the water surface, or even exposed. In fact, scour in the main current adjacent to a large bar may cause the water surface there to drop slightly so that the top of the bar is exposed. This has been observed in the Rio Grande and in other natural channels. If the banks are not stable erosion occurs where the high-velocity water impinges, and deposition occurs on the opposite bank. The ultimate development is a meandering stream if other factors such as slope, discharge, and size of bed material are compatible.

The alluvial bars which form on the bed of an alluvial channel is a type of roughness element that plays a very significant role in river mechanics and channel geometry. These bars are much larger than the ripples and dunes illustrated in fig. 7.1.9. Their amplitude ranges from very small to as large as the average depth of flow. Their widths range from $\frac{1}{2}$ to nearly $\frac{1}{3}$ channel width, depending upon the size of the system, and they may be several hundred feet long.

The three most common and accepted types of bars are: point bars, alternate bars, and middle bars (see fig. 7.1.8).

The position, shape, and magnitude of the alternate bars is a function of channel alignment, bed material, and width–depth ratio. These bars normally occur in the straight reaches or crossings between two consecutive bends. Conditions in the upstream bend usually dictate on which side of the channel the first alternate bar will form. Thereafter the sequence is fixed, because the second bar must be on the opposite side of the channel, and so on. In the next bend downstream the secondary circulation developed within it is usually, unless the radius of the bend is very large, of sufficient magnitude to terminate the sequence of alternate bars. Within each bend the pool forms adjacent to the outside bank and a point bar forms on the inside bank. Consequently, the alternate bars are in a sense locked in between the two bends, and their physical characteristics vary with the type of bed material, the width–depth ratio of the channel, and the characteristics of the bends. The number of bars between two bends may increase as the discharge decreases, and vice versa. The flow meanders around these alternate bars. In fact, this flow phenomena is probably the way meanders are initiated and develop with time. For example, consider the successive stages of the development of a meander in sand. Initially the channel may be straight, but the meandering of the thalweg rapidly initiates the development of the meander. Another significant point is that the wavelength of the meander is essentially constant for constant discharge throughout the development of the meander, even though the amplitude increases significantly. The rate of movement of the alternate bars is mostly just a change in size and shape as flow conditions change and the bars alter their geometry to suit the new flow conditions. However, if the bends of the channel are moving, then this allows an additional freedom of movement for the bars. Leopold [1964] has documented that bars move very slowly – of the order of a foot or so per day. Simons has verified the foregoing information on bars by observing their development and movement with time in canals and rivers, and has studied their detailed movement in the sand-bed flumes at Colorado State University. He observed that as a specific weight of the bed material decreases the rate of both change of shape and movement downstream rapidly increases.

Various methods for predicting form of bed roughness have been developed. One of the more useful methods was presented by Simons and Richardson [1963].

12. Tractive force

The tractive-force theory is formulated on the basis that stability of bank and bed material is a function of the ability of the bank and bed to resist erosion resulting from the tractive force exerted on them by the moving water.

Consider the free body of a segment of the full width of the channel as shown in fig. 7.1.10.

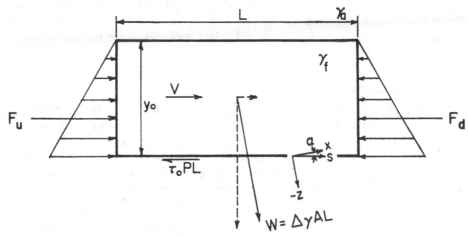

Fig. 7.1.10 Free-body diagram of segment of open-channel flow.

Equating the forces parallel to the flow yields

$$F_u + W \sin \alpha = F_d + \tau_0 p L \tag{31}$$

where W is the weight of the entire segment of fluid;

p is the wetted perimeter – that is, the length of the cross-sectional boundary which is in contact with the fluid flowing in the channel;

F_u and F_d are the upstream and downstream hydrostatic forces acting on the free body where $p = \gamma y$ – since the flow is uniform, $F_u = F_d$;

τ_0 is the average boundary shear which is retarding the flow;

A is the cross-sectional area of the flow;

L is the length of the free body segment;

α is the angle which the channel slope makes with the horizontal; and

$\Delta \gamma$ is the difference between the specific weight γ of the fluid flowing and the specific weight γ_a of the ambient fluid, normally the air.

The product $\Delta\gamma AL$ may be substituted for the weight W and the equation re-arranged to solve for the boundary shear

$$\tau_0 = \Delta\gamma\, \frac{A}{wp}\, \sin\alpha - \Delta\gamma\, RS = \gamma\, RS \tag{32}$$

where R is called the hydraulic radius which is the area A divided by the wetted perimeter p; and

S is $\sin\alpha = dz/ds$, which for relatively flat slopes may be considered more conveniently as $\tan\alpha = dz/dx$.

Equation (31), it may be noted, evaluates the boundary shear in terms of the static characteristics of the geometry and the fluid.

A tractive force theory was clearly presented and illustrated by Lane [1953] to assist with the design of channels for conveying essentially clear water in coarse non-cohesive materials and where bank stabilization is to be achieved by armour plating with coarse non-cohesive material. For a review of the design of alluvial channels in accordance with regime and other concepts, refer to Lacey [1958] and Simons and Albertson [1963].

13. Sediment transport

Knowledge of sediment transport in alluvial channels is just as important as knowledge of resistance to flow. The ability of a stream to transport bed material

TABLE 7.1.3 Variation of concentration on a dry-weight basis of total bed material load with regimes of flow and forms of bed roughness

		Total bed material load (p.p.m.)	
Regime of flow	Forms of bed roughness	Median diameter of bed material 0·28 mm	Median diameter of bed material 0·45 mm
Lower flow regime	Ripples	1–150	1–100
	Dunes	150–800	100–1,200
Transition	Zone in which dunes are reducing in amplitude with increasing shear stress	1,000–2,400	1,400–4,000
Upper flow regime	Plane	1,500–3,100	——
	Standing waves	3,000–6,000	4,000–7,000
	Antidunes	5,000–42,000	6,000–15,000

is relatively small when the form of bed roughness consists of ripples and/or dunes. In the upper regime of flow the streams are capable of carrying much larger volumes of sediment per unit volume of water (see qualitative data in Table 7.1.3 suggested by Simons and Richardson [1963]). Some of the more useful concepts for estimating bed material discharge have been presented by Einstein [1950], Colby and Hembree [1955], Colby [1964], Simons *et al.* [1965], and Bishop *et al.* [1965]. For a more detailed treatment of the sediment problems encountered in designing and operating irrigation canals constructed in alluvium, refer to Simons and Miller [1966].

14. Summary

The basic concepts of fluid mechanics applicable to flow in open channels has been presented. For a more detailed treatment of hydraulics and fluid mechanics, see Albertson *et al.* [1960] and Albertson and Simons [1964] and other fluid-mechanics texts. Also, many valuable concepts pertinent to the design of hydraulic structures associated with the conveyance and distribution of water have been presented by the U.S. Bureau of Reclamation [1960, 1963, 1964].

REFERENCES

ALBERTSON, M. L., BARTON, J. R., and SIMONS, D. B. [1960], *Fluid Mechanics for Engineers;* (Prentice Hall, Englewood Cliffs, New Jersey), 568 p.

ALBERTSON, M. L. and SIMONS, D. B. [1964], Fluid Mechanics; In Chow, V. T., Editor, *Handbook of Applied Hydrology*, (McGraw-Hill, New York), Chapter 7, 49 p.

BAZIN, H. [1897], Etude d'une nouvelle formule pour calculer le debit des canaux decouverts; *Annales des Ponts et Chaussees*, Memoire No. 41, Vol. 14, Ser. 7, 4me trimestre, pp. 20–70.

BHOWMIK, N. G. [1965], *The Hydraulic Design of Large Concrete-lined Canals;* Thesis. Colorado State University, Fort Collins, Colorado.

BISHOP, A. A., SIMONS, D. B., and RICHARDSON, E. V. [1965], Total bed-material transport; *Proceedings of the American Society of Civil Engineers, Journal of the Hydraulics Division* 91 (HY2), 175–91.

COLBY, B. R. [1964], Discharge of sands and mean-velocity relationships in sandbed streams; *U.S. Geological Survey Professional Paper* 462-A, 47 p.

COLBY, B. R. and HEMBREE, C. H. [1955], Computations of total sediment discharge, Niobrara River near Cody, Nebraska; *U.S. Geological Survey Water Supply Paper* 1357, 187 p.

EINSTEIN, H. A. [1950], The bed-load function for sediment transportation in open channel flows; *U.S. Department of Agriculture Technical Bulletin* 1026; 71 p.

ELLIS, G. H. [1916], The flow of water in irrigation canals; *Transactions of the American Society of Civil Engineers*, Paper no. 1373, 1644–88.

GARBRECHT, G. [1961], Flow calculations for rivers and channels; Die Wasser-Wirtschaft, (Stuttgart), Parts I & II, 40–5 and 72–7. (U.S. Bureau of Reclamation Translation 402.)

HAUSHILD, W. L., SIMONS, D. B., and RICHARDSON, E. V. [1961], The significance of fall velocity and effective diameter of bed materials; *U.S. Geological Survey Professional Paper* 424-D, 17–20.

LACEY, G. [1958], Flow in alluvial channels with sandy mobile beds; *Proceedings of the Institution of Civil Engineers*, 9, 145–64.

LANE, E. W. [1953], Progress report on studies on the design of stable channels by the Bureau of Reclamation; *Proceedings of the American Society of Civil Engineers*, **79**, 1–31.

LEOPOLD, L. B. and EMMETT, W. W., [1963], *Downstream pattern of River-bed scour and fill;* U.S. Geological Survey paper prepared for presentation of the Federal Interagency Sedimentation Conference, Jackson, Mississippi.

LEOPOLD, L. B. and MADDOCK, T. [1953], The hydraulic geometry of stream channels and some physiographic implications; *U.S. Geological Survey Professional Paper* **252**, 56 p.

MANNING, R. [1889], On the flow of water in open channels and pipes; *Transactions of the Institution of Civil Engineering of Ireland*, **20**, 161–207. (Supplement, 1895, **25**, 179–207).

SIMONS, D. B. and ALBERTSON, M. L. [1963], Uniform water conveyance channels in alluvial material; *Transactions of the American Society of Civil Engineers*, **128**, 65–106.

SIMONS, D. B. and MILLER, C. R. [1966], Sediment discharge in irrigation canals; *Proceedings of the International Committee on Irrigation and Drainage, 6th Congress*, (New Delhi, India) Quest 20, Rep. 12, 20275–307.

SIMONS, D. B. and RICHARDSON, E. V. [1962], Resistance to flow in alluvial channels; *Transactions of the American Society of Civil Engineers*, **127**, 927–52.

SIMONS, D. B. and RICHARDSON, E. V. [1963], Forms of bed roughness in alluvial channels; *Transactions of the American Society of Civil Engineers*, **128**, 284–302.

SIMONS, D. B., RICHARDSON, E. V., and HAUSHILD, W. L. [1963], Some effects of fine sediment on flow phenomena. *U.S. Geological Survey Water Supply Paper*, 1498-G, 46 p.

SIMONS, D. B., RICHARDSON, E. V., and NORDIN, C. F. [1965], Bedload equation for ripples and dunes; *U.S. Geological Survey Professional Paper*, 462-H, 9 p.

U.S. BUREAU OF RECLAMATION [1960], *Design of Small Dams* (U.S. Government Printing Office, Washington, D.C.), 611 p.

U.S. BUREAU OF RECLAMATION [1960], *Design of Stable Channels with Tractive Forces and Competent Bottom Velocity;* Sedimentation Section, Hydrology Branch, Bureau of Reclamation, Denver Federal Center, Denver, Colorado.

U.S. BUREAU OF RECLAMATION [1963], *Hydraulic Design of Stilling Basins and Energy Dissipators* (Supplement of Documents, Washington, D.C.), Engineering Monographs **25**, 114 p.

U.S. BUREAU OF RECLAMATION [1964], *Design Standards No. 3: Canals and Related Structures;* Commissioner's Office. Denver Federal Center, Denver, Colorado.

II. Hydraulic Geometry

G. H. DURY

Department of Geography, University of Sydney

Hydraulic geometry is the graphical analysis of the hydraulic characteristics of a stream channel. These include width, depth, slope, discharge, velocity, bed material, and load. Discharge can be regarded as the only fully independent variable. Bed material is an independent variable in so far as its properties are determined by outcropping bedrock in the floor or sides of the channel; and it can also be regarded as an independent variable, since its properties are controlled by the geology of the catchment. Loose material, however, will be sorted during transit, and will also undergo reduction of size and change of shape. The remaining characteristics interact upon one another in a complex fashion. Whether they are to be classed as dependent or independent in a particular context may depend on the point of view. It will be convenient here to consider the whole series by groups, and to introduce at an early juncture the frequency-characteristics of discharge.

1. Interrelationships of discharge, velocity, depth, and width

Discharge is by definition the amount of water passing through the cross-section in unit time, expressed, for instance, in cubic feet per second. In what now follows depth is mean water depth and width is water surface width.

Consider an ordinary stream flowing in a single alluvial channel; and imagine to begin with a condition of low flow. Then, when discharge increases, so will depth and width increase. Part of the increase in depth will be accounted for by scour if the bed material can be shifted, although scour may not at once begin when stage rises (fig. 7.II.1). Part at least of the increase in width will result simply from the rise of water-surface level, but part at some sections will involve the temporary removal of bank material. Velocity also increases as discharge increases, in response to the enlargement of the cross-section and the associated reduction of average friction on the flowing water.

Variations in these and other characteristics at a given cross-section are called *at-a-station variations*. Although the foregoing verbal summary of their inter-relationships is self-evident enough, there is more to the matter than this. Velocity, depth, and width can all be expressed as power functions of discharge (fig. 7.II.2). The prospect of bringing order into the variations of these four characteristics is, however, probably less important here than the indicated general inference, namely, that values of width, depth, and velocity cannot be

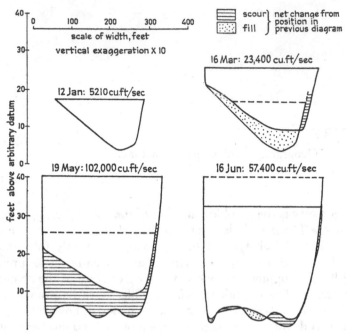

Fig. 7.11.1 Changes in size and shape of channel on the Colorado River at Grand Canyon, Arizona, during the passage of the December 1940–June 1941 flood (Freely adapted from Leopold and Maddock, 1953).

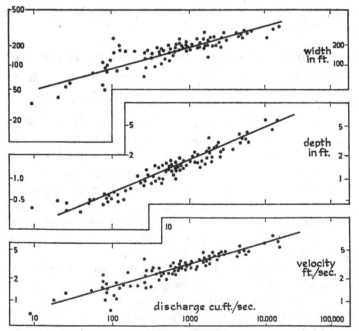

Fig. 7.11.2 Variation of width, depth, and velocity with varying mean annual discharge in the downstream direction: Powder River system, Wyoming and Montana (Adapted from Leopold and Maddock, 1953).

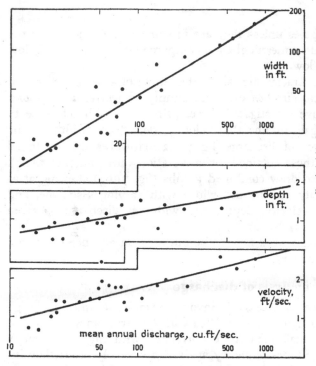

Fig. 7.11.3 Variation of width, depth, and velocity with varying discharge at-a-station: Powder River at Locate, Montana (Adapted from Leopold and Maddock, 1953).

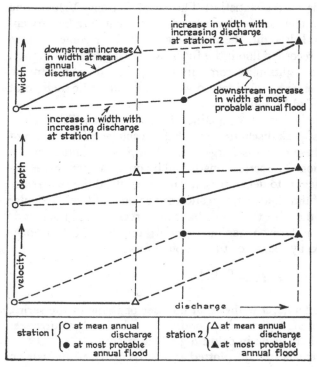

Fig. 7.11.4 Graphs of hydraulic geometry, showing at-a-station and downstream variations combined: diagrammatic, all scales logarithmic; Station 2 is downstream of Station 1 (Adapted from Leopold and Maddock, 1953).

used for comparative purposes unless they are in some way connected to the scale of discharge. It would be meaningless to compare one cross-section at low flow with another at high flow.

Variations of hydraulic characteristics along the length of a stream are called *variations in the downstream direction* or, more simply, *downstream variations*. Leopold and Maddock have investigated these, principally with reference to mean annual discharge, again finding that velocity, depth, and width can be expressed as power functions of discharge (fig. 7.II.2). However, the functional values usually differ from those obtained for at-a-station variations (fig. 7.II.3). It thus becomes possible to draw combined graphs (fig. 7.II.4) showing at-a-station variations in velocity, depth, and width with varying discharge, and downstream variations in velocity, depth, and width at discharge of given frequency.

Considerations of frequency need now to be outlined, in advance of further discussion of hydraulic geometry.

2. Magnitude-frequency analysis of discharge

Analysis of discharge-frequency necessarily involves magnitude, since very high discharges are rare (great magnitude correlates with low frequency) while low discharges are common (small magnitude correlates with high frequency). Several techniques of magnitude-frequency analysis are widely used: they include that developed by E. J. Gumbel, which will be outlined here.

Gumbel analysis depends on the statistical Theory of Extreme Values, one proposition of which is – crudely put – that records will eventually be broken. However severe an historical drought, there will sooner or later come a worse one; no matter how high a recorded flood, the future will bring one still higher. But although floods and droughts may come in spells, they do not come in cycles. In order to assess their time-relationship we need a non-cyclic method of analysis, such as that provided by Gumbel.

The operation of this method is best explained by an example. Table 7.II.I lists in column 1 the highest peak discharge recorded in each year for a series of years at a single station. These peak discharges are called *floods*, whether or not they actually cause inundation. The series constituted by one peak per year is the *annual series*. The requirement is to define for each recorded flood the *recurrence-interval*, the average span of time between two successive floods of that particular magnitude. In column 2 of the table the floods have been rearranged in descending order of magnitude, and given ranking (serial) numbers. Recurrence-intervals are now calculated by means of the simple equation

$$r \cdot i \cdot = \frac{n+1}{r}$$

where $r \cdot i \cdot$ is recurrence-interval, n is the total number of items in the series, and r is the ranking order of a particular flood. The calculated recurrence-intervals for the series listed are given in column 3 of the table.

Observed magnitudes of floods are now plotted against recurrence-intervals on

TABLE 7.II.I Annual floods on the Wabash River, at Lafayette, Indiana, 1924–57

Source: U.S. Dept. of Interior open-file Report

	1		2	3
Year	Peak discharge (ft³/sec)	Ranking Order	Discharge (ft³/sec)	Recurrence-interval (years)
1924	59,800	1	131,000	35
1925	63,300	2	93,500	17·5
1926	57,700	3	90,000	11·7
1927	64,000	4	74,600	8·8
1928	63,500	5	74,400	7·0
1929	38,000	6	73,300	5·8
1930	74,600	7	67,500	5·0
1931	13,100	8	64,000	4·4
1932	37,600	9	63,500	3·9
1933	67,500	10	63,300	3·5
1934	21,700	11	63,300	3·2
1935	37,000	12	62,000	2·9
1936	93,500	13	59,800	2·7
1937	58,500	14	58,500	2·5
1938	63,300	15	57,700	2·3
1939	74,400	16	52,600	2·2
1940	34,200	17	50,600	2·06
1941	14,600	18	46,600	1·95
1942	44,200	19	44,200	1·84
1943	131,000	20	41,900	1·75
1944	73,300	21	41,300	1·67
1945	46,600	22	41,200	1·59
1946	39,400	23	38,400	1·53
1947	41,200	24	38,000	1·46
1948	41,300	25	37,600	1·40
1949	62,000	26	37,000	1·34
1950	90,000	27	35,300	1·29
1951	50,600	28	35,000	1·25
1952	41,900	29	34,700	1·21
1953	35,000	30	30,000	1·17
1954	16,500	31	21,700	1·13
1955	35,300	32	16,500	1·09
1956	30,000	33	14,600	1·06
1957	52,600	34	13,100	1·03

special graph paper – Gumbel paper (fig. 7.II.5). If the data conform strictly to the Theory of Extreme Values, then the plotted points lie on a straight line. In practice, a certain amount of scatter is usual, and some plots give lines which are inflected either upwards or downwards. Nevertheless, they can still be used to read off magnitudes against recurrence-intervals, and vice versa. Against the recurrence-interval of 50 years is read off the magnitude of the 50-year flood, the peak discharge expectable as an annual maximum once in a 50-year span. Emphatically, this does *not* mean at regular intervals of 50 years. There is nothing to prevent the occurrence of two 50-year floods in two successive years. Furthermore, some 50-year spans must inevitably include 100-year floods (on the average, one such span in two) 500-year floods (one in ten), and 1,000-year floods (one in twenty).

For preference, the period of record should be twice as long as the greatest recurrence-interval for which flood magnitude is obtained. Few series of records are as long as 100 years, but for many stations the magnitudes of the 25-year,

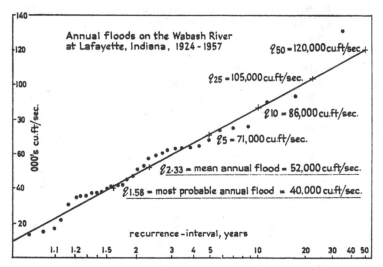

Fig. 7.11.5 Example of a Gumbel graph: scales simplified and graph face omitted (cf. Table 7.11.1).

10-year, and 5-year floods can be satisfactorily defined. These floods come within the practicable limits of much engineering work, flood control, and flood alleviation. For present purposes, lesser recurrence-intervals and lesser floods are of main interest. The 2·33-year flood is *the mean annual flood*; the 2-year flood is the *median annual flood*, equalled or exceeded in one year in two; and the 1·58-year flood is the *most probable annual flood*. These are symbolized respectively as $q_{2.33}$, $q_{2.0}$, and $q_{1.58}$.

Both below and in a later section it will be necessary to discuss discharge in relation to stage – specifically, in relation to discharge at the bankfull stage. Bankfull discharge is symbolized as q_{bf}. If it were possible it would be useful to locate bankfull discharge on the magnitude-frequency scale, but complications occurring on natural rivers, quite apart from those due to engineering works, make the desired act of location very difficult. But the matter can be approached in another way. Peak floods below the median value disappear from half the record of annual peaks, since they are by definition equalled or exceeded in one year in two. Similarly, peaks equivalent to the most probable annual flood vanish from more than half of the annual series. Now when account is taken of flood peaks smaller than the annual maxima it can be shown that, in addition to being the most probable annual maximum, the discharge equivalent to $q_{1.58}$ is likely to occur once every year. These seem good grounds for suggesting that $q_{1.58}$ = natural bankfull discharge. It is possible to take the matter further, saying that natural bankfull discharge = channel-forming discharge, on the grounds that analysis of such matters as sediment delivery shows that channels are shaped not by events of great magnitude and low frequency but by those of modest magnitude and high frequency. In addition, it is at bankfull stage that water is in contact with, and is acting on, the whole perimeter of the channel.

3. Interrelationships of slope, discharge, and velocity

Downstream channel slope on most rivers tends to decrease from source to mouth, giving longitudinal profiles a concave-upward form. The basic reason for a downstream decrease of slope is a downstream increase in channel efficiency.

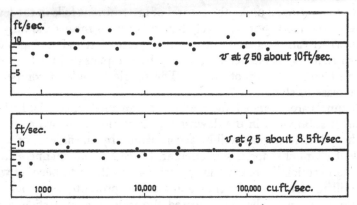

Fig. 7.11.6 Velocity-relationships on the Yellowstone–Missouri–Mississippi system, at the five-year and fifty-year floods (Adapted from Leopold, 1953).

This increase is a response to the increase in channel size which results from the entry of tributaries. The combined effects of relief distribution, basin shape and size in relation to the layout of the stream network, and runoff characteristics frequently ensure that the concave-upward profile tends to be approximately logarithmic. Alternatively, the logarithmic form can be regarded as the outcome of statistical probability.

Leopold began the study of velocity in relation to slope with an analysis of rivers in the U.S. Mid-west, for which he used mean discharge as a basis of reference. This discharge corresponds to below-bank stages. Leopold showed that, at mean discharge, velocity through the cross-section can actually *increase* downstream. Later work on the Yellowstone–Mississippi revealed that there is no tendency for velocity to change along the length of this system, despite a downstream decrease in slope from 100 to 0·5 ft/mile, either at the 50-year or at the 5-year flood (fig. 7.11.6).

It appears to follow that constant velocity along the length of the stream is attained somewhere between the below-bank stage of mean discharge and the modest overbank stage of the 5-year flood. Theoretical analysis of stream behaviour at bankfull combines with pilot empirical studies of velocity at $q_{1.58}$ to suggest that constant downstream velocity is attained at bankfull or at the most probable annual flood, according to which is in question. That is to say, channel slope tends to be adapted to promote constant velocity downstream at channel-forming discharge.

These observations and this conclusion make nonsense of the assertion – unfortunately still a common one – that streams rush hurriedly down their mountain courses but move sluggishly across the plains. This could be so only if there were no downstream change of channel size.

4. Interrelationships of slope, width, depth, and bed material

Many longitudinal profiles are far from regular: they include reaches where slope increases downstream. Profiles consisting of segments of concave-up curves may have been developed under the control of intermittent falls in the level of the sea relative to the land. But sea-level change is by no means the only possible cause of downstream steepening of gradients and of segmented profiles. Any factor which reduces channel efficiency may be compensated for by some equal but opposite factor or group of factors. The readiest mode of compensation is often an increase in slope.

Large channels are, as stated, more efficient than small channels. If therefore a single channel subdivides in the downstream direction into two or more lesser channels the subdivision may well be accompanied by an increase of slope. The most efficient channel shape is semicircular. Since alluvial channels are usually shallower than semicircular, anything which causes them to become wider and shallower will reduce their efficiency and tend to promote a compensating increase in slope. Schumm has investigated the influence of particle-size in the materials of the beds and banks of alluvial channels, finding that a high proportion in the silt and clay grades produces a low width/depth ratio, whereas a low proportion of silt-clay is associated with great width in relation to depth. The two kinds of shape are, respectively, efficient and inefficient.

At the other end of the range of possible shapes are channels which are mere slots in bedrock, excavated vertically down belts of weakness but confined by strong vertical walls. These, too, are markedly inefficient, and likely to appear on longitudinal profiles as reaches where slope increases downstream.

Whether in alluvium or in bedrock, channels range considerably in their frictional drag on the flowing water. It is partly for this reason that the well-known equation of Manning, which relates velocity to size, shape, and slope, also includes a roughness factor. Roughness in this context includes not only physical obstruction such as that produced by irregularly jutting masses of rock in place but also the effect of particle-size in the beds and banks. To this extent it overlaps with the properties investigated by Schumm. In addition, roughness includes the frictional drag produced by material in transit. Once again, the greater the roughness, the greater the slope required to promote a particular velocity at a given discharge.

It follows from all this that any downstream increases of slope which may be detected by survey or from accurate maps cannot be fully understood unless their possible causes have been checked in the field.

5. Grade

In its application to rivers, the concept of grade has often proved elusive and sometimes confusing. It calls for a balance between the capacity to do work (in particular, to transport load) and the amount of work to be done – e.g., the quantity of load supplied. A graded profile is supposed to be produced when capacity and performance are matched. However, the whole of the very bulky early debate on grade is rendered largely meaningless by lack of reference to magnitude-frequency of discharge. Use of the concepts of magnitude-frequency, and especially of channel-forming discharge, directs attention not so much to grade as to steady-state conditions. In any event, it has been pointed out that slope is determined with reference to a whole series of additional variables, so that a graded profile, however defined, need not be a smooth continuous curve throughout.

6. Load in relation to material of the bed and banks

It is useful for a number of purposes to separate material in transit into some-what arbitrary fractions, according to its mode of progression along the channel. *Bedload* consists of material too coarse to be supported in the flowing water for any appreciable time. It moves by sliding, rolling, or saltating, continually coming to rest and thus reverting to bed material. *Suspension load* is carried in the body of the current. The degree of coarseness of particles varies with velocity, and especially with the degree of turbulence. Streams at bankfull can be assumed to have turbulent flow, with rising and descending eddies. At a velocity of 3 ft/sec, which is modest for flow at bankfull stage, transport of sand is approximately uniform through all parts of the cross-section, and a stream can deliver 10 tons per day for every foot of filled channel width at a given range of depth.

Some writers separate off as *wash load* that part of the suspended load which, although consisting of solid particles, is so fine-grained that its settling velocity is very small or nil. This portion is carried in colloidal suspension. *Solution load* is by definition transported in the dissolved state.

Most of the available information on transport of load by rivers deals either with suspended load or with solution load. Movement of bedload is remarkably difficult to measure in field conditions. On the other hand, a great deal is known about the tendency of particles larger than sand size to become reduced in di-mensions with progression downstream, and also about the critical stress needed to move particles of a given size. It is fully clear that bedload movement tends to increase as velocity increases: that is – up to the bankfull stage at least – as discharge increases. Transport in solution varies widely in its effectiveness. It is controlled at least in part by the geology of catchments and by climate. In some arid (salty) and humid-tropical environments it can exceed transport in suspension, just as it can in catchments where the rocks are markedly soluble. The relationships of solution-load to discharge are more complicated than those of bedload, for solution-load concentration is likely to be low at times of high discharge, whereas total transport in solution is similarly low at times of low discharge. Generally speaking, the greatest effectiveness of transport in solution seems likely to be attained at stages intermediate between low and very high.

Suspended load has been well studied. At many stations on many rivers it

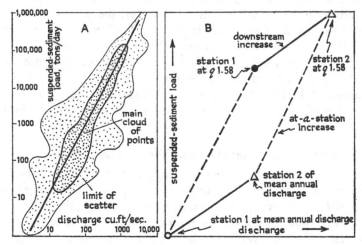

Fig. 7.11.7 (A) Variation of suspended-sediment load with varying discharge, Powder River at Arvada, Wyoming (Adapted from Leopold and Maddock); (B) Combined at-a-station and downstream variations of suspended-sediment load with varying discharge; diagrammatic, all scales logarithmic: Station 2 is downstream of Station 1 (Compare Fig. 7.11.2) (Adapted from Leopold and Maddock, 1953).

increases as a power function of discharge, increasing as a rule far more rapidly than discharge does (fig. 7.11.7(a)). As for width, depth, and velocity, so for suspended-sediment load, a combined graph can be drawn to show at-a-station and downstream variations (fig. 7.11.7(b)). Like the concentration of solution-load, the concentration of suspended-sediment load is apt to be less at times of very high flow than at times of moderate flow, simply because the suspended material is disseminated through a great volume of water. Magnitude-frequency analysis of sediment transport shows that the greatest total transport is effected by discharges of modest magnitude and high frequency, similar to those recognized as responsible for shaping the channels themselves.

REFERENCES

LEOPOLD, L. B. [1953], Downstream change of velocity in rivers; *American Journal of Science*, 251,606–624.

LEOPOLD, L. B. and MADDOCK, T. JR. [1953], The hydraulic geometry of stream channels and some physiographic implications; *U.S. Geological Survey Professional Paper* 252, 57 p. (The basic reference on hydraulic geometry.)

LEOPOLD, L. B., WOLMAN, M. G. and MILLER, J. P. [1964], *Fluvial Processes in Geomorphology* (W. H. Freeman and Co., San Francisco and London), 522 p. (This incorporates much relevant work, especially in Chapters 6 and 7.)

WOLMAN, M. G. [1955], The natural channel of Brandywine Creek, Pennsylvania; *U.S. Geological Survey Professional Paper* 271, 56 p.

WOLMAN, M. G. and MILLER, J. P. [1960], Magnitude and frequency of forces in geomorphic processes; *Journal of Geology*, **68**, 54–74.

III(i). The Human Use of Open Channels

ROBERT P. BECKINSALE
School of Geography, Oxford University

The human use of rivers and streams depends on the nature of the rivers as well as on the needs and customs of the riverine societies. Most modern methods of river use are merely the technical refinements of usages practised in some advanced early civilizations long before the Christian era.

Today the problems of the human use of open channels with regimes fall under nine main interrelated facets: flood control; irrigation and drainage; water power; flotability; navigation; water supply; fishing and wild-life conservation; recreation and religion; and water-pollution (fig. 7.III(i).1). In the following short discussion, together with Chapter 10.III, which treats the first three facets most dependent upon the regime variations of discharge, little more can be attempted than to demonstrate these aspects from notable examples so as to illustrate the present state of man's response to the socio-economic opportunities offered by rivers.

1. Flotability

The crudest human use of rivers, the uncontrolled floating of objects, is still very popular. In large areas the timber, pulp, and paper industries depend mainly upon river floatways for the transport of their raw material. This is especially so in the northern forests of Canada, Scandinavia, and the U.S.S.R., where the river regimes (D) (see Chapter 10.1) have a spate in spring or early summer. Many early sawmills grew up inland near waterfalls, and in, for example, Norway a few mills still use direct water-power. With the coming of steam-power and later of hydroelectricity the chief pulp-processing mills tended to be sited at the junction of rivers or at their mouths in more convenient locations for markets and export. Because typical extra-tropical timbers float easily, it is usual in many countries to pile the logs in winter near floatways, into which they can be rapidly pushed at the thaw. As in much of Scandinavia the thaw progresses inland and upward, a fairly continuous supply of logs can be sent to the river mouth in the floating season. In most lumbering regions various forms of mechanical transport are used to take logs to the mill, but the prime method is still to drag logs to a floatway. In Norway in 1960 nearly 10,000 miles of floatway handled 33 million logs; in Sweden a public-floatway system of over 21,000 miles, with a staff of nearly 40,000 persons, deals with about 170 million logs annually, and the number handled has risen to 208 million (in 1949). In Finland about 25,000 miles

Fig. 7.III(i).1 Model for multiple-purpose integrated river basin development.

1. Multiple-purpose reservoir.
2. Recreation; swimming, fishing, camping.
3. Hydroelectric station.
4. Municipal water supply.
5. City and industrial waste treatment plant.
6. Pump to equalizing reservoir for irrigation.
7. Diversion dam and lake.
8. High-level irrigation canal.
9. Levees for flood control.
10. Erosion control: stream dams and contour terracing.
11. Regulating basin for irrigation.
12. Wildlife refuge.
13. Low-level irrigation canal.
14. Gravity irrigation.
15. Contour ploughing.
16. Sprinkler irrigation.
17. Community water treatment plant.
18. Navigation: barge trains, locks, etc.
19. Re-regulating reservoir with locks.
20. Farm pond with pisciculture.

(Adapted from *A Water Policy for the American People*, Vol. I, X)

of floatway are in use, and the busiest carry up to 3 million m³ of timber annually for the national production of nearly 4½ million tons of pulp and 1½ million tons of paper (fig. 7.III(i).2).

The mass floating of logs causes problems for hydroelectric operators;

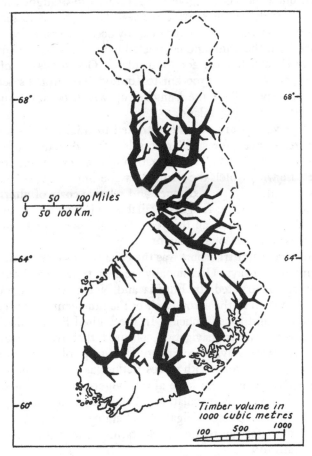

Fig. 7.III(i).2 Chief timber floatways in Finland, 1960.
The blank areas in parts of the south-west occur largely because here short-haul road transport has replaced river floating (Adapted from *Atlas of Finland*, 1960 and Millward, 1965).

similarly, dams hinder log floating and complicated legal disputes have been common. In Finland the eventual solution was to compel the hydrosite constructor to build a by-pass trough for logs and to compensate the log-floating company for 1½ times its direct material losses caused by damming the river. Difficulties of another kind occur in the tropics, where many valuable hardwoods do not float when green. Teak is felled some time before being dragged to the waterfront, a task often performed in Burma by elephants,

M

2. Navigation

Treaties concerning navigation on the River Po date back to A.D. 1177 and on the Rhine to A.D. 1255. For thousands of years there was little difference between sea and river craft, but as ocean-going vessels increased in draught their navigation inland was increasingly restricted to estuaries. Today the Amazon is the only unaltered waterway usable for long distances by ocean liners. Here floating docks a short distance from the bank and connected to it by a swing bridge suffice for ocean-going vessels at Manaus, 2000 km inland. Ocean vessels of up to 14 ft draught can ascend a further 1700 km to Iquitos. The world's second longest natural waterway may well be the Yangtze, up which ocean-going vessels can ascend to Hankow, 1,000 km inland.

Elsewhere large vessels are usually restricted to tidal reaches and lakes, and navigation upstream, even for smaller vessels, often demands dredging, locks, canalization, and river control. The utilization of a waterway for traffic depends heavily on the ease with which the channel can be improved, on its connections with other water routes, on the existence and relative costs of alternative means of transportation, and on the freights available. AF and CF river regimes are especially suited to navigation all the year. D regimes suffer from seasonal ice and other regimes from seasonal drought.

In many countries water transport was the chief means of moving heavy goods until the coming of railways. Occasionally steamer traffic was flourishing when the railways were constructed. In Eurasia and North America the survival of waterway transportation in the second half of the nineteenth century was often a question of national and social policies. In the United States the Constitution, and since 1824 the laws on navigation, ensured that the improvements and maintenance of the waterways were financed out of public funds. All navigable waterways were free from tolls, with no restrictions as to user. Similarly, in more recent times the Soviet planners have always vigorously promoted river transportation and hydroelectric installation. They have, for example, allocated enormous funds for the Greater Volga Project and given generous tariff inducements in favour of waterway traffic. These tariffs offset the more direct routes followed by the railways.

Today river transport is especially important in industrialized countries with bulk cargoes and in some densely peopled lands, such as China and South-East Asia, where a large population of boat owners, often living partly or entirely on the water, combine habitations with trading or fishing. The latter seems to lack statistics; the former may be best exemplified from North America and Europe.

In the United States the minimum depth of 6 ft was increased to 9 ft in 1930, and since 1944 was raised to 12 ft on the lower 1,200 km of the Mississippi. By 1947 about 28,000 miles of the 65,000 miles of so-called navigable waterway in the country had been improved, and the waterway movement totalled 31,500 million ton-miles on the rivers and 112,000 million ton-miles on the Great Lakes. In 1964 the corresponding figures were 144,253 million ton-miles on rivers and 105,912 million ton-miles on the Great Lakes. The most impressive increase was on the Mississippi system, including great tributaries such as the Ohio. Here, in spite of competition from pipelines, the freight carried expanded from 115 million short tons (30,382 million ton-miles) in 1950 to over 226 million short tons (80,087 million ton-miles) in 1964. Much of this increase was due to improvements in navigability and in barge techniques.

Navigation on the Great Lakes is closed by ice for about four months annually, but it has long been the world's busiest system of inland water transport and is unsurpassed for size of boats and length of hauls. After 1932, when the Welland Canal (by-passing Niagara Falls) was reconstructed, lake carriers of 20,000–25,000 tons drawing 27 ft could navigate between all the Great Lakes. The Saulte-Ste-Marie canal between Lakes Huron and Superior was regularly used by over 110 million tons of cargo annually. But rapids on the lower St Lawrence closed the system for ocean-going traffic. Between 1953 and 1958 this was corrected by the construction of the St Lawrence Seaway, whereby a series of locks, connected with hydroelectric projects, lifts vessels 69 m to Lake Ontario. Thence the existing Welland Canal, which is considered a part of the Seaway, lifts vessels a further 100 m to Lake Erie. Lake carriers of up to 25,000 tons now carry iron ore from the lower St Lawrence to upper lake ports, and ocean vessels of up to 8,000 tons travel 2,700 km inland. Chicago, at nearly 180 m above sea-level, and Duluth-Superior now each deal direct with over 4½ million tons of foreign cargo annually, irrespective of their large lake trade. Traffic on the Seaway, as distinct from that on the four upper lakes, rose to 67 million short tons of cargo in 1966, and of this amount 14 million tons came from overseas. The chief commodities by volume were iron ore, wheat, bituminous coal, corn, manufactured iron and steel, fuel oil, barley and soya-beans, all ideally suited to bulk handling.

In Europe, which is about the same size as the mainland United States, navigation is favoured by the low-lying plain that extends from the Netherlands to the Urals, but is hindered by political fragmentation. The busiest waterways are the Rhine, Elbe, Danube, and Volga. The Rhine takes 4,000-ton barges to the Ruhr and 2,000-ton boats as far as Basel, while many of its tributaries and feeder canals take barges of 1,000–1,500 tons. In 1965 cargo handled on the Rhine amounted to 223 million tons or a nearly 13% above that in the former peak year 1961. The international freight consisted of 16,050,000 tons carried on the Netherland's section; 79,599,000 tons (27·5 millions downstream and 52 millions upstream) over the German–Dutch frontier; and 12,310,000 tons between West German and French ports and Basel (Switzerland). To these amounts must be added the internal traffic of 59,048,000 tons between West German ports, and 56 million tons between Netherland's ports. The total freight movement amounted to 43,821 million ton-kilometres. Table 7.III(i).1 shows the length of navigable inland waterway (river and interconnecting canal) and certain freight characteristics of various countries in Europe in 1966. In the last column, A relates to all freight journeys while B excludes freight carried by short road hauls.

Except in Britain, in 1965 and 1966 the growth of water traffic in Europe was well maintained and was not seriously affected by new pipeline developments for crude petroleum. At Strasbourg, for example, decrease in crude-oil traffic was offset by the rise in the transport of refined petroleum products. The upper Rhine trade to and from Basel reached an all-time record of 8·8 million tons in 1965, while the cargo traffic on the newly finished canalized Moselle amounted to 3·4 million tons. The Moselle is being improved upstream of Metz, and by the early 1970s will be connected with the Saône, so providing a Rhine–Rhône waterway.

TABLE 7.III(i).1

Country	Total tonnage (1,000 tons)	Total ton/km (millions)	Average length of haul (km)	Total length of navigable waterway in regular use (km)	Percentage total natural freight movement carried by waterway (ton/km) A	B
Belgium	79·6	5,392	68	1,595	26*	
France	93·5	12,652	135	7,677	8	9
West Germany	207·9	44,826	216	4,496	26†	30
Netherlands	197·8	25,240	128	6,044	64	
U.S.S.R.	279·0	137,582	493	142,700	6	
United Kingdom	7·8	195	25	1,345	0·2	

* Percentage as for 1961. † Percentage for 1965.

The Danube, which traverses Alpine Europe, is being extensively regularized. The projects include thirteen dams with hydroelectric installations on the Austrian section alone, and a large barrage at Sip below the Iron Gates which will dam back the water far upstream of the rapids, and will provide locks 300 m long and 34 m wide for navigation. Since 1932 a Rhine–Danube link has been undertaken by a company with a State concession to exploit the local water-power and use the net earnings to finance canal building. By 1967 about forty-three hydrostations with a total annual production of 2,300 million kWh had been constructed in the Main basin, and their earnings had facilitated loans sufficient to complete the first 20 miles of the 120-mile Rhine–Danube connection. Boats of 1,500 tons can now reach Bamberg on Main, and will, it is planned, reach Nuremberg by 1970 and Regensburg on the Danube by 1981. Then boats of that size will be able to travel 2,125 miles from Rotterdam to the Black Sea, and tonnage at Danube ports (21 million tons of cargo loaded and 22·5 million tons unloaded in 1962) will greatly increase. The freight carried on the river in 1966 weighed 45 million tons, of which 28 million was internal traffic.

The third great expansion of the European waterway network has been in the U.S.S.R., where the hauls are long, once the rivers have been regulated against strong summer floods. In 1937 the Moscow–Volga canal provided an adequate water supply and good water route for the capital. In 1952 the lower Volga–Don canal linked the Volga with ocean transport. Then, by a series of huge dams and reservoirs, already discussed, the Volga and Don were controlled, and connecting waterways and canals leading northward were improved, so that by 1965 a Volga–Baltic waterway was opened. The guaranteed navigable depth is 9½ ft on the main rivers and 8 ft on the Kama. Vessels of up to 5,000 tons can now navigate between the White Sea, Baltic, Caspian, and Black Sea. In 1966 the Soviet river and canal systems operated 138 billion ton-miles of freight traffic, or over four times the ton-mileage of 1950. The freight movement on the Volga system alone in 1960 was estimated at 40 billion ton-miles, with cargo traffic of up to 17 million tons between Kubyshev and Volgograd. The statistics given for the U.S.S.R. in the above Table include the transport of timber, which in 1966 amounted to 69,700,000 tons and 27,600,000 ton-km.

3. Domestic and industrial uses and the problem of open-channel pollution

In most countries the domestic and industrial consumption of water increases rapidly with the growth of population, of industrialization, and of the standard of living and hygiene. Even in areas with a long-established piped supply, modern domestic appliances cause the water consumption to rise rapidly. In 1830 the British domestic user managed on less than 4 gallons (18 l.) per day, whereas by 1960 he needed over 60 gallons (270 l.) daily. In the United States each person uses on an average about 140 gallons a day, and the inhabitants of some large cities need over 180 gallons. In contrast, at Karachi the average *per capita* daily consumption is probably about 20 gallons.

Recently, owing to depletion of ground aquifers, cities have turned increasingly to rivers and overground reservoirs for an adequate water supply. The Romans built magnificent aqueducts for their Mediterranean cities. Nearly two millennia later long-distance aqueducts have become common. Probably the most striking modern scheme supplies the gold-mining towns in the Western Australian desert. Here between 1898 and 1903 a reservoir was constructed on the Helena River near Perth, and a steel pipe, mostly 30 in. in diameter, was laid from it 346 miles to Coolgardie and 351 miles to Kalgoorlie. In 1913 Los Angeles obtained a supply from a reservoir 233 miles away in the Sierra Nevada. Today this is supplemented by an aqueduct from Lake Mead on the Colorado River, a distance of 266 miles, of which 108 miles are in tunnels through mountain ranges. Recent schemes, such as that of Whyalla in South Australia (233 miles of pipeline; 1,558 ft total lift), demonstrate the distance that municipalities are prepared to go for water.

In some of these long-distance projects the water is totally lost to the source drainage basin, whereas in most smaller schemes a large proportion of the water extracted from a river is returned as sewage effluent to the same drainage basin. It is common in densely peopled areas for municipalities extracting water and discharging effluent to be spaced at intervals along the same river. This practice is safe provided the discharge of the river remains adequate and all domestic and industrial effluents are properly treated before they re-enter it.

The World Health Organization has laid down standards of desired quality for domestic water supplies. To try to improve community water supplies, which are unhygienic or unpiped and inadequate in vast areas, it co-operates with various international development associations to provide funds, surveys, and training. Thus in 1965 water surveys were made under its guidance in, for example, Ecuador and Panama, and a loan made for the development of the Johore River as a supply for Singapore.

Industrial consumption of water is usually on a bulkier scale than that of domestic users. For steam-raising and cooling purposes coal-fired electricity plants need about 600–1,000 tons of water for each ton of coal burned. The water requirement for direct cooling for a station of 2,000 MW is about 1,000 million gallons (4,500 million l.) per day, a quantity sufficient for the domestic uses of a large conurbation. The manufacture of a ton of steel requires about 250 m³ of water, of a ton of sulphate woodpulp about 240 m³, of a ton of soda woodpulp 320 m³, and of a ton of woollen and worsted fabrics about 580 m³. However, most of this water is returned into circulation and compared with irrigation, industrial and domestic users are relatively small absolute consumers of water. Yet they cause one of the greatest difficulties of water supply. Often open channels are ultimately the recipients of effluent from sewage works and industrial plants, which may result in unsafe and unsightly pollution.

In most countries waste disposal is less advanced than water supply. In Britain cholera epidemics led to a royal commission on water supply and drainage in 1843, but improved supplies led to so great an increase in the effluent from sewers and industrial concerns that by 1870 river pollution also required public investigation. Here and in many other districts of the world some rivers are discoloured and polluted sufficiently to discourage fish life and all domestic and recreational uses.

Pollution, or quality of water, is partly a question of the specific use for which the water is needed. Thus to encourage fish life, low dissolved oxygen content, and concentrations of toxic metals, ammonia and so on must be avoided; for domestic use the absence of harmful bacteria, of objectionable colouring matter, and of organic constituents with an offensive taste or smell are also important; for cooling water in thermal electric stations, the prime needs are low temperatures, relative freedom from suspended solids, and a tendency not to promote slime and corrosion. As yet very little is known about 'the chronic physiological effects' on public health of minute amounts of chemical substances contained in water supplies. The difficulty of avoiding pollution in areas where heavy artificial fertilizing and hormone spraying are practised is only just being realized.

The main remedies against excessive river pollution today are to increase the rate of discharge ('Dilution is the solution to pollution'); to increase the rate of re-aeration, by increasing turbulence by building weirs etc. and constructing purification lakes or reservoirs; to improve the flora, the clarity of the river, and cleanliness of its bed; and, inversely, to increase the purity of effluents entering the river. These ideas include a detailed monitoring system for river-water quality, whereby users can be warned immediately of the precise quality of the supply and, if necessary, undesirable water can be periodically flushed out of a drainage system. The building of purification reservoirs seems especially promising and could be combined with schemes for flushing river-beds. This latter idea, the sudden increase in flow, appeals strongly to people interested in the preservation of migratory fishes.

Open channels bring economic and human problems of a kind other than those caused by the direct consumption or use of the water. In warm climates with no resting season for hydrophytes water plants multiply and soon block shallow channels. In cooler climates the water weeds are removable by a single annual cutting, but in warm regions repeated cuttings by machine or hand are necessary. In some areas chemical sprays are becoming popular. In others, where the vegetation is suitable, herbivorous fishes, such as Chinese grass carp, puntin carp, and goramy, have been introduced. The great weed nuisance in much of the tropics is the beautiful water hyacinth (*Eichornia crassipes*), which is still spreading. In the Sudan, where it has only reached the southern watercourses, hormone spraying from boats and aircraft is among the control methods. Biological controls used locally include, first, the seacow or manatee, which is successful in parts of Guyana but is so popular with the natives as a source of meat that it is becoming scarce; and, second, certain species of snail which feed on the hyacinth.

However, these snails act as hosts for the fluke (*Bilharzia*) which causes *bilharziasis* (Schistosomiasis). At a stage in its cycle the fluke leaves the snail and moves freely in water; it then attaches itself to, and eventually enters the blood stream of, human beings that come in contact with it. A serious deterioration of the intestines and liver of the affected person then occurs. The fluke breeds in the human body before returning, as excreta, to the water snail. Bilharziasis is endemic in many parts of the tropics, and is second to malaria among the world's parasitic diseases. Unlike malaria, it is spreading, and today probably affects 150 million people. It increases in frequency near irrigation projects, especially where seasonal water supplies are replaced by perennial. In parts of Egypt its incidence rose from 5 to 75% of the adult population between 1948 and 1963 (*World Health*, World Health Organization, July–August 1964, p. 27).

The main needs to prevent bilharziasis is to control the host snails and human hygiene. The former could in some areas be at least partly achieved by lessening the shallow-water margin of watercourses and reservoirs, and also by raising and then lowering the water level to expose the snails. This latter method was successfully used in the T.V.A. scheme to control mosquitoes; marginal vegetation growth was retarded by maintaining high water during the early growing season, and at a critical time the water level was greatly lowered, so exposing the immature mosquito larvae.

The harmful inhabitants of open channels in warm regions also include carnivores ranging in size from crocodiles and alligators to the South American scourge, the small bloodthirsty piranha (*Serrasalmo*). However, these objectionable denizens are more than compensated for by the abundance of edible species.

4. Open channels and freshwater fisheries

In all countries fishing is one of the most popular recreations, and tens of millions of amateur fishermen catch a fish or two in their leisure hours. Professional and part-time freshwater fishermen and fish-farmers are commonest in monsoon Asia, where fish is an indispensable part of the diet and fish sauces, such as the *prahoc* of Cambodia, flavour almost every meal.

Of the world's fish catch in 1966, freshwater and diadromous (migratory) species provided about 7,910,000 metric tons or 13·9%. Of this the diadromous fishes supplied 1,760,000 tons, the main species being salmons, trouts, smelts, capelin, and other *Salmonidae* (1,180,000 tons); shads, milkfish, and similar species (520,000 tons); freshwater eels (41,000 tons); and sturgeon (about 16,000 tons). The surprising growth in the catch of capelin in 1966 (521,000 tons, of which Norway caught 379,000 tons and Iceland 124,900 tons) should not be allowed to hide the long-standing importance of salmons. Among anadromous species (which migrate up rivers to spawn) the salmons are supreme. In a single year at sea they can put on up to 10 lb in weight. They are caught in rivers and in tidal estuaries, and their importance can be judged from the catches on the Pacific coasts and rivers in 1966, when over 260,000 metric tons were landed in and off the United States and Canada, and about 190,000 tons in and off Japan and the eastern U.S.S.R. In that year the Scottish catch was 1,300 tons.

These diadromous fishes depend for their life cycle on a freshwater channel with a depth and discharge of healthy water sufficient to enable them to reach their spawning grounds. They happen to frequent the coastal rivers of glaciated areas, where hydroelectric dams are most common. The obstruction to migration upstream caused by modern dams is insurmountable, except perhaps for eels (which migrate to the sea to spawn), while the injuries or fatalities caused by turbines to small fry migrating downstream may be over 50% of those passing through. Consequently, there has been a constant disagreement between fish conservationists and water-resource developers, and today elaborate precautions are taken to allow fish migrations to continue unharmed. The normal practice is to build fish ladders or other forms of stepped waterfalls which the migratory fish use to by-pass the dam. Sometimes these devices prove relatively ineffective and waste too much water. Other methods involve the stripping of spawn and the artificial reproduction of fry in hatcheries and the restocking or artificial seeding by collected spawn of the upstream spawning grounds. These, and a better river flow in spring to encourage the fish, are the methods being used, for example, on the Volga system, where the huge dams have disturbed the conditions for semi-migratory roach, bream, etc., and obstructed the route to the spawning grounds for migratory sturgeon, whitefish, and some forms of herring. The U.S.S.R. (22,000 tons in 1962; 15,100 tons in 1966) produces nearly all the world's commercial catch of sturgeon, a large fish which yields a fine flesh as well as caviare and isinglass. In 1932–6 about 75% of the sturgeon catch came from fisheries on the Volga and north Caspian, but yields here declined markedly in the 1960s. On the other hand, the vast new reservoirs above the Volga–Don dams are becoming important fishing grounds for non-migratory species, and nearly one-fifth of Soviet Russia's entire fish production (5,350,000 tons in 1966) comes from fish farms and inland waterways.

The advantage of manipulating reservoir releases so that a surge of fresh water encourages and enables anadromous fish to migrate from estuaries to their spawning grounds has also been demonstrated on the Roanoke River in North Carolina. Here striped bass apparently responded to such surges, which also brought the additional benefits of improving the stream bed for fish food organisms. The reservoir release, however, may be a loss to hydroelectric production, as only a volume sufficient to reach the estuary is likely to affect the maximum number of fish.

Caution may be necessary in some localities after sunny weather has stratified the water in deep reservoirs, causing the bottom layers to become low in oxygen content. This bottom water, if passed through turbines to the river downstream, may be harmful to fish life there. Such a possibility can be avoided by selective withdrawal of the better-oxygenated surface layers of the reservoir and by artificial aeration below the dams.

No such difficulties occur in connection with fish that spawn without migrating appreciable distances. These non-migratory freshwater species, which yielded a total catch of 6,150,000 metric tons in 1966, are ideally suited to lakes, reservoirs, and pisciculture generally. Provided they have adequate food, and water with an oxygen saturation above about 75% and reasonably free of excreta, these fish will thrive in still or almost still water. The protein-producing potential of rivers and freshwater ponds is enormous, especially in the tropics. A suitable reservoir stocked with carp will produce 200–400 lb of fish per acre per year in central Europe; 500–600 lb in Alabama; and up to 1,000 lb in parts of Monsoon Asia. In Java the milkfish (*Chanos chanos*) and other species reach marketable size in less than one year; here and elsewhere in south-eastern Asia spawn or fry are often bought by the thousand for stocking tanks, where they will grow to 5 or 6 lb in nine months. River fisheries and pisciculture, such as keeping carp on paddy-fields, play a large part in the economy of monsoon countries. In India 447,500 tons or 36% of the total national fish catch in 1966 came from rivers and fresh-water reservoirs. Then the annual freshwater catch of Indonesia was 281,400 tons (28% national total catch); of Pakistan 231,800 tons (56% total); and of Cambodia 125,000 tons (76% total). In pisciculture there are fascinating specialities: for example, in 1966 Japan produced mainly by aquiculture 19,800 metric tons of eels (48% world total) and Denmark 15,100 tons of trout (52% world total commercial production). In addition, where malaria is prevalent fish play a vital role in keeping down mosquitoes, and no stagnant or slow-moving water should be without edible species. Buffalo fish are useful denizens of the ricefields of the Mississippi.

5. Open channels and the preservation of wild life

The preservation of biological species, including rare species of fish, concerns the planner of water resources. This is often a question of prohibiting or removing industrial pollution and of protection also against other human interference and predators. The reintroduction of locally extinct or failing species often brings satisfactory results once the river habitat has been restored, if necessary, to its original quality. Multiple-purpose basin planning usually incorporates the provision of lakes which act incidentally as resting places for migratory birds such as geese. In areas where the streams dry up seasonally river life can be greatly increased by the building of dams, which may provide a small dry-season flow or at least provide a refuge during the drought. This has been done successfully on some of the small trout streams of the Sierra Nevada in California.

6. Social uses of open channels

In a few countries some springheads and rivers acquire great religious significance. The ice cave near Gangotri (10,300 ft above sea-level), which is the source of the Bhagirati, and the junction of this headstream with the other main headstream of the Ganges at Devaprayag are famous pilgrimage places. The chief pilgrimage centre, however, is lower down at Allahabad, where the Jumna joins the mature Ganges. Here the bathing festival of Magh Mela is usually attended by 250,000 Hindus, and every twelfth year the special festival of Kumbh Mela is attended by 1 million Hindus, who wash away their sins in the sacred river.

In nearly all countries rivers and lakes are important directly for recreation in addition to fishing for sport and subsistence. Bathing, swimming, and washing, personal and sartorial, are universal uses in warm lands. In societies where leisure is common and fresh air a cult or a necessity the construction of dams and reservoirs often makes recreation perennial rather than seasonal, or at least lengthens its season and renders it safer and easier than on unregulated rivers. Reservoirs become the scene of boating, sailing, water ski-ing and, where free of pollution, swimming, while their shores provide camping grounds for tourists. In some countries, especially the United States, recreation is an important component of many water projects, and in most industrialized societies the appreciation of the value of recreation as a source of pleasure and income has greatly increased. Fifty years ago the Hoover Dam (Lake Mead) project catered on a large scale for boating, sailing, and swimming in the hot sunny climate. In 1933 the Muskingum Watershed Conservancy multiple-purpose scheme covering about 8,038 square miles of territory included fourteen reservoirs, which provided 16,000 acres of lake and 365 miles of shoreline. Each lake or reservoir was encircled by a landscaped strip at least 100 ft wide perpetually reserved for public use. Within fifteen years about $2\frac{1}{2}$ million visitors were coming annually. Recently a scheme for a 37-mile long reservoir on the Delaware River above Tocks Island allowed about one-third of the total estimated cost for purposes of recreation and provided for a capacity of over 10 million visitor-days annually.

Many of the great dams themselves become noted tourist attractions, for example, the Glen Canyon dam with its tourist viewpoints and the Aschach dam on the Danube in Austria.

REFERENCES

GENERAL

UNITED STATES GOVERNMENT [1950], *A Water Policy for the American People*, 3 vols., (Washington, D.C.).

FLOTABILITY AND NAVIGATION

Atlas of Finland [1960] (Helsinki). (See plate 24.)

Atlas of Sweden [1953–68] (Stockholm).

CANADIAN GOVERNMENT [1967], *Traffic Report of the St. Lawrence Seaway, 1966* (Queen's Printer, Ottawa).

MEAD, W. R. [1958], *An Economic Geography of the Scandinavian States and Finland* (Univ. of London Press), 302 p.

MILLWARD, R. [1964], *Scandinavian Lands* (Macmillan, London), 488 p.

ROM, V. Y. [1961], The Volga–Baltic waterway; *Soviet Geography*, **2** (9), 32–43.

TAAFE, R. N. [1964], Volga River transportation; In Thornan, R. S. and Patton, D. J., Editors, *Focus on Geographic Activity* (McGraw-Hill, New York), pp. 185–93.

UNITED NATIONS [1967], *Annual Bulletin of Transport Statistics for Europe* (New York).

UNITED STATES [1967], *Annual Statistical Abstract* (Washington, D.C.).

VENDROV, S. L. *et al.* [1964], The problem of transformation and utilization of the water resources of the Volga River and the Caspian Sea; *Soviet Geography*, **5** (7), 23–34.

DOMESTIC AND INDUSTRIAL WATER USES AND POLLUTION

ISAAC, P. C. G., editor [1967], *River Management* (Maclaren, London), 258 p.
See particularly the following chapters:

BRIGGS, R. *et al.*, The monitoring of water quality, pp. 38–55.

HOUGHTON, G. U., River-water quality criteria in relation to waterworks requirements, pp. 153–67.

LESTER, W. F., Management of river water quality, pp. 178–92.

LOVETT, M., Control of river quality, pp. 193–8.

MERCER, D., The effects of abstractions and discharges on river-water quality, pp. 168–77.

WOLMAN, A., editor [1962], *Water Resources;* National Academy of Science, Publication 1000-B, National Research Council (Washington, D.C.).

WORLD HEALTH ORGANIZATION [1963], *International Standards for Drinking Water;* 2nd edn. (Geneva).

OPEN CHANNEL AND FRESHWATER FISHERIES

FOOD AND AGRICULTURE ORGANIZATION OF THE UNITED NATIONS [1967], *Yearbook of Fishery Statistics*, 1966, Vol. 22 (Rome).

ISAAC, P. C. G., editor [1967], *River Management* (Maclaren, London), 258 p.
See particularly the chapters by:

BRAYSHAW, J. D., The effects of river discharge on inland fisheries, pp. 102–18.

HULL, C. H. J., River regulation, pp. 86–101.

III(ii). Rivers as Political Boundaries

ROBERT P. BECKINSALE
School of Geography, Oxford University

1. Rivers as internal administrative boundaries

Rivers are commonly used as administrative boundaries within a state. Streams, being occasionally impassable or awkward to cross, form recognizable limits for minor administrative units such as parishes in England, many of which date back to the eighth and ninth centuries A.D. Yet in England the larger administrative units, such as counties or shires, are rarely bounded by rivers, largely because they were conceived as being centred upon a defensive town situated on a waterway. The main exception is the Thames, which acted as a defensive barrier between the kingdoms of Wessex and Midland Mercia and which is today a county boundary for most of its length. When a river acting as a boundary changes its course, naturally or with artificial aid, the original boundary line is retained and becomes of considerable geomorphic interest, as can be seen from the lower Dee and Dove (fig. 7. III(ii).1).

Similar domestic river boundaries are more common in states or provinces colonized in recent times by Western powers. In Australia the Murray between New South Wales and Victoria, and in Canada the Ottawa are inter-provincial boundaries. But the use of internal river boundaries reaches its maximum in the United States, where, for example, the Delaware, Potomac, and Savannah east of the Appalachians and the Ohio and Mississippi west of the Appalachians are state boundaries for most of their length. This method of using ready-made river lines would no doubt have caused endless interstate friction had not the Federal Government financed all improvements for navigation.

2. Rivers as international boundaries

In monsoon Asia the national territories of paddy-oriented nations tend to extend across floodplains and rivers, and the only notable river boundary is the middle Mekong (Thailand–Laos), here rapid and gorge-like. Even in arid Asia national territories usually include both sides of a river. Thus in the U.S.S.R. the lower Amu-Dar'ya (ancient Oxus) is avoided as a boundary except for a short stretch, although its steeply incised upper course acts generally as a boundary between Afghanistan and Tadzhikistan-Badakshan. In fact, the only great international river boundary in Asia is the Amur and Argun rivers between Manchuria and the U.S.S.R.

In Europe the Rhine was a defence line in Roman times, but international

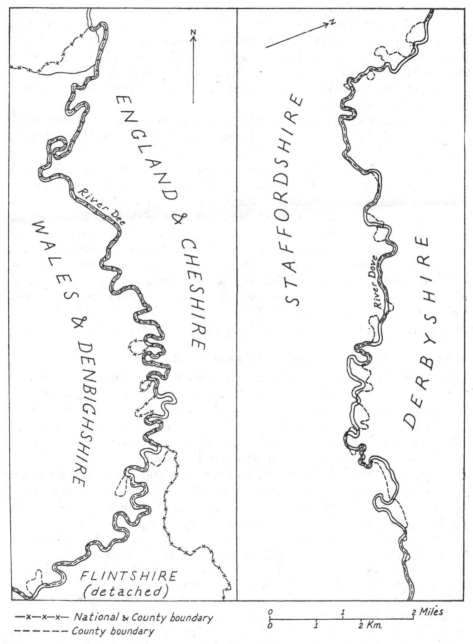

Fig. 7.III(ii).1 Administrative boundaries as evidence of changes in a river's course: the lower Dee and lower Dove, England. Embankments or levees are not shown.

river boundaries are uncommon. Apart from short stretches such as the Torne River (Sweden–Finland), the only notable examples are the middle Rhine from Konstanz to Lauterbourg and the lower Danube in and below the Iron Gates, except for a stretch in Romania. Yet on the Swiss–German section of the Rhine there are four sizeable Swiss enclaves on the German side of the river (Früh [1939], pp. 494–8), and in the rift-valley section downstream of Basle much of its barrier nature formerly consisted of marshes and braidings. Today the river is canalized between levées, and the French have built a lateral canal and cultivated appreciable areas of lowland (*ried*). On the lower Danube large expanses of marsh remain on the north bank.

Unlike politicians in the Old World, statesmen in the Americas often seized on distant river lines as international boundaries. Although in North America the lower St Lawrence, a great entry route, is entirely in Canada, from Cornwall or Massena on the middle St Lawrence westward for the next 1,715 miles to the north-westernmost point of the Lake of the Woods, the United States–Canadian boundary consists almost entirely of water. This tremendous length of water line largely explains why no less than 55%, or 2,198 miles, of the total United States–Canadian boundary follows rivers or lakes. The United States–Mexican boundary is even more concerned with inland water. Of its 1,905 statute miles, about 1,210 miles are along the Rio Grande and a further 20 miles along the lower Colorado (Boggs, 1940). In South America international river boundaries are also common, as for example, on long stretches of the Putumayo (Colombia–Peru); Guaporé (Brazil–Bolivia); Uruguay (Argentina–Uruguay); Pilcomayo (Argentina–Paraguay); and Parana (Paraguay–Brazil–Argentina).

3. Problems of international river boundaries

The advantage that a river is easily recognizable as a boundary is in well-populated districts usually offset by three main drawbacks. First, most rivers naturally change their courses, especially where they meander and suffer violent floods; second, improvement of a river channel for navigation, etc., and withdrawals of water from it for irrigation, etc., affect both banks; and third, flood-plains and flat riverine land attract settlers to both banks.

The legal problems that international river boundaries give rise to are mainly concerned with definition and demarcation. The problems of definition involve either the physical definition (agreed recognition) of a watercourse or the definition of the boundary line along a mutually agreed-upon river.

The physical definition or recognition of a river for a boundary is easy once the territory is accurately surveyed and mapped and the river names are traditionally or legally accepted. If, however, in a recognized river the waterline is highly unstable difficulties of definition may constantly arise. In this case uncertainties and future problems are best resolved by restricting, with the approval of all riparian states, the channel by means of engineering constructions. This has been done on the Rhine in its marshy, braided rift-valley section and on the Rio Grande. Near the latter when the United States–Mexican treaties were signed in 1848 and 1853 there were few settlers and little economic interest

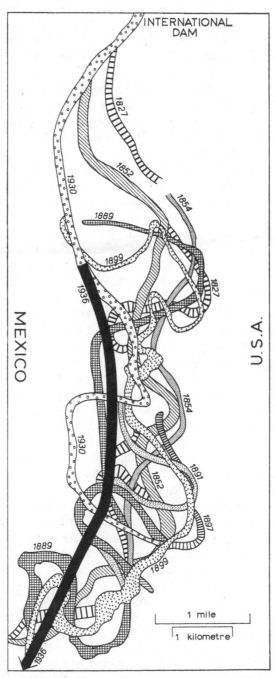

Fig. 7.III(ii).2 The Rio Grande near El Paso–Ciudad Juarez, showing some of the main shifts in the major channel from 1827 to 1936 (Adapted from Boggs, 1940).

except in navigation. The settlement and variety of water uses increased rapidly, but the channel remained highly unstable due to excessive silting and frequent cut-offs of meanders during violent floods (fig. 7.III(ii).2). Between 1907 and 1933 the river bed near El Paso was raised 8 ft by silting. In the following years, under an International Boundary Commission with wide powers, a flood-control reservoir was built at Caballo, New Mexico, and the river bed straightened (from 155 to 86 miles between El Paso and Fort Quitman) and endyked so that silting virtually ceased. The parcels of land cut off by the new international boundary along the rectified channel were exchanged in equal proportions of about 3,500 acres on each side. On the basis of benefits received, the United States paid 88% of the total costs (Boggs, 1940).

In the above cases the river itself was not in doubt, whereas in partially explored areas the river may need definition or mutual recognition before demarcation can begin. There exists today a dispute about a frontier between Guyana and Surinam, two former European colonies, which involves 5,800 square miles in the triangle between the Courantyne River (named as the boundary) and the so-called New River. Guyana claims that the New River is really a tributary of the Courantyne, whereas Surinam claims that it is in fact a continuation of the Courantyne wrongly named New River on the original European maps (*The Times*, 1968).

The problem of determining which river formed the master stream of a drainage basin was argued fully in the Argentine–Chile Frontier controversy of 1966 in which the author acted as hydrological adviser to the Republic of Chile (Foreign Office, 1966). The dispute originated largely in a Boundary Award of 1902 when the territory involved was largely unexplored. This Award stated that the international boundary ran from a fixed boundary post 16 up the Rio Encuentro from its junction with the Rio Palena (or Carrenleufú) to its source on the western slopes of the Cerro de la Virgen. Later it was discovered that the river joining the Palena opposite boundary post 16 rose at a considerable distance from the Cerro de la Virgen. In the recent arbitration it was accepted that the lower course of the river opposite boundary post 16 was the Rio Encuentro. However, above this lower course, which is partly in a deep gorge, the river consists of two main headstreams: one, hereinafter called A, which rises in a cirque high up on the mountain flanks to the east and approaches the main valley-floor at a sharp angle; and another, hereinafter called B, which rises at a lower altitude and drains mainly the northern part of a broad glaciated valley that continues the general direction of the lower gorge. As the tributaries and their sources lacked any traditional names, the arbitrators had to decide which tributary was the main channel or continuation of the lower Encuentro on other historical evidence and on scientific grounds. Scanty historical evidence showed that a few observers considered channel A was the main channel. Modern scientific hydraulic principles left no doubt that A was the master stream. Consequently, the arbitrators decided that channel A was the Rio Encuentro.

The scientific arguments used by Chile to show that channel A was the master stream were that, compared with channel B, it had:

1. greater length (exact measurements were given in all the following linear, areal, and volumetric measurements);
2. greater discharge (based partly on gauge measurements and partly on scientific precipitation/altitude relationships);
3. greater drainage basin;
4. greater geomorphic age and incision (being in existence in its upper course when stagnant ice still occupied the valley floor now drained by channel B);
5. at and near their junction, channel B had a greater gradient of bed and of water level than channel A;
6. strong similarities of bed load existed only between channel A and the lower course below the junction of channel A and channel B.

The arbitrators considered that items 1–3 were the principal criteria to be applied in a problem of this kind and that items 4–6 did not contribute significantly to its solution.

Argentina laid much stress on the lineal continuity of channel B with the lower trunk river, and the Court admitted that this could be an important factor and quoted Strahler's use of it as equal to 'longest total stream length as a factor, when continuing a single-channel profile headward into channels of lower order (Strahler, 1964). However, the Court recognized Chile's contention that the lineal continuity referred to was primarily continuity of valley-form (of the glaciated trough), whereas more importance should be attached to 'the continuity of the general force of the river', which was more evident in channel A.

Argentina introduced the Strahler method of stream order-numbering to show that at their junction the two channels were of the same order, whereas Chile suggested that on the Horton system channel A was the master channel. However, the Court did 'not consider that the two different methods of order designation (applied as they are to maps of different degrees of accuracy and on different scales)' helped to resolve the problem. However, the author retains the opinion that Horton's method of 1935 and 1945 (Leopold, Wolman, and Miller [1964], pp. 134–5) would normally provide a satisfactory solution to such a problem, as it does select and extend to its headsource the chief order stream in a drainage basin, whereas Strahler's 1952 simplification of Horton's method was deliberately designed to avoid making that decision.

In its conclusions the Arbitration Court ordered part of the new boundary to 'follow the thalweg of the Encuentro', although fieldwork had shown that in this mountainous drainage basin there is in fact usually no thalweg in the strict sense, as most of the river channels have almost flat beds lined irregularly with large pebbles and boulders. As, however, will now be shown, it is usually very difficult to define and demarcate precisely a river boundary.

In water, especially where flowing, marks are often impracticable to fix or to maintain, and are often thought unnecessary, as they are liable to be a danger or nuisance to both parties when the river or lake is used for certain lawful purposes, such as navigation and fishing. Water boundaries are usually demarcated by reference to land marks, and are often in practice marked in detail only on the official reference maps. The demarcation, however, should be based on a precise definition. Such boundaries follow either

1. the shore;
2. the median line;
3. the thalweg; or
4. some arbitrarily selected line.

The fourth type occurs today, for example, as a parallel of latitude on Lake Victoria between Tanzania and Uganda and as azimuths or straight-line courses on the United States–Canadian water boundaries.

The first type is rare, and probably the chief existing example is on the Shatt al Arab, where in 1914 the boundary was drawn along the low-water line of the left or Persian bank, thus depriving Persia of free access to the river fairway. Later, Iraq succeeded to Turkish rights, and there were many incidents between Persian and Iraqi officials on the Shatt al Arab before a Treaty of Friendship was signed. The agreement provided for the inclusion of anchorages in the Shatt al Arab off both Abadan and Khorramshahr in Persia and the general retention of the rest of the 1914 frontier.

The third type, the thalweg or deepest channel, was commonly used in international agreements when navigation interests predominated. The definition here refers to the cross-section of a river's bed, although in some early treaties thalweg and the 'middle of the main channel' were assumed to be coincident. The concept envisages an 'uninterrupted line determined by the deepest places in the bed' (Kaeckenbeeck [1918], p. 176; Haataja [1927]), and although the thalweg is not completely stable, it offers 'greater stability than the middle line of a stream'.

The second type of definition or demarcation, the middle line or median line, could entail great inconvenience, as it changes its position with changes in the water level and in the shape of the bed (e.g. the Rhine between Germany and Switzerland). Moreover, the deep-water channel (thalweg) might be entirely on one side of the middle line, so depriving one riparian state of beneficial possession. Unfortunately the median line failed to develop precise concepts. It was variously defined as:

1. the 'middle' of a watercourse or a line at all points equidistant from each bank;
2. a line paralleling the general line of the banks and dividing the horizontal surface area of the water into two equal parts;
3. in the case of a lake, a line along the mid-channel dividing the navigable portion and being at all points equidistant from shoal water on each shore (*International Waterway Comm. compiled reports, p. 578*).

It was not, however, until 1930 that the following exact definition was proposed by Boggs [1940, pp. 181–2]. 'A median line is a line every point of which is equidistant from the nearest point or points on opposite shores of a lake or river' (see also Boggs [1937] and Burpee [1938]). Ambiguities were now virtually impossible, although islands may still cause difficulties.

The United States–Canadian water boundaries exemplify the increasing precision in definition and demarcation. By the treaty of 1783 (Art. 2) the boundary followed the 'middle' of twelve lakes, rivers, and inland straits. However, curved lines proved impracticable, and in 1908 the International Commission was empowered to replace them by 'a series of connecting straight lines'. The boundary from Cornwall on the St Lawrence to the mouth of the Pigeon River on Lake Superior now consists of 270 straight-line courses (although the two of these that follow parallels of latitude mathematically speaking are not *straight*; that is are not *azimuths*). Smaller, deeply indented lakes and smaller winding rivers are more difficult to demarcate. Thus, the international boundary between the mouth of the Pigeon River and the north-westernmost point of the Lake of the Woods now consists of 1,796 straight-line courses. Farther east the boundary in the 'centre of the main channel or thalweg' of the St Croix river now comprises 1,008 straight-line courses, and that along Hull's stream consists of 766 straight-line courses in a distance of 26·6 miles. No wonder Boggs [1940, p. 54] calls this 'one of the best-marked frontiers in the world'.

4. International boundaries across river channels

Most great and many smaller rivers are international, although international boundaries commonly follow watersheds. The boundary that cuts most ruthlessly across rivers and drainage basins is that between Canada and the United States from Lake of the Woods to the Pacific which follows parallel 49° N for about 1,257 miles. In this section waterway problems have been almost continuous and have been successfully settled by an International Joint Commission. A similar commission on the United States–Mexican boundary on the Rio Grande was equally successful, and under a treaty of 1945 was given wider powers of collaboration 'in order to obtain the most complete and satisfactory utilisation' of the water available.

Examples of integrated water-resource schemes elsewhere are becoming less rare. The Nile Waters Agreement has resulted in, for example, the Owen Falls Dam at the outlet of Lake Victoria, the Sadd-el-Aali barrage and, indirectly, in several Sudanese schemes. Here the working arrangements are implemented by the irrigation departments of Egypt (U.A.R.) and of the Sudan Republic.

A United Nations' *Report* [1958] stresses 'the inadequacy of international law' in respect to multiple-purpose, integrated-basin development schemes and sets out the principles drafted by the International Law Association at Dubrovnik in August 1956 as being likely 'to aid adjustment and agreement'. Since these principles were suggested at least two notable international drainage-basin development agreements have been successfully negotiated.

The Indus Water Treaty of 1960 originated largely because the north-eastern part of the Indus basin was sub-divided politically in 1947, leaving the upper courses in India and the lower courses of the chief tributaries mainly in Pakistan (United Nations [1966], pp. 47–66). At the time of partition, of the annual flow of

the Indus system (about 207,500 million m³ or 168 million acre-feet) about 40%
ran out to sea at the delta, 16% was lost through seepage, 39% used in irrigation
canals in Pakistan, and 5% for similar purposes in India. The latter diverted
water mainly from the three eastern rivers (Ravi, Sutlej, and its right bank
tributary the Beas), which have a total annual discharge of 40,350 million m³.
But Pakistan was already using 14,800 million m³ from these eastern rivers,
whereas India, to meet all her local projects, needed all told about 38,000 million
m³. On the other hand, the three western rivers, Chenab, Jhelum, and Indus,
have four times the discharge of the eastern rivers and can be easily used in the
plains only by Pakistan.

After thirteen years of negotiations, greatly aided by the World Bank, the
Treaty agreed that the water of the three eastern rivers shall be for unrestricted
use by India and that of the three western rivers for unrestricted use by
Pakistan. Thus Pakistan will in the next ten or thirteen years make good the
14,800 million m³ it withdrew from the eastern rivers by extra withdrawals from
those in the west. In the meanwhile India will supply to Pakistan from the eastern
rivers the deficit between 14,800 million m³ and the extra amounts withdrawn by
Pakistan from the western rivers. From these western rivers Pakistan will receive
about 167,000 million m³ minus a relatively small proportion which will be
diverted to India for various purposes. India may develop hydroelectric stations
(to some specific design criteria) and build storage reservoirs of about 4,500
million m³ total capacity on the headstreams of the western rivers in India for
general purposes, particularly flood control. The Treaty is implemented by a
Permanent Indus Commission and contains mutual obligations on exchanging
hydrologic data, etc., operating the river work, maintaining river channels,
floating timber, pollution and procedures about disputes. It provides 'an ad-
mirable instance of international co-operation' (United Nations [1966], p. 66).

Its counterpart in the New World is the Columbia River Treaty of 1964
between the United States and Canada. This 'major milestone in the history of
international river development in North America' (Sewell, 1966) required
twenty years of studies and negotiations. The instigation came partly from
proposals to build in the United States section of the Columbia dams which
would back up water into Canada (cf. the Sadd el Aali dam and the Sudan). The
opportunity for integrated basin development led in 1944 onwards to studies of
the whole Columbia River basin in the United States and Canada. Emphasis,
however, was put on hydroelectric generation and on flood control (fig.
7.III(ii).3). The *Report* of 1959 produced three alternate schemes of maximum
development, each providing about 17 million kW of firm power and about
50 million acre feet of water storage at a total cost of about $4 billion. As about
23 million acre feet of this storage would be developed in Canada, where the head
or fall was relatively small, its storage value would lie mainly in supplies to
power-stations downstream in the United States, where also flood-control
benefits would be greatest. Consequently, it seemed that some inducement or
compensation should be given to Canada to develop these storage units. This
concept of 'downstream benefit sharing' caused difficulties and led to a narrowing

Fig. 7.III(ii).3 The main engineering projects on the Columbia River drainage system (From Sewell, 1966).

Only the larger natural lakes are shown. The ultimate installed hydroelectric capacity of some of the larger stations is: Grand Coulee 5,574,000 kW; John Day 2,700,000 kW; Mica 2,000,000 kW; The Dalles 1,743,000 kW; Chief Joseph 1,728,000 kW; Wanapum 1,330,000 kW; Priest Rapids 1,262,000 kW; McNary 986,000 kW.

in the treatment of the integrated river-basin problem. According to Sewell [1966, p. 150], the negotiators aimed at reaching agreement and at making a benefit/cost profit rather than producing an overall scheme that would yield the greatest net benefits. Some of the difficulties arose because the United States' stretch of the river was already partly developed and their negotiators wished for an unfinished project in an advanced stage of design to be included in the international agreement. On the Canadian side there were difficulties due, for example, to Province–Federal politics and to vague alternative sources of hydroelectric power.

Eventually the Treaty was signed by Canada in 1964. Under its provisions 15·5 million acre feet of storage will be built on the headwaters of the Columbia River in Canada at three sites: Arrow Lakes (7·1 million live storage), Duncan Lake (6·4 million), and Mica Creek (7 million, rising later to 12 million live storage). This storage will be used mainly for increasing the winter flow of the Columbia and in reducing flood damage. The benefits will be shared by the two states. In 1964 the power benefits were estimated at 2·6 million kW capacity and 13 billion kWh output, but these have since been reassessed at 2·75 million kW and 15 million kWh. The flood-control benefits were assessed at $126 million. Canada elected to sell her share of the increased power to the United States for the next thirty years for a lump sum of $253·9 million (U.S.) ($273·3 million Canadian), which was enough to pay for the construction of the three storage dams and a considerable proportion of the cost of installing generating units at the Mica project (dam height 645 ft; ultimate installed capacity 2 million kW, for use exclusively in Canada). Canada was also paid $64·4 million (U.S.) as her share of flood-control benefits. The United States had the option to build the Libby dam on the Kootenay River in Montana, the reservoir of which will stretch several miles into Canada (fig. 7.III(ii).3). Either country is allowed to take water for domestic uses at any time and Canada to divert up to 1·5 million acre feet from the Kootenay to the Columbia for power purposes after twenty years. No doubt, as the parties hoped, the Treaty will stand out as 'an example of large-scale international co-operation, hopefully to be imitated elsewhere'. However, the negotiations showed that a truly comprehensive approach is difficult to put into practice and, in the opinion of some, that a higher proportion of the investigation costs should have been spent on economic, as distinct from engineering, studies.

REFERENCES

ADAMI, V. [1927], *National Frontiers in Relation to International Law*; Translated by Behrens, T. T. (Oxford University Press, London). (See especially pp. 199–210 and 218–19.)

BOGGS, S. W. [1937], Problems of water-boundary definition; *Geographical Review*, **27**, 445–56.

BOGGS, S. W. [1940], *International Boundaries* (Columbia University Press, New York), 272 p.

BURPEE, L. J. [1938], From sea to sea; *Canadian Geographical Journal*, **16**, 3–32.

DEPARTMENT OF EXTERNAL AFFAIRS, CANADA [1964], *The Columbia River Treaty: Protocol and Related Documents* (Ottawa).

FAUCHILLE, P. [1925], *Traité de droit international public;* Vol. I, Part 2 (Paris).

FOREIGN OFFICE [1966], *Award . . . for the Arbitration of a Controversy between the Argentine Republic and the Republic of Chile;* Reference S. O. Code No. 59–163 (H.M.S.O., London).

FRÜH, J. [1939], *Géographie de la Suisse;* Vol. 2 (Librarie Payot, Paris).

GLOS, E. [1961], *International Rivers: A Policy-Oriented Perspective* (Singapore).

GRIFFIN, W. L. [1959], The use of international drainage basins under customary international law; *American Journal of International Law,* **53,** 50–80.

HAATAJA, K. [1927], Questions juridiques surgies lors de la révision de la frontière finlandaise entre le golfe de Bothnie et l'océan Glacial; *Fennia,* **49,** 1–46.

INTERNATIONAL COLUMBIA RIVER ENGINEERING BOARD [1959], *Water Resources of the Columbia River Basin;* (Ottawa).

INTERNATIONAL JOINT COMMISSION [1959], *Report on Principles for Determining and Apportioning Benefits* . . . (Ottawa).

JONES, S. B. [1945], *Boundary Making* (Washington).

KAECKENBEECK, G. [1918], *International Rivers* (London). (See especially p. 176.)

LAPRADELLE, P. DE [1928], *La frontière: étude de droit international;* Vol. 1, Part 2 (Paris).

LEOPOLD, L. B., WOLMAN, M. G. and MILLER, J. P. [1964], *Fluvial Processes in Geomorphology* (Freeman, San Francisco and London), 522 p.

PRESCOTT, J. R. V. [1965], *The Geography of Frontiers and Boundaries;* (Hutchinson's University Library, London).

SEWELL, W. R. D. [1966], The Columbia River Treaty: Some lessons and implications; *The Canadian Geographer,* **10,** 145–56.

SMITH, H. A. [1931], *The Economic Uses of International Rivers* (P. S. King and Son, London).

STRAHLER, A. N. [1964], Quantitative geomorphology of drainage basins and channel networks; In Chow, V. T., Editor, *Handbook of Applied Hydrology,* Section 4-II.

The Times [1968], News Item, 2 February, p. 6 (London).

UNITED NATIONS [1958], *Integrated River Basin Development* (New York).

UNITED NATIONS [1966], *A Compendium of Major International Rivers in the ECAFE Region;* Water Resources Series No. 29 (New York).

CHAPTER 8

Snow and Ice

I. The Hydrology of Snow and Ice

MELVIN G. MARCUS

Department of Geography, University of Michigan

1. Introduction

Snow and ice are significant elements of the world hydrological system, which occur subject to tremendous variations in space and time. Snow or ice are present in the atmosphere, in lakes and rivers and oceans, on the land, and even beneath the earth's surface. Sometimes their appearance is brief and local, as in the sudden snow flurry which coats the earth with a quickly melted veneer of white, or the violent hailstorm that brings to heated, summer landscapes a contradictory deluge of ice. In other places snow and ice dominate the earth's surface. The ice caps of Antarctica and Greenland and the frozen floes of the Arctic Ocean seem, from the human view at least, hostile and permanent features of our world.

Yet even the great ice sheets are transitory features when viewed in the broad sweep of earth history. Snow and ice are, after all (in a necessary statement of the obvious), simply solid water. As such, they prevail or disappear in response to variations of heat flow within the earth–atmosphere energy system. Change is a constant condition of nature, and nowhere is this more apparent than in the advancing and retreating tides of snow and ice. Man has witnessed these changes through each day, each season, each millennium; for the history of man coincides with one of the rare geological epochs – comprising less than 10% of the earth's history – when snow and ice abound.

Considering that twentieth-century man lives in a relatively active phase of an ice age, it is significant that we know surprisingly little about snow and ice phenomena and the processes associated with them. It is, in fact, only in the last three decades that more than a handful of researchers have focused their attention on solid water phenomena. Nevertheless, we have learned enough to paint a broad, if incomplete, picture of the characteristics, distributions, and processes relating to snow and ice. The following sections briefly cover those subjects, but the treatment is necessarily selective and generalized for reasons of space.

2. Properties of snow and ice

Water is a mineral, and may be defined in terms of its physical characteristics. Most properly, *water* refers to hydrogen dioxide (H_2O) in solid, liquid, or gaseous states. Common usage, however, distinguishes the solid and gaseous forms as ice and water vapour respectively. An unfortunate semantical problem exists when we describe liquid water. This is because we not only use the term water in a general and encompassing sense but also to describe the substance in its liquid state. For the purposes of this article the latter definition will be used; but when quantities of ice or vapour must be described [liquid] water-equivalent units will be used.

A. Physical properties

Crystal form

Ice is characterized by crystals of the hexagonal system, and commonly takes on a variety of prismatic, pyramidal, or dipyramidal forms. Hexagonal symmetry occurs, and is especially prominent in the aggregates of ice crystals which form snowflakes. Crystal size, form, and aggregate structure are highly variable and depend greatly on mode of formation and local environment. Crystals in a glacier, for example, may vary from less than 1 mm to over 1 m in length.

Density

Density relationships between ice and liquid water are anomalous; only a few other substances experience expansion during crystallization. Distilled water has a density of 1·0 at 4° C and 760 mm pressure, which is the standard base for density measurements. At 0° C ordinary ice has a density of 0·92. Depending upon the presence of impurities or gas bubbles and the structural organization of crystals, ice densities vary considerably in nature. Glacier ice, for example, has an average density of 0·84–0·85, while maturely developed lake ice has a density of 0·89–0·90. Sea ice, because of the presence of impurities, has a density of 0·91–0·93. Snow densities are especially variable and dependent on degree of compaction. Density extremes for newly fallen snow range from 0·004 to 0·33, although most meteorological services accept and record an average density of 0·10.

The significance of water's anomolous density behaviour is obvious. Were densities to increase inversely with temperature, ice would sink to the bottom of lake and ocean basins. Since very little of this ice would melt seasonally, there would be a progressive accumulation of ice from depths upward in the earth's lake and ocean basins.

Energy relationships

Energy and water are inseparable and interacting elements of the environment, and changes of state between water and ice account for major energy fluxes. Ice melts to water at 0° C; sea ice at −2° C. Melting ice absorbs 80 g calories of energy for every gram of water melted, an equal amount being released when the process is reversed. The specific heat of ice, depending on temperature, is approximately 0·50, or about half the specific heat of water.

Other properties

Ice is colourless to white to pale blue, and varies from transparent to translucent. Hardness varies with temperature. At −5° C and −44° C, for example, ice has a hardness of 1·5 and 4·0 respectively.

B. Types of snow and ice

The variety of forms taken by snow and ice are too numerous and specialized of definition to enumerate here. It can be generally stated, however, that there are three principal zones of occurrence of snow and ice in the hydrological cycle, each characterized by familiar snow or ice types:

1. Snow or ice which occurs and originates directly in the atmosphere from sublimation or the freezing of previously condensed droplets
Ice clouds and precipitation types such as snow, sleet, and hail are examples. Since these crystals are either precipitated to earth or return to vapour, their existence in the atmosphere is relatively brief.
2. Snow or ice which occurs on the earth's surface
Examples exist at all scales. Frost, rime ice, and glaze ice occur when atmospheric moisture crystallizes directly on the earth's surface. Sea, lake, and river ice result from the freezing of water already present at the surface, while snow cover, névé, firn, and glacier ice result primarily from the accumulation and subsequent metamorphosis of precipitation.
3. Ice which occurs beneath the earth's surface
Most familiar, perhaps, is interstitial ice in soil or detritus. Such ice may be ephemeral, seasonal, or – as in the case of permafrost – relatively permanent.

3. Distribution of snow and ice

In recent geological time – some 10,000–20,000 years ago – continental glaciers covered about 32% of the earth's land surface. Much of that ancient ice has receded and disappeared, and man pretends that he lives in a warm, non-glacial age. Yet the evidence is startlingly contradictory (fig. 8.1.1 and 8.1.2). Some 10% of the land surface remains covered by glaciers, while 7% of the ocean surface is coated by pack or sea ice at their maximum winter extent. An additional 22% of the earth's land surface is underlain by continuous or discontinuous zones of permanently frozen ground. Seasonal snow coats the continents throughout the mid-latitudes; and even at the Equator, the higher mountain summits are capped by snow and ice. The glacial age is not over; it has only diminished in intensity.

Ice and snow remain perennial features of landscape wherever conditions favour survival. The polar latitudes account for most permafrost ice, pack ice, and glacier cover. For example, over 99% of glacier ice is found in Antarctica, Greenland, and islands of the Arctic Archipelago. But alpine regions favoured by high winter precipitation and short, cool summers also support tens of thousands of glaciers and perennial snowfields. Outstanding in this respect are the monsoon-affected Himalayas and, standing in the path of marine air mass flow, the glacierized mountain belts of north-western North America, western South America, and Scandinavia.

4. Snow and ice as an input–output system

Snow and ice which occur at the earth's surface may be schematically viewed as a simple input–output subsystem within the hydrological cycle; that is, the

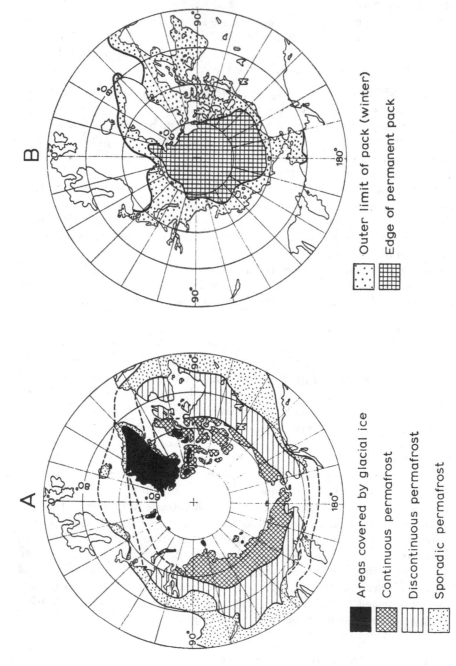

A

B

Outer limit of pack (winter)

Edge of permanent pack

Areas covered by glacial ice

Continuous permafrost

Discontinuous permafrost

Sporadic permafrost

Fig. 8.1.1 Northern hemisphere extent of present glacial ice and permafrost (A), and of pack ice (B) (Partly after Black, R.F., *Bulletin of the Geological Society of America*, Vol. 65, 1954).

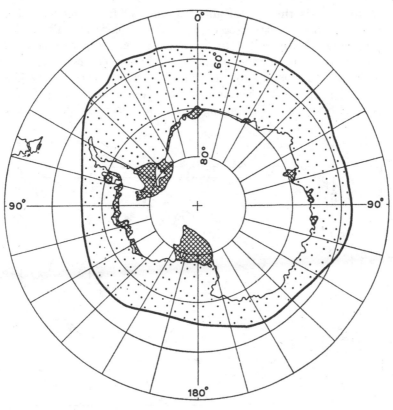

Approximate seaward limit of ice shelf

Approximate present northern limit of pack ice at annual maximum

Fig. 8.1.2 Southern hemisphere extent of present ice shelve and pack ice (the central white area indicates the Antarctic ice cap).

growth or diminution of ice and snow cover is a response to net differences between *accumulation* (water input) and *ablation* (water output). Thus, snow or ice covers are open-ended systems whose life spans are dependent upon interacting heat and moisture fluxes across the system interfaces. Glaciers and seasonal snow cover are two widespread and significant examples of this principle.

A. Glacier mass balance

The *mass balance* or *hydrological budget* of a glacier is the net quantity of water gain or loss occurring in a glacier over time – usually a period of one or more glacier balance years. A balance year is the time interval between the formation of two consecutive summer surfaces, where *summer surface* is defined as the time when minimum mass occurs at the site. Mass balance terms vary with time and can be defined seasonally: (1) the *winter season*, which begins when the rate of accumulation exceeds the rate of ablation, and (2) the *summer season*, which begins when the ablation rate exceeds the accumulation rate. Thus, the glacier budget year does not coincide with the calendar year, but begins and ends in late summer or autumn for most temperate and subpolar regions.

Mass balance relationships for any point on an idealized valley glacier are illustrated in fig. 8.1.3 after recent definitions proposed by the International Commission of Snow and Ice. During the balance year, the glacier experiences

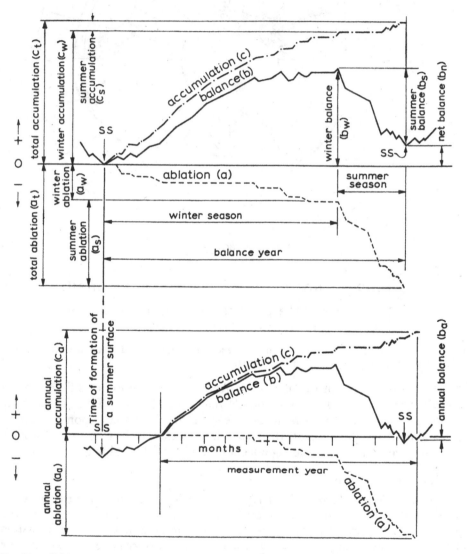

Fig. 8.1.3 Mass balance terms as measured at a point on a glacier (From International Commission of Snow and Ice, 1969).

cumulative accumulation c and ablation a. The balance b throughout the budget year is shown by the solid line and given as $b = c + a$ (where c is always positive and a is always negative). Using additional values from fig. 8.1.3, *net balance* is given as $b_n = c_t + a_t = b_w + b_s$. In this example, since water input is greater than water loss ($c_t > a_t$), the sample point experienced a positive balance year. An equilibrium condition would exist if $c_t = a_t$; a deficit budget if $c_t < a_t$. Aerial mass balance quantities can be similarly calculated by integrating point values over the entire glacier surface.

In fig. 8.1.4 the total glacier is shown at the end of the budget year. Net water gain is included in the snow-covered accumulation zone, and net deficit is encompassed by dashed lines in the ablation zone. If the glacier is in equilibrium the water equivalents of ice flowing across the cross-section under firn line m will

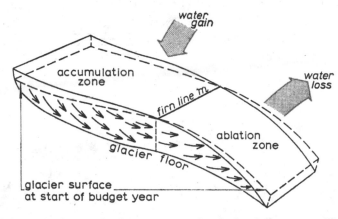

Fig. 8.1.4 Schematic changes in the geometry of a glacier during an equilibrium budget year.

equal ablation and accumulation. Thus, the glacier will retain its original shape and volume. For positive or negative budgets a resultant adjustment of size, shape, and firn line position would occur.

Glacier mass budget can be determined by a number of methods, but the most frequently used technique involves sampling ablation and accumulation over the glacier surface. Figure 8.1.5 illustrates a sampling pattern used on Place Glacier, British Columbia, by Canadian glaciologists. Ablation and/or accumulation is measured at stake positions during the ablation season. Additional soundings of snowpack depth are also accomplished along designated profiles, and variations in density are determined from samples taken in the pits. The data is then integrated to give water-equivalent values of the accumulation, ablation, and budget terms. It is important that field work continue to the end of the glacier year if net budgets are to be calculated. In some cases records of stream discharge below the glacier snout are maintained to check ablation calculations. Such discharge observations must, of course, be corrected for rainfall and snow melt in the catchment area above the glacier.

N

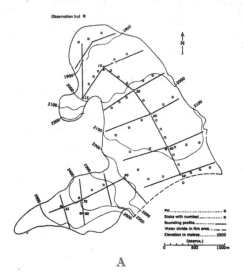

A

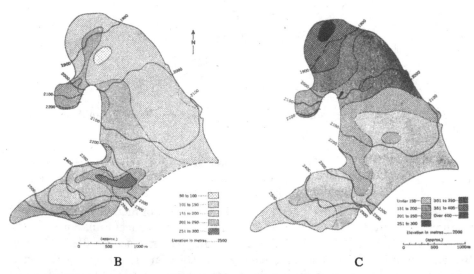

B   C

Fig. 8.1.5 Observations on the Place River Glacier, British Columbia, 1964–5 (From Østrem. G., 1966, Mass balance studies on glaciers in western Canada; *Geographical Bulletin*, Vol. 8 (1), pp. 81–112).

A. Location of stakes, pits, and sounding profiles 1964–5.
B. Accumulation map 1964–5.
C. Ablation map 1965, based on readings from 52 stakes.

In summary, glaciers may be viewed as an open-ended and relatively long-termed storage element in the hydrological cycle. Minor fluctuations of glacier systems occur continually, but larger changes are cumulative over decades and centuries. Climatic change is primarily responsible for these mass budget variations, but the interrelationships between climate and glaciers is exceedingly complex and poorly understood. We know that a glacier may grow or diminish in response to changing energy and moisture fluxes across its surface, but we still have difficulty factoring out the relative importance of individual climatic factors. Changes in planetary energy, moisture, and momentum patterns undoubtedly influence mass budgets, but it is also true that local environmental conditions – topography, land–water relationships, landform orientation, etc. – cause microscale and mesoscale variations in climates. Thus, the relative significance of climatic factors such as radiation, wind, cloud cover, and precipitation may differ appreciably from glacier to glacier.

B. Seasonal snow cover

In principle, the seasonal snow cover responds to energy and moisture fluxes as do glaciers. Inputs and outputs determine the mass budget, and the only distinctions are in scale and time. Unfortunately, we can only guess at the amounts of water which are exchanged as the seasonal snow line advances and retreats; no precise broad-scale or global measurements have been made. The seasonal snow cover has, however, a significant impact on human activities. Agriculture, transportation, and flood control are only a few of the activities which are influenced by the size and duration of the snow cover. This information is, in fact, of such importance that we may expect in the near future – through use of satellite sensors – daily reports on the budget, condition, and migration of seasonal snow.

REFERENCES

DYSON, J. L. [1962], *The World of Ice*; (Alfred Knopf, New York), 292 p. (An interesting and superbly illustrated book treating all aspects of snow and ice.)

FRASER, C. [1966], *The Avalanche Enigma*; (John Murray, London), 301 p. (A fine general source on snow characteristics, avalanche mechanics, and the impact of avalanches on human activity.)

INTERNATIONAL COMMISSION OF SNOW AND ICE [1969], Mass balance terms; *Journal of Glaciology*.

SHARP, R. P. [1960], *Glaciers* (The University of Oregon Press, Eugene, Oregon), 78 p. (An excellent account of glacier structure, flow, and mass budget, made easily understandable for the non-specialist.

WEEKS, W. and ASSUR, A. [1967], *The Mechanical Properties of Sea Ice*; Cold Regions Scientific and Engineering Series, Part II-C3, U.S. Army Cold Regions Research and Engineering Lab., (Hanover, New Hampshire), 80 p. (An excellent technical summary of lake and sea ice properties.)

II(i). The Geomorphology and Morphometry of Glacial and Nival Areas

IAN S. EVANS

Department of Geography, Cambridge University

Glacial and nival hydrology is particularly sensitive to temperature and radiation, and hence latitude, altitude, and aspect are especially important influences on glacial and nival landforms. The latitudinal sequence polar ice-cap, polar desert, tundra, boreal forest is matched by the altitudinal zonation nival, sub-nival, alpine, sub-alpine, although high altitudes differ from high latitudes in the relationship of seasonal and diurnal fluctuations and in radiation, wind, and weather. In addition, slope direction and relative position on a slope are very important in mountain areas, and complicate the pattern of zonation. For example, whereas small glaciers are zonal forms found at the snowline for their aspect, large glaciers are azonal, and a large actively fed glacier in a steep valley may transgress down into a much warmer zone. Slope gradient and aspect affect the surface receipt of precipitation and radiation. This leads to asymmetry of nival balance (affecting intensity of glaciation) and of temperature (affecting freeze–thaw cycles, chemical weathering, and availability of meltwater). The greatest direct radiation at a given angle of latitude is received by an equatorial-facing slope of similar angle, and the least by steep north-facing slopes. This effect is most marked in middle latitudes and in clear conditions. Since west-facing slopes are heated in the afternoon when the air temperature is highest, south-west-facing slopes tend to be warmest and north-east-facing slopes coolest. This probably explains the general mid-latitude tendency for glaciers and cirques to face preferentially north-eastward. Wind drifting from smooth summits and plateau areas produces snow accumulation on lee slopes, and in the west-wind belt this encourages east-facing glaciers and cirques. However, since falling precipitation travels with the wind, its incidence is greater on a surface perpendicular to its direction of travel (i.e. of westerly aspect). In rugged terrain this effect roughly counterbalances wind drifting, leaving the radiation factor dominant.

1. Cirques

A cirque is defined morphologically as a large steep-sided hollow, open on the downstream side but essentially closed upstream by a steep, arcuate headwall below a divide; its floor slopes more gently than its sides, and also more gently

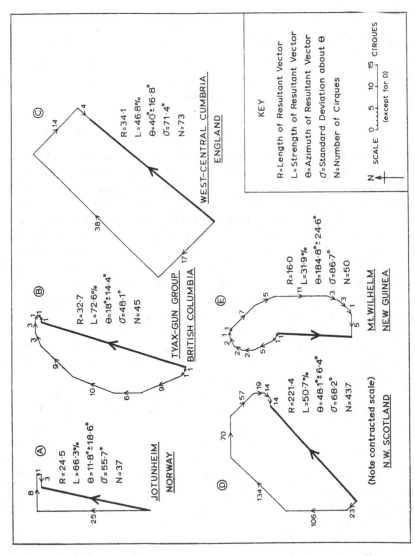

Fig. 8.II(i).1 Analysis of azimuthal distributions of cirque aspects. A and B result from asymmetry of solar heating, whereas C and D have probably been affected in addition by westerly winds, so that they have weaker azimuthal concentrations. E is a still more dispersed distribution from an equatorial region where solar heating is more uniform.

than the succeeding slope downstream. Little ice flows, or flowed, in from out-side the cirque. The convex change in slope at the downstream end of a cirque floor is known as a threshold (or sill). Especially if it is ice-moulded, this feature helps to distinguish glacial cirques from comparable forms developed in non-glacial areas of favourable structure, where mass movement broadens valley heads faster than stream cutting deepens them. Also a cirque floor does not normally coincide closely with a stratigraphic surface.

The influence of aspect is greatest in areas of 'marginal glaciation', where the regional snowline is not far below mountain crests. Its effect has remained im-portant even where the glaciation intensified subsequently. Glaciers facing north-east aided the development of cirques by rapidly removing frost-wedged debris and simplifying accumulation basins. This extended north-east-facing valleys by pushing back divides until they were located far to the south-west of range centre-lines. If later lowering of the snowline permitted glacier development on south-west-facing slopes as well the same opportunities for catchment accumula-tion and cirque development were often not available there, the divide could not be pushed back, and the majority of cirques continue to face north-east. Figure 8.II(i).1 shows cirque-aspect statistics for several groups of cirques influenced chiefly by radiation, and for others which are influenced in addition by wind. Even within cirques, north-east-facing slopes tend to be steeper, so that cirques facing other azimuths may be internally asymmetric. Azimuthal distributions of gradient are complicated by the development within cirques of gently sloping floors as well as steep walls. A range in altitude which includes both shows greater gradient variance at the preferred azimuth of cirques. Mean gradient is affected when the altitudinal range includes mostly walls or mostly floors. The altitude of cirque floors is an altitude of minimum gradient and maximum surface area, especially for preferred cirque azimuths.

In plateau areas cirques are often isolated and easy to delimit. But in more thoroughly dissected terrain their coalescence creates difficult problems. If several hollows, separated by distinct spurs, coalesce to share the same threshold they may be considered individual cirques if the floor can be partitioned easily between them (fig. 8.II(i).2(a)). If, however, the spurs are much smaller, minor protrusions from a continuous backwall, the feature is a single cirque (fig. 8.II(i).2(c)). A distinction between 'valley-side' and 'valley-head' cirques is not useful, since all cirques inevitably concentrate drainage and are necessarily the heads of small valleys. A more meaningful dichotomy is between 'closed' cirques (*cirques en fauteuil*), whose floors are essentially basins cut in rock, and 'open' cirques (*cirques en van*), whose floors slope generally outwards, but at an angle less than that of the slopes below and above. Open cirques are more common well above the snowline, in major glacial source areas of high relief; whereas closed cirques are more characteristic of areas of marginal glaciation, and their floor altitudes have been shown to bear an apparently reasonable relation-ship to the regional snowline. Alternatively, on weak rocks such as shales cirques are commonly open; whereas on massively jointed igneous rocks, lime-stones, and sandstones they more often have rock basins or at least flat floors.

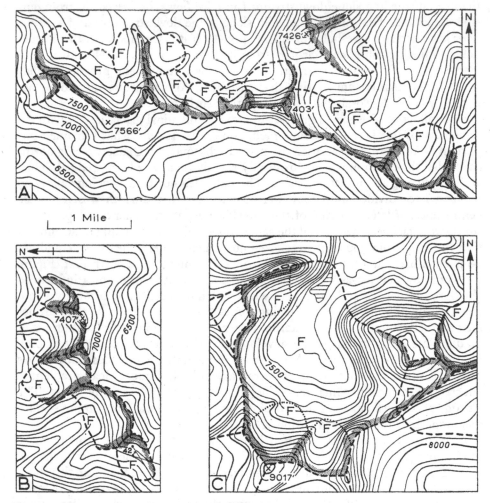

Fig. 8.11(i).2 Cirque definition and variation, in the northern part of the Bridge River District, south-western British Columbia (Heights in feet). Delimitations are based on ground and air photographs (Maps reproduced by permission of the British Columbia Department of Lands, Forests and Water Resources).

A. Coalescent individual cirques at the sources of Eldorado and Taylor Creeks, Tyax–Gun Group. Cirque floors (F) are adjacent but distinct. Contours are shown as continuous lines, cirque outlines as dashed lines, and the upslope margins of the cirques are shaded.

B. Shallow cirques south-east of Eldorado Creek, showing (*bottom*) the definition of cirque plan closure (i.e. the range in azimuth of the longest contour: 42°).

C. A large diversified cirque, deep both in plan and in profile, at the source of the southern fork of Blue Creek, Shulaps Range. The plan closure is 271°. There are several tributary cirques, each with weakly developed floors and thresholds, now almost obscured by detritus.

It seems that strong rock is necessary to sustain a threshold with a reversed slope.

Cirque geometry can be described quantitatively, firstly, with reference to closure in long profile (i.e. in vertical section) as the difference between the maximum backwall slope and the minimum outward floor slope (which may be negative, i.e. reversed) (fig. 8.11(i).3); and secondly, the closure in plan may be defined as the range in azimuth of the longest contour within the cirque. A 'plan closure' less than 90° indicates a cirque which is only a shallow indentation in the mountainside (fig. 8.11(i).2(b)), while one in excess of 180° denotes a cirque closed

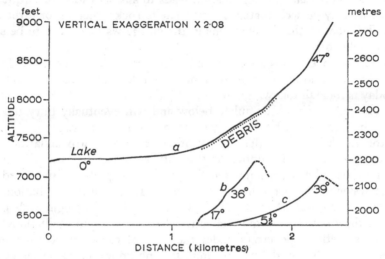

Fig. 8.11(i).3 Cirque long profiles.

A. Large cirque shown in Fig. 8.11(i).2(c). This has quite a large profile closure which is at least 47° (i.e. 47°−0°).
B. The poorly developed westernmost cirque shown in Fig. 8.11(i).2(b). This has the relatively small profile closure of 19° (i.e. 36°−17°).
C. A simple north-east-facing cirque in the Tyax–Gun Group. Profile closure is 33½° (i.e. 39°−5½°).

on more than three sides (fig. 8.11(i).2(c)). These two parameters are the most important measures of cirque shape. Dealing with a single cirque long profile is troublesome, firstly, in the need to define a cirque 'centre-line', and secondly, in that the profile selected may not be characteristic of the cirque as a whole. At the loss of some detail, this may be avoided by constructing a clinographic curve showing the altitudinal variation of mean gradient within the cirque. Profile closure can be expressed as the difference between maximum and minimum altitudinal mean gradient.

The measurement of cirque dimensions is more difficult, since width, length, and especially depth are sensitive to the influence of external topography. Cirque area is one simple expression of cirque size; the length of the longest unbroken contour within a cirque is another, increasing with elongation but less affected by cirque outline. There is no obvious upper limit to cirque size, except that cirque coalescence takes time, and pre-glacial topography rarely favoured early formation of large cirques; cirque growth in area is necessarily slower when opposed by other growing cirques on several sides. While cirque size may, of course, vary with length of glaciation (largely controlled by altitude relative to snow lines), the passage of time does not have the cardinal importance often assumed. Tectonic environment and pre-glacial topography account for most of the differences in 'stage of glacial dissection'. Hobbs's view that cirques normally subdivide and become more intricate as they develop is now generally rejected in favour of the view that spur elimination leads to simplification of cirque form. Irregularities may be accentuated if they relate to rock structure, or if the cirque glacier subdivides so that indentations in the cirque wall continue to be sapped while spurs do not.

While cirque rock basins are produced solely by glacial erosion, the development of the surrounding steep walls is more complex. The process observed on them today is rockfall following loosening by the expansion of water freezing in cracks. The fallen rocks accumulate below and will eventually bury the whole cliff, which developed when a glacier carried away such fragments and permitted the attack by freeze–thaw to continue. A more controversial possibility is that the glacier actually undercut the cliff, as might the sea. Freeze–thaw action might have been accelerated at the glacier margin by an increased availability of water in the bergschrund or rimaye. Temperature oscillations there would be reduced, but when abundant water is available a single cycle of deep freeze and thaw is probably more effective than many rapid oscillations. Meltwater may sometimes penetrate under the ice and cause freeze–thaw at considerable depths, followed by 'plucking', the incorporation of loosened blocks into the moving ice.

2. Nivation

It has been suggested that cirques develop from nivation hollows. Although nivation is important early in the transformation of a steep gully or landslide scar into a cirque, most nivation hollows are on a much smaller scale, and lack the initial relief to accumulate sufficient snow to develop into cirques. They are generally found on gentle slopes, especially where a waste mantle has accumulated, or on weak rocks, where they are commonly elongated parallel to rock structure. Though not proven, their causal relation to snowbanks is reasonably inferred from their spatial association with long-lasting snowpatches, where the concentrated availability of meltwater favours frost disintegration of the waste mantle, as well as solifluction, earthflows, and surface wash on bare areas uncovered late in summer. Eluviation of fine material through the waste mantle permits the development of closed depressions.

Avalanches are a quite different manifestation of nival action. Their importance in areas of high relief and snowfall has not been adequately recognized. When a snow cornice breaks and sweeps down a cliff it cleans off a considerable amount of rock already loosened by freeze–thaw. Such slab avalanches are most common on lee slopes, and permit them to remain steep as they retreat. But slush avalanches are also very effective erosive agents, even on much gentler slopes. Hence, even windward or sunny slopes can be indented by dendritic networks of shallow gullies ('chutes'). These are also followed by streams and rockfalls, but avalanches are probably the major factor in their genesis.

3. Glacial erosion

Erosion by valley glaciers depends upon rock erodibility, the amount of basal debris, the 'hydraulic geometry' of the glacier channel, and basal ice conditions, such as velocity, viscosity, and temperature. These relationships are very complex, and an understanding of how glaciers erode is a long way off. Landforms related to past glaciation suggest that erosion increased where glaciers became either thicker or faster. Fast-flowing glaciers can achieve considerable abrasion, especially if they slide over their bed; however, it is plucking which seems to account for most glacial erosion. This is greatly facilitated if glacier ice can freeze on to a loosened block: refreezing has been observed at shallow depths, but is more difficult to envisage at depth where pressure is more even and, in a temperate glacier, ice is kept at pressure melting point. Changes in pressure through time are important in causing oscillations between 'cold' and 'temperate' ice. Sections of frozen bed material may thus be attached to the glacier sole and removed from their setting at a time of basal melt (i.e. increased pressure). This may explain the lifting of very large blocks of chalk, found in the till of Norfolk and Denmark. In other cases the push of the ice and especially of basal moraine may pry out loosened fragments. On a large scale, the $4\frac{1}{2}$-km-long limestone island of Osmussar in north-west Estonia seems to have been turned through $16°$, over a shale base.

Glacial erosion of incoherent or jointed rock is more comprehensible than the continued erosion of hard, massive rocks to great depths. The development or opening of joints parallel to the surface is favoured by expansion due to unloading (i.e. the removal, by erosion, of the weight of superincumbent rock). It has recently been found that horizontal stresses in the earth's crust are several times greater than vertical loads. Hence when a deep trough is cut its walls expand laterally into it, releasing compression in rocks at this level, but increasing it in lower layers below the trough. This excessive stress concentration probably causes up-arching into the trough, fracturing the rock so that it can be removed by the glacier. As the trough is deepened the stress concentration becomes even greater, counterbalancing the weight of an increased thickness of ice. In this process we may at last have an explanation of why glaciers can cut deep rock basins in hard rock, and why erosion increases with ice thickness so long as ice is confined in a channel.

It is often suggested, even now, that glaciers merely transport material already loosened and debris fallen from the slopes above. While these two sources are important, they completely fail to account for the excavation of deep rock basins, which are the most distinctive result of glacial sculpture. Such basins and troughs cannot be due to glacial removal of rock deeply rotted in the Cenozoic, since their distribution relates to the former glacier system and their walls are generally of sound rock; their pattern is quite different from that of lows in a 'weathering front'. Furthermore, the great quantity of 'rock flour' carried by streams draining from glaciers demonstrates how much abrasion and corrasion is proceeding beneath even small glaciers. The scepticism of some workers relates to the small quantity of moraine currently found in some glaciers. It has been estimated that the Mer de Glace removes only a 'few dozen' cubic metres/year, which lowers its basin by a few microns; that the Glacier de Saint-Sorlin (Grandes Rousses), on weaker rock, lowers its basin 0·1 mm/year; and that there is a lowering of 0·05 mm/year in East Antarctica, which is more rapid than fluvial denudation in areas of low relief, but much less than that in semi-arid or mountain areas. On the other hand, the glacial rock flour carried by streams draining Muir Glacier, Alaska, corresponds to an average loss of 19 mm/year from the area beneath the glacier. Apart from the great approximations involved in such estimates, it is unreasonable to expect modern glaciers, however large, to erode as much as more extensive past glaciers. To the extent that valleys currently glaciated have been modified to suit larger past glaciers, little contemporary modification is either likely or necessary. Thus the most recent Yosemite Glacier in California cut only 220 m into loose sediments filling the 550-m-deep closed basin which was cut into granite by its larger predecessor. Similarly, several Alpine glaciers are floored by a layer of old moraine or proglacial deposits often as thick as the overlying ice. The floor of the present glacier channel reflects the bedrock surface, since both have been cut by similar glacier systems.

The reality of glacial erosion is confirmed most strikingly by the way that channels of former glaciers, as reconstructed on the basis of their deposits, striations, and ice-moulded forms, are well adjusted to their estimated ice discharge. As in fluvial hydraulic geometry, the effect of an increase in discharge (mean velocity times cross-sectional area) is shared out among velocity, width, and depth, all of which tend to increase. Where two glaciers join, their surface must slope continuously downstream (if flow is to continue), and if the bed inherited from fluvial action slopes gently, glacier depth cannot increase downstream; hence velocity increases excessively, increasing erosion until an equilibrium depth is reached, with an increased bed slope at the confluence. This effect is most marked near the source of a glacier, where ice from several cirques joins to form a trunk glacier, leading to an abrupt increase in discharge and a steep drop to a 'trough-head'. A closed rock basin is more likely to form in the ablation zone, where ice moves towards the glacier surface to compensate for the the net surface loss by ablation. Where the surface slopes gently, or near the snout, there may actually be a vertically upward flow component. The downstream reduction in discharge may then be accommodated by a rise in the glacier floor. Hence the largest rock basins are often just above glacier termini, and that of Sogne Fjord (Norway) is at least 1,150 m deeper than its terminal threshold. The rotational flow of cirque glaciers is a special case of this phenomenon when a cirque basin is cut beneath the ablation zone, where ice-flow direction changes from downward to upward. Similarly, where a valley broadens, especially at the margin of a mountain range, the sudden increase in glacier width produces a decrease in depth and velocity, encouraging the formation of a reversed slope closing a rock basin.

Other closed rock basins, so characteristic of glaciated regions, are due to differential rock erodibility; in particular, basins are cut where joints are more closely spaced. Some basins coincide with shear belts of shattered rock, others with weak dykes. Indeed, the locations of basins formed under an ice-sheet are controlled principally by rock structure.

Where a small glacier joins a large one, the result is a negligible step beneath the trunk glacier but an abrupt step beneath the small one (which becomes inset). This is left as a 'hanging valley', but in reality it is a 'hanging paleochannel', for the surfaces of the two glaciers were originally accordant. In many ways glacial hydraulic geometry is similar to fluvial hydraulic geometry, except for the larger cross-section of the channels due to the fact that ice is more viscous than water. Thus the parabolic shape of a glaciated valley (i.e. paleochannel) compares with the cross-section of a river channel, but glacier channels show less variation from this mean form because their laminar flow tends to reduce, rather than exaggerate, channel asymmetry. Laminar flow, with a lack of cross-currents, also leads to the smoothing of sharp pre-glacial valley curves, so that glacier troughs are either straight or curving broadly. There is no evidence that glaciers meander.

As glaciers grow, they cease to be confined in the previous valley system. Ice streams overflow low points in divides to escape to valleys where the ice surface is lower. Rapid flow wears down the divides, creating 'through valleys'; and commonly a rock basin is cut across a former divide, where ice flow was more constricted and therefore needed a deeper channel. Greater ice surface slope on the down-stream side permits greater erosion than where the ground slopes upstream; hence the low point which becomes the post-glacial divide is displaced 'up-ice', often a considerable distance. Glacial 'transfluence' of divides becomes more widespread as ice overwhelms the landscape, and a mountain ice-sheet produces radiating valleys by deepening suitably oriented valley sections and integrating them by divide breaching. In areas of low relief glacial erosion is less concentrated and an ice-sheet does not carve new valleys in geologically homogeneous plains, except that deep basins may result where flow is locally concentrated by a topographic obstruction.

Deglaciation produces rapid landscape modification as fluvial and slope processes begin readjusting the glaciated landscape to a new fluvial equilibrium. Unstable, newly exposed rock walls collapse, and gorges (usually initiated by sub-glacial streams) are cut into oversteepened slopes. Renewed glaciation encounters a changed situation and makes further changes towards its own equilibrium, so that erosion by repeated glaciation may well exceed that by a single long glaciation.

4. Streamlined and depositional forms

Ice tends to mould large bed obstacles into streamlined forms, by erosion and deposition combined in various proportions. The most distinctive form is the drumlin, of the order of a kilometre long and less than half as broad; in plan rounded upstream and pointed downstream; and composed commonly of till, but sometimes of stratified drift or even bedrock. The long axis is aligned in the direction of ice flow, and the upstream end is steeper. Drumlins are probably formed principally by erosion of older glacial deposits, and their elongation may depend on a force/resistance ratio (i.e. it increases with basal ice velocity and decreases with till resistance). Drumlins are usually found in large groups (within which elongation tends to be consistent) not far upstream from terminal moraines, and were formed especially beneath glacier lobes with divergent flow. Though similar forms ('crag and tail') are produced by till deposition in the lee of a bedrock obstacle, drumlins are not necessarily related to such obstacles, but rather to a regular instability at the ice/till interface.

In the ablation zone, and especially during glacier wastage, meltwater is a powerful geomorphic agent. At the glacier bed it moves under considerable and rapidly changing pressure, and is a very effective erosive agent. Sudden discharges caused by the bursting of subglacial barriers, for example, as water in ice-dammed lakes reaches a critical level and drains out under the ice, are particularly erosive. Major subglacial meltwater channels may be distinguished from subaerial ones in having 'up-and-down' long profiles and being discordant with local topography, as when they are cut through spurs. On the glacier surface, meltwater washes fine-grained material out of supraglacial moraine, which is deposited as loose 'ablation till', distinct from the compact 'lodgement till' deposited subglacially under pressure and with fine material retained. Meltwater deposits some material as coarse foreset beds in subglacial channels, forming linear gravel ridges (eskers) when the ice walls melt away, but much is taken farther to form great depositional plains of outwash alongside and downstream from the glacier, mixing with an increasing amount of non-glacial material. Near the glacier, outwash is not only coarser and less well sorted but also disturbed by loss of support when the ice melts, so that it can be distinguished as 'ice-contact stratified drift'. During glacier wastage outwash may cover much of the near-stagnant glacier tongue, accumulating in meltwater channels, crevasses, and other hollows in the ice. Ice melting produces inversion of relief: 'kames' are the positive forms once surrounded by ice, while 'kettles' are the negative ones, where a block of ice was surrounded by drift deposits. Lateral accumulations between a glacier and a valley-side become 'kame terraces'. Where the transition from active to near-stagnant ice is rapid, thrust-planes often carry debris to the surface, where differential ablation produces linear ridges. From these ridges, debris slumps down over the glacier snout and on to the proglacial outwash plain as sheets of 'flow till'. In this way, till and outwash can be interbedded without any readvance of the ice margin. On valley glaciers much till is deposited by ablation at the ice margin, building an end moraine. The inside of this is a cast of the glacier margin, the outside often a talus at the angle of rest. Such a moraine may rim the glacier as far upstream as the firn line, but it is better developed where ablation is greater; it is often lacking on the shady side of the valley. The lateral parts of the moraine are often better preserved, since meltwater streams breach the terminal moraine at the glacier snout: complete arcuate ridges are typical of the smallest glaciers, which produce less meltwater. In lowlands the till component is often small, and the moraine is a series of deltas, the head of an outwash plain, or a bulldozed ridge of various deposits. End moraines are formed by the culminations of readvances, rather than pauses in retreat. Once formed, their influence may keep the glacier margin stationary for a longer period, by containing minor oscillations.

REFERENCES

AHLMANN, H. W. [1919], Geomorphological studies in Norway; *Geografiska Annaler*, **1**, 1–148 and 193–252.

ANDREWS, J. T. [1963], Cross-valley moraines of north-central Baffin Island: a quantitative analysis; *Geographical Bulletin*, **20**, 82–129.

BLACHE, J. [1952], La sculpture glaciaire; *Revue de Géographie alpine*, **40**, 31–123.

BOULTON, G. S. [1967], The development of a complex supraglacial moraine at the margin of Sørbreen, Ny Friesland, Vestspitsbergen; *Journal of Glaciology*, **6** (47), 717–35.

CHARLESWORTH, J. K. [1957], *The Quaternary Era with special reference to its glaciation* (London), 1,700 p.

CHORLEY, R. J. [1959], The shape of drumlins; *Journal of Glaciology*, **3** (25), 339–44.

CLAYTON, K. M. [1965], Glacial erosion in the Finger Lakes region (New York State, U.S.A.); *Zeitschrift für Geomorphologie*, n.f. **9**, 50–62.

COTTON, C. A. [1942], *Climatic Accidents in Landscape-Making* (Christchurch, New Zealand), 354 p.

DAHL, R. A. [1965], Plastically sculptured detail forms on rock surfaces in northern Nordland, Norway; *Geografiska Annaler*, **47**A, 83–140.

DAVIS, N. F. G. and MATHEWS, W. H. [1944], Four phases of glaciation, with illustrations from south-western British Columbia; *Journal of Geology*, **52**, 403–13.

EMBLETON, C. and KING, C. A. M. [1968], *Glacial and Periglacial Geomorphology* (London), 608 p.

EVANS, I. S. [In preparation], *Measurement and Interpretation of Asymmetry in Glaciated Mountains.*

FLINT, R. F. [1957], *Glacial and Pleistocene Geology* (New York), 553 p.

GEIGER, R. [1965], *The Climate Near the Ground* (Cambridge, Mass.), 611 p.

GILBERT, G. K. [1904], Systematic asymmetry of crest lines in the High Sierra of California; *Journal of Geology*, **12**, 579–88.

LEWIS, W. V., Editor [1960], *Norwegian Cirque Glaciers;* Royal Geographical Society Research Series, No. 4, 104 p.

LINTON, D. L. [1963], The forms of glacial erosion; *Transactions of the Institute of British Geographers*, **33**, 1–28.

LLIBOUTRY, L. [1965], *Traité de Glaciologie*, t.2. *Glaciers-variations du climat-sols gelés* (Paris), pp. 429–1040.

MATTHES, F. E. [1900], Glacial sculpture of the Bighorn Mountains, Wyoming; *United States Geological Survey, 21st Annual Report*, Part 2, 167–90.

MATTHES, F. E. [1930], Geologic history of the Yosemite Valley; *United States Geological Survey Professional Paper* 160, 137 p.

RAPP, A. [1960], Recent development of mountain slopes in Kärkevagge and surroundings, North Scandinavia; *Geografiska Annaler*, **42**, 65–200.

SEDDON, B. [1957], Late-glacial cwm glaciers in Wales; *Journal of Glaciology*, **3** (22), 94–9.

SISSONS, J. B. [1967], *The Evolution of Scotland's Scenery* (Edinburgh), 259 p.

TRICART, J. and CAILLEUX, A. [1962], *Le modelé glaciaire et nival* (Paris), 508 p.

II(ii). Periglacial Morphometry

BARBARA A. KENNEDY

Department of Geography, Cambridge University

1. Introduction

It is virtually impossible to find a universally acceptable definition of the term 'periglacial': for example, should it be restricted to the description of glacier-margin zones, or would this be too pedantic? The present discussion will take a highly pragmatic view and equate 'periglacial' areas with those in which the ground at depth is perennially frozen: these are the permafrost zones. Such a definition will include some high-altitude areas of frozen ground, in addition to the main high-latitude regions, but will exclude alpine meadows where the ground is completely thawed each summer. It should be stressed that the major characteristic of such periglacial regions is not the number of freeze–thaw cycles but the depth of frost penetration and length of time that the surface temperatures are below 32° F.

Although calculations have been made of the fraction of the world's water at present 'stored' in glaciers and ice-caps, no such estimates exist for that proportion which is removed from the hydrologic cycle as frozen ground water in regions of continuous and discontinuous permafrost. Figure 8.II(ii).1 shows the relative extent of ice-sheets and permafrost for the northern hemisphere (which contains the major areas of frozen ground), and it is clear that the volume of water involved in the latter regions is considerable.

With some exceptions (notably in the north-eastern U.S.S.R.) most of the areas of present periglacial climate were covered by continental ice sheets during the Pleistocene (see fig. 8.II(ii).1), and many of their landforms are therefore those common to all regions of glacial erosion and deposition. In addition, many coastal areas of frozen ground show clear effects of recent glacial 'rebound' in the form of flights of strandlines: the eastern coast of Baffin Island is a case in point.

Regarding other landforms, the opinion of geomorphologists is sharply divided. Peltier [1950, p. 221] has proposed a distinct periglacial cycle, resulting from the action of 'intense frost shattering, solifluction and congeliturbation'. The continued operation of these processes – upon an area initially possessing strong relief – is considered to result in 'extensive surfaces of cryoplanation with slopes less than 5°' which grade laterally into even lower-angle 'congelifractate-covered surfaces of downwastage or lateral planation' (Peltier [1950], p. 225).

Some workers have accepted the view of a distinct, periglacial assemblage of landforms: for example, Suslov [1961, p. 137] considers that the presence of

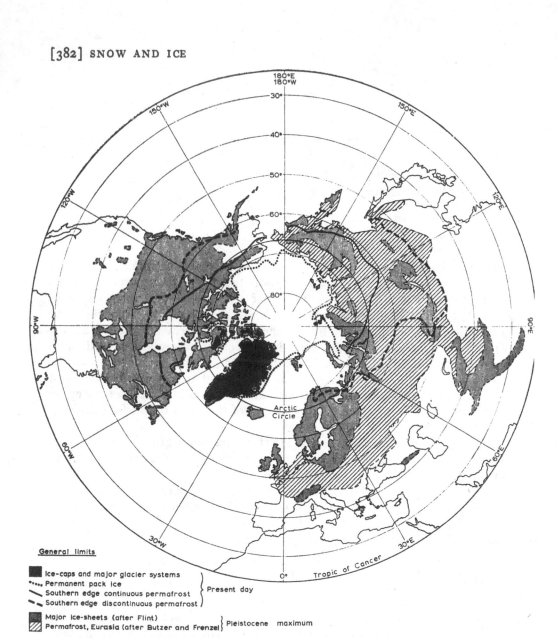

Fig. 8.11(ii).1 Past and present distribution of ice sheets and frozen ground in the northern hemisphere.

(The differences between some of these distributions and those shown in Fig. 8.1.1 reflect current differences of opinion.)

permafrost will affect the entire character of relief, though one should emphasize that this is in relation to eastern Siberia, an area of comparatively low relief which was not entirely glaciated. In much of northern North America, on the other hand, workers have felt it necessary to stress how little the landscape corresponds to the Peltier ideal (Bird [1953], p. 36; Mackay [1958], p. 26).

Rather than attempt to reconcile these views, the divergence of which may well result from historical differences between areas, the present discussion of periglacial landforms will be limited to those features which are uniquely present in regions of frozen ground. With the exceptions outlined below, the landforms of periglacial areas tend, perhaps disconcertingly, to resemble those of the other major fluvial regions of the earth's surface, both in their complexity and types. If one considers the vast extent of permafrost, as outlined in fig. 8.11(ii).1, and the range of climatic regimes, lithology, and relief to be found in such areas, the diversity of landforms is scarcely surprising.

2. Individual landforms

A. Pingos

'Pingo' has become the generally used term to denote certain ice-cored mounds or 'hydro-laccoliths' which are common in some low-lying permafrost areas of predominately fine-grained sediments.

A major concentration of pingos is found around the mouth of the Mackenzie Delta, N.W.T., Canada, where they have been intensively studied by Mackay [1963, pp. 69–94].

In form, pingos are conical hills of between 10 and 150 ft in height, with basal diameters ranging from 100 to 2,000 ft. The maximum heights are attained by those pingos with intermediate diameters between 500 and 700 ft, as are the steepest side slopes (up to 45°).

The occurrence of pingos is very closely related to the presence of drained lake-beds and marsh-filled channels, and it appears that the central ice-lens is formed, as shown in fig. 8.11(ii).2, by the migration of water as the permafrost front advances inwards from the old shorelines. The classification into 'closed' or 'open-system' pingos is related to the absence or presence of a continuing source of water after the formation of the initial core.

Although many pingos are long-established features of permafrost landscapes, all collapse in time; either because a change in climate destroys the permafrost altogether or because the over-lying sediments are cracked and the ice-core exposed to insolation. A collapsed pingo will be represented by an almost circular depression with a raised rim.

In themselves, pingos can scarcely be termed 'significant' landforms, yet in those permafrost areas in which they are found they create the nearest thing to a unique periglacial landscape.

B. Ground-ice slumps

Where sections of permafrost underlain by thick layers of ice are exposed to direct lateral undercutting – as along a river bank or sea coast – severe slumping ensues.

As the ice content of the ground may be anything from 50% upwards, the slump mass is liable to extreme wastage, and the chief feature which distinguishes these forms from slumps of other kinds is that the debris 'toe' is markedly

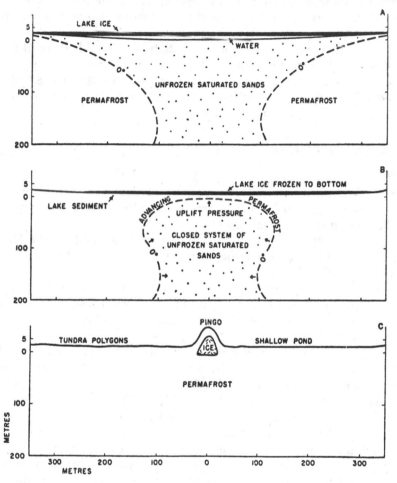

Fig. 8.11(ii).2 Schematic origin of a pingo (From Mackay, 1963).

In diagrams A, B, and C a vertical exaggeration of 5 × has been used for the height above zero in order to show the lake ice and the open pool of water.

(A) A broad shallow lake has an open pool of water in winter with a frozen annulus around it. No permafrost lies beneath the centre of the lake.

(B) Prolonged shoaling has caused the lake ice to freeze to the bottom in winter and induced downward aggradation of permafrost. Infilling has raised the lake bottom a small distance. The deepest part of the lake, which has the thinnest permafrost, is gradually domed up to relieve the hydrostatic pressure.

(C) The pingo ice-core, being within permafrost, is a stable feature. The old lake bottom is occupied by tundra polygons and shallow ponds. Because of scale changes in the diagram, the volume of the ice-core should not be construed as showing a direct relationship to the initial volume of unfrozen material.

underfit when considered in terms of the volume of the slump scar. The slumps themselves are arcuate-headed, and their sides may be bordered by levées of hardened mud.

C. Oriented lakes

Two of the three major occurrences of oriented lakes in North America are found in permafrost areas: near Point Barrow, Alaska, and in the Mackenzie Delta, N.W.T. (the third region is that of the Carolina 'bays'). The critical factor governing the development of such lakes in the North at the present time appears to be the strength of winds along the Arctic coast and the lack of barriers in the tundra landscape which will break their fetch.

In form, oriented lakes are elliptical in outline, with a triangular 'deep' at their centre. Lengths of the long axes may range from 100 ft to 2 miles, and the most common length:width ratio is 2:1 (Mackay, 1963).

The observation that such lakes tend to be oriented north–south, at right angles to the prevailing winds, has been shown to agree with forms derived from theoretical calculations of circulation patterns in two dimensions.

D. Ice-wedge polygons

These polygons are found in many low-lying areas of permafrost and are formed by the junction of ice wedges which initiate in tension cracks developed in the surface materials. In size, ice-wedge polygons may range from 5 to 100 ft across (fig. 8.11(ii).3), and they may possess either high or low centres, depending upon their stage of development.

E. 'Naleds' or icings

The build-up of hydrostatic pressure in an unfrozen aquifer in permafrost may become so great that water bursts through the overlying beds and freezes on the ground surface. Such icings are, obviously, winter features, but on melting they leave devastated areas comparable to those created by the passage of small avalanches.

3. Valley-side features

Although it is impossible to distinguish a 'periglacial valley form', some valleys within permafrost areas may possess minor characteristics which can be specifically related to the climatic regime.

A. Solifluction features

The process of solifluction – or downhill creep of soil with a very high water and ice content above the permafrost table – has frequently been cited as a dominant and characteristic feature of periglacial regions, but within any one area the evidence for its operation is usually given by very minor features of the landscape.

Among the forms attributable to solifluction are low, arcuate lobes and terraces and soil or vegetation stripes.

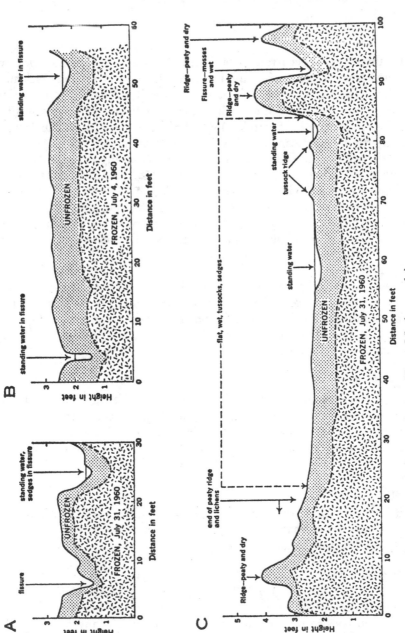

Fig. 8.II(ii).3 Cross-sections of tundra polygons (From Mackay, 1963).

A. High-centred peaty polygon.
B. Polygon subject to periodic inundation.
C. Low-centred polygon.

B. Nivation and snow-bank hollows

These concave niches in valley sides are not restricted to permafrost areas, but they are probably more common in such regions than elsewhere.

In size and appearance, nivation hollows may vary considerably. In some cases the back walls are bare, whereas the footslope, which may be vegetated, is rilled and frequently grades into a solifluction terrace downslope. In other cases the whole niche may be vegetated, giving no direct evidence of the processes at work.

The influence of the presence of such hollows upon slope form is also various. In most areas it would seem that profiles on which niches develop are steepened, but the degree of steepening depends upon the position of the niche on the profile and the proportion of the total length which it occupies. In areas of very low precipitation – notably Banks Island, N.W.T. – there is evidence to suggest that strong development of nivation hollows leads to a general decline in the angle of the slope above: possibly this relationship arises from the unusual concentration of moisture represented by the snow patch, which provides favourable conditions for solifluction on the back wall. Whatever the mechanism, slopes in the Kellet drainage, Banks Island, which are the site of persistent snow banks are, on average, 8° less steep than those without marked nivation hollows.

The most pronounced nivation hollows are found on north- or north-east-facing slopes, where insolation is weakest and late-lying snow banks consequently favoured.

C. Asymmetrical valleys

There is evidence to suggest that the steepening of south- and west-facing slopes, unrelated to structural controls, is peculiar to certain permafrost environments, though not all asymmetric valleys currently developing in such areas are of this type.

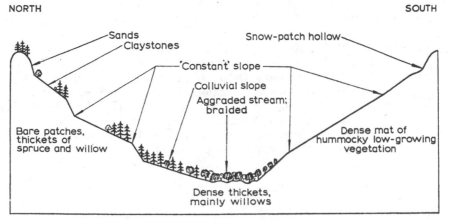

Fig. 8.11(ii).4 Cross-section of a typical asymmetric valley, Caribou Hills, North West Territories, Canada (From Kennedy and Melton, 1967).

Figure 8.II(ii).4 shows an idealized cross-section of a typically asymmetrical valley at the foot of the Caribou Hills, N.W.T. In this area south-facing valley sides have average maximum angles of 28° and are significantly steeper than the comparable sections of north-facing profiles, which average only 23°. The development of this asymmetry is strictly limited to the lower valleys, which have a relative relief of up to 200 ft and may reach a quarter of a mile in width. In the narrow, relatively shallow upper valleys north-facing slopes are significantly the steeper.

It appears that the development of periglacial asymmetry, in this sense, is controlled fairly strictly by valley dimensions. Only in deep, wide valleys does the combination of maximum insolation, greatest depth of active layer, and large downslope gravity component on south- and west-facing slopes lead to rapid, large-scale movement of surface material and steepening of the valley sides.

In narrower valleys, or those less incised, the steepening of north- and east-facing profiles is jointly controlled by the presence of large snow-patch hollows and slumping along the plane of the relatively shallow permafrost table, as the result of basal undercutting by streams.

REFERENCES

BIRD, J. B. [1953], *Southampton Island;* Memoir 1, Geographical Branch, Department of Mines and Technical Surveys (Ottawa), 84 p.

BIRD, J. B. [1967], *The Physiography of Arctic Canada* (Baltimore), 336 p.

BROWN, R. J. E. [1960], The distribution of permafrost and its relation to air temperatures in Canada and the U.S.S.R.; *Arctic,* 13, 163–77.

BUTZER, K. [1964], *Environment and Archaeology* (London), 524 p.

FLINT, R. [1957], *Glacial and Pleistocene Geology* (New York), 553 p.

FRENZEL, B. [1968], The Pleistocene vegetation of northern Eurasia; *Science,* 161, 637–49.

KENNEDY, B. A. and MELTON, M. A. [1967], Stream-valley asymmetry in an arctic environment; *Research Paper 42, Arctic Institute of North America,* Montreal, 41 p.

MACKAY, J. R. [1958], *The Anderson River Map – Area, N.W.T.;* Memoir 5, Geographical Branch, Department of Mines and Technical Surveys, (Ottawa), 137 p.

MACKAY, J. R. [1963], *The Mackenzie Delta Area, N.W.T.;* Memoir 8, Geographical Branch, Department of Mines and Technical Surveys, (Ottawa), 202 p.

PELTIER, L. C. [1950], The Geographical Cycle in Periglacial Regions; *Annals of the Association of American Geographers,* 40, 214–36.

SUSLOV, S. P. [1961], *The Geography of Asiatic Russia;* Translated by Gershovsky, N. O. (San Francisco and London), 594 p.

III. The Economic and Social Implications of Snow and Ice

J. ROONEY

Department of Geography, Southern Illinois University

Snow represents both a valuable resource and a menacing natural hazard. Although its economic utility is difficult to evaluate precisely, we can make some meaningful estimates of its contribution to agriculture, recreation, domestic and industrial water supply, and even to our less-tangible aesthetic needs. For example, most of the irrigated agriculture in the western United States, notably in the Central Valley of California, is supported by meltwater from snowpack; while in the past twenty years the number of Americans who ski has risen from 50,000 to approaching 4 million, representing only one aspect of the rapidly increasing economic benefits associated with snow-oriented recreation.

Knowledge concerning the equally significant negative social and economic impact of snow and ice is probably more sketchy. It is believed that the costs of snow-caused disruption and loss of life, plus the funds spent to combat the hazard, amount to at least 1 billion dollars a year in the United States.[1] During the period 1958–66 the U.S. Weather Bureau directly attributed more than 500 deaths to severe snowstorms, and many more persons died in traffic accidents resulting from more minor storms or prematurely from some form of over-exertion, such as snow clearing.

1. Snow as a water-supply source

Snow is the principal source of water not only in many of the world's mountainous areas but also in the densely settled plains and valleys adjacent to them. It is estimated that about one-half of the streamflow in the western United States is of snowpack origin. Considerable study has been devoted to snow-zone management in the Sierra Nevada, and it is believed that California is representative of many other areas characterized by marked contrasts in elevation and vegetation. Approximately 51% of California's streamflow originates in the

[1] According to the U.S. Weather Bureau, damage attributed to the snow hazard in the United States ranged from an absolute minimum of $1,502,550 in 1964 to a maximum of $738,841,500 in 1958. However, these estimates are based on the high and low estimates of the U.S. Weather Bureau for those storms which were deemed severe enough for publication in *Storm Data*, U.S. Weather Bureau, Department of Commerce, Washington, D.C. They probably represent no more than 20% of the damage attributable to snow and ice each year in the United States.

snowpack zone, 32% comes from the lower forest zone, and the remainder from the foothills and lowland areas.

One of the primary aims of any water-management strategy is to reduce the annual variability of supply, and it is here that the snow zone takes on particular significance. Not only is the yield from the snow zone the most important source of water, but in the most critical dry years it is far more dependable than sources in other zones. For example, in the southern Sierra Nevada less than one year in ten is a dry year (defined as having less than one-half the mean annual precipitation) in the snow zone, whereas two to three are dry in the coniferous forest zone, and more than four in ten are dry in the foothill zone. The variability of water supply from the snow zone is further reduced by the low evaporation from the snowpack.

The economic impact of snow-zone water in both California and the whole of the western United States has consequently been tremendous. Cash-farm income in California ranks first in the United States, and nearly all crops are irrigated. A comparison of the flourishing cotton, vegetable, fruit, beet, grain, and potato agriculture of the irrigated sections of the San Joaquin Valley with the poor grazing lands of its dry western foothills is all that is necessary for one to appreciate the value of the snowpack as a water-supply source.

2. Recreation

Participation in recreation associated with snow and ice has been expanding faster than most other forms of recreational activity, with ski-ing the most prominent beneficiary. In the United States 'ski-ing activity days' have been increasing at the rate of 15% per year, and there has been an annual increase of 20% in the ski-ing population. According to the Ski Trade Association, almost 3 million 'serious' skiers spent approximately $1 billion in connection with the sport during 1967–8 alone, and over 100,000 winter-sports enthusiasts were flown to Denver by a single airline in that year.

To accommodate this ballooning United States' demand nearly 1,200 ski resorts are now in operation, over half of them equipped with chair lifts. Most of the spectacular ski areas are located in the sparsely settled western section of the country, necessitating extensive travel and adding to the general economic significance of the ski industry (figure 8.III.1). Another economic multiplier effect has been the development of sophisticated residential estates in association with ski resorts, such as the flourishing complexes at Jackson, Wyoming, and Snow Mass at Aspen, Colorado, the latter involving a planned $75 million investment.

It should not be thought that all the economically important recreational aspects of snow are limited to ski-ing. Snowmobiling is another activity which has experienced a phenomenal expansion during the past five years. Snowmobile production in North America has grown from 15,000 units in 1963 to 175,000 in 1967, with many of the owners participating in economically significant racing and endurance events, especially in the states of Michigan and Wisconsin. Sledding and tobogganing are enjoyed by millions, and have been so perfected by some as to warrant inclusion as an Olympic event. Ice skating and hockey are avidly pursued by many, and professional hockey is now one of the major spectator sports in North America. Additional snow and ice recreational activities include ice-boating, curling, snow-shoeing, and sled-dog racing, all of which are both socially and economically important in many areas.

The snow-based recreation industry of western Europe is even more important than that of North America, both absolutely and in relative economic terms. France alone, for example, has between 1 and 1½ million skiers (which

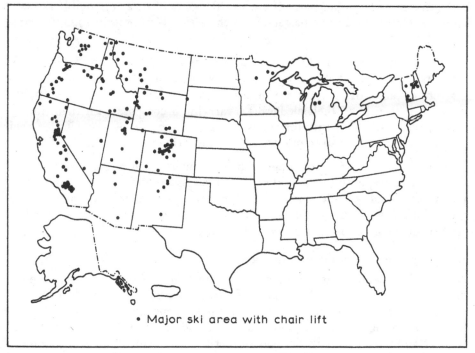

• Major ski area with chair lift

Fig. 8.III.1 National Forest ski areas in the United States (Source: U.S. Department of Agriculture, Forest Service, PA 525).

figure is expected to surpass 2 million by 1970), some 200 winter-sports resorts with 2,000 hotels containing 45,000 bedrooms. There are 40 aerial ropeways, 150 cabin or chair lifts, 800 ski lifts, 150 ski schools, and 2,000 ski instructors in the country. New centres are being opened every year, one of the most recent being the elaborate resort of La Plage in Savoy, which was begun from scratch in 1961 and can now accommodate more than 6,000 vacationers at a time. A large proportion of those taking skiing holidays in France are foreigners, with 40% of the holidays being taken at Christmas, 25% at Easter, and 35% in between the two. The economic significance of this invisible tourist export is even more important in Switzerland, where the number of nights passed by foreign tourists increased from the already-high figure of 15 million in 1912 to 19

million in 1966, the latter representing only 59% of the total tourist accommodation. In 1966 foreign tourists spent 2,900 million Swiss francs on Swiss holidays,
which compares with the 8,110 million francs earned by the country's manufactured exports in 1960. One in ten of the Swiss working population is employed
in transportation, hotels, and catering.

However, the aesthetic benefits of snow are difficult to assess in financial
terms. How much is a White Christmas worth, or a drab winter landscape re-
vitalized by a blanket of white? In those areas where prolonged snow-cover is
common winter seems to take on a special meaning, with the pattern of life
being distinct from other parts of the world. Some of these intangible assets of
snow were captured in this passage from the *Atlas Maritimus* in 1728, referring
to winter in Finland:

> '... the inhabitants look abroad ... to travel and carry on their needful
> affairs, and without troubling themselves about night or day, sea or land,
> rivers or lakes, dry land or wet, the face of the world being all smooth and
> white, they ride on their sledges ... carrying a compass with them for the
> way, wrapt in warm furs for the weather and a bottle of aquavita for their
> inside, with needful store of dried bread and dried fish for their food'.

Traces of this attitude remain and help to explain the mixed reactions which
still occur in response to snow. The negative impact of snow in today's urban
world can be fully understood only after a careful analysis of such positive
psychological associations, for it is often the aesthetic 'benefits' of snow which
prevent optimum effort being directed toward its control.

3. The negative aspects of snow and ice

Perhaps the most obvious detrimental economic effect of snowfall lies in the disruption of communications and allied services, and it is at once clear that, although where snow is common authorities may spend more money in the long
run in combating it, it is in regions where snowfall has a high variability that the
greatest and most expensive individual disruptions occur. A classic example of
snow disruption took place in Britain during the abnormally snowy winter of
1962–3. On average, southern Britain can expect not greatly in excess of 10 days
with snow lying per year, the most likely period for it (2 chances in 7) being
about the second week in January. Snow began to fall on 26 December and continued intermittently until 23 February, during which there were 35 consecutive
days with maximum temperatures not exceeding 2·7° C, giving the coldest
winter for the country as a whole since 1829–30 and for some regions since
1740. The prolonged snowfall at one time blocked 95,000 miles of road in
southern England, directly caused the deaths of 49 people, and was responsible
for incalculable economic loss.

Railroads provide some of the best comparative data on snow disruption,
partly because their traffic is so well documented and partly because they are so
susceptible to it. Besides blocking lines, freezing snow jams points and moving
parts of signals, obscures signal lights, breaks telegraph wires, and falling snow

may completely obscure visibility. In regions used to coping with the effects of snow costs are quite modest and the disruption at a minimum. The New York Central R.R., for example, normally spends 0·74% of its operating expenses on snow and ice removal (this figure rose to only 1·18% in 1945 as a result of the famous Buffalo–Albany blizzard), and of the normal winter train delays on the Chicago, Milwaukee, St Paul, and Pacific R.R., only 12% are due to weather conditions (mostly to blizzards, snow drifting, and snow-plough delays in the northern Rockies section). Compared with these figures, the heavy snowfalls in Britain during 1940 put 1,500 miles of track out of service for 12–72 hours, and

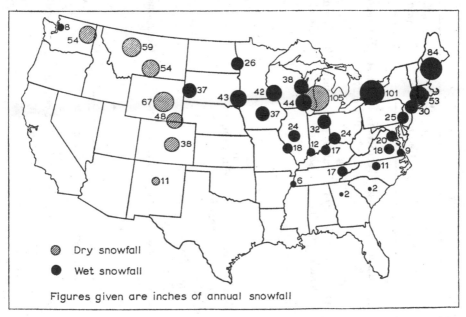

Fig. 8.III.2 Average snowfall at the thirty-five U.S. cities chosen as study sites for snow hazard. Figures give mean annual snowfall in inches (Dry snowfall<0·1 in. of water per inch of snow; wet snowfall>0·1 in. per inch).

during the bad winter of 1947 one trans-Pennine route was closed by snow for two months. Where severe snow and ice occur only occasionally, the disruption may be on a huge scale, as with the ice storm of 30 January to 2 February 1951 in Alabama, Louisana, and Arkansas, in which the principal damage was the breaking of telegraph wires (one stretch of line, for example, suffered 3,500 breaks in 111 miles) and which was not permanently restored for over two years.

An accurate assessment of the troublesome effects of snow in urban areas urgently requires some type of categorization. To this end I have separated urban snow disruption into two parts: internal, when inter-change within the city is hampered; and, external, when conditions affect the interaction between a community and its surrounding area. Measurement was based on a five-order hierarchy of disruption which categorized impact from paralysis (first order) to

TABLE 8.III.1 Hierarchy of disruptions: internal and external criteria

Activity	1st order (paralysing)	2nd order (crippling)	3rd order (inconvenience)	4th order (nuisance)	5th order (minimal)
INTERNAL					
Transportation	Few vehicles moving on city streets	Accidents at least 200% above average	Accidents at least 100% above average	Any mention	No press coverage
	City agencies on emergency alert, Police and Fire Departments available for transportation of emergency cases	Decline in number of vehicles in CBD	Traffic movement slowed	Traffic movement slowed	
		Stalled vehicles			
Retail trade	Extensive closure of retail establishments	Major drop in number of shoppers in CBD Mention of decreased sales	Minor impact		No press coverage
Postponements	Civic events, cultural and athletic	Major and minor events Outdoor activities forced inside	Minor events	Occasional	No press coverage
Manufacturing	Factory shutdowns Major cutbacks in production	Moderate worker absenteeism	Any absenteeism attributable to snowfall		No press coverage
Construction	Major impact on indoor and outdoor operations	Major impact on outdoor activity Moderate indoor cutbacks	Minor effect on outdoor activity	Any mention	No press coverage

Communication	Wire breakage	Overloads		Any mention	No press coverage
Power facilities	Widespread failure	Moderate difficulties	Minor difficulties	Any mention	No press coverage
Schools	Official closure of schools; Closure of rural schools	Closure of rural schools; Major attendance drops in city schools	Attendance drops in city schools		No press coverage
EXTERNAL[a]					
Highway	Roads officially closed; Vehicles stalled	Extreme-driving-condition warning from Highway Patrol; Accidents attributed to snow and ice conditions	Hazardous-driving-condition warning from Highway Patrol; Accidents attributed to snow and ice conditions	Any mention, for example, 'slippery in spots' warning	No press coverage
Rail	Cancellation or postponement of runs for 12 hours or more; Stalled trains	Trains running 4 hours or more behind schedule	Trains behind schedule but less than 4 hours	Any mention	No press coverage
Air	Airport closure	Commercial cancellations	Light plane cancellations; Aircraft behind schedule owing to snow and ice conditions	Any mention	No press coverage

[a] Warnings are the key to this classification. They provide excellent indicators because they are widely publicized.

minimal (fifth order), with transportation the most critical variable (Table 8.III.1). Paralysis disruptions may occur in either or both of these situations. With respect to internal activity, the complete restriction of mobility with its myriad ramifications is normally the most serious problem that can be attributed to snow, since most functions characteristic of urban areas require movement from one section of the city to another. The isolation of a city from its surrounding area, or vice versa, would represent a paralysis external disruption. This type of measurement has provided considerable understanding of the disruption producing roles of the physical paramenters (the physical snow environment), and insight concerning the implications of snow-hazard perception and adjustment.

Generalizations regarding the snow hazard were based on observations, taken at thirty-five U.S. cities with annual snowfall ranging from 1 to 105 in. (fig. 8.III.2). Information concerning disruptions was primarily obtained from daily newspaper coverage and public records. From this material it was possible to classify the impact of all snow days under study during a ten-year period 1956-65.

The salient relationships between the physical snow environment and disruption can be summarized as follows:

A. Annual snowfall

1. The frequency of higher-order (paralysis, crippling, and inconvenience) disruption increases with annual accumulation, however, not without exception. When paralysis disruptions are considered separately the correlation with annual snowfall is weak.
2. Disruption, though it increases with annual snowfall, does so at a diminishing rate. Inversely, the intensity of disruption (disruption per inch) decreases as average annual snowfall increases.
3. Snow of lower moisture content tends to produce less difficulty than the wet variety. Disruption per unit of snow at sites where the moisture content was lower than 0·10 per inch (chiefly cities in the Mountain West) was significantly less.

B. Individual storms

1. At the individual storm level the relationship between depth of snow and curtailment of human activity is very strong. Furthermore, intersite comparisons revealed in general that as annual accumulation increases the amount of snow necessary to produce paralysis conditions increases as well. Thus, to take the extremes, the average paralysis disruption at Muskegon (105 in,) was associated with a 14-in. storm; whereas in Greensboro (11 in.), 4½ in. produced those circumstances.
2. A significant statistical difference exists between the impact of snow in association with winds of 15 miles per hour or more, and that accompanied by wind of lesser velocity.
3. Both the time and the rate of snowfall proved critical at the single storm level. The time of fall often means the difference between a nuisance and a

crippling disruption, while the rate of accumulation has a profound impact on the effectiveness of snow-control operations.

4. The effect of air temperature proved difficult to isolate, perhaps because temperature is so casually related to the amount of snow and the rate at which it accumulates.

Disruption resulting from snow and ice, like that resulting from floods or hurricanes, is only partially attributable to the physical properties of the hazard. Of equal or greater importance may be the adjustments which communities and individuals make to modify the hazard, or to reduce its impact.

TABLE 8.III.2 Results of correlation

	Coefficients of correlation					
	Magnitude of disruption (internal and external)			Intensity of disruption (internal and external)		
	1st order, paralysing	2nd order, crippling	3rd order, inconvenience	1st order, paralysing	2nd order, crippling	3rd order, inconvenience
Annual snowfall	0·23	0·73	0·87	−0·48	−0·56	−0·71
Annual snowdays	0·15	0·64	0·81	−0·48	−0·62	−0·69

1. 0·45 Significant at the 0·01 level of probability.
2. 0·35 Significant at the 0·05 level of probability.

A comparison of the disruption patterns with the snow environments of all thirty-five cities revealed some interesting relationships. A very strong positive correlation was obtained by comparing the sum of the three highest orders of disruption magnitude with annual snowfall (Table 8.III.2). Using this combination, differences in annual snowfall account for nearly 76% (i.e. $0·87^2 \times 100$) of the variation in disruption. However, by combining only the highest two orders of disruption and comparing them with snowfall, it appears that the snow environment explains only 53% (i.e. $0·73^2 \times 100$) of the variation.

The relationship between first-order disruption magnitudes only and snowfall is in marked contrast with the coefficients obtained through the summations referred to above. Paralysis situations tend to be ubiquitous regardless of mean annual snowfall or snowdays. Disruption at cities which average less than 20 in. a year is in many cases more frequent and more severe than the disruption at places recording considerably greater accumulations. For example, St Louis, Richmond, Knoxville, Louisville, Evansville, and Greensboro experience more paralysis occurrences than communities with twice to three times as much snow. To cite a specific case, St Louis (18 in.) has recorded over 60% more first-order and 40% more second-order disruptions than Milwaukee (44 in.). As

o

a result, the relationship between paralysis and annual snowfall is statistically insignificant, accounting for less than 6% of the variation. Measurements of disruption intensity (i.e. disruption per 10 in. of snow) also provide substantiating evidence concerning these relationships (Table 8.III.2.).

A generalized pattern of snow-caused disruption in the United States based on the data from the thirty-five study sites tends to emphasize the significance

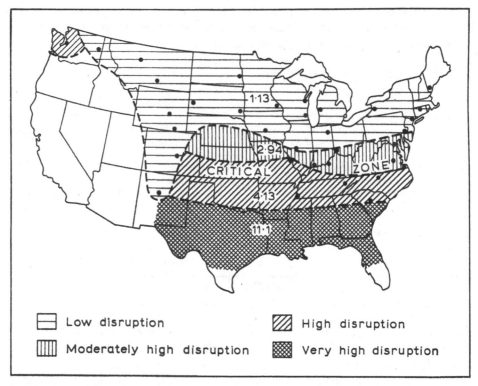

Fig. 8.III.3 A generalized pattern of snow-caused disruption in the United States. Figures indicate average number of first- and second-order disruptions per 10 in. of snowfall for the sites within each zone. Shading gives the relation of disruption to snowfall, the low disruption indicating a high degree of adjustment to it, and vice versa.

of the human environment (fig. 8.III.3). Four disruption zones are conspicuous. The northernmost zone is one marked by reliable annual snowfall, a high level of adjustment, and low disruption related to the amount of annual snowfall. It is in this area that the most sophisticated public snow-control programmes exist and where individuals have taken the initiative to protect themselves and their property from the snow hazard. It is also here where behavioural patterns have been conditioned through more frequent exposure to snow and ice, the result being a less emotional, and perhaps more routine response to most snow situations.

South of the high-exposure zone is a narrow transitional area where disruption

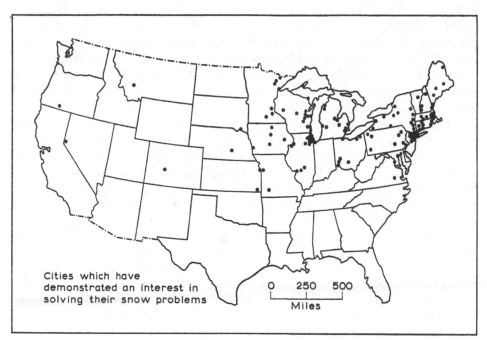

Fig. 8.III.4 Distribution of snow hazard interest in the United States (Compiled from *The American City*, 1950–67).

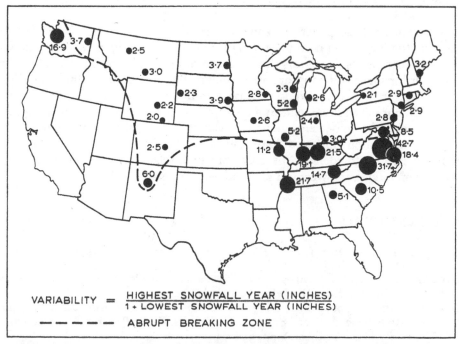

Fig. 8.III.5 Pattern of snowfall variability in the United States. Proportional circles represent the variability index and illustrate the abrupt breaking zone.

is moderately high in relation to snowfall and where inter-site variation is substantial. This zone and the wider belt to the south are characterized by marginal adjustment and relatively high disruption per inch of snow. Both areas exhibit a critical and abrupt scaling down in the level of adjustment, which has produced a general state of unpreparedness for snow. Many of the cities within this area experience a greater *absolute* number of paralysis and crippling disruptions than do their northern 'snow belt' counterparts. The fact that adjustment wanes rapidly within the transition zone is partly documented by the pattern of snow-hazard interest (fig. 8.III.4). Interest in adjustment decreased abruptly south of the east–west line approximating the 20-in. snow isohyet – the same area in which both the disruption per unit of snow and the annual snow variability increase!

Annual variability of snow may well explain much of the observed patterns of adjustment and disruption (fig. 8.III.5). A comparison of expected variability over a ten-year period for cities with accumulations between 10 and 20 in. points out the degree of uncertainty associated with low-snowfall areas. Cities in this 'normally' meagre to moderate snowfall zone are confronted with a degree of uncertainty unknown in the north. Here the decision-makers must cope with a snow environment characterized by a 2- or 3-in. accumulation in one year and as much as 30–40 in. in the next. It is clear that community adjustment is not a direct function of annual snowfall, but rather there seems to be a critical accumulation to which cities respond that is explained largely by community decision-making and perception.

In conclusion, the pattern of snow-caused disruption in the United States suggests the immediate need to systematize decision-making in regard to snow-control programmes. It appears that planning has generally been formulated in a haphazard manner, with appropriations being allocated via the political market-place. Even in high snow-exposure areas planning has been geared to some preconceived normal environment. As a result, high-snowfall cities are not equipped to cope with the five-, ten-, and fifty-year storms, while low-exposure communities have often embraced 'hope and pray' strategies. A substantial research effort is needed to identify the costs of snow-related disruption before the benefits of various snow-control alternatives can be assessed.

REFERENCES

AMBASSADE DE FRANCE [1968], *The French Tourist Industry* (Service Presse et d'Information, London).

ANDERSON, H. W. [1963], Managing California snow lands for water; *U.S. Forest Service Research Paper* (Berkeley, California).

ANDERSON, H. W. [1966], Integrating snow zone management with basin management; In Kneese, A. V. and Smith, S. C., Editors, *Water Research* (Johns Hopkins University Press, Baltimore), pp. 355–73.

BAUER, H. [1968], *All About Switzerland* (Swiss National Tourist Office).

BELL, C. [1957], *The Wonder of Snow* (New York), 269 p.

CHAMPION, D. I. [1947], Weather and railway operation in Britain; *Weather*, **2**, 373–80.

HAY, W. W. [1957], Effects of weather on railroad operation, maintenance, and construction; *Meteorological Monographs*, **2** (9), 10–36.

LIEBERS, A. [1963], *The Complete Book of Winter Sports* (New York), 228 p.

MEAD, W. R. and SMEDS, H. [1967], *Winter in Finland* (Hugh Evelyn, London), 144 p.

ROONEY, J. F. JR. [1965], *The Urban Snow Hazard: An Analysis of the Disruptive Impact of Snowfall in Ten Western and Central United States Cities* (Clark University, Worcester, Massachusetts), 150 p. (Also from University Microfilms, Ann Arbor, Michigan.)

ROONEY, J. F. JR. [1967], The urban snow hazard in the United States: An appraisal of disruption; *Geographical Review*, **57**, 538–59.

RUSSELL, J. A., Editor [1957], Industrial operations under extremes of weather; *Meteorological Monographs* 2 (9).

THOMPSON, J. C. [1959], The snow probability factor; *The American City*, **74**, 80–3.

CHAPTER 9

Short-term Runoff Patterns

I. The Flood Hydrograph

JOHN C. RODDA

Institute of Hydrology, Wallingford

1. The hydrograph

Where the quantity of runoff from a river basin is measured continuously, a graph of flow as a function of time is obtained. This discharge hydrograph expresses the sequence of relationships that occur between runoff and the other components of the basin water balance, together with their adjustments to the physical characteristics of the basin.

During dry periods the flow of a river decreases exponentially. This base flow

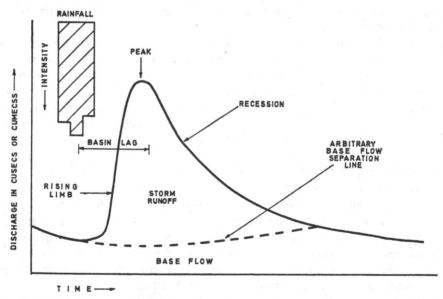

Fig. 9.1.1 Components of a hydrograph.

(fig. 9.1.1) continues until rain falls, when the rate of discharge increases rapidly to a peak or crest. This point is arrived at when the quantity of water draining to the gauging station from the basin has reached a maximum. The peak is usually attained shortly after rain has ceased, and thereafter flow is largely determined by the amount of water in storage. This stored water will be held in the soil and in the bedrock, and will have resulted from recent infiltration and percolation from earlier rain. The rate of withdrawal from storage controls the shape of the recession limb of the hydrograph, this relationship between storage and discharge being expressed as:

$$Q_t = Q_0 K^t \qquad (1)$$

where Q_t is the discharge at time t after some instant when the discharge was Q_0, K being a recession 'constant', which is less than unity. Further rain during recession can cause two or more peaks, but without it, discharge decreases until the extra water in storage due to the recent rain has been depleted and the flow is approaching its original volume.

Thus the discharge hydrograph consists of a series of irregular saw-tooth-shaped fluctuations superimposed on a gently undulating section. These two components are usually defined as storm runoff (or the flood hydrograph) and base flow, the latter being attributed to ground-water discharge. A further analysis of the hydrograph has been made by separating storm runoff according to the path to the river. The initial rise in the hydrograph is attributed to channel precipitation – water falling directly on to the connected water surfaces of a basin– while the bulk of the increase in discharge is said to be caused by surface runoff. Interflow is the third component, and this water, which moves laterally through the upper soil horizons until it is intercepted by a stream channel, is particularly important during recession.

These concepts may be useful for descriptive purposes, but their scientific basis is limited. For example, surface runoff can only occur when the rainfall intensity exceeds the infiltration rate, yet in many storms this never happens, no surface runoff results but the hydrograph rises sharply. Indeed, for most hydrographs it is difficult to distinguish between storm runoff and base flow, let alone the other components, so most modern approaches concentrate on this two-fold division, employing it in unit hydrograph analyses.

2. Hydrograph shape and magnitude

The shape and dimensions of the hydrograph are controlled by a variety of factors, many being interrelated. These factors can be divided into two main groups; those of a permanent nature and those that can be classed as transient (fig. 9.1.2). The first group in general represent the characteristics of the basin, while the second is associated with climate and related features. Of course, one or two factors fall into either category, and there are others that may change from one group to the other – usually as a result of man's activities.

While the factors that influence the characteristics of the hydrograph are for the most part readily recognizable, relatively few studies have aimed at establishing quantitative relations between the flood and these factors. Yet a precise method of predicting the size and shape of the flood hydrograph would be invaluable to the hydrologist, in place of the somewhat dubious means that he is forced to employ at times. Such a method could be utilized for flood-warning purposes and to provide the basic information for the design of spillways, culverts, bridges, and similar structures. For flood-warning purposes, however, it would be necessary to know *when* a flood is likely, as opposed to the

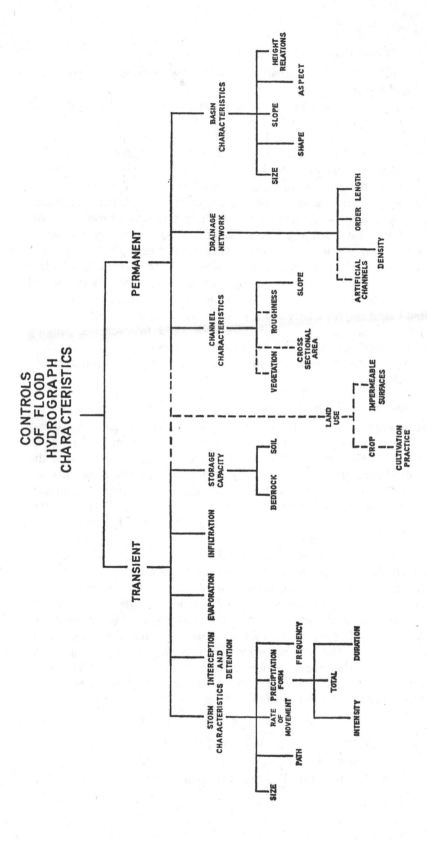

Fig. 9.I.2 Controls of flood hydrograph characteristics.

design need, which requires information on *how often* a flood of a particular magnitude would occur.

Some floods are caused by dam bursts, earth movements, and high tides, but these are rare by comparison with the floods due to intense rain or rapid snow melt. In the case of rain-induced floods the path of the storm in relation to the alignment of the basin, the size of the storm, and its rate of movement are important, as well as the intensity and amount of rain. Once the rain has reached the surface of the basin, the rate and amount of runoff will be influenced by such factors as the current evaporation and infiltration rates, the soil moisture status, and the type of land use. Movement of this water into rills and brooks and the passage of the flood wave to the gauging stations are governed by a series of factors, from the basin morphometry to the hydraulic and biological characteristics of the channel system. These factors control not only the form of the hydrograph but also the time interval between the rain and the flood. Basin lag and time of concentration are indices of these time-response characteristics of a basin; the first being the time between the centre of mass of rain and centre of mass of runoff, and the second, the time taken for water to reach the gauging station from the most distant point in the basin. However, even though these features are important, far more attention has been given to estimating the magnitude of the flood peak.

3. Factors influencing the magnitude of the flood peak

The interdependence of factors, the difficulties of quantifying and identifying them correctly, and the problems of establishing meaningful statistical relationships between variables have to be overcome in determining the controls of the flood peak. Interdependence is probably the major obstacle, because virtually all the controls are related in some way to one another. Hence the usual approach is to select as variables factors that on physical grounds are least likely to be interdependent. There are difficulties in quantifying some of the factors, such as vegetation or land use, while in many cases measurements of others are simply not available (e.g. infiltration rates and rainfall intensities). Often the only material that can be obtained, other than river-flow records, is what can be derived from map analyses or from aerial photographs. This emphasizes the importance of quantitative geomorphology to this branch of hydrology; although few of the morphometric properties that can be assessed have been shown by objective methods to be important controls. Basin size, basin and channel slope, and various measures of the channel system can be determined from maps. These and similar factors have been demonstrated to be significant and have been incorporated into a range of 'flood formulae' by means of standard statistical techniques producing expressions of the following type:

$$Q_t = aX_1{}^b \, X_2{}^c \, X_3{}^d \qquad\qquad (2)$$

where Q_t = the T-year annual peak discharge;

    $a \ldots d$ = regression coefficients;

    X_1 etc = the factors controlling the flood peak.

A. Basin area

Other factors being equal, the larger the size of the basin, the greater the amount of rain it intercepts and the higher the peak discharge that results. This rather obvious conclusion has been the basis for a large number of flood formulae in the general form:

$$Q = CA^n \tag{3}$$

where Q = peak discharge;

A = basin area;

C = a constant that varies according to the land use or topography of the basin;

n = a constant that has a range from 0·2 to 0·9, depending on climate to some extent.

One of the many examples of this type of approach is a study of the mean annual flood (Q_m) in a number of basins in England and Wales, where the area–discharge relationship was determined as:

$$Q_m = CA^{0·85} \tag{4}$$

However, there are a number of other factors that are partly dependent on basin area and must be in some way accounted for by the constants in this type of relationship. For example, larger basins are usually less steep than smaller ones, and this applies to other factors, such as channel slope and rainfall intensity. As a consequence, when these additional characteristics of the basin are employed the inclusion of basin area with them usually gives the most meaningful result.

B. Basin shape

Basin shape is of obvious importance in influencing peak flow and other hydrograph characteristics, although it is a feature which is difficult to express numerically. However, a number of shape indices have been developed, some of the best known being the form factor and Miller's circularity ratio. The former is the ratio of average width to axial length of the basin, while the latter demonstrates the circularity of a basin by:

$$R_c = \frac{A_b}{A_c} \tag{5}$$

where A_b area of basin;

A_c area of the circle having the same length of perimeter as the basin.

This expression has a value of unity for a circular basin, while for two basins of the same size the flood potential would be considered greatest for the one with the smallest circularity coefficient. This index was employed in analysing the flow from a number of Appalachian basins, but it was found to have a low correlation with peak discharge. On the other hand, in the same study, peak discharge was found to be highly correlated with the longest length of basin as measured from the head of the basin to the stream gauging station. This and other indices of shape have been criticized on the grounds that they do not approach the ideal pear-shaped basin – the lemniscate being put forward as offering a better comparison.

C. Basin elevation

The altitudinal extent of the basin above the gauging station exercises direct and indirect control over the magnitude of the flood peak. With basin slope and several additional factors, it determines the proportion of runoff, and indirectly it influences a number of other important controls, such as precipitation, temperature, vegetation, and soil type. However, it is difficult to compute a single term which gives a meaningful measure of basin elevation. Indeed, several studies have shown the various indices that have been devised to have no significant relation to the size of the flood peak.

D. Basin slope

Slope, like elevation, is an obvious control of peak discharge, but again it is a factor which is difficult to interpret meaningfully. Some methods of slope assessment are extremely involved and require measurement of length of all contours in a basin, or counting the number of intersections between contours and a grid overlay. Others are relatively simple, but the importance of these basin slope indices has been difficult to establish, whereas measurements of channel slope have been proved significant. One slope index (S) devised for a study of fifty-seven British drainage basins showed little significance, even when log S was correlated with log Q_m. However, when slope was combined with basin area, as below, a coefficient of multiple correlation of $+0.93$ resulted:

$$Q_m = 0.074 A^{0.74} S \qquad (6)$$

E. The drainage network

The several characteristics of the flood hydrograph hinge on the efficiency of a basin's drainage system. A quick rise to a high peak is the mark of a well-developed network of short steep streams. Conversely, a minimal response to intense rain usually reflects an incipient channel system. How a particular basin relates to these extremes of development can be assessed in terms of linear aspects of the drainage network, the areal relationships of the system, and the various channel gradients.

Linear aspects of the channel system are expressed in terms of stream order, bifurcation ratio, stream length, and length of overland flow. Other than longest length of stream channel, none of these measures, by themselves, have been shown to exercise extensive control over the flood peak. On the other hand, their inclusion with other factors has reduced the error of estimate of peak flow, and this also applies to areal relationships and channel gradients. For a study of New England floods, ninety-three slope factors were computed, main-channel slope being found the most significant variable. In the same study peak flow showed no relation to drainage density, once channel slope and storage area had been taken into account. On the other hand, the drainage density term improved the prediction equation in a flood study for the United Kingdom, and it has been shown to correlate with the mean annual flood in other studies.

Apart from channel slope, no other measures of the hydraulic nature of the channel have been included in flood peak studies, although roughness and wetted perimeter would be important terms. Similarly, there is an absence of expressions for the character of the in-channel and riverine vegetation, features that can also influence the nature of the hydrograph.

F. Climatic factors

In latitudes where snow melt combines with rain to produce a spring flood maximum, snow depth and temperature are important controls of peak flow. Elsewhere, the magnitude of the flood is related to the rainfall that provokes it, coupled with the current storage capacity of the basin. There are difficulties in expressing this rainfall, because each storm is typified by a differing set of magnitude, duration, and intensity relationships, as well as those of frequency, distribution, and areal extent. One way of avoiding these difficulties is to employ the basin's mean annual rainfall as an index of its flood susceptibility. This was the basis for a second part to the study of floods in Britain, where the slope factor (S) was replaced by mean annual rainfall (R) to give:

$$Q_m = 0.009A^{0.85}R^{2.2} \tag{7}$$

However, this term is not a sufficiently realistic measure of the flood-producing rain, and several more likely ones have been determined. These include various intensity–duration parameters and one of rainfall frequency. For example, in the New England study of floods, referred to previously, the intensity index showing the highest correlation with peak discharge proved to be the daily maximum rainfall of the same frequency as the flood. Further improvements could be made by the inclusion of factors indicative of antecedent conditions or the availability of storage, but such terms have rarely been employed.

G. Vegetation and land use

Speculation about the effects on runoff of felling a forest, or one of the other land use changes brought about by man's activities, has continued for many centuries. Only at the end of the nineteenth century were experiments commenced to determine what hydrological differences resulted from alterations in land use, but since then over a hundred of these experiments must have been conducted in various parts of the world. In fact, none of the other controls of runoff have received comparable attention. The classical approach is to alter the land use on one of a pair of otherwise identical basins after an initial calibration period, then to ascribe differences in runoff patterns to the contrasting land usage. The basin maintained in its original state acts as a control, so that extraneous influences, such as climatic change, can be identified. This is the disadvantage of the single-basin approach, because the land use change that is made after the calibration period can be confused with trends in rainfall or the long-period variations in another element of climate. Results from these experiments are often not representative of larger areas, and few, if any, vegetation factors have been employed as variables in assessing peak discharge. Nevertheless, the information gained from these studies, which are as yet far from complete, is of considerable interest and importance (Penman, 1963; Sopper and Lull, 1967) from the floods point of view and in terms of water resources.

H. Comprehensive formulae for estimating the flood peak

Various measures of basin morphometry and climate have been combined in formulae for assessing the T-year peak discharge. All such formulae include a basin area factor, and most contain some index of rainfall intensity and frequency, in addition to differing measures of several morphometric characteristics. The efficacy of each extra term is demonstrated by a reduction in the standard error of estimate and a rise in the coefficient of multiple correlation in the examples shown below.

Author	No. of basins	Location	Formula
Potter	51	Allegheny–Cumberland Plateau	$\log Q_{10}{}^1 = -1\cdot4 + 0\cdot17 \log A^1 - 0\cdot55 \log T + 0\cdot93 \log P + 0\cdot45 \log S$
Morisawa	15	Appalachian Plateau	$\log Q_{10}{}^1 = -8\cdot96 + 0\cdot54 \log A^1 + 0\cdot72 \log t + 4\cdot24 \log P - 0\cdot29 \log S$
Benson	164	New England	$\log Q_{2\cdot33} = 0\cdot4 + 1\cdot0 \log A + 0\cdot3 \log Sl - 0\cdot3 \log St + 0\cdot4 \log F + 0\cdot8 \log O$
Rodda	26	United Kingdom	$\log Q_{2\cdot33} = 1\cdot08 + 0\cdot77 \log A + 2\cdot92 \log R_{2\cdot33} + 0\cdot81 \log D$

$Q_{10}{}^1$ = peak discharge (cusec/acre) for a ten-year recurrence interval.
$Q_{2\cdot33}$ = peak discharge (cusec) in the mean annual flood.
A^1 = basin area (acres).
P = rainfall intensity factor.
T = topography factor.
S = rainfall frequency factor.
t = topography factor combining measures of relief, circularity, and first-order stream frequency.
A = basin area (square miles).
Sl = main channel slope (ft/mile).
St = % of surface storage area plus $0\cdot5\%$.
F = average January degrees below freezing (° F)
O = orographic factor.
$R_{2\cdot33}$ = mean annual daily maximum rainfall (in.).
D = drainage density (miles/square mile).

Even the inclusion of three or more independent variables covering a wide range of climate and basin characteristics leaves considerable differences between observed and predicted peak discharges. Indeed, it has been suggested that one year of discharge records is more valuable for predicting the T-year flood than innumerable estimates of flood magnitude–frequency relations. It could be that the problem of prediction will never be solved by application of the basic physical laws, even where these are fully understood, because the physical system is far too complex. Some hope of success is offered by simplification of the system and its input through the use of quantitative models to simulate the major processes and interactions within the basin.

4. Model simulation of basin discharge characteristics

The use of models in the study of various aspects of the hydrograph has been common for some time. The model system consists of a limited number of parts, analogous to the most important features of the prototype, and it operates by transforming numerical input data into a quantitative representation of the behaviour of the prototype. Parameters in the model, which are obtained initially by comparison with the prototype, are adjusted so that the two outputs match. This optimizing technique is not a simple procedure, because a number of parameters have to be adjusted, and it is not completely clear which mathematical methods of optimization are best. However, if all basins could be described fully by an expression of the type

$$Q = PX_1{}^q \tag{8}$$

in which P is the percentage run-off and q is a rainfall intensity factor, then for one particular basin the rainfall and discharge records would be scrutinized and the optimum values for the parameters P and q sought by a sequence of adjustments, to improve the agreement between calculated and observed peak discharges – the agreement being measured objectively. This two-parameter model is obviously far too simple a representation of a basin, and it ignores many of the physical processes involved in runoff. Yet the parametric approach has the advantage of not requiring detailed measurements of the entire range of processes involved, because where a significant physical effect is neglected, the model will show this by its failure. The use of digital computers in these simulation studies permits quite complex models to be employed, but there are other examples where parametric methods are employed, such as co-axial correlation and the unit hydrograph.

A. Co-axial correlation

The co-axial method of graphical correlation has been widely applied for the prediction of the total volume of storm runoff. A range of factors have been incorporated in the diagrams and in the example shown (fig. 9.1.3). A four-parameter model was adopted for computing the storm run-off in the River Thames at Teddington.

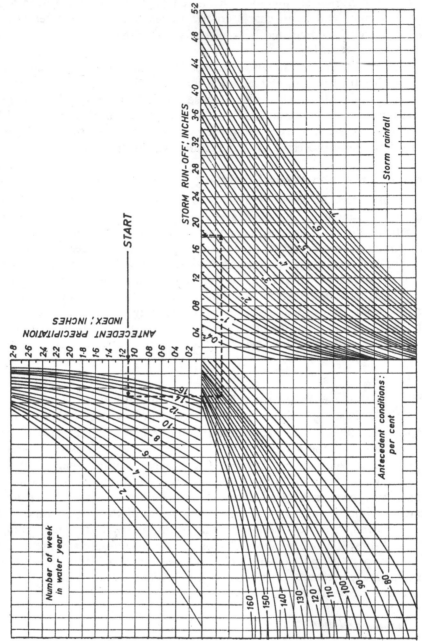

Fig. 9.1.3 Coaxial diagram for computing storm runoff for the River Thames at Teddington (After Andrews, 1962).

B. *The unit hydrograph hypothesis*

The unit hydrograph (Sherman, 1932) is a simplified concept of the behaviour of a basin in converting rainfall to stream flow. It is based on the premise that the storm runoff is derived by a linear operation from the rainfall that is effective in causing the runoff and that the system is time invariant. For a given drainage basin, the T-hour unit hydrograph is the storm runoff due to a unit volume of effective rain generated uniformly in space and in a time T – the volume of the effective rainfall commonly being taken as 1 in. or 1 cm over the drainage area. It is assumed that the runoff from effective rainfalls of the same duration, produced by isolated storms on the same basin, causes hydrographs

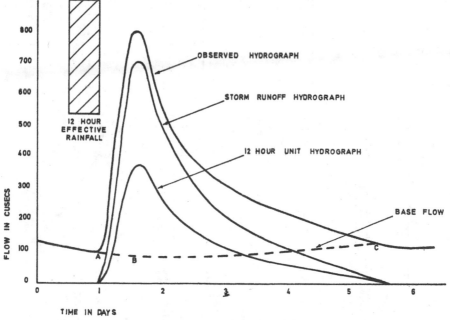

Fig. 9.1.4 Derivation of a hypothetical unit hydrograph.

of equal length in time. Another assumption is that ordinates of the unit hydrograph are proportional to the total volume of direct runoff from falls of rain of equal duration and uniform intensity, irrespective of the total volume of rain. In other words, for two storms of 15 hours duration, the first producing 2 in. of effective rainfall and the second 3 in., if the storm runoff from the first storm passes the gauging station in ten days, then this time will be approximately the same for the second storm. Also, if on the sixth day after the first storm 15% of the total storm runoff passes the gauging station, then this will apply to the second storm. It is evident that for natural basins these assumptions about the unit hydrograph cannot be justified completely, but for many purposes the results produced by the theory are acceptable.

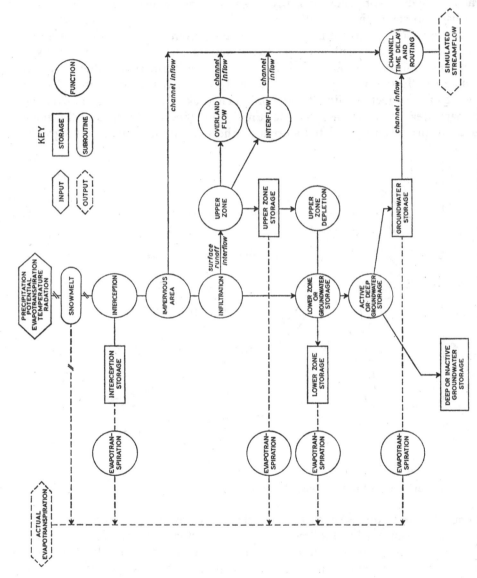

Fig. 9.1.5 Stanford Watershed Model IV flowchart (After Crawford and Linsley, 1966).

To derive a unit hydrograph for a particular basin, records of rainfall and discharge are examined for an isolated storm with reasonably uniform rainfall. For such a storm (fig. 9.1.4) the base flow (ABC) is separated from the remainder of the hydrograph so that the volume of storm runoff can be determined. In the hypothetical example shown this volume is 85,400,000 ft³, which is equivalent to a depth of 1·86 in. over an imaginary basin of 20 square miles. Then by dividing each ordinate of the storm runoff hydrograph by 1·86 the hydrograph resulting from 1 in. of runoff is obtained – the unit hydrograph.

Empirical relations between unit hydrographs and basin characteristics have been aimed at, in a similar manner to the way in which the controls of peak discharge have been sought. Formulae have been developed which provide values of time and magnitude of peak, time of base of the unit hydrograph, and basin lag time from measures of characteristics of the basin. In one study lag time (m) and its second and third moments (m_2 and m_3) were correlated with area (A), length of main channel (L), and the slope of the catchment (S); the best prediction equations being:

$$m_1 = 27 \cdot 6 A^{0 \cdot 3} S^{-0 \cdot 3} \tag{9}$$

and
$$m_2 = 0 \cdot 41 L^{-0 \cdot 1} \tag{10}$$

using such methods, synthetic unit hydrographs can be predicted for basins with no records, but caution should be exercised in applying them, because they can involve appreciable errors.

C. Digital models

This is a relatively new method for investigating the behaviour of systems, and it has only been applied to simulating hydrological relations within the last ten years. The other approaches considered so far deal with particular aspects of runoff and its controls, but the quantitative model must continuously simulate all the processes involved in the basin water balance. Digital-computer programmes reproduce components of the entire physical system, and parameters in the programmes can be altered to represent any set of circumstances. Time scales can be compressed, and the behaviour of a basin over a period of years can be reproduced in several minutes. Most models carry out two main operations: converting rainfall and potential evapotranspiration into runoff (fig. 9.1.5), then transforming volumes of runoff into discharge hydrographs. There are appreciable differences between models, however, particularly in simulating basin storage and in the methods of determining actual evapotranspiration from potential.

This is a very rapidly developing field, where a considerable effort is being made. Like all other approaches that aim at an understanding of the controls of the flood hydrograph, its limitations lie mainly in the veracity of the basic data.

REFERENCES

ANDREWS, F. M. [1962], Some aspects of the hydrology of the Thames Basin; *Proceedings of the Institution of Civil Engineers*, **21**, 55–90.

BENSON, M. A. [1962], Factors influencing the occurrence of floods in a humid region of diverse terrain; *U.S. Geological Survey Water Supply Paper 1580-B*.

CARLSTON, C. W. [1963], Drainage density and streamflow; *U.S. Geological Survey Professional Paper 422-C*, 8 p.

COLE, G. [1966], An application of the regional analysis of flood flows; *The Institution of Civil Engineers, Proceedings of the Symposium on River Flood Hydrology* (London), pp. 39–57.

CRAWFORD, N. H. and LINSLEY, R. K. [1966], *Digital Simulation in Hydrology: Stanford Watershed Model IV*; Department of Civil Engineering, Stanford University, Technical Report No. 39, 210 p.

MORISAWA, M. E. [1959], Relation of quantitative geomorphology to stream flow in representative watersheds of the Appalachian Plateau Province; *Office of Naval Research Project NR 389-042, Technical Report 20, Department of Geology, Columbia University, New York*, 94 p.

NASH, J. E. [1960], A unit hydrograph study, with particular reference to British catchments; *Proceedings of the Institution of Civil Engineers*, **17**, 249–82.

NASH, J. E. and SHAW, B. L. [1966], Flood frequency as a function of catchment characteristics; *The Institution of Civil Engineers, Proceedings of the Symposium on River Flood Hydrology* (London), 115–36.

NASH, J. E. [1967], The role of parametric hydrology; *Journal of the Institution of Water Engineers*, **21**, 435–74.

PENMAN, H. L. [1963], *Vegetation and Hydrology*; Commonwealth Bureau of Soils, Harpenden, Technical Communication No. 53, 124 p.

POTTER, W. D. [1953], Rainfall and topographic factors that affect runoff; *Transactions of the American Geophysical Union*, **34**, 67–73.

RODDA, J. C. [1967], The significance of characteristics of basin rainfall and morphometry in a study of floods in the United Kingdom; *UNESCO Symposium on Floods and their Computation* (Leningrad).

SHERMAN, L. K. [1932], Streamflow from rainfall by the unit-graph method; *Engineering News-Record*, **108**, 501–5.

SOPPER, W. E. and LULL, H. W., Editors [1967], *International Symposium on Forest Hydrology: Proceedings of a National Science Foundation Seminar* (Pergamon Press, London), 813 p.

II. Relation of Morphometry to Runoff Frequency

G. H. DURY

Department of Geography, University of Sydney

By strict definition, channel morphometry is the measurement of channels, while channel morphology is the study of channel shape. By extension, both terms are used to connote shape characteristics, whether in plan or in the vertical plane. Before the relation of morphometry to discharge frequency can be examined, it is necessary to define in this context the terms *pattern* and *habit*.

1. Channel pattern

This is the trace of a channel in plan, as shown, for instance, on vertical air photographs or as represented on maps. Meandering patterns and braided patterns occur widely. *Habit* in this sense can be substituted for *pattern*, as when a stream is said to have a meandering or a braided habit; but the study of habit as a mode of stream behaviour goes beyond the two dimensions of the plan view. It involves the shape of the channel in cross-section, and also the long-profile of the bed.

No comprehensive classification of channel pattern has yet been produced. The following provisional classification, including a number of well-represented types of which some are illustrated in fig. 9.11.1, is most unlikely to be complete:

1. Meandering
2. Braided
3. Straight
4. Straight-simulating
5. Deltaic-distributary
6. Anabranching
7. Reticulate
8. Irregular

Morphologic studies of channels have hitherto been concentrated on natural channels which are meandering, braided, or straight-simulating, while laboratory work has dealt with meandering, braided, and straight channels. These types will be chiefly discussed in what follows, meandering channels being first used to explain some of the principles involved.

Meanders are sinuous bends. In practice, a complete range of intermediate patterns links straight channels to channels which are highly sinuous and obviously meandering, while a second or possibly a variant continuum reaches

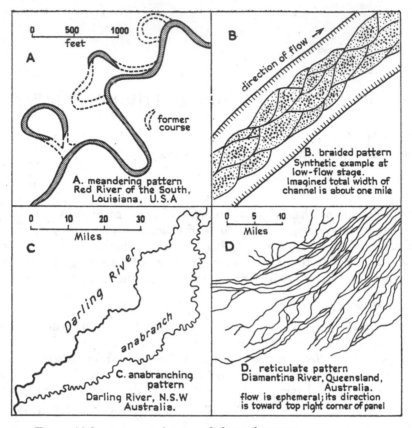

Fig. 9.11.1 Four widely represented types of channel pattern.

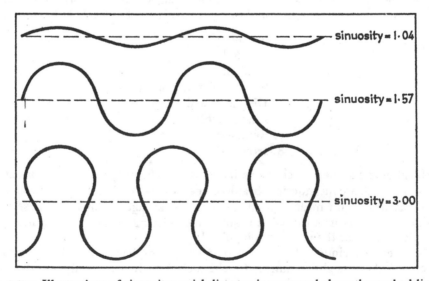

Fig. 9.11.2 Illustrations of sinuosity: axial distance is measured along the pecked lines.

from straight channels to channels which are merely irregular in plan. If it is required to discriminate between channels which meander and other single channels which do not some arbitrary fix on the scale of sinuosity is normally used. Sinuosity is the ratio of channel distance to axial distance (fig. 9.11.2): a sinuosity of at least 1·5 is used by some workers as a criterion of meandering.

In addition to sinuosity, the attributes of a meandering channel include amplitude, wavelength, and radius (fig. 9.11.3). Amplitude, which at one time commanded a great deal of attention, is not greatly considered nowadays. Radius is a somewhat crude measure of channel curvature, since meander bends tend to assume the form of sine-based waves rather than that of arcs of circles. Wavelength is fundamentally significant, especially in relation to bedwidth and

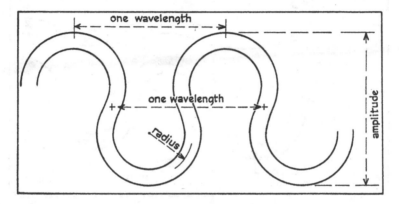

Fig. 9.11.3 Some elements of meander geometry.

to discharge. Variations in the width/depth ratio ensure that the wavelength/ width ratio is also variable; but on many streams the wavelength is between 8 and 12 times the bedwidth, with the modal value between 9 and 10 times bedwidth.

A meandering channel is asymmetric in cross-section at bends, the greatest depths occurring near the outer banks. Between bends it is more symmetrical, and shallower. The deeps at bends are *pools*, while the shallows between bends are *riffles*. Meandering appears to begin with the establishment of a pool-and-riffle sequence. Straight laboratory channels, shaped in homogeneous materials, can deform into pool-and-riffle when water is fed straight into them at constant discharge. We infer that straight uniform channels are unstable. A characteristic spacing from one pool to the next in a flume experiment is about five bedwidths. As meanders form, alternate pools migrate to alternate sides, giving the approximate wavelength of two inter-pool spacings, or 10 bedwidths, as in nature (fig. 9.11.4).

Field experiments with marked sediments show that riffles tend to be eroded somewhat at low stages, the material removed from them being fed into the pools. When discharge increases and stage rises to banktop or thereabouts the pools are scoured anew, the sediment excavated from them lodging on riffles farther downstream. That is, the fullest expression of the pool-and-riffle sequence is promoted by discharge at or near bankfull. This circumstance reinforces what was said in an earlier section about channel-forming discharge.

Bedwidth on meandering channels, and on single channels generally, is accordingly measured between banktops, unless water-surface width is specific-

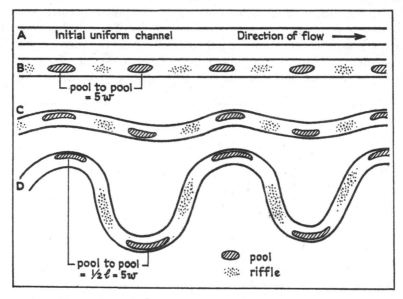

Fig. 9.11.4 A, uniform channel, no pool-and-riffle; B, straight channel with pool-and-riffles; C, slightly sinuous channel; D, meandering channel.

ally in question. A well-established relationship exists between bedwidth, w, and bankfull discharge, q_{bf}, in the form

$$w \propto q_{bf}^{0.5}$$

or, numerically, with w in feet and q in ft³/sec, approximately

$$w = 3q_{bf}^{0.5}$$

Meander wavelength, l, has already been described as an approximate linear function of bedwidth, i.e.

$$l = kw$$

or, numerically,
$$l = 10w$$

Independent analysis produces the expectable result that

$$l \propto q_{bf}^{0.5}$$

or, in the same units as above,

$$l = 30q_{bf}^{0.5}$$

The values of the numerical constants are affected in specific cases by variation of the width/depth ratio of the channel.

The connection between channel size and meander size (wavelength), on the one hand, with value of discharge, on the other, takes this discussion a long way from the outmoded proposition that meandering is caused by obstacles. Precisely the opposite is true. Obstacles, including variation in the cohesiveness of alluvium, distort and even suppress meanders, as is well shown by studies of the lower Mississippi. Meander patterns can be produced on a cathode-ray screen by means of electrical analogue devices; when suitable resistances are introduced into the circuits distorted bends and cut-offs appear which can be matched from the traces of actual rivers.

Just as the development of pool-and-riffle indicates that a uniform bed-slope is unstable, so does the side-to-side swing which produces meanders indicate that a straight trace is unstable. It can be proved theoretically that if a channel deviates from straightness a meandering trace is in the statistical sense the most probable trace. It minimizes the variability of water-surface slope downstream, the variability of bed shear, and the variability of friction. Hence, once a meandering pattern has been established, it is likely to persist, unless some really powerful disturbing factor comes into operation. But although a great deal is known about the behaviour of meandering streams and about their statistical properties, it is still not possible to define the ultimate cause of meandering.

A number of penetrating descriptive and analytic studies of braided channels are on record, although not a great deal has been done in relation to channel-forming discharge. The braided pattern is best displayed at low stages, when the characteristic multiple bars are revealed, interlaced by the dividing and re-combining minor channels. At common low stages the bars may cover, perhaps, 80% of the streambed.

Braided channels have much greater width/depth ratios than have meandering or other single channels. Thus, the highest flow velocities, which occur close to the water surface, are also close to the bed in braided channels. This is to say that the gradient of velocity from surface to bed is steep, and that shear stress on the bed is powerful. Braided patterns are associated with mobile streambeds, which are typically deformed into the roughly diamond-shaped bars which separate the minor low-stage channels. Given mobile bed material of the sand grade or coarser, anything which promotes widening and associated shallowing of the channel is likely also to promote braiding.

One possible cause of widening, shallowing, and braiding is weakness of the banks. If these are incohesive, collapsing so rapidly that they impel the channel to become wider and shallower, braiding can result. A variant on this situation is provided by outwash streams of glacial meltwater, which, because they are filling in the valleys that they occupy, are virtually without banks in the ordinary sense. At the opposite extreme come channels with a high proportion of fine (silt-clay) material in their sediment, where the width/depth ratio is low and where braiding is improbable.

A particularly illuminating case of braiding is that where a meandering stream converts to the braided habit, either in a given reach or in the downstream direction. It can do this if its floodplain alluvium is underlain by coarser material, such as glacial outwash or older alluvium deposited by a former larger stream, and if one or more pools are especially deep at unusually sharp distorted bends. The bar thrown up immediately downstream of a deep pool and sharp bend splits the channel into two. Each of the divided channels is less efficient than the single channel which it partly replaces, simply because it is smaller. The loss of efficiency can be compensated by a steepening of slope: the stream deepens

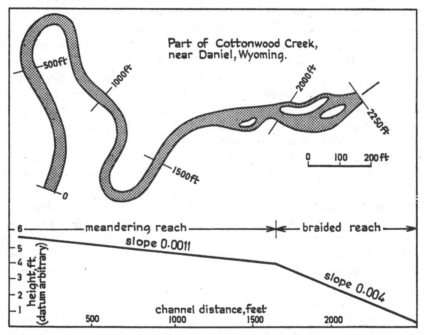

Fig. 9.11.5 An example of changing pattern and changing slope (Adapted from Leopold and Wolman, 1957).

each of the two lesser channels, helping the bar to dry out at most stages and permitting it to be colonized by plants. Repeated subdivision in this fashion can convert a meandering into a braided trace.

Because of the inefficiency of its total channel, a braided stream will have a steeper slope than a meandering stream of equal channel-forming discharge (fig. 9.11.5).

Truly straight channels scarcely exist outside the laboratory, except where a headwater stream is firmly held on the line of a fault. Straight-simulating channels are discussed below. The class of streams with irregular traces is a ragbag into which is stuffed patterns not otherwise classified. Deltaic-distributary streams, occurring in the special conditions indicated by their title, will not be further discussed here, except for the comment that some are notably meandering, as in certain deltas of South-East Asia, while others, including some of the Mississippi passes, are just as notably straight.

Anabranching and reticulate channel patterns are widely represented on the pediplains and alluviated basins of inland Australia, but are unlikely to be confined to these localities. Work on anabranching channels has been restricted mainly to the alluvial plains of the Murray and Murrumbidgee, where the anabranches – offshoots – rejoin the original trunk or unite with a next-neighbouring trunk, sometimes after a distance of tens of miles. Little rigorous work has yet been done on reticulate channels, which resemble braided channels in being complexly subdivided at low stages; but they differ from braided channels in being typified by markedly ephemeral flow, and in having a distinctly open network. They occupy whole valley floors as opposed to channel beds. Anabranching and reticulation alike seem to be responses to very gentle transverse slopes, which allow the current to subdivide freely at times of high stage. Both kinds of pattern are widely enough represented for them to constitute distinct classes.

2. Floodplains

By definition, floodplains are liable to inundation. They are best understood in the context of meandering streams. Shifting meanders work over the valley-bottom alluvium, eroding their outside bends but depositing on the insides and forming a depositional strip along the length of the valley. Some floodplains are composed chiefly of point-bar deposits laid down in multiple crescents on the inner sides of bends, with nil to minor additions of sediment from floodwaters, but others may be constructed mainly by overbank deposition.

Another kind of difference is that some floodplains possess levées – natural raised banks bordering the channel – whereas elsewhere levées are absent. The lower Mississippi is a noteworthy levée-builder, causing backswamps to form on its outer floodplain and obstructing the entry of tributaries (fig. 9.II.6). On very large streams such as this it is usual to find routeways, and even settlement, concentrated on the levées, out of reach of some floods and away from the ill-drained backswamps. The lower Thames, by contrast, like the Cotswold feeders of the upper Thames, is typically devoid of levées, with a floodplain sensibly flat in cross-section: on such a floodplain there is nothing to choose between one location and another, in respect of wetness or of freedom from flood danger.

Differences between channel-side accretion and overbank deposition, and between levée-building and the lack of it, are not yet fully understood, but appear to be related to the calibre of sediment transported through the channel, and to the relative proportion of bed load and suspended load. The reasoning involved in this interpretation can be used to classify braided channels as of the depositing bedload type.

If a meandering stream on a floodplain is cutting down very slowly or not at all it can be expected to reach bankfull stage once every year, and to inundate its floodplain in about two years in three. Many present-day meandering streams

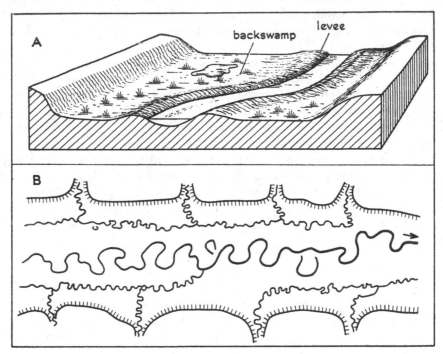

Fig. 9.11.6 A, levee–backswamp association; B, deferred tributary junctions.

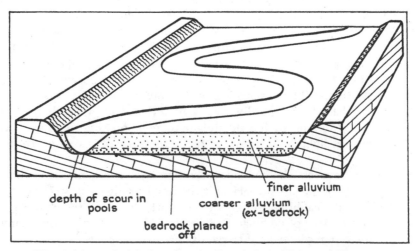

Fig. 9.11.7 Floodplain underlain by planed-off rock floor (Compare Fig. 9.11.9).

are, however, suspected to be slightly incised, so that their frequencies of bank-full flow and of inundation are reduced below the expectable level. Other streams, engaged in raising their floodplains, cause unusually frequent inundation.

If the working-over of the floodplain by shifting meanders keeps pace with any downcutting, then the base of the floodplain will correspond to the maximum depth of scour in pools (fig. 9.11.7). In actuality, the floodplain base is often cut across older alluvium rather than across bedrock, for reasons which will appear under the next subhead.

3. Meandering valleys

Large numbers of streams occupy meandering valleys, where steep crescentic slopes on the outsides of bends oppose gentler lobate slopes on the insides, but

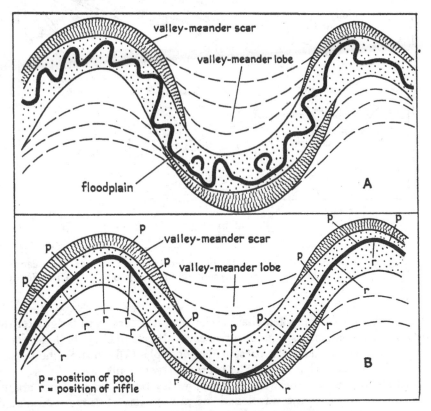

Fig. 9.11.8 A, manifestly underfit stream; B, underfit of the Osage type. Both portrayed as occupying ingrown meandering valleys.

where the stream channel is shaped wholly in alluvium and describes curves far smaller than those of the valley (fig. 9.11.8a). The steep outer slopes of the valley are called *meander scars*, the gentler inner slopes are *meander lobes, spurs,* or *slip-off slopes*.

The combination of scars and spurs belongs to trains of *in-grown meanders* which have increased in sinuosity during downcutting. The increase in sinuosity is not infrequently documented by crescentic patches of terrace on the slip-off slopes; and where the terraces are well enough preserved to reconstruct much of the sequence of downcutting it usually appears that the ingrowing stream was very little sinuous when it began to incise. The curves of the valley, that is to say, were produced largely or entirely during the course of incision.

Meandering streams on floodplains in the bottoms of meandering valleys are called *manifestly underfit*: manifestly because their condition is self-evident, and underfit because their bends are significantly smaller than the bends of the valley.

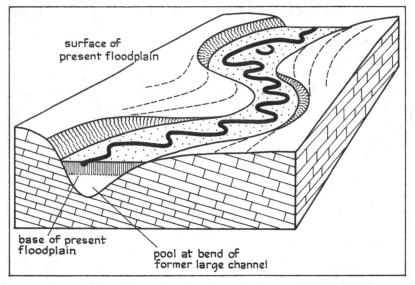

Fig. 9.11.9 A common condition of manifestly underfit streams in ingrown meandering valleys: the fill beneath the base of the present floodplain is older than the floodplain alluvium.

The valley bends were cut when channel-forming discharge was far greater than it now is. On some manifestly underfit streams the channel of the large former stream is preserved beneath the present-day floodplain (fig. 9.11.9), the present-day floodplain constituting merely the topmost layer of fill.

According to location, the wavelength of valley bends ranges from about 5 times to about 10 times the wavelength of present-day stream meanders; the 5/1 ratio is widely represented. Calculation based on the connection $l \propto q_{bf}^{0.5}$ between meander wavelength and bankfull discharge gives a 25-fold increase of bankfull discharge for a 5-fold increase of wavelength. When allowance is made for additional hydraulic characteristics, such as channel size, channel shape, channel slope, and roughness, the requirement comes down to about a 20-fold increase in bankfull discharge.

Now although in special conditions underfitness may be due to river capture, the cessation of meltwater discharge from glaciers, or the cessation of overspill from ice-dammed lakes, most streams which are now underfit have had their channel-forming discharges reduced by climatic change. It is highly likely that increase, reduction, and renewed increase of discharge occurred repeatedly during the Pleistocene, but little is known of any but the last main episode of shrinkage. This, possibly itself interrupted by an increase, took place between about 12,000 and 9,000 years ago. This was the time of the last major transition from high-glacial to interglacial conditions. When allowance is made for the reduced air temperatures of the time it can be shown that the swollen channel-forming discharge required to shape valley meanders could be produced by an increase in mean annual precipitation to $1\frac{1}{2}$ or 2 times its present value. Not only did the rise of temperature and the reduction of precipitation to present-day levels cause streams to become underfit: they also dried up the former pluvial lakes of arid lands.

Since the last main episode of stream shrinkage and channel filling, lesser fluctuations of temperature and precipitation have produced lesser episodes of renewed channelling and subsequent filling.

Not all meandering valleys are occupied by underfit meandering streams, a circumstance which until recent years has prevented the recognition of underfitness as general throughout vast areas. It is now known that a second type of underfit stream, the *Osage type* (fig. 9.11.8(*b*)) is quite common. This type is named after the Osage River of Missouri, U.S.A., where its qualities were first recognized. Although contained in a meandering valley, an underfit of the Osage type lacks stream meanders; but it possess a pool-and-riffle sequence, with the pools and riffles spaced appropriately to the existing channel at intervals of about 5 bedwidths. Because the existing channel is smaller than that of the former stream which cut the valley bends, the pool-and-riffle spacing is too close for these bends and for the curves which they impose on the existing channel: that is, pools occur more numerously than do channel curves. The stream has reduced its bedwidth and has correspondingly reduced its pool-and-riffle spacing, without going on to develop meanders. It behaves as if it were straight, apart from being enforcedly inflected round the curves of the valley. Thus, channels of underfits of the Osage type are assigned to the straight-simulating type of pattern.

The apparent wavelength/bedwidth ratio on streams of this kind is in many cases about 40/1, far larger than the rough 10/1 obtained for alluvial meandering channels; but in actuality the ratio on underfits of the Osage type is L/w, that between the wavelength of the large former stream and the bedwidth of the shrunken present-day stream. Whereas manifestly underfit streams can be recognized on sight, underfits of the Osage type can only be presumptively identified by a high apparent ratio of wavelength to width: their existence can only be proved (or disproved) by survey of the bed-profile and an investigation of the relationship between bedwidth and pool-and-riffle spacing.

P

REFERENCES

DURY, G. H. [1965], General theory of meandering valleys; *U.S. Geological Survey Professional Paper* 452. (Examines underfit streams.)

DURY, G. H. [1966], Incised valley meanders on the lower Colo River, New South Wales; *Australian Geographer*, **10**, 17–25. (Names the Osage type.)

LANGBEIN, W. B. and LEOPOLD, L. B. [1966], River meanders: Theory of minimum variance; *U.S. Geological Professional Paper* 422-H.

LEOPOLD, L. B. and WOLMAN, M. G. [1957], River channel patterns: Braided, meandering and straight; *U.S. Geological Survey Professional Paper* 282-B. (The basic reference on channel patterns.)

LEOPOLD, L. B. and LANGBEIN, W. B. [1966], River meanders; *Scientific American*, **15** (6), 60–9. (A simple, but valuable, account of meanders.)

LEOPOLD, L. B., WOLMAN, M. G. and MILLER, J. P. [1964], *Fluvial Processes in Geomorphology* (Freeman, San Francisco and London), 522 p. (This contains a very complete summary up to 1964.)

WOLMAN, M. G. and LEOPOLD, L. B. [1957], River flood plains: Some observations on their formation; *U.S. Geological Survey Professional Paper* 282-C.

III. Human Response to Floods

W. R. DERRICK SEWELL

Department of Economics, University of Victoria

1. Man's affinity for floodplain occupance

One of the most conspicuous features of human settlement pattern is man's affinity for riverine locations. Throughout history he has been attracted to the lands adjacent to rivers. Today a very considerable proportion of the world's population lives in such areas. There are some fairly obvious reasons why this should be so. River valleys often contain deposits of rich alluvium, providing the basis for the development of a thriving agricultural industry. Some of the world's great civilizations have developed in the bottom lands of major rivers, notably along the banks of the Tigris and Euphrates, the Nile, the Indus, and the Yangtze. River valleys are often transportation corridors, providing access for roads and railways. They also provide level land for the construction of houses and factories. For certain activities a riverine location is essential, particularly those which depend upon the river for transportation or for large quantities of water for processing. Other activities which have a greater freedom of locational choice may also be attracted to the river because of the activities already there. Various aesthetic considerations also provide an attractive force. Riverine locations carry a prestige value, for example, for private homes.

Settlement beside a river, however, can be a mixed blessing, for once in a while the river may overflow its banks and exact a heavy toll of property losses, income losses, and sometimes losses of life as well. In some cases man has learned to live with such periodic inundations of the floodplain and has turned them to economic advantage. The case of the Nile is perhaps the most famous. 'Egypt,' said Herodotus some 2,400 years ago, 'is the gift of the Nile.' His statement is still true today. In late June each year the lower Nile, swollen by tropical rains and the melting snows of its upper reaches, begins to rise. By late September its floodplain has become a large lake. As the floodwaters begin to recede in October they deposit a rich residue of silt, which revitalizes the soil. Through the construction of a network of canals, dykes, reservoirs, and ditches, the Egyptians have developed an agricultural system which is geared to the annual inundation of the Nile's floodplain.

In the case of the Nile floods are a critical input into the economy. In most cases, however, floods are regarded as a burden rather than as an advantage. Often they cause huge losses of property and income, and sometimes large losses of life as well. There are records of floods in China which have caused more than 1 million deaths at a time. Property losses can also be staggering. In 1951, for example, the Kansas River in the United States overflowed its banks and caused damages exceeding $1·5 billion. Flood losses in the United States have exceeded $1 billion several times in recent years. Effects of floods in the lesser-developed countries are even more serious, for recovery is much more difficult. Several times in recent years, for example, hundreds of thousands of people have been left homeless in India, Pakistan, Korea, and China, and their sources of food and livelihood have been severely damaged.

Despite the huge losses that have been experienced, floods have not discouraged settlement in river valleys. On the contrary, there is substantial evidence that occupance of floodplains in many parts of the world is increasing. The Yellow River and the Yangtze River in China have overflowed their banks many times in the past four thousand years, and millions of people have been drowned as a result, yet the peasants continue to flock into the floodplain. In this case the reasons are fairly obvious: there are few alternative opportunities for earning a living or for growing the food needed by a burgeoning population.

Occupance seems to be increasing, however, even in those cases where economic necessity is not involved. In some instances the costs of living or working in a place outside the floodplain might be lower than for those living on the floodplain itself, yet others flock in to join them. In the United States it has been estimated that at least 12% of the population lives in areas subject to periodic inundation (White *et al.*, 1958). A similar proportion of the Canadian population also lives in floodplains (Sewell, 1965). In both cases the proportion seems to be increasing. Floodplain occupance seems to be growing at a more rapid rate than overall population increase. Flood losses, therefore, seem destined to continue to mount.

Given these trends, what are the possible courses of action that can be taken to reduce the impending toll of flood losses?

2. Possible adjustments to floods

There are several possible adjustments to floods. Briefly, these might be grouped into the following categories:

    A. Accepting the loss
    B. Public relief
    C. Emergency action
    D. Structural changes
    E. Flood proofing
    F. Regulation of land use
    G. Flood insurance
    H. Flood control.

Each of these adjustments has advantages and disadvantages. Each of them is appropriate for some situations but not for all. Generally several adjustments are tried before a final selection is made. Often a combination of several adjustments is chosen. Some of the adjustments can be adopted by individuals: others, however, depend upon group action. What are the characteristics of the various adjustments, in what circumstances are they most appropriate, and what are the implications of their adoption for public policy?

A. Accepting the loss

The most common adjustment, perhaps, is to accept the loss. This is certainly the case where people are too poor to do anything else or are unaware of any alternative course of action. In the more developed countries, however, it is seldom the result of a conscious decision, particularly when a flood has been experienced in the past. Usually an attempt will be made to find means for offsetting future losses.

B. Public relief

One of the more common alternatives to accepting the loss is to rely upon public relief. An immediate reaction to the announcement of a flood disaster is the establishment of a relief fund to assist flood victims. Sometimes such funds are purely voluntary, and calls are made for contributions from people in adjacent communities, from the country at large, or even overseas. Often these voluntary relief funds gather very substantial amounts of money. Another type of relief fund is provided by the Government. Typically there is no set policy for determining the amount to be granted, and often there is considerable debate on the matter. A third type of relief is that administered by the Red Cross and other similar voluntary organizations. Friends and relatives may also offer assistance to flood victims, providing them with food and temporary accommodation.

The principal justification for the various forms of relief is that they help to ease the immediate distress, and to aid the initial rehabilitation. It is sometimes a very useful adjustment to floods. Generally, however, it tends to become regarded as a right rather than as a charitable gift. As a consequence, it tends to remove the incentive to avoid future flood losses, and therefore encourages persistent human occupance of the floodplain.

In the United States the Federal Government has allocated funds for relief of flood victims, through gifts, low-cost loans, deferred payments, and subsidies of various kinds. Similar types of flood relief have been provided in Canada and in several other countries. Results of studies in North America seem to suggest that the granting of public relief in the past constitutes a major reason for persistent human occupance of floodplains, particularly when the provision of such relief is made without any obligation on the part of the recipient to undertake measures to reduce his vulnerability to future flood losses (Sewell [1965], p. 70).

C. Emergency action and rescheduling

Potential losses of property and income can be reduced by various types of emergency action or by rescheduling of activities. Emergency action consists mainly of removing persons or property from the area subject to flooding, and flood fighting. Some communities have come to rely almost exclusively on this form of action. Each year when the floodwaters begin to rise they make preparations to evacuate the area. Sometimes the actual evacuation is undertaken by the individuals themselves, but generally the local authority or the central government assumes responsibility for overall organization. Massive evacuation programmes are undertaken along the Lower Mekong, the Indus, the Mississippi, and other major rivers each year.

Efforts are also made to reduce the impact of floods by flood fighting, such as by building temporary dykes along the river or outside a building, elevating goods and equipment from the reach of the floodwaters (such as by removing them to an upper floor of the building), or by protecting equipment with plastic sheeting or grease. Here again there are opportunities for both individual action and group action. Generally, however, action taken beyond the individual's dwelling or work-place depends upon organization by government authorities.

Another form of emergency action that is undertaken in some places is the rescheduling of operations. Business managers, for example, may so organize their production schedules that they avoid having damageable goods in the flood-prone area at the time when the flood is expected. In some cases they may schedule vacations to avoid losses of income caused by the closing of the plant. Rescheduling can be carried out fairly easily in some activities, but not in all. Many types of agricultural activity, for example, are restricted as to the time they can be undertaken, although there is sometimes a certain amount of latitude as to the timing of planting or harvesting. There are often opportunities in transportation and in manufacturing industries, however, for rescheduling. Potential losses can be reduced, for example, by making arrangements for sending passengers and goods over routes outside the floodplain, and by arranging for production of certain goods and services at plants outside the floodplain during the flood period.

The effectiveness of emergency action usually depends upon the extent of preparation before the flood occurs. The floodplain occupant must know how probable it is that there will be a flood and what its effects are likely to be. They must also be aware of the kinds of emergency action they might take to reduce potential losses. In addition, they must be given adequate warning of the onset of the floods to enable them to put their plans into operation. Key elements in emergency action, therefore, are the provision of information about the likelihood and the potential effects of floods, and the development of a flood-forecasting and warning system.

Emergency action seems to be an appropriate adjustment in those situations where the flood-to-peak interval is greater than one hour. (The flood-to-peak interval is the period from the time the river reaches flood stage to the time it reaches its maximum stage.) When this interval is less than 1 hour only limited kinds of emergency action can be taken. When it is greater than 1 hour many opportunities for reducing losses present themselves. Very substantial opportunities exist when it is more than one day. In the case of the 1952 flood on the upper Mississippi, for example, the flood-to-peak interval was 180 hours (7½ days). Many floodplain dwellers took action to reduce potential losses, such as disconnecting utility lines, raising their houses above flood level by jacking them up and supporting them on concrete-block piers, removing belongings to upper floors, and sealing windows and doors with temporary barricades.

Emergency action can substantially reduce potential flood losses. In the United States, for example, it has been estimated that such action taken following flood warnings appear to have reduced losses by at least 5% and sometimes by as much as 15% (White [1939]; U.S. Select Committee, [1959], p. 7). Such action, however, is not a panacea. It is most effective when the flood duration is short, where flood velocity is low, and where the frequency of flooding is high. Interest in emergency action tends to lag when the interval between floods is long. Its success hinges upon floodplain occupants being able to interpret information about floods and their being able to select the appropriate adjustment. There is evidence that floodplain occupants are not always able to do so efficiently. Moreover, there is sometimes resistance to taking any action, even though a flood warning has been issued.

Like public relief, emergency action tends to encourage persistent human occupance of floodplains. It has the advantage, however, that it encourages individuals to take action to reduce personal losses.

D. Structural changes

Another way of reducing potential flood losses is to modify building structures to repel floodwaters. Among the various types of structural change are construction of walls with impervious materials, closure of low-level windows and other openings, and underpinning of buildings. In some cases buildings can be built on stilts. This enables buildings to perform several functions. Cars can be stored between the stilts and moved during the flood period. In certain instances land fill is a practical proposition. On Annacis Island in British Columbia a massive programme of land fill has been undertaken to reduce potential losses from flooding of the Fraser River. Factories have been built upon this fill.

Structural change is appropriate where the duration of flooding is short and where the velocity is low. It appears to be most effective when the depth of flooding is less than 3 ft. It is possible, however, to build structures to withstand depths of floodwater in excess of 15 ft. Modifications can be made to existing structures, or they can be incorporated into new ones. It is usually much less expensive, however, to build them into new structures. Land fill can help to reduce the impact of flooding. It is an especially attractive adjustment when undertaken prior to construction in an urban or industrial area. It generally becomes prohibitively expensive once such development has taken place.

Structural change and land elevation can be undertaken by individuals or by groups. Thus far governments have played only a minor role in encouraging the adoption of these adjustments. There are signs, however, that they are likely to become fairly widespread in the next few years in the United States, particularly as a result of changes in Federal Government policies relating to flood management (U.S. Government, 1966). Various incentives might be offered to encourage the adoption of these measures, including grants of low-cost loans to those who are willing to undertake structural change or the withholding of mortgages unless the structure is built to withstand a specified flood.

Structural change and land elevation tend to encourage persistent human occupance. They do offer, however, a means of reducing potential losses, and they do place part of the burden on the floodplain occupant.

E. Flood proofing

Flood proofing is essentially a combination of structural change and emergency action. It does not necessarily involve evacuation. Rather it concentrates on the adoption of certain measures that can be put into action as soon as a flood warning is received. Among the various types of flood proofing are the installation of removable covers, such as steel or aluminium bulkheads, over doors or windows, or the installation of sump pumps and elevated outlet pipes to remove water which seeps into basements and interiors of buildings. In stores counters can be placed on wheels to facilitate rapid removal. There are, of course, many other possibilities for flood proofing (Schaeffer, 1960).

Flood proofing offers considerable opportunities for reducing flood losses. Many factories and stores in the Golden Triangle of Pittsburgh, for example, have adopted this adjustment and have found it a very effective means of dealing with floods. Flood proofing is now an integral part of the TVA flood-control programme. To be effective, however, it requires a well-organized flood information system. Like other adjustive actions to floods, flood proofing tends to foster persistent human occupance of floodplains, yet similar to emergency action and rescheduling and structural changes, it does place part of the responsibility for taking action on the shoulders of the individual.

F. Regulation of land use

The land in the floodplain has a wide variety of potential uses, ranging from urban and industrial development, through agricultural and recreational uses, to leaving it in its virgin state. Potential losses tend to vary with the type of use, being highest on land used for urban or industrial purposes, and lowest on land set aside for agriculture or recreation. There would be no losses, of course, on land that is not used at all. The immediate conclusion might be to keep all development out of the floodplain. This would be neither realistic nor necessarily the economically most sensible course of action. Ideally, an attempt should be made to determine which activities can afford to locate in the floodplain and still pay the 'natural tax' of flood losses (Renshaw, 1961). If the activity cannot afford the latter, then it should not be allowed to use the floodplain. A set of regulations based upon this concept would encourage potential floodplain occupants to examine carefully locations outside the floodplain as well as inside the floodplain.

Figure 9.III.1 illustrates the manner in which floodplain land might be allocated among competing alternative uses. Each use has a rent-earning capability, determined by the returns it can produce after paying the costs of hiring the various factors of production. Assuming that the total land in a given floodplain is *OW* acres, and there are three competing uses, how can the area be allocated efficiently among them? The lines *FQ, DU*, and *BW* indicate the returns that could be earned in urban, agricultural, and recreational uses, respectively. On this basis *ON* acres would be used for urban purposes, *NS* for agriculture, and *SW* for recreation.

Flood losses, however, constitute a cost and must be taken into account in calculating net returns. Such losses result in a reduction of net returns, as represented by the dotted lines, *EP, CT*, and *AV*. It will be observed that urban land use is reduced to *OM*, agricultural land use to *MR*, and recreational use to *RV*. *VW* acres are then abandoned. This indicates that once flood losses

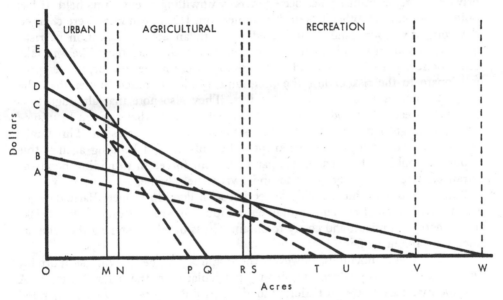

Fig. 9.III.1 The allocation of floodplain land among competing uses.

are taken into account, some urban occupants would move elsewhere rather than absorb the loss, and that land shifts into uses where losses per acre are less.

There are a wide variety of means of regulating floodplain occupance, notably through statutes, ordinances, subdivision regulations, government purchase of property, and subsidized relocation. Each of these methods have been used to varying degrees in North America and elsewhere (Murphy, 1958). Generally, government action is required to formulate and enforce the regulations. In some cases local authorities have enacted such regulations, but it often requires the action of a senior government to make this type of adjustment effective. Local authorities generally hesitate to enact regulations because they fear a neighbouring municipality will not do so, and therefore will attract activities which might otherwise have located in the first municipality. Co-operation between local authorities or central government enactment and enforcement appears to be a prerequisite for successful regulation.

Regulation of land use has a number of advantages. The most important perhaps is that it encourages careful weighing of the costs against the benefits of floodplain occupance. It forces consideration of the relative advantages of being in the floodplain versus location elsewhere. It offers a valuable complement to other types of adjustment, such as emergency action, flood proofing, structural change, flood control, and flood insurance.

G. Flood insurance

Another possible response to flood problems is to insure against the losses which they cause. Thus far, however, it has been adopted to only a minor extent. The private insurance industry has been generally unwilling to enter this field. It has pointed out several difficulties in this connection. If uniform rates were charged the company would find itself loaded up with an adverse selection of risks because people in the highly flood-prone areas are the ones most likely to take out a policy. On the other hand, if an attempt were made to charge rates proportionate to the risk of loss the premiums would be much higher than the property owners would be willing to pay. They also note that although it is possible to estimate flood frequencies, it is conceivable that a given insurance company might have to pay out claims several years in succession. This might result in the company going bankrupt. The only way of hedging against this problem would be for several companies to join forces in providing flood insurance, or for the Government to underwrite the scheme.

There are other difficulties in connection with flood insurance. Basic data for the determination of fair and equitable premiums for areas of varying flood risks are sometimes difficult and costly to obtain. Costs of administering the scheme, therefore, might be fairly high.

Despite these difficulties, however, there appears to be growing support for insurance as an adjustment to floods, particularly in the United States. A Presidential Task Force on Federal Flood Control Policy recently recommended that serious consideration be given to flood insurance, sponsored if necessary, by the Federal Government (U.S. Government, 1966). The Task Group noted the advantages of such insurance, particularly the fact that it provides an incentive for floodplain occupants to reduce damage potentials. In this way they would reduce their premiums. Flood insurance also shifts the burden of flood losses on to those who are responsible for them, the floodplain occupants.

Flood insurance tends to encourage increased occupance of the floodplain, but it does so selectively. Only those activities that can afford to pay the premiums can afford to continue occupancy (Krutilla, 1966).

H. Flood control

Finally, man may adjust to floods by trying to control them. Two main lines of action are possible: one in the land phase (flood abatement) and the other in the channel phase (flood protection) (Hoyt and Langbein, 1955). Examples of flood abatement are the modification of cropping practices, terracing, gully control, bank stabilization, and revegetation. In the United States the Department of

Agriculture has undertaken major programmes of this type aimed at controlling the development of floods. Projects focused on forest replanting, soil-erosion control, and improvement of farming methods have been carried on with flood abatement as one of their major objectives. Many millions of dollars are allocated each year by the Federal Government for this purpose. Similar programmes have been sponsored in Canada by the Prairie Farm Rehabilitation Administration, the Maritimes Marshlands Rehabilitation Administration, the Department of Agriculture, and the Department of Forestry, as well as by various provincial government agencies. Generally, flood control is one of several objectives of these programmes. Typically, other objectives include agricultural readjustment, soil conservation, and the preservation of fauna and flora.

Flood-protection programmes are concerned with the channel phase of floods. Their objectives are to control the flood once it has formed and to minimize the damage it causes by regulating its flow or directing it away from damageable property. It may involve the construction of control works, such as dykes, floodwalls, or dams and reservoirs, or the undertaking of channel improvements and dredging. Flood protection is one of the most widespread of all adjustments to floods, both in the more advanced and in the lesser developed countries.

In some cases the provision of flood protection is allied with the development of projects for other purposes. Flood control, for example, is one of several objectives of the Tennessee Valley Authority Scheme, as it is in the Mississippi River and the Columbia River schemes. In this way flood control can often be provided at a much lower cost than if it were furnished on a single-purpose basis.

For some years there has been a controversy in the United States as to whether flood abatement or flood protection is the more efficient adjustment to floods. This has come to be referred to as the 'upstream–downstream' controversy, or the conflict between 'big dams and little dams' (Leopold and Maddock, 1954). On the one side are those who argue that the best way to deal with floods is to control them where they form. They suggest that deforestation and devegetation are major causes of floods, and so the most appropriate course of action is to plant trees in the place of those that have been cut down, to improve the vegetative cover in areas subject to soil erosion, and to phase out farming practices that contribute to such erosion. They see dams as a possible means of controlling runoff but feel that these are most effective in the headwaters rather than downstream. They point out further in this connection that upstream reservoirs are less likely to take out of production agricultural land or forest land. The most vigorous supporters of these views have been the Department of Agriculture, particularly through its Forest Service and its Soil Conservation Service, and various private conservation groups.

On the other side are those who believe that the most efficient way to deal with floods is to control them close to the place where they are likely to do the most damage. They point out that it takes several reservoirs in the upstream region to do the work that one large reservoir farther downstream might do. They also note that the contributions of programmes of reforestation and soil conservation to runoff control may be insufficient to deal with the flood problem. The major benefits of such programmes, they claim, are in the regions where they are undertaken. Big dams and various downstream control works are seen as the much more effective alternatives. The major proponents of the latter view are the U.S. Corps of Engineers, an agency with a long tradition of constructing large-scale engineering works to deal with flood problems.

There is some truth in both sets of arguments. Flood abatement and flood protection are not necessarily alternatives. Often they can be complementary parts of an overall programme of adjustment to floods. It is worth noting in this connection that the Department of Agriculture in the United States has recently increased its emphasis on engineering works in its programmes, and it is now building big dams as well as small ones. By the same token, the Corps of Engineers, is tending to encourage the adoption of adjustments in addition to engineering works in dealing with flood problems.

Flood abatement and flood protection can be undertaken by individuals, but generally they require co-operative action, and usually government sponsorship. Both tend to foster increased human occupance of the floodplain. Flood protection in particular tends to develop a false sense of security among floodplain occupants. Floodplain occupants may take the construction of a dyke or dam to mean that there will never be any more flooding. Consequently, more and more people move in, and activity in the floodplain intensifies. When a flood of greater magnitude than that which the dykes or reservoirs were designed to control eventually comes along much greater damage is done than if no protection had been provided at all. Flood protection, therefore, needs to be supplemented by other measures which control the increase of potential flood losses.

3. Limitations to the range of choice

The types of response to floods outlined in the foregoing discussion constitute a Theoretical Range of Choice from which a flood manager could select an appropriate course of action (White, 1964). Typically, however, only a few of these possibilities are taken into account in flood-management decision-making. The result may be that adjustment is much less efficient than it could be if the whole range was considered.

One of the major foci of research on flood problems in recent years, particularly at the University of Chicago under the leadership of Gilbert White, has been the factors which tend to limit the range of choice. Results of these studies suggest that two main sets of factors are involved: the flood manager's perception of the nature and magnitude of the flood problem and his perception of alternative responses to that problem; and various social guides which tend to encourage the consideration of some responses and to discourage the consideration of others. Kates [1962] has shown that there are wide differences in the perceptions of individual floodplain occupants as to the nature and magnitude of the flood problem, and that their perceptions often differ considerably from those of the engineer or technician. He notes that floodplain occupants often perceive the flood hazard and its potential effects rather imperfectly. As one might expect, those who have experienced a flood in the area in the past tend to have more accurate perceptions of the hazard than those who have not had such experience. But it does not necessarily follow that even when there is accurate perception of the hazard that there will be effective action to deal with it. Some floodplain occupants may feel that they will not suffer any damage in the future, and even if they do, it will not be serious. From the evidence gathered so far it seems that action is most likely when several flood events have been experienced and when the losses have been severe. For the most part there tends to be apathy about the flood hazard. Action seems to await a crisis to provide the necessary trigger.

One reason why the floodplain occupant is often unconcerned about flood problems is that they may be only one of many problems that concern him in his daily life. As a result, he devotes only a small part of his time to dealing with such matters. Typically, he is unaware of the wide range of actions he can take to reduce potential flood losses. Often he places great faith in adjustments that are considered to be of limited value by the technician, such as the clearing of brush or debris from the river channel or last-minute flood-fighting efforts.

Various social guides, such as law, historical precedent, jurisdictional constraints, and government policies, condition to an important extent the adjustments that are chosen. Typically, the questions will be asked – what has been done in the past, and whose responsibility should it be – rather than what is the best course of action in this case? Sometimes legislation facilitating government action is drafted in fairly restrictive terms, focusing upon only one or two types of adjustment. In some cases only flood relief or flood protection are seen as possible responses to flood situations, and these are written into the legislation. Consequently, such other alternatives as flood insurance, flood proofing, etc., may not be considered at all in actual decision-making, even though they may be much more effective in dealing with flood problems than the latter.

Gilbert White and his colleagues have concluded from their studies that such limitations on the range of choice account in part for the continuous increase in flood losses in the United States. More than $7 billion has been spent on various adjustments in the United States since 1936, but flood losses have continued to mount. Government policy and historical precedent have tended to favour a concentration on flood protection, emergency action, and flood-relief payments. Unfortunately, these adjustments attack the effects rather than the causes of flood losses. Generally too, they have tended to encourage a transfer of responsibility for action from the individual to the public at large.

A tangible result of the studies has been some significant shifts in U.S. government policy relating to flood problems. Important among these shifts have been a broadening of the approach to include adjustments in addition to flood prevention, emergency action, and flood relief. Consideration is now being given to policies which would broaden the practical range of choice to include flood proofing, structural change, flood insurance, and other alternatives that have promise of reducing flood losses. Particular attention is also being paid to courses of action which would encourage private individuals to consider the risks of floodplain occupance and to take appropriate measures to curb flood losses.

4. The Lower Fraser Valley: a case study

The Lower Fraser Valley in British Columbia provides an opportunity to examine the changing pattern of responses to a major flood problem. For more than a century this 800-square-mile, wedge-shaped area has been the main focus of settlement and economic development in the province. Today it provides a home for more than one million people, and it contains the larger part of the province's manufacturing, commercial, and agricultural activity. It is also a

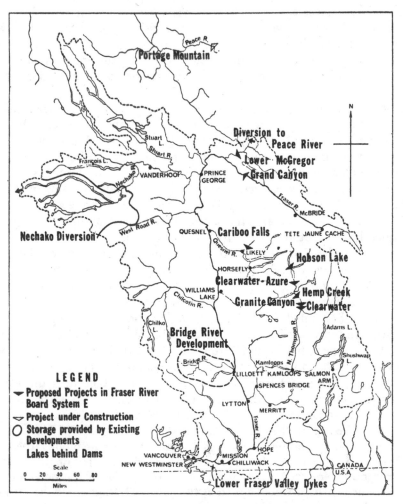

Fig. 9.III.2 The Fraser River basin. Scheme of hydroelectric power and flood control recommended by the Fraser River Board.

major transportation corridor, linking the Pacific Coast with eastern Canada, and northern North America with the United States (fig. 9.III.2).

A considerable portion of this settlement and economic activity is located in the floodplain of the Valley. More than 10% of the Valley's population, most of its agricultural activity, a large segment of its manufacturing and commercial activity, two transcontinental railways, and the major highways are in areas that are subject to inundation by major floods.

The concentration of population and economic development in the floodplain is in part a reflection of the economic advantages it offers, particularly its fertile alluvial soil, its flat terrain, and its access to other areas. It does not follow, however, that the present pattern is a result of a conscious weighing of these advantages against the costs of floodplain location. On the contrary, it is probable that many of the people who live there are unaware of the flood hazard, and so do not take it into account in their decisions. Occasionally, however, there is a major flood and the costs of floodplain occupance are brought home to them in dramatic, and sometimes catastrophic, fashion. This was indeed the case in 1948.

A. The 1948 Flood

The 1948 Flood was by far the worst disaster in British Columbia's history (Hutchinson, 1950). It was also one of the most costly floods ever to occur in Canada. There had been a heavy snow pack during the previous winter, and the spring was late in coming. Suddenly at the beginning of May temperatures in the Interior of the province began to rise, and they remained high for several weeks. The snow began to melt rapidly, and this resulted in disastrous flood conditions downstream. The danger mark on the Mission gauge, some 50 miles from the mouth of the river is 20 ft. By 25 May the river had reached 18·8 ft on the gauge (Fraser River Board, 1963). The next day the dykes at Mission broke. There were breaks at several other points downstream too. For more than thirty days the Valley was in a state of siege by the flood waters. Vancouver was cut off from the rest of Canada, except by air.

The flood resulted in huge losses of property and income throughout the Valley. No complete assessment of these losses was made, but those for which compensation was paid totalled more than $20 million. Some 55,000 acres of agricultural land were inundated. Farm buildings were damaged or destroyed. Fields in the Valley were covered with several feet of mud and debris. More than 2,000 people were left homeless. The effects of the flood extended far beyond the Valley itself, particularly because of the interruption of communications. Recovery was slow, and it was costly. Residents and politicians were resolved that such a disaster must never occur again. A thorough investigation of the flood problem was called for, leading to a set of recommendations for action.

B. Evolution of adjustment to floods in the Lower Fraser Valley

The early settlers in the Lower Fraser Valley experienced several severe floods. Some of them moved elsewhere, others decided to bear the loss in the hope that they would not be inundated again. A few, however, built dyking systems (Sewell, 1965). The first dykes were built by private landowners to protect their property. Later a number of Dyking Districts were formed to provide flood protection on a communal basis. The early dykes proved to be a disappointment, technically and financially. Many of them collapsed, and most of them fell into disrepair because it became impossible to collect enough money to maintain them. Eventually the Government felt obliged to take over the dyking system, to maintain it and to extend it where necessary. But public ownership did not solve the financial difficulties nor did it overcome the technical problems. The debt continued to mount and the dykes continued to fall down. The poor condition of the dyking system was in large part responsible for the huge losses that occurred in the 1948 Flood.

Besides the construction of flood-protection works, attempts were made to deal with the flood problem by emergency action. Individuals and communities had devised methods of flood fighting and evacuation over the years, and these had often proved to be effective, particularly in minor, short-term inundations. Evacuation and flood fighting in the 1948 Flood, however, were required on such a large scale that they could not be left to private initiative. Carefully planned and co-ordinated efforts were clearly required. Responsibility for organizing evacuation and temporary relief was assumed by the provincial and federal governments.

Up to 1948 flood relief had been a matter that was left largely to private initiative. flood victims either drew on their own resources or depended upon assistance from relatives and friends. In 1948, however, the senior governments decided that such aid would be inadequate to facilitate recovery and offered assistance for relief and rehabilitation.

Adjustment to floods in the Lower Fraser Valley up to 1948, therefore, was characterized by the adoption of relatively few measures, mainly bearing the loss, flood protection, and emergency action. Over the years there was a shift in responsibility from the private individual to the Government. The relative infrequency of major floods had led to a relaxing of vigilance on the maintenance of the dyking system. Nothing was done to control occupance of the flood-plains.

The 1948 Flood led to some important shifts in public policy. First, the Federal and Provincial Governments offered financial assistance on a fairly large scale for relief and rehabilitation. Although no promise was made that such aid would be forthcoming in the future, it is evident that this established a precedent. Many of those who now live in the floodplain believe that financial assistance would be provided in the event of another flood. Second, the two governments agreed to search for a more permanent solution to the flood problem. As an initial step they established in 1949 an engineering board – known initially as the Dominion-Provincial Board, Fraser River Basin, and subsequently as the Fraser River Board – to undertake a thorough investigation of the causes of the problem and to recommend measures that should be adopted to deal with it. It presented its Final Report in 1963.

C. The Fraser River Board's proposed scheme

The Fraser River Board proposed as a solution to the flood problem the construction of a scheme of hydroelectric power and flood-control development, estimated to cost about $400 million, and the reconstruction of the dyking system in the Lower Fraser Valley, costing some $5 million (fig. 9.III.3). The scheme would control a Design Flood, which has a discharge of some 600,000 ft^3/sec at Hope (fig. 9.III.4) and a chance of occurrence of one in 150 years (fig. 9.III.5) This control would result in a reduction of potential flood losses, valued at some $75·3 million per annum (fig. 9.III.6). The scheme would also generate about 785,000 kW of firm power. Revenues from the sale of this power would be sufficient to cover the entire cost of the scheme. The only costs of controlling the floods, therefore, would be the costs of repairing the dyking system.

So far little has been done to implement the Board's recommendations. There have been discussions between the Federal Government and the British Columbia Government, but no agreement has yet been reached. Even if an agreement had been reached to construct the proposed scheme, however, it is not clear that it would lead to a long-run reduction in flood losses in the Lower Fraser Valley.

Three main factors have delayed the implementation of the Fraser River Board's proposed scheme: the uncertainty as to which level of government is responsible for initiating action to deal with flood problems; the pre-emption of the power market by other projects under construction in British Columbia; and opposition from recreation interests. Traditionally, flood problems have been regarded as a matter of local responsibility in British Columbia. In 1948, however, the senior governments indicated that they were willing to give assistance in dealing with such problems, both by providing funds for relief of flood victims and by undertaking an engineering investigation. But no commitment was made by either the Federal or Provincial Government beyond that. The Federal Government feels it is unable to initiate action because flood problems are beyond the jurisdiction granted to it under the British North America Act. The Provincial Government, however, feels that the proposed solution is beyond its financial capabilities. Moreover, the local authorities seem unwilling to take any remedial action themselves because they feel the senior governments have assumed overall responsibility for dealing with it!

Another reason for the delay in action has been the pre-emption of power markets. While the studies of the Fraser River were being undertaken, investigations of other rivers in the province were also under way, notably those relating to the Columbia River and the Peace River. Huge hydroelectric power schemes are now being built on the latter rivers which will provide power for the Province and for export to the United States for the next fifteen to twenty years. Unfortunately, the planning relating to the Fraser River was not phased in with that relating to the other two rivers, and so the possibilities of an integrated scheme involving all of them were not considered.

The proposed Fraser River scheme has also been delayed because of opposition

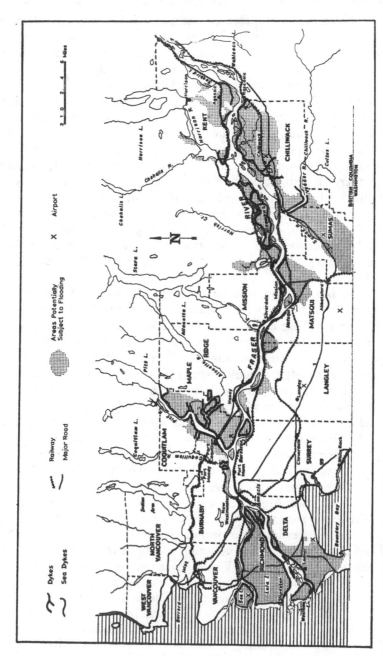

Fig. 9.III.3 The lower Fraser Valley dyking systems.

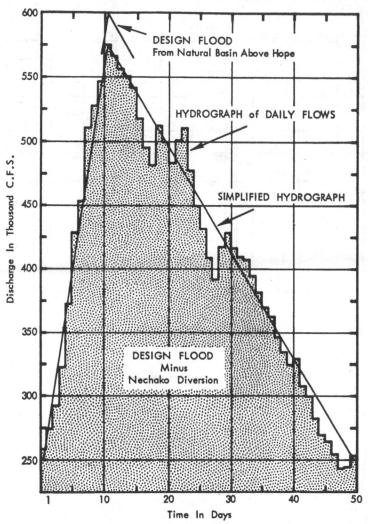

Fig. 9.III.4 Design flood hydrograph at Hope, British Columbia.

by salmon fishing and recreation interests. The river is one of the world's largest remaining sources of salmon. Salmon fishing is one of the oldest industries in the province and still makes an important contribution to employment and income in British Columbia. The landed value of the catch from the Fraser River is about $20 million per annum. Some 10,000 fishermen and shore workers are employed on a full-time or part-time basis in obtaining the catch and processing it. In addition, there is a rapidly growing sports fishery in the province which provides recreation opportunities and an increasing source of income for vendors of fishing equipment and supplies and for guides.

Construction of dams on the Fraser River would interfere with the migration and spawning of the salmon runs, and so the commercial fishermen and the

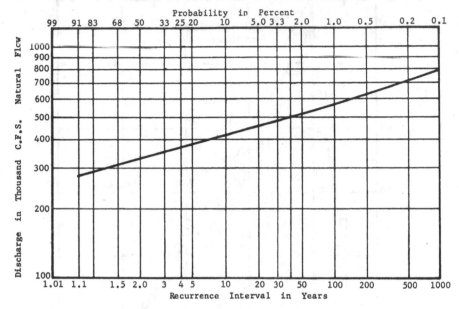

Fig. 9.III.5 Annual peak discharges of the Fraser River at Hope, British Columbia.

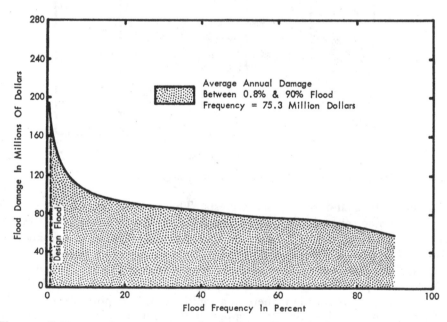

Fig. 9.III.6 Damage frequency curve for the lower Fraser Valley.

sports fishermen have been solidly against proposed adjustments to the flood problem involving such structures. In an effort to accommodate this opposition, the Fraser River Board decided to restrict its selection of sites for dam construction to streams in the headwaters of the river where there would be only minimal effects on the salmon runs. Such a location resulted in a scheme that was considerably less efficient both technically and economically than others that could have been developed. The storage reservoirs are far from the major areas of flooding, thus requiring a larger number of projects than would have been the case with near storage. The power plants associated with them are also far from major load centres, and there are no opportunities for large-scale economies of power production at the projects selected. Power costs, therefore, are not competitive with those of other sources in the province.

Despite the efforts of the Board to accommodate the fishing interests, there has been opposition to the proposed scheme. Some of them claim that the construction of any structure will have adverse effects elsewhere in the river system, and therefore should not be allowed. There is also opposition from recreation interests, who feel that the construction of some of the proposed reservoirs will destroy 'the unique natural beauty' of the areas involved. They are particularly opposed to the projects recommended for the Clearwater River. Unfortunately, these are among the most economical of those recommended by the Board, and their elimination would seriously undermine the whole scheme, both technically and economically.

The proposed scheme would control a flood equal to that which occurred in 1894, the largest one on record. But it would not protect against all floods. The success of the scheme, therefore, would hinge in part upon the enactment and enforcement of floodplain zoning, aiming to control the increase in flood-loss potential. Unfortunately, the Fraser River Board had little to say about floodplain zoning, other than suggesting that 'the responsible authorities examine the use of lands in the floodplain with a view to restricting the use of those lands to developments that would suffer least from flooding'. The lack of a clear definition as to who the responsible authorities are inevitably means that the problem is unlikely to be studied! And without such a study regulations are unlikely to be forthcoming.

Reliance is placed on the local authorities to enact floodplain regulations. It seems, however, that such regulations will only be enacted and enforced if *all* local authorities do so. Without such uniform action individual local authorities will fear that they will lose industrial and urban development to other areas that do not enact regulations.

D. The Fraser River experience in a wider context

Adjustment to floods in the Lower Fraser Valley has been characterized by four main features. First, it has generally been a reaction to crisis, with short periods of feverish activity punctuated by longer intervals of inaction and apathy. The consequence has been a piecemeal approach which has been corrective rather than preventive in character. Second, there has been little innovation in flood policies over the years. Reliance has continued to be placed on a narrow range of adjustments – mainly flood protection, emergency action, and flood relief. The Fraser River Board considered other possible adjustments, such as flood-plain zoning, but gave them only cursory attention, possibly reflecting the uncertainty as to which level of government would implement such measures but also reflecting a bias that characterizes most engineering reports, a bias towards adjustments that involve the construction of control works.

A third feature has been the increase in the share of costs of flood protection assumed by the general public. Today the senior governments pay more than half the costs of flood protection in the Valley, the remainder being borne by the local areas. A consequence of the shift of the financial burden has been to make individual floodplain occupants less concerned about the flood hazard. The extra costs of floodplain location are not brought home as vividly as they would be if he had to carry most of the cost. Moreover, the present policy of providing flood relief without any obligation on the part of the floodplain occupant to take remedial action has further reduced his concern about the problem. In a sense, the provision of such relief has furnished an incentive to further increases in the flood-loss potential.

A further feature has been the tendency to consider the flood problem in isolation from other problems of developing the Lower Fraser Valley. Consequently, while some government policies have tried to reduce potential flood losses, others have been encouraging developments in the floodplain which would increase such losses. The Water Resources Branch, for example, has been busy building and repairing dykes in an effort to curb losses, while the Department of Highways has been building highways through the floodplain, and the Department of Industrial Development has been promoting development in locations subject to flood hazard!

The Fraser River experience offers some lessons for policies relating to flood management elsewhere. First, it suggests that a concentration on a narrow range of alternatives may lead to inefficient adjustment. Sometimes, when the suggested course of action is blocked, no action is taken at all. Second, it shows that there are dangers in removing completely from the individual the incentive to consider the risks he is running by occupying the floodplain. The greatest danger is that activities will move into the area that might be more efficiently located elsewhere. To the extent that their flood losses are covered by public expenditures, their location in the floodplain is publicly subsidized. The adoption of such adjustments as floodplain zoning, flood proofing, and flood insurance would encourage the individual to consider the relative advantages of floodplain location versus location elsewhere, and at the same time reduce the potential burden on the public purse. Finally, it emphasizes that the lack of a clear definition as to who is responsible for dealing with flood problems generally means that there will be inaction. Floodplain dwellers will assume that the Government is dealing with the matter, while government agencies assume that it is beyond their terms of reference. Meanwhile the flood loss potential continues to grow. Catastrophe is the inevitable result.

REFERENCES

FRASER RIVER BOARD [1963], *Final Report on Flood Control and Hydro-Electric Power in the Fraser River Basin* (Queen's Printer, Victoria, British Columbia).

HOYT, W. A. and LANGBEIN, W. B. [1955], *Floods* (The Ronald Press, New York).

HUTCHINSON, B. [1950], *The Fraser* (Clark Irwin and Co., Toronto).

KATES, R. W. [1962], *Hazard and Choice Perception in Flood Plain Management*; University of Chicago, Department of Geography Research paper No. 78.

KRUTILLA, J. V. [1966], An economic approach to coping with flood damage; *Water Resources Research*, **2**, 183–90.

LEOPOLD, L. B. and MADDOCK, T. [1945], *The Flood Control Controversy* (The Ronald Press, New York).

MURPHY, F. C. [1958], *Regulating Flood Plain Development*; University of Chicago, Department of Geography Research Paper 56.

RENSHAW, E. F. [1961], The relationship between flood losses and flood control benefits; In White, G. F., Editor, *Papers on Flood Problems*, University of Chicago, Department of Geography Research Paper No. 70, pp. 21–45.

SCHAEFFER, J. R. [1960], *Flood Proofing: An Element in a Flood Damage Reduction Program;* University of Chicago, Department of Geography Research Paper No. 65.

SEWELL, W. R. D. [1965], *Water Management and Floods in the Fraser River Basin*; University of Chicago, Department of Geography Research Paper No. 100.

U.S. GOVERNMENT [1966], Executive Order No. 11296, August, 1966, U.S. 89th Congress Second Session, House of Representatives, Report of the Task Force on Federal Flood Control Policy, *House Document 465*, Washington, D.C.

U.S. SELECT COMMITTEE ON NATIONAL WATER RESOURCES [1959], *River Forecasting and Hydrometeorological Analysis*; Committee Print No. 25.

WHITE, G. F. [1939], Economic aspects of flood forecasting; *Transactions of the American Geophysical Union*, **20**, 218–33.

WHITE, G. F. et al. [1958], *Changes in Urban Occupance of Flood Plains in the United States*; University of Chicago, Department of Geography Research Paper No. 57.

WHITE, G. F. [1964], *Choice of Adjustment to Floods*; University of Chicago, Department of Geography Research Paper 93.

CHAPTER 10

Annual Runoff Characteristics

I. River Regimes

ROBERT P. BECKINSALE

School of Geography, Oxford University

The regime of a river may be defined as the variations in its discharge. In its widest sense the regime involves all occurrences and is portrayed by a curve based on continuous or hourly observations. Such curves, however, present complicated problems of analysis and for some purposes the discharge variations are better expressed by graphs of mean monthly flow. When used for critical purposes, such as the delimitation of hydrological regions, the ideal seasonal regime hydrograph (station-model) would show additionally, for each month and for the year as a whole:

1. the mean flow;
2. the mean maxima and minima; and
3. the absolute maximum and minimum.

It would also be helpful to insert (not as a curve) the absolute *daily* maximum and minimum recorded during the period.

However, for general comparative purposes and for global classifications the monthly means seem adequate and, indeed, such simple data is still not available for large areas. On this broad scale comparison is facilitated by expressing the mean monthly value either as a ratio of the mean monthly flow for the year (taken as 1) (as in figs. 10.1.5 and 10.1.6), or as a percentage of the mean annual flow (taken as 100) (as in figs. 10.1.2 and 10.1.3). If the actual quantities are stated as one of the ordinates this method does not lose much in practical utility, particularly if the mean annual total flow is also stated somewhere on the regime hydrograph. Only by insisting on actual quantity as well as on comparative ratios will the possibilities of inter-regional water exchanges be kept constantly before civil engineers and planners.

1. Factors controlling river regimes

Seasonal variations in the natural runoff of a drainage basin depend primarily on the relationships between climate, vegetation, soils and rock structure, basin morphometry, and hydraulic geometry. Of these only rock structure and, to a lesser extent, basin size can be strictly independent of climate. It should be stressed that the features of basin morphometry and hydraulic geometry are only of direct relevance to the seasonal regimes of large river basins.

A. Climate

The direct climatic control over river regimes lies in the difference between the monthly sequence of precipitation (positive) and the values of insolation or solar radiation (negative), controlling evaporation. Normally water-vapour pressure is highest and rainfall most abundant in the summer half-year when evaporation and evapotranspiration are also greatest (fig. 10.1.1(b)). A few restricted areas, however, have most precipitation in the winter half-year because locally moist maritime airflow, associated with frontal depressions, is strongest then. The direct relation between rainfall and insolation is also departed from in subsident or anticyclonic air masses, and in sunny tropical and sub-tropical regions with prolonged upper-air (cT) subsidence (or even with local lee-wave subsidence behind high topographic barriers) potential evaporation often far exceeds precipitation and permanent streams are absent.

In continental areas with snow-cover in the cold season, lower air subsidence creates a shallow anticyclonic air mass (cP), beneath which chances of precipitation remain negligible until a warmer air mass intrudes. Thus frigid cP air masses, with very low moisture content, have almost the same impediment on precipitation as hot subsident cT air masses do, but their low evaporation rate and snow retention ensure that they have eventually a *positive* effect on runoff. Beneath cold cP air masses soils usually freeze to a depth of several metres, and when the snow melts in spring runoff may for a while be very great from above the still-frozen sub-soil. In such climates vast areas are underlain by permafrost, above which the soil thaws out in summer. The top of the permafrost layer is usually deeper in river valleys, but in extremely cold northerly areas it remains close to the surface all the year.

The direct effect of climate on a river's surface is also significant, as it includes its freezing as well as the direct channel precipitation it receives and, more important, the evaporation it loses – both of which increase when the channel widens, anastomoses, or winds excessively. These channel-precipitation and evaporation factors vary in potency with the increase in channel area, and from region to region. In humid basins such expansions of channel surface give a net channel *gain* of the water-surface area increase multiplied by the difference between the annual precipitation and the annual floodplain surface runoff (which would have flowed directly into the channel in any event). In arid and semi-arid regions the evaporation increase would probably be equivalent to the increase in water-surface area multiplied by the local evaporation rate, since extra channel length or width would usually be associated with channel shallowing and warming. Such losses can be enormous. Thus on the upper Niger, where the river enters upon and flows north-eastwards over a sedimentary plain (fig. 10.1.2), its channel spreads over a wide area, and evaporation soon lowers its volume from 1,545 m³/sec at Koulikoro to 1,146 m³/sec at Mopti about 300 miles downstream.

B. Vegetation

Except in deserts, the influence of climate cannot be divorced from that of vegetation, which must be viewed as thermally driven chains of cells that conduct solutions from the soil to aerial surfaces, and thence to the atmosphere. Hence plant growth normally leads to great losses of water from the soil. The extent to which plants initiate other processes which may offset these transpiration losses is small. The turbulential uplift of a moist airflow athwart the edge of a tall forest may slightly increase rainfall. Similarly, tall plants extract from air masses at or near condensation point (cloud and fog) rain-drip which is of slight significance locally. But as a rule vegetation greatly decreases runoff, and the less the vegetation, the more abundant and rapid the runoff will be. A close plant-cover, at least for a while, modifies the violent effects of heavy rains and nullifies the hydrological effects of very light showers.

Vegetation actually growing in and on river channels may be regarded as a kind of surface roughness which may markedly reduce the capacity of the channel and retard the flow. In shallow streams the growth of hydrophytes raises the surface-level either all the year or in the warm season, and in some large tropical rivers (such as those in the sudd-hindered Bahr-el-Ghazel) a considerable retardation of flow and loss of discharge occur directly due to vegetation.

C. Soils and rock structure

From the point of view of regimes the chief properties of soils are their permeability and their water-holding capacity. These factors work in conjunction with climate, vegetation, and relief to control the amount of sub-surface water that eventually reaches the streams either as springs or as seepages. Hydrographs are often drawn to show the proportion of total runoff that originates as sub-surface flow and ground water. For example, the monthly sequence of the soil–water balance of the River Havel is shown in fig. 10.1.1(b).

Some soils and rock structures, unless coated by a continuous clay cover, are highly permeable and have a large water-holding capacity. Among the chief of these are some varieties of porous limestones and of coarse-grained sandstones, and certain well-jointed igneous rocks, notably basalt. Such rock formations, up to a certain limit (which may never be reached locally), tend to even out ground-water discharge, especially during dry seasons. In time of prolonged or heavy rain, after the underground storage capacity becomes almost full or the water-table rises high above spring-heads, this moderating influence often lessens and depends largely on the relative speed, directness, and convergence of underground flow towards the main exits.

D. Basin morphometry and hydraulic geometry

The shape and gradient of landforms greatly influences runoff, which as a rule increases in amount and rapidity with increase of slope. Conversely, flat areas, especially where marshes and lakes occur, tend to accumulate water and to

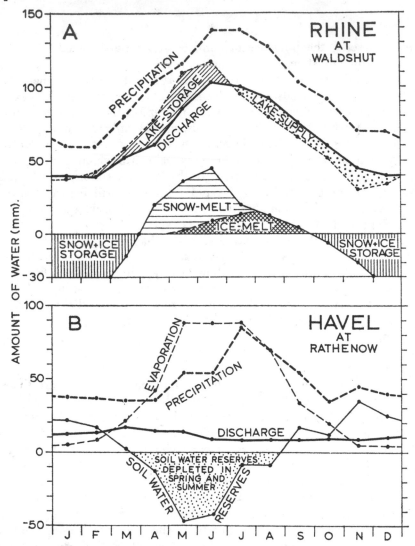

Fig. 10.1.1 Annual water-balance of two rivers.

A. Upper Rhine at Waldshut, showing influence of Alpine snow and ice, and of Lake Constance.

B. River Havel, a tributary of the Elbe, showing relation between precipitation, evaporation, runoff, and soil-water reserves (Adapted from Wundt, 1953).

modulate the regime downstream. The regulating effect of lakes is abundantly documented. Figure 10.1.1(*a*) indicates the moderating influence of Lake Constance on the flow of the Rhine downstream at Waldshut.

The influence of alluvial floodplains resembles that of lakes only during floods, when enormous quantities of water are absorbed and stored above and below ground. The classic example is the lower Yangtze-Kiang and the Hwai Ho in lowland China where the large, shallow, permanent lakes are conjoined in times of great summer floods by vast temporary inundations over 100,000 km² in extent. In hydrological regions, where the warm-season evapotranspiration exceeds the rainfall, the typical alluvial floodplain deprives the main river of a great deal of water and continues to do so until a change of weather allows the soil to be recharged with moisture. During periods of soil-moisture deficiency many

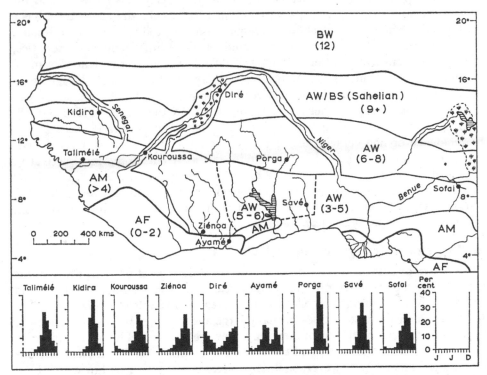

Fig. 10.1.2 Hydrological regions and characteristic river regimes of West Africa (Adapted from Ledger, 1964 and *International Atlas of West Africa*, Plate 10).

Thick solid line denotes approximate boundary of major hydrological region; thin solid lines show subdivisions of major region; thin pecked line shows minor variant of a sub-division, in this instance AW (5–6) or Dahomean sub-type. Numbers denote approximate length in months of dry season. Symbols are explained in the text. BS (arid steppe) and BW (desert) are areas where permanent streams cannot originate.

floodplain rivers contract rapidly to their low-water channel, leaving large areas of bed exposed. However, many rivers, especially those flowing on fine alluvium, tend to contract relatively little, because their channels are naturally (and in places artificially) puddled with clay particles. But for this puddling, most riverine plains near sea-level would be constantly waterlogged or flooded, as they lie well below mean river-level.

The regime of a river will also be affected by the geometry of its drainage basin. Thus high convex landforms in the tropics and sub-tropics favour the formation of local orographic uplift cells during spells of intense daylight insolation which often cause heavy convectional rainfall on the upper mountain slopes. This is a frequent cause of violent spates in mountain torrents on tropical highlands and islands.

Channel characteristics, such as size and shape of the cross-section and slope and roughness of bed, are partly the result of the regime, and have only a minor influence upon it. On the other hand, the morphometry of a basin, which involves the size, shape, stream-pattern, and orientation of a drainage basin, has a more decisive influence upon both the regime and its speed of reaction to climatic factors. Very small basins tend to show rapid reactions and violent characteristics, to be hypersensitive to brief downpours, during which overland flow is more important than channel flow. As basin size increases, the channel storage effect becomes increasingly dominant. In sizeable drainage basins the basin shape and the stream pattern may either modulate or accentuate the regime. This is largely a question of the convergence or non-convergence of tributaries and of the coincidence or non-coincidence of times of arrival of flood and of low-water. The total length of channel is also significant, as it affects the local arrival of seasonal variations from upstream. This is well shown on the Niger, where the upstream high water (September–October), after travelling over 2,000 km largely across a flat swampy plain, forms a flood on the middle course from January to March (fig. 10.1.2).

2. Types of river regimes, or hydrological regions

The above discussion reveals the importance of the size of units in the analysis and classification of river regimes. Whereas the river regimes of small and moderately sized basins may closely reflect regional runoff controls, especially climate, the main watercourses of many large and complex basins often acquire regimes unrepresentative of the territory they are crossing. The lower Colorado and lower Nile are obvious, and the lower Rhine and lower Rhône less obvious examples. There are, however, such large areas of the world within which local and regional river regimes reflect the regional climatic rhythm that some form of areal differentiation into river-regime types or *hydrological regions* seems desirable. Notable attempts at identifying hydrological regions have been made recently, particularly for Italy, France, and West Africa (fig. 10.1.2). This areal differentiation will progress in accuracy and coverage as hydrological observations increase, and will, without the need for extrapolation, eventually be based entirely on local measurements, showing the influence of both regional climatic and non-climatic factors on runoff. It will also be possible to distinguish all rivers with regimes markedly different from that of the hydrological region they are crossing.

At present these complexities are shown by inserting on maps hydrographs for selected points (fig. 10.1.3). Such a method is excellent where stations abound, but these data for a station on a river represent more than the discharge at that location; they relate also to the hydrological area providing the runoff or the total environment that a wise engineer would not ignore. Thus on large-scale maps station regime hydrographs *and* hydrological regions need to be inserted. As the scale of the map decreases, however, it becomes more and more difficult to show sufficient hydrographs and increasingly convenient to equate characteristic river regimes with generalized hydrological regions. If these regional

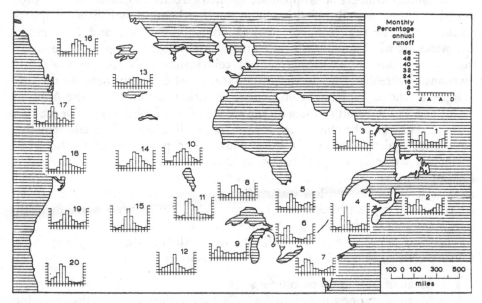

Fig. 10.1.3 Regime hydrographs of typical rivers in Canada and the northern United States (After Bruce and Clark, 1966 and Langbein and Wells, 1955).

General hydrological regions are given in Fig. 10.1.4. The rivers are: 1. Gander R., Newfoundland; 2. St. Mary's R., Nova Scotia; 3. Hamilton R., Labrador; 4. Upsalquitch R., New Brunswick; 5. Harricanaw R., Quebec; 6. Saugeen R., Ontario; 7. Susquehanna at Harrison, Pa; 8. English R., Ontario; 9. Pecatonica R., Illinois; 10. Saskatchewan R., Manitoba; 11. Assiniboine R., Manitoba; 12. Republican R., Nebraska; 13. Yellowknife R., N.W.T.; 14. N. Saskatchewan R., Edmonton, Alberta; 15. Yellowstone R., Montana; 16. Yukon R., Dawson; 17. Skeena R., Usk, B.C.; 18. Fraser R., Hope, B.C.; 19. Columbia R., The Dalles, Oregon; 20. Kings R., Piedra, California.

regimes are designated by a shorthand nomenclature (initials) it is possible to indicate the following, even on small-scale maps:

1. the boundaries of the hydrological regions;
2. the nature and components of the river regimes, however complex; and
3. any special local influences, quite apart from regional ones.

The climatic terminology used by Köppen seems adaptable for this purpose and is desirably genetic. In the world classification used below it is assumed that Köppen's terms retain their climatic meaning:

A = tropical rainy climates; all months with mean of over 18° C.
B = dry climates with an excess of potential evapotranspiration over precipitation.

Q

C = warm, temperate rainy climates.

D = cold, snowy climates; the mean temperature of the coldest month being not more than −3° C.

The rainfall symbols of Köppen are applied directly to runoff and are promoted to be the second capital letters in the shorthand, thus F denotes appreciable runoff all the year. W marked winter low-water, and S summer low-water. It will be noticed that the major hydrological regions of the world (fig. 10.1.4) show a remarkable general coincidence with Köppen's climatic divisions. Where discrepancies occur they are in part due to the need for revising Köppen's scheme in the light of modern climatic statistics (as in the Amazon basin) and as the result of modern knowledge of the distribution of vegetation types. In fig. 10.1.4 the boundaries of many of the divisions will remain tentative until more hydrological details are available. The possibility of further subdivision of larger hydrological regions is illustrated in fig. 10.1.2. An attempt is also made in fig. 10.1.4 to distinguish between tropical areas where the river regimes have no marked low water (AF) and those where the tropical rivers experience an appreciable low-water season (AM), but which is not sufficiently severe nor sufficiently prolonged to allow them to be classified as AW.

The third category of letters designates (in small type) temperature regimes, which obviously also have relevance to hydrological regimes:

a = Mean of warmest month over 22° C.

b = Mean of warmest month under 22° C; more than four months averaging over 10° C.

c = One to four months averaging over 10° C, and mean of warmest month under 22° C.

d = Mean of coldest month under −38° C.

A. Megathermal regimes (A)

1. AF: EQUATORIAL DOUBLE MAXIMA. In some equatorial areas the heaviest rains occur in spring and autumn following the equinoxes, and there is no dry season. Regions exhibiting these double maxima are the main valley of the Congo lying athwart the equator, probably parts of Indonesia and of the upper Amazon, and a coastal strip in eastern Brazil. In these cloudy areas the surface insolation is rarely excessive and allows the abundant rainfall to maintain a high discharge all the year; however, the coastal parts of West Africa that experience this regime seem to have a brief but marked low-water stage.

The widespread invasions of summer maritime (mT) air masses give much of the tropics and sub-tropics a strong summer maximum of runoff. After the rains the rivers dwindle to a marked minimum in late winter and early spring when insolation is great. The duration of high water varies with the length and intensity of the rainy season, and as a rule decreases rapidly inland, except where rising relief intervenes. Probably several hydrological subdivisions are neces-

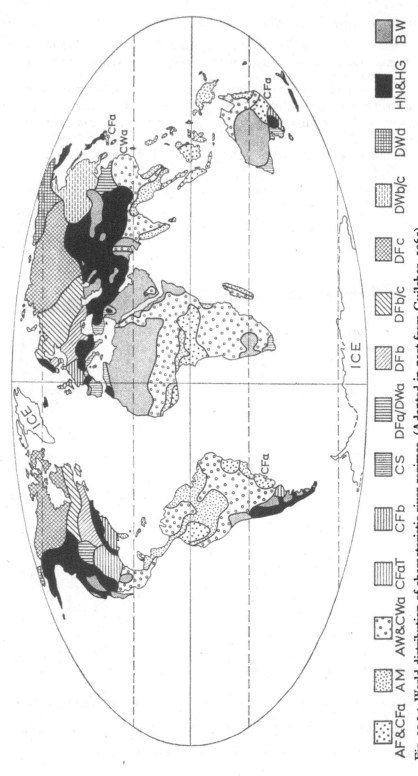

Fig. 10.1.4 World distribution of characteristic river regimes, (Adapted in part from Guilcher, 1965).

Symbols are explained in the text and illustrated in Figs. 10.1.5 and 10.1.6. BW denotes desert and other dry areas (BS) where streams cannot originate.

Mountain zones in some areas of Asia and the western United States should be shown as watercourses crossing desert (BW) and dry steppe (BS) regions, as is done for southern South America, but such details could not be attempted at this scale.

sary, based on the length of the low water (figs. 10.1.2 and 10.1.5), but information is usually inadequate for this at the present. Suggested divisions are:

2. AM: TROPICAL STRONG SINGLE MAXIMUM WITH A SHORT LOW WATER PERIOD OF UNDER THREE OR FOUR MONTHS. The rainier parts of West Africa and vast areas of the Amazon basin and of monsoonal South-East Asia belong to this heavy rainfall type.

3. AW: TROPICAL SINGLE MAXIMUM WITH A LONG LOW WATER. Where the

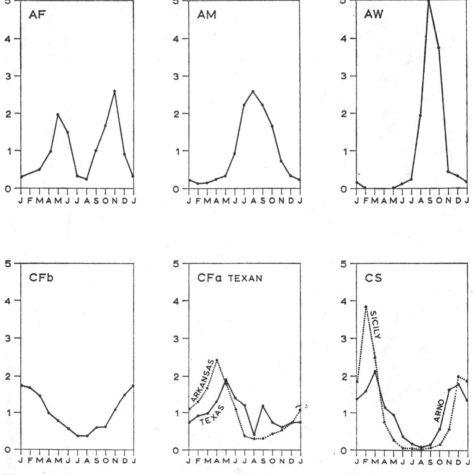

Fig. 10.1.5 Characteristic river regimes controlled mainly by rainfall and warm season evaporation.

AF. Lobaye R., a northern tributary of the Congo; AM, Lower Irrawaddy; AW, Pendjari R., a tributary of the Volta (See Fig. 10.1.2); CFb, Thames R., England; CFa Texan Buffalo R., Arkansas and Guadalupe R., Texas; CS. Arno R. and Imera Meridionale, Sicily.

All graphs show monthly coefficient of mean monthly flow for whole year.

rainy season shortens and annual falls lessen, the low-water period increases from four or five to six or seven months, and most small rivers become almost or quite dry. With increasing aridity the dry season extends to eight or nine months or more, and the hydrological characteristics degenerate into semi-arid steppe where permanent streams cannot originate and favoured watercourses have an episodic flood only once or twice a year, and ultimately into desert (BW), characterized by very infrequent flash floods.

B. Mesothermal regimes (C)

In warm subtropical and mild mid-latitude climates the regimes in two hydrological regions closely resemble those in the tropics and might well be grouped with them. They are:

1. CFa: WARM SUB-TROPICAL DOUBLE MAXIMA, which resembles the AF regime and occurs in all-the-year rainfall coastlands of eastern South America and eastern Australia about latitude 30°.
2. CWa: WARM SUB-TROPICAL WITH STRONG SUMMER MAXIMUM AND WINTER MINIMUM, which prolongs the hydrological regions AM and AW outside the tropics in monsoonal South-East Asia.

Elsewhere in C climates at least three distinctive hydrological regions can be delimited (fig. 10.1.5):

3. CS: STRONG SUMMER MINIMUM. In areas with a so-called Mediterranean climate summer usually brings clear skies and intense surface insolation. Under high temperatures and drought the local rivers dwindle or dry up unless fed by snow-melt or karst storage. In other seasons moist westerly airflow with active frontal uplift tends to prevail and steep orographic barriers can cause severe floods.
4. CFa/b: ALL-THE-YEAR FLOW WITH SLIGHT WARM-SEASON MINIMUM. In some of these areas, particularly on exposed seaboards, rainfall may be most in winter but the regime is remarkably even and drops to a slight minimum in late summer and early autumn. Most of this hydrological region is essentially CFb, especially in Europe.
5. CFaT: ALL THE YEAR FLOW WITH SPRING MAXIMUM AND WINTER MINIMUM. In a large area west of the lower Mississippi the precipitation is least in winter and most in summer. The maximum runoff occurs in May, or more rarely in April, and the minimum in August or September. In the south the summer rainfall is sufficient to cause the total warm season flow to exceed slightly that of winter, but farther north these conditions are reversed. Perhaps in the classification suggested it would be better to replace T, denoting Texas the typical regional location, by the number of the calendar month of maximum flow—i.e. CFa⁵ or CFa⁴.

In the three hydrological regions just described snow falls occasionally in the coldest months, but, except on hills, it seldom lasts more than a few consecutive days, and very seldom causes a flood, except in early spring.

C. Microthermal regimes (D)

Where one or more months experience mean temperatures below —3° C, snow-cover normally lasts for a month or more. As the frigidity of the winter half-year increases so does the proportion of the cold-season precipitation that falls as snow and is lost to the winter runoff and added to the spring flood. The actual depth of snow tends to decrease, and the depth and duration of freezing of soils and rivers to increase, with the severity of winter temperatures. Everywhere most precipitation falls in summer, and the annual totals are mediocre or small, being usually under 1 m in eastern Canada, 600 mm in Europe, and 400 mm in Siberia. Six main hydrological regions can be distinguished, three of which occur mainly in eastern Asia (fig. 10.1.6):

1. DFa/DWa: SUMMER PLUVIAL MAXIMUM; WINTER NIVAL MINIMUM. In north China and in the vicinity of the state of Kansas in North America the river regimes show a marked summer maximum (coincidental with the pronounced summer rains) and a winter minimum due to relatively cold, dry, snowy weather. Locally, snow-melt causes a brief secondary maximum in spring.

2. DWb/c: STRONG SUMMER PLUVIAL MAXIMUM; LONG WINTER NIVAL MINIMUM. North of DWa regions, in the eastern Asia coastlands draining mainly to the Sea of Okhotsk, the frost-bound period lasts for six or seven months. On rivers such as the Amur the snow-melt in May or June causes a moderate flood which, often after a slight recession, rises in August or September to a main maximum due to summer rains (fig. 10.1.6).

3. DWd: STRONG SUMMER PLUVIO-NIVAL MAXIMUM: PROLONGED COLD SEASON MINIMUM. In north-eastern Siberia the total annual precipitation averages well under 250 mm, and in parts under 100 mm. The maximum is strongly concentrated in summer, as the severe winters allow very little snowfall. Permafrost remains near the surface all the year, and in winter most rivers freeze solid except in deep pools, so that cold-season runoff is practically nil. With the thaw in June the flood is moderate (although truly excessive compared with the negligible winter runoff) and merges directly into the runoff from the July–August rains to form then a single pluvio-nival maximum.

In the colder parts of Europe and of Asia and North America not described above the hydrological regimes are dominated by winter snow and by summer evapotranspiration, although maximum rainfall occurs in summer. Vast areas were affected by thick ice sheets during the Quaternary glaciations and have an immature drainage, with numerous lakes and swamps that greatly moderate the warm-season flow. Three major characteristic regimes are distinguishable:

4. DFa/b: MODERATE PLUVIO-NIVAL OR NIVO-PLUVIAL SPRING MAXIMUM; SLIGHT SUMMER MINIMUM. In much of New England and the southern parts of the St Lawrence basin, and in a broad belt from southern Sweden to the

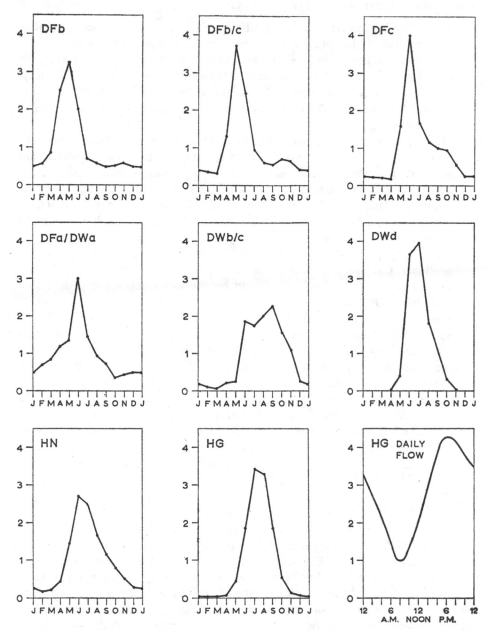

Fig. 10.1.6 Characteristic river regimes controlled mainly by cold season snowfall and warm season rainfall.

DFb Dnepr R. at Kremenchug; DFb/c Volga R. at Kuybyshev; DFc Yenisey at Igarka; DFa/DWa Republican R. near Bloomington, Nebraska; DWb/c Amur R. at Komsomol'sk; DWd Indigarka R.; HN Reuss R. at Andermatt; HG Massa R. at Massaboden. HG DAILY FLOW, South Cascade Glacier stream during a fine warm spell (After Meier and Tangborn, 1961).

All graphs, except the last, show monthly coefficient of mean monthly flow for year.

Black Sea in Europe east of the Elbe basin, the river regimes are either pluvio-nival or nivo-pluvial. In the former a moderate spring spate (March or April) is followed by a slight low water in August or September. In the latter the spring maximum is slightly stronger and comes in April or May, while the main winter flow may be only slightly greater than the summer minimum. Most of this region is definitely DFb, particularly in Europe.

5. DFb/c: STRONG NIVAL SPRING MAXIMUM; SECONDARY AUTUMN PLUVIAL MAXIMUM. Over large areas in eastern Canada, European Russia, and northern Scandinavia a variant of the nivo-pluvial regime dominates. A strong snow-melt maximum (usually in May) directly follows a cold-season minimum (January–March) and is itself followed by a small secondary maximum in late autumn (October–November) when a moderate rainfall is not offset by high evapotranspiration (figs. 10.1.3 and 10.1.6).

6. DFc: VIOLENT NIVAL SPRING MAXIMUM; STRONG WINTER MINIMUM. In northern Canada and the northern expanses of western and central Siberia (draining to the Arctic Ocean by the lower courses of great rivers such as the Ob, Yenisey, and Lena) a classic lowland nival regime prevails. A severe low water lasts from December to late April or early May, during which a small discharge persists beneath the carapace of ice on rivers and lakes. In May and June a sudden thaw occurs simultaneously over wide areas and causes a violent June maximum. The mean monthly flow of the lower Yenisey at Igarka in June (78,000 m³/sec) is exceeded only on the Amazon. The decline of the flood is less rapid, as it is moderated slightly by the effect of summer and early autumn rains. A feature of the northward-flowing Siberian rivers is that the upper and middle courses usually thaw out between late April and mid-May, whereas their mouths remain frozen until early June. The resultant floods spread out over wide areas and floating ice menaces banks and structures.

D. Mountain regimes (Hohenklima: above 1,500 m)

Widespread forms of microthermal hydrological regimes occur on high mountains outside the polar ice-caps, and it is proposed to designate these HN (*nivale* or Highland Snow) and HG (*gletscher* or Highland Ice). Because mean shade temperatures decrease upwards on an average about 6° C per 1,000 m, there is an elevation on most high mountains when precipitation begins to accumulate as snow. The margin of the snow-cover is lowest in winter, except in the tropics, where it lowers in the wet season. In the warm season the snow melts back and, where the height is sufficient, retreats to a permanent (*firn*) snowline. During this ascent the melt-water spate of the rivers lasts as long as the snow cover. Where ice forms, it persists as glaciers, sometimes far below the permanent snow-line, and ice-melt in summer may long sustain a river's flow. On lofty tropical and sub-tropical mountains the thin atmosphere allows on clear high-sun days intense solar radiation which causes a rapid melt-water runoff. Such streams often grow daily to a late-afternoon flood and dwindle fast after nightfall as temperatures fall rapidly below freezing-point (fig. 10.1.6). Travellers should cross them at first light, thereby avoiding the afternoon spate and the chance in hot weather of flash floods due to convectional downpours near the summits.

The mountain slopes between the lower and upper seasonal positions of the snowline may experience rain in the warmer months. Pardé has classified regimes with an appreciable snow-cover influence on the basis of a *coefficient of nivosity* which expresses the percentage of the warm-season flow contributed by melt-water. For Alpine mountains in Switzerland and Savoy the coefficient was the basis for the following subdivisions:

6–14%: pluvio-nival
15–25%: nivo-pluvial
26–38%: transition to nival
29–50%: pure nival to nivo-glacial
51% and over: glacial (rising to 67% on the Massa basin which is nearly 70% ice-covered).

Probably the enormous snowfalls on mountains in western North America would yield higher coefficients of nivosity than occur in Alpine Europe. But irrespective of the critical statistics used, it seems that these lower mountain slopes must be considered as a gradation of DFa/b (pluvio-nival), DFb (nivo-pluvial), and DFb/c (transition to nival) lowland regimes above a CFa/b base. However, the HIGHER slopes can hardly be considered DFc and DFd, as at some height the thinness of the mountain atmosphere introduces the unique excessive diurnal range during high-sun periods. As a result of rapid night cooling, above the firn-line the main daily temperatures normally remain below zero all the year. Even on very snowy mountains where large snowfalls may persist when mean summer temperatures are well above zero, frosts can still be expected at night. Because of this great diurnal range and proneness to night frost, the pure nival (or nivo-glacial) and the glacial hydrological regions on mountains should be designated by HN and HG respectively.

Since the effect of the climate on HN and HG river regimes depends on the presence of snow or ice, it is difficult to give altitude limits to these hydrological regions, except for broad latitudinal zones. There are, however, great differences between the effect of mountain-snow in the tropics and in high latitudes. In the tropics the temperature difference between the seasons is relatively small and the high basal temperatures ensure a high snowline all the year. Thus seasonal snow-melt areas are small and the local firn-line (HG) lies at about 4,600–6,000 m. Away from the topics the seasonal temperature differences rapidly increase, and both the permanent and temporary snowlines rapidly lower, the latter being at or near sea-level in D climates. In the same way the area affected by warm season snow-melt on mountains also greatly increases.

In most sub-tropical and cool-latitude mountains the deeply dissected relief and wide variations on local snowfall cause the HN and HG hydrological regions to be intricately interconnected. In the western Alps the mean altitude of the nivo-glacial basins varies from about 1,400 to over 2,000 m. The HG regions are usually situated above these altitudes, and commonly have 15–20% or more of their basins ice-covered. Here the river regimes normally show a single summer maximum culminating in July (or rarely in June), whereas throughout nivo-glacial regions the rivers commonly reach their maximum in June (fig. 10.1.6). These slight variations in peak flow, as with those of any lowland regime, could be indicated by adding the calendar number of the month of high water to the symbols. Thus HG⁶ equals June and HN⁵ May. In the southern hemisphere HG¹ (January) and HG² (February: ultra-glacial) occur in New Zealand and southern Chile. Worth indicating also is the liability of HG regions to glacial melt-water debacles, for which the symbol j (from the Icelandic *jokulhlaup*) seems suitable.

Indeed, for all regimes the genetic shorthand nomenclature could, where necessary, be made more explicit by adding the calendar month of high-water, in addition to symbols for special local influences such as:

j. jokulhlaup
l. moderated by lake storage
p. moderated by porous catchment
v. retarded by vegetation in channel, and
i. lowered by irrigation abstraction.

Acknowledgement. The author is especially indebted to Professor Maurice Pardé for great help and encouragement over a long period of time.

REFERENCES

Excellent summaries with good bibliographies are:

GUILCHER, A. [1965], *Precis d'Hydrologie, marine et continentale* (Masson, Paris), pp. 267–379 (Also with good section on lakes.)

KELLER, R. [1962], *Gewässer und Wasserhaushalt des Festlandes* ('Teubner, Leipzig), 520 p.

PARDÉ, M. [1949], *Potamologie* (University of Grenoble) (Roneographed), 2 vols., 336 p.

PARDÉ, M. [1955], *Fleuves et Rivières* (Colin, Paris), 3rd edn., 224 p.

PARDÉ, M. [1961], *Sur la puissance des crues en diverses parties du monde* (Geographica, Saragossa), 293 p.

WUNDT, W. [1953], *Gewässerkunde* (Berlin and Heidelberg), 320 p.

Excellent shorter summaries, also with bibliographies are:

BRUCE, J. P. and CLARK, R. H. [1966], *Introduction to Hydrometeorology* (Pergamon, London) (especially pp. 33–56).

CHOW, V. T., editor [1964], *Handbook of Applied Hydrology* (McGraw-Hill, New York), 1,418 p. (Section 14 on 'Runoff', by Ven Te Chow, and Section 16 on 'Ice and Glaciers', by Mark F. Meier.)

ROCHEFORT, M. [1963], *Les Fleuves* (P.U.F., Paris) ('Que sais je?'), 128 p.

AF, AM, AW and CFa Regimes.

Among regional monographs and articles are:

HURST, H. E. [1952], *The Nile* (Constable, London), 326 p.

LEDGER, D. C. [1964], Some hydrological characteristics of West African rivers; *Inst. Brit. Geogr.*, **35**, pp. 73–90.

LOCKERMANN, F. W. [1957], *Zur Flusshydrologie der Tropen und Monsunasiens* (Bonn) (Roneographed), 619 p.

ROCHEFORT, M. [1958], *Rapports entre la pluviosite et l'écoulement dans le Brésil subtropical et le Brésil tropical Atlantique* (Paris), 279 p.

RODIER, J. [1963], *Bibliography of African Hydrology* (UNESCO, Paris), 166 p.

RODIER, J. [1964], *Régimes hydrologiques de l'Afrique Noire à l'Ouest du Congo* (Paris).

C and D Regimes

IONIDES, M. G. [1937], *The Régime of the Rivers Euphrates and Tigris* (Spon, London), 278 p.

MASSACHS ALAVEDRA, V. [1948], *El regimen de los rios peninsulares* (Barcelona), 511 p.

PARDÉ, M. [1964], Les régimes fluviaux de la péninsula Ibérique; *Revue de Géographie de Lyon*, 129–82.

For *Italy* there is an excellent map in:

A. R. TONIOLO, *Atlante Fisico Economico d'Italia*, 1940, Map 9, Idrografia Terr estre.

Europe: annual review of hydrological literature in:

Revue de Geographie de l'Est, (Nancy), 1961, onwards by RENÉ FRÉCAUT.

U.S.A.:

LANGBEIN, W. B. and WELLS, J. V. B. [1955], The water in the rivers and creeks; In *Water*, U.S. Department of Agriculture Yearbook, pp. 52–62.

LANGBEIN, W. B. *et al.* [1949], Annual runoff in the United States; *U.S. Geological Survey Circular* 52.

Glacial: HG Regimes

MEIER, M. F. and TANGBORN, W. [1961], Distinctive characteristics of glacier runoff; *U.S. Geological Survey Professional Paper* 424.

For further references to debacles, see:

MEIER, M., In Chow V. T. [1964], Section 16, pp. 30–2. and

BECKINSALE, R. P. [1966], *Land, Air and Ocean;* 4th edn. (Duckworth, London), pp. 327–8 and 339.

Vegetation

CHOW, V. T. [1959], *Open-channel Hydraulics* (McGraw-Hill, New York), pp. 102–5.

WARD, R. C. [1965], Evapotranspiration from the Thames floodplain; in Whittow, J. B. and Wood, P., Editors, *Essays in Geography* (University of Reading), pp. 145–67.

II. Climatic Geomorphology

D. R. STODDART

Department of Geography, Cambridge University

The rise of climatic geomorphology dates from the period of exploration of the new tropical empires of France and Germany at the end of the nineteenth century: with its concern with the unusual and spectacular landform, the subject still bears the mark of this early work by scientific explorers. Prominent among these were Von Richthofen in China, Passarge, Jessen, Walther, and Thorbecke in Africa, and Sapper in Central America and Melanesia. This rapid expansion of knowledge of hitherto remote parts of the globe followed closely on the rediscovery in the 1860s and 1870s of fundamental geomorphic principles in the arid west of the United States, and their codification by Davis in the cycle-of-erosion concept from 1883 onwards.

In assimilating the new data into accepted theories, Davis treated the landforms of non-temperate climatic regions as deviants from the 'normal' scheme. In Germany each climatic region was thought to have its own assemblage of characteristic landforms and sequences of development. In more recent years German workers such as Büdel and Louis have continued to refine the concept of distinct morphoclimatic regions, while increasingly aware of the importance of climatic changes; while the French have viewed climate as but one, though a major, control of landscape morphology.

There are three central themes which require discussion in any treatment of climatic geomorphology. They are: (1) the view that landforms differ significantly betwen different climatic areas; (2) that these differences are the result of areal variations in climatic parameters and their effect on weathering and run-off; and (3) that, though considerable, climatic changes in Quaternary times have not disguised the climate–landform relationship. Each of these is attended by considerable difficulties, and can only be treated in broadest outline here.

1. Reality of distinctive climatogenetic landforms

While a general distinction has long been drawn between the landforms of arid, glacial, and humid temperate lands, surprisingly little objective morphometric evidence exists on landform variation within the wide range of fluvial conditions. Partly this results from the poor quality of topographical mapping over many parts of the earth's surface: Eyles [1966] has shown that Malayan maps, for example, are inadequate for many morphometric purposes. Hence morphometric work on maps often fails to demonstrate significant differences in landforms between climatically diverse areas. Viti Levu, Fiji, forms a good example: the south-east windward side of the island is wet, forested, and deeply dissected, with rainfalls often exceeding 3 m; the leeward side is under grassland, with rainfalls often less than 0·5 m. Measurements of drainage density, hypsometric integral, and Horton parameters for the Fiji 1:50,000 maps has failed to demonstrate significant differences in form between various rainfall and vegetation groupings. The Fiji maps are of good quality by comparison with many tropical areas.

Peltier [1962] adapted a less intensive approach, and sampled topographic

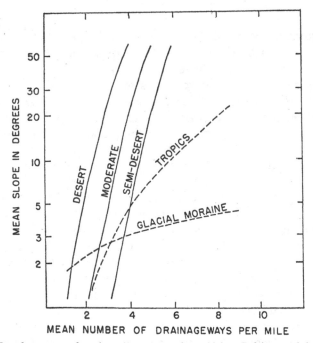

Fig. 10.11.1 Morphometry of major climatic regions (After Peltier, 1962).

maps on a world basis by randomly selecting geographical co-ordinates, correcting for latitude so that high-latitude areas were not over-represented. Because of the great variability in the scale of available maps, he measured the maximum difference in height within a 100-square-mile area at each sampling point, and used this to calculate mean relief in ft/sq. mile and mean slope in degrees. He then classified his sampling locations into major climatic groups (tundra, microthermal, mesothermal, and tropical), and also calculated the mean number of drainageways per mile at each station as a measure of topographic texture. When mean slope is plotted against mean number of drainageways per mile by climatic regions (fig. 10.11.1), the curves for desert, mesothermal, and microthermal areas are almost parallel, but with sideways displacement in response to changing runoff, whereas the curves for the tropics and for glacial morainic country are notably anomalous.

This suggests that first-order distinctions may be made between glaciated, tropical, and all other fluvial landscapes, and that second-order differences exist between desert, semi-desert, and temperate fluvial landscapes. The second-order differences could be interpreted in terms of Langbein and Schumm's sediment–yield curve, with erosion limited at lowest rainfalls by lack of water and at higher rainfalls by vegetation growth, reaching a maximum in the semi-arid lands. Two points need to be made here: first, the coarseness of Peltier's measures needs to be emphasized, for it raises the major problem of the scale at which climatic effects become apparent in landforms; and second, the fact that in Langbein and Schumm's scheme vegetation plays a major intermediary role between climate and landform.

Most identifications of climatically-controlled landscapes are made on two levels: (a) generalized impressions of whole landscapes, and (b) the recognition of peculiar landscape components. In the first case an illusion of differences in landform is certainly given by differences in the vegetation cover. In the Viti Levu case already mentioned the heavy green tropical forest of the wetter side, concealing the ground surface, contrasts with the brown and yellow appearance of the bare grasslands, where details of form are clearly apparent. On air photographs the forest would tend to conceal much fine dissection and the grass cover to reveal it, thus leading to underestimation of drainage density in the wet lands and to exaggeration in the dry. In the field, however, vegetation differences, reflecting climate, often suggest morphometric differences which may be uncritically accepted.

The recognition of peculiar landform components thought to be diagnostic of particular climates has been the chief tool of the climatic geomorphologist, in the absence of quantitative work on landform geometry. Several examples of such diagnostic landforms may be mentioned.

A. Surface duricrusts (Laterites, silcretes, and calcretes)

Laterites are often taken to indicate humid tropical conditions, though most laterite pavements are found in subtropical areas such as Bihar and the Guianas, and are frequently if not invariably related to past geomorphic conditions. Typically laterite pavements outcrop on interfluves between incised streams, under conditions of stripping of surface soil. Such crusts are often interpreted as of Tertiary age, or as having been under continuous formation since the end of the Mesozoic. Exposures of silcretes and calcretes similarly are often related to past rather than present climatic conditions.

B. Inselbergs

The term inselberg has been loosely used since Bornhardt's explorations in East Africa for steep-sided residual hills rising above low-angle plains in semi-arid areas, and Cotton has termed this assemblage the 'savanna landscape'. Inselbergs have been described over a wide variety of climatic conditions, from humid subtropical in Georgia, North America, to humid tropical in the Guinea coastlands, south India, and Brazil, and to desert areas in western North America,

Mauretania, and south-west Africa. While these examples cover a wide variety of forms, it is possible to argue that inselbergs are lithologically or structurally controlled azonal features resulting from the combination of sharp areal variability in the resistance of rocks and of efficient debris-transporting mechanisms. Many inselbergs on resistant rocks are so large that they certainly predate Quaternary climatic changes: hence present climates are not necessarily those in which the inselbergs were formed.

C. Pediments

Pediments as classically described are smooth low-angle slopes surrounding desert mountains, formed by processes of either sheet-flooding or lateral stream migration as transportational or erosional successors of back-wearing mountain slopes. The controversy over the origin of pediments has tended to obscure their diversity of form, to which Tuan drew attention in south-east Arizona. The model of pediment formation by slope retreat was remarkably similar to one of Penck's models of slope evolution, which Penck himself believed to be tectonically rather than climatically controlled, and later workers, such as Lester King, have suggested that, far from being diagnostic of aridity and thus atypical, the pedimentation process is in fact the universal mode of slope evolution. Davis himself, in his 1930 paper on 'Rock floors in arid and humid climates', came close to this position. Many arid-zone pediments are clearly polycyclic, developed during the complex sequence of Pleistocene pluvials and interpluvials: many appear to be being destroyed under present climatic conditions, rather than being formed by them.

D. Tors

Tors resemble inselbergs in many respects apart from scale. In Britain Linton has interpreted the Dartmoor tors as indicative of Tertiary deep tropical weathering of an inhomogeneous rock, with subsequent stripping of the weathered mantle to reveal piles of corestones. Palmer and Nielson showed the importance in this case of Pleistocene frost-shattering and mass-movement under periglacial conditions; but no one would suggest that the morphologically similar features of, for example, eastern Nicaragua and Rhodesia originated periglacially.

In each of these cases of the climatic interpretation of distinctive landforms two conclusions stand out: first, that form is an ambiguous guide to origin, and is often more complex than first generalizations would suggest, and second, that most supposedly diagnostic forms are older than Pleistocene climatic fluctuations and cannot be assumed to be genetically related to the climates in which they are now found. Landscapes differ, of course, in the degree of genetic ambiguity attached to them: desert dunes are quite clearly related to arid conditions, and even coastal dunes are largely absent in the humid tropics. Fossil dunes have been successfully used by Grove and others to reconstruct the former extent of African deserts in the Pleistocene. At the opposite end of the ambiguity scale, by contrast, are such forms as summit convexities on slopes (interpreted both paleoclimatically and structurally by different workers in Arizona) and V-shaped and saucer-shaped river valleys, described by Louis as typical of different climatic zones in Tanzania but found adjacent to each other in the New Guinea uplands by Bik.

Though landscapes often look different in different parts of the world, we must conclude that, with the exceptions of extreme cases, such as the glacial and arid, landscape assemblages of other climatic zones have yet to be unequivocally interpreted in climatic terms, and that many of the type-landforms of particular climatic zones are themselves of dubious significance. This is not to deny that climatically controlled landform differences exist, though morphometric confirmation of this is scanty; but it is to assert that the climatic inputs and geomorphic outputs in denudation systems are so little known that one cannot be inferred from the other.

2. Climatic controls of geomorphic significance

Because of the dearth of morphometric data, most treatments of climatic geomorphology concentrate either on the climatic characteristics associated with particular landscape types or on the definition of climatic zones within which

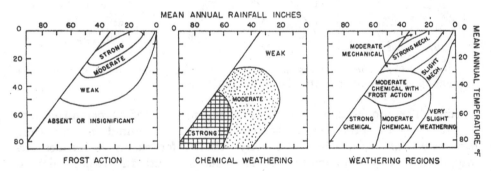

Fig. 10.11.2 Climatic control of frost action, chemical weathering, and weathering regions (After Peltier, 1950).

distinct landform assemblages might be expected to develop. Peltier [1950], in the best-known scheme of morphogenetic regions, distinguishes nine, of which six have been generally recognized in the literature (glacial, periglacial, 'moderate', savanna, semi-arid, arid) and three have yet to be worked out (boreal, maritime, selva). W. M. Davis had recognized only the 'normal', arid, and glacial cycles, and in spite of working in the humid tropics, he did not add a separate scheme for low latitudes.

Peltier's scheme (Table 10.11.1) uses two climatic parameters, mean annual temperature and mean annual rainfall: the morphogenetic regions are defined in terms of dominant processes varying areally and in combination, and not in terms of landform geometry (fig. 10.11.2 and 10.11.3). It is surprising that few attempts have been made to refine this scheme in twenty years. Apart from wind action in arid lands and ice action in cold ones, regions are differentiated in terms of water availability, both in channels and on slopes. A first refinement would thus be to replace mean annual rainfall by a measure of the availability of water for geomorphic work, for example, in crude terms, rainfall less evapotranspiration, or rainfall less potential evaporation. Calculations involving these measures have been made by Chorley and by Tanner. The more refined the attempts to give precision to climatic limits, however, the greater the problem of what the limits are for, and the more acute the difficulties of local variability. Standard meteorological measures are not necessarily those of greatest geomorphic significance, and Visher has experimented in mapping a range of climatic parameters of presumed geomorphic value. Many of these are not simply additive in their effects, and the problem of generalizing morphogenetic regions from them

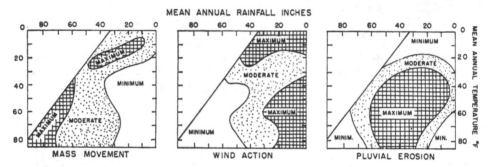

Fig. 10.11.3 Effectiveness of mass movement, wind action, and pluvial erosion under different climatic conditions (After Peltier, 1950).

remains. The problems of delimitation using more complex parameters are illustrated by the problems of the boundaries of the humid tropics: using temperature and atmospheric humidity criteria, for example, Garnier has mapped as parts of the humid tropics areas such as southern Arabia, which on vegetational criteria are manifestly not.

The climate at the ground, and even more the climate in the soil, normally differ considerably from the climate of the Stevenson screen. Thus in the humid tropics the interposition of a 30–50-m layer of vegetation between atmosphere and lithosphere means that the climate of the open air affects geomorphic processes only indirectly. Detailed studies of forest climate have been made in Uganda, Nigeria, Colombia, and the Guianas. Roughly two-thirds of the rainfall reaches the ground as rainfall: the rest is intercepted by leaves, evaporated, or channelled down trunks. The finest rains are filtered out and may not reach the ground at all, and the mean droplet size is markedly increased. Temperature and humidity variations are greatly reduced under the forest canopy, and in the soil temperature is invariant at quite small depths. Direct insolation at the surface is replaced by complex patterns of sunfleck and shadow. Comparative data on ground climate are available from many temperate forests, but are lacking for many areas (Geiger, 1965).

Data are simply not available to map the distributions of these geomorphic climates, and thus we generally substitute other distributions. The composition and distribution of vegetation was used by Köppen in his search for significant climatic boundaries, and vegetation is a major criterion in the morphoclimatic maps of Tricart and Cailleux. Landform, vegetation, and climate are, of course, complexly interrelated, and it is questionable whether vegetation is not as ambiguous as other measures in the search for meaningful criteria.

TABLE 10.11.1 Morphogenetic regions

Morphogenetic region	Estimated range of average annual temperature (° F)	Estimated range of average annual rainfall (in.)	Morphologic characteristics
Glacial	0–20	0–45	Glacial erosion Nivation Wind action
Periglacial	5–30	5–55	Strong mass movement Moderate to strong wind action Weak effect of running water
Boreal	15–38	10–60	Moderate frost action Moderate to slight wind action Moderate effect of running water
Maritime	35–70	50–75	Strong mass movement Moderate to strong action of running water
Selva	60–85	55–90	Strong mass movement Slight effect of slope wash No wind action
Moderate	38–85	35–60	Maximum effect of running water Moderate mass movement Slight frost action in colder parts No significant wind action except on coasts
Savanna	10–85	25–50	Strong to weak action of running water Moderate wind action
Semi-arid	38–85	10–25	Strong wind action Moderate to strong action of running water
Arid	55–85	0–15	Strong wind action Slight action of running water and mass movement

Source: Peltier [1950], 215.

Recognition of the interdependence of these controls leads, however, to important insights, which are often most readily apparent when the system is deranged. This is well demonstrated in the humid tropics, in the comparison between erosion under forest, and on bare ground, or between forests and savanna. Rougerie found sediment yields from experimental plots in the Ivory Coast to be roughly fifty times greater from bare ground than under forest. The effects of forest clearing during shifting cultivation, or of forest replacement by

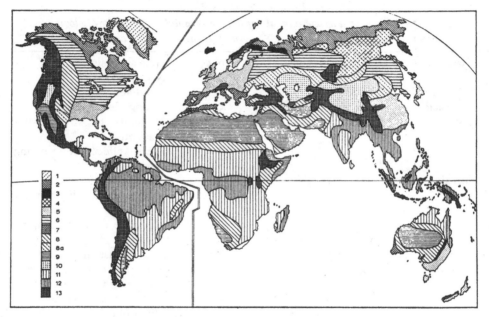

Fig. 10.11.4 World morphoclimatic regions (After Tricart and Cailleux, 1965, Fig. 49).

1. Glaciated regions.
2. Periglacial regions with permafrost.
3. Periglacial regions without permafrost.
4. Forest on Quaternary permafrost.
5. Mid-latitude forests: with maritime climate or lacking severe winter.
6. Mid-latitude forests: with severe winter.
7. Mid-latitude forests: Mediterranean type.
8. Semi-arid steppes and grasslands.
8a. Semi-arid steppes and grasslands: with severe winter.
9. Deserts and degraded steppes: without severe winters.
10. Deserts and degraded steppes: with severe winters.
11. Savannas.
12. Intertropical forests.
13. Azonal mountainous regions.

savanna grassland, either caused by man's activities or by climatic changes, would thus be of major geomorphic importance. Rougerie also found that a simple sediment–yield curve relating erosion and rainfall could not be constructed, for the amount eroded by a given rainfall varied with antecedent conditions, both seasonal and short-term. Rains at the beginning of the wet season gave higher sediment yields than later rains, though different trends were apparent with different rainfall intensities.

Rougerie's work highlights another major problem in defining climatic parameters: that of periodicities and magnitudes. In simplistic terms it is possible to draw contrasts between, for example, the sparse but occasionally torrential rainfalls of the deserts, the strongly seasonal and high-intensity rainfalls of the savannas, and the continuous high-intensity rains of the humid tropics, and to infer geomorphic consequences from them. But the more we learn of rainfall characteristics, the less they appear to conform to the assumed model (Beckinsale, 1957; Peel, 1966), and it is doubtful whether any useful purpose is served by elaborate deductive reasoning about what might be happening.

On a world scale, therefore, it is possible to describe the distribution of certain attributes of climate of geomorphic importance, such as thunderstorm incidence,

TABLE 10.11.2 Tricart and Cailleux's morphoclimatic zones

---

1. Cold zone
    (a) Glacial domain
    (b) Periglacial domain

2. Forested zone of middle latitudes (modified by man, with glacial and periglacial survivals):
    (a) Maritime domain, mild winter (strong survival of glacial and periglacial forms).
    (b) Continental domain, severe winter (Quaternary permafrost may survive).
    (c) Mediterranean domain, dry summers.

3. Arid and sub-arid zone of low and middle latitudes:
    (a) Rainfall distinction: steppe and desert.
    (b) Winter temperature distinction into cold and warm regions.

4. Intertropical zone, differentiated by rainfall seasonality:
    (a) Savannas.
    (b) Forests.

Source: Tricart and Cailleux [1965, pp. 268–88].

frost frequency, and tropical cyclones (Common 1966); it is also possible to map the zonal distributions of soils and of vegetation. Several workers have combined such criteria with those of relief to produce maps of morphoclimatic regions, in which the critical though only inferred parameters must be weathering-climate and availability of water for transportation of debris. Table 10.11.2 lists the morphoclimatic regions recognized by Tricart and Cailleux [1965, pp. 268–88], and fig. 10.11.4 shows their distribution. Apart from the arid and glacial lands, such maps do not record the occurrence of different kinds of processes, only the variety of combinations of the same processes: the differences are of degree and not of kind. Further, it does not follow that in such maps we are also mapping the boundaries of landform systems, many of which may not be related to present climates.

3. The problem of climatic change

Since the time of Agassiz, evidence has accumulated that not only has climate changed drastically and rapidly over large parts of the earth during the Quaternary but stratigraphic evidence shows that climatic changes have occurred during most of geological time. These changes pose a critical problem in climatic geomorphology, for it cannot be assumed that the landforms found in any given climate have developed in response to it. How, then, can the links between climate and landform be identified?

Albrecht Penck was aware of this problem, but believed that the glacial shifts of the climatic belts were minor, with changes limited to marginal zones. In the central parts of the deserts the humid tropics and other zones he considered Pleistocene changes to have been minimal. It is true that the most spectacular evidence of recent climatic change comes from the margins of the deserts, where relict dune fields now vegetated indicate former dry conditions, and old shorelines and lake deposits wetter phases. Thus Grove has described an ancient erg in Hausaland, Nigeria, and evidence for a 'Mega-Chad' which may have existed only 10,000 years ago. Comparable forms are known from the western United States and from Australia. The sensitivity of the semi-arid belts to even small changes in water availability can be explained in terms of Langbein and Schumm's sediment–yield curves and the role of changing vegetation cover, though the effects of both rainfall and temperature changes are difficult to disentangle.

There is, however, growing evidence, much of it still ambiguous, that considerable climatic changes affected even the interior of the deserts, and possibly other climatic zones as well. Pollen analysis in the Sahara has shown that both Mediterranean and Guinean floral elements invaded the interior deserts in the late Pleistocene, and similar work in East Africa and Angola had suggested vertical shifts of the vegetation belts of several hundred metres. It has been objected that we know so little of the synecology of the present vegetation that we cannot legitimately infer climatic changes from scattered pollen records, but many lines of evidence taken together lead to a strong presumption of major changes, perhaps with considerable regional variability. What happened to the rain-forests in the Pleistocene is one of the yet unsolved problems of climatic geomorphology. It is often argued that because of their species composition and diversity the tropical forests must have undergone only marginal fluctuations at their latitudinal and altitudinal boundaries. The appearance of the present forest may be a poor indicator of antiquity, however. Charcoal layers under supposedly undisturbed Nigerian forests, archaeological remains under Central American and South-East Asian forests, demonstrates the rapidity of regeneration after disturbance. Relict desert sands in Central Africa extend as far north under rain-forest as the mouth of the Congo, though now restricted to the Kalahari.

Quaternary climatic changes were essentially short-term changes of great intensity. Though the length of the Quaternary is itself in doubt, there is now some agreement on a figure of 2 million years, rather than the much shorter figures previously accepted. 'Pre-Quaternary' glaciations now being described from Iceland and elsewhere are dated at 5 million years, and may make an even longer Quaternary necessary. During this time there were at least four, and probably more, major climatic fluctuations, apparently synchronous in high latitudes all over the globe, and a very large number of minor fluctuations both in glacials and interglacials. If it is accepted that particular climate–vegetation conditions generate specific sets of landforms, then the effect of these Quaternary changes would depend on the rate at which landforms adjust to new equilibrium conditions. Such rates will be a function of not only the magnitude and duration of the processes involved, but of the nature of the pre-existing landforms and the resistance of the rocks. Except in the case of ice action, it is likely that many climatic conditions existed for so short a time that they were morphologically significant only in areas of weak rocks and considerable relief.

In a sense the Quaternary has been so complex that the greater part of it has been ignored, especially as later changes tended to obliterate the effects of earlier ones. Much effort has gone into reconstructing the climates and landforms of the Holocene and of Wisconsin times, and only recently has comparable attention been given to the geography of the last interglacial. Little is known about earlier times, covering the great part of the Quaternary. There is the added complication that this period was one of major tectonic movement, with rifting and vulcanicity in East Africa, and uplift of the Andes in South America and comparable mountains in South-East Asia.

In spite of their intensity, therefore, it is possible that Quaternary climatic changes were less important as landscape-forming agents, because of their short duration, than earlier climatic conditions. The cooling of the Tertiary in north-west Europe has long been known on geological grounds, and warmer sea temperatures at the end of the Mesozoic have been deduced from isotopic measurements on fossil foraminifera. Baulig and many others, particularly in southern Europe, have described deep weathering profiles, laterite-like deposits, and low-angle surfaces which they ascribe to humid tropical conditions during the Tertiary. Strakhov [1967] has attempted to summarize the evidence for Tertiary climatic changes on a world scale. In most areas, however, present landscapes are complex mosaics consisting of small areas inherited from Tertiary conditions, and tracts of forms developed during the Quaternary complex climatic conditions.

Taken together, the evidence suggests that climatic changes have been so continuous in the last 50 million years, and so rapid in the last 2 million years, that equilibrium landforms can rarely have been developed. Davis considered his 'climatic accidents' to be infrequent replacements of one type of climate and landscape by another, but his concept may have to be modified to take account of continuous climatic changes and the constant readjustments of denudation systems to them.

4. Conclusion

Climatic geomorphology rests on the assumptions that different climatic inputs in the denudation system will result in different landform outputs, and that equilibrium landforms will differ between climatic regions in such measurable parameters as drainage density, maximum slope angle, and slope form, as well as in the development of characteristic type-landforms. Many of these differences have yet to be demonstrated, and some at least of the supposedly characteristic forms may have diverse origins. Climate itself is an elusive quantity to define, and most previous workers have used such coarse measures as mean annual rainfall and mean annual temperature. What is required is, first, data on water availability and temperature conditions at the ground surface and in the weathered mantle, where areal movements and weathering changes take place, and second, some information on the water and sediment outputs from the system, through analysis of river discharges, hydrographs, and sediment yields.

In the absence of such data judgement must be suspended on the reality of climatic control of fluvial landforms, though clearly the arid and glacial landforms differ fundamentally from those of more humid areas. In the most intensive analysis yet made of discharge/erosion relationships in contrasting climatic areas, Douglas has concluded that climate as such is of small significance, at least as traditionally understood, and that we are dealing simply with different combinations of the same processes. Similar sediment yields could, he believes, be provided by similar rainfall seasonality conditions under what appear to be very different climatic conditions. Most of the analyses so far made of climatic–morphologic cycles have been highly deductive: and it is, of course, possible to construct internally coherent and logically consistent schemes relating climatic characteristics to weathering processes and rates, vegetation types, run-off characteristics, erosion rates, and landform development. Such schemes require more intensive testing in the field than they have yet received.

If landform development is viewed as the adjustment of form to process in denudation systems, the emphasis of climate at the expense of other factors, particularly lithology and structure, is bound to be distorting. It is possible that at certain scales of landform development climatic parameters are important independent variables, whereas at others they are dependent or not significant. The investigation of such scale-linkages in the correlation structure of diverse denudation systems is the main task facing climatic geomorphology.

REFERENCES

BECKINSALE, R. P. [1957], The nature of tropical rainfall; *Tropical Agriculture*, **34**, 76–98.

COMMON, R. [1966], Slope failure and morphogenetic regions; In Dury, G. H., Editor, *Essays in Geomorphology*, (Heinemann, London) pp. 53–81.

EYLES, R. J. [1966], Stream representation on Malayan maps; *Journal of Tropical Geography*, **22**, 1–9.

GEIGER, R. [1965], *The Climate Near the Ground* (Harvard University Press, Cambridge, Mass.), Revised edition.

LANGBEIN, W. B. and SCHUMM, S. A. [1958], Yield of sediment in rela tion to mean annual precipitation; *Transactions of the American Geophysical Union*, **39**, 1076–84.

PEEL, R. F. [1966], The landscape in aridity; *Transactions of the Institute of British Geographers*, **38**, 1–23.

PELTIER, L. C. [1950], The geographic cycle in periglacial regions as it is related to climatic geomorphology; *Annals of the Association of American Geographers*, **40**, 214–36.

PELTIER, L. C. [1962], Area sampling for terrain analysis; *Professional Geographer*, **14** (2), 24–8.

STODDART, D. R. [1968], Climatic geomorphology: review and reassessment; *Progress in Geography*, 1.

STRAKHOV, N. M. [1967], *Principles of lithogenesis;* Volume 1. Tomkeieff, S. I. and Hemingway, J. E., Editors, (Oliver and Boyd, Edinburgh and London).

TRICART, J. and CAILLEUX, A. [1965], *Introduction à la géomorphologie climatique;* (S.E.D.E.S., Paris).

III. Human Responses to River Regimes

ROBERT P. BECKINSALE

School of Geography, Oxford University

Although flood control, irrigation, and the generation of water-power are carried on under a wide range of hydrological conditions, they are especially important where wide seasonal fluctuations of discharge occur. Here the period of high discharge poses particular flood-control problems, as well as providing sufficient water, when adequately impounded, to carry on irrigation and water-power generation during the dry season. It is important to recognize that few single-purpose projects are now being initiated, and that it is the multi-purpose schemes which represent the most sophisticated response by man to seasonal variations of discharge.

1. Flood control and streamflow routing

The need for flood control usually follows the settlement of people on the rich soils of a floodplain. The paddy cultivation of lowland monsoon Asia and the ancient irrigation systems of the Nile and Euphrates were dependent mainly upon river regimes with a violent warm-season flood. Records at the Roda gauge near Cairo date back to A.D. 640. Here and elsewhere little could then be done about exceptional low-water years, whereas high floods could be partly controlled by constructive levées. On the lower courses of many monsoon rivers (with AM; AW regimes) the natural levées were soon strengthened artificially and often resulted in a natural raising of the channel bed. Eventually over vast areas the rivers were flowing along ridges well above the adjacent floodplains. No doubt increasing deforestation and cultivation near the watersheds added to the silt content of the streams and to deposition in the lowland channels. Also the appreciable rise in sea-level since about 12,000 B.P. has greatly decreased the gradients of coastal plains.

Given similar regimes, the problems of flood control tend to be least on rivers with sufficient gradient to scour their beds and greatest on rivers incapable of preventing channel deposition. A river with a reasonable gradient can be controlled by dams, adequate storage reservoirs, dykes, and channel improvements. This is exemplified on a small scale by the Rhône just above Lake Geneva, where the river floods violently from snow and ice-melt in early summer and then spreads over a wide bed, much of which is exposed in the low-water season. In 1895 an attempt was made to improve the reach between Brigue and Lake Geneva by constructing eleven flood-dykes and several jetties to restrain the river to a narrower central channel. But the jetties proved to be wrongly sited, and the river-bed silted, causing the dykes to be overtopped and breached by floods. From 1928, after a scientific study, the hydraulic characteristics of the central channel were improved by making a dyked and graded bed. In many parts the river now began to scour, and only exceptional floods now overflow into the area between the channel dykes and a parallel line of external dykes. With the aid of drainage devices, the former floodplain has been transformed from marshes to fertile agricultural land (UNESCO, 1951a).

The floods at Florence in 1966 demonstrate the violence of Mediterranean (CS) river regimes. Owing to channel silting and artificial levée building, the Arno even in normal high water rises above the level of much of the adjacent city. Only extensive channel dredging, dyke strengthening, and an elaborate system of reservoir control near the watersheds will ensure that the riverine parts of the city are not flooded occasionally every few centuries.

On rivers with very low gradients and heavy silting, flood control today is largely a question of distant headstream control, coupled with storage reservoirs, levées or embankments, channel improvements, flood spillways, pumping stations, and flood forecasting. The problems of flood control have been overcome on a small scale on the Waal in the Netherlands, where the river regime is moderate. Here main dykes enclose a major (winter) bed 1 m to 2 km wide. The polders between these master dykes and the minor (summer) bed are protected against high water in summer by dykes only 3 or $3\frac{1}{2}$ m above the mean summer river-level. When the river floods most of the drainage is carried by the summer bed, but the winter bed acts as a storage reservoir for water in excess of the capacity of the central channel.

On a greater scale the lower Mississippi–Yazoo, with a floodplain of 75,000 km², demonstrates the problems of flood control of a meandering river with a high silt content and moderately violent regime. The natural levées are usually 3–4·5 m above the swamps 3–5 km back. Under the incentive of rich bottom land for cotton and corn, the levées have been strengthened artificially and now extend for more than 3,500 km and allow the channel flood-level to rise 7·5 m above that of the adjacent plain. Channel shortening by means of cut-offs, dyke strengthening, and the construction of flood spillways leading to areas where flooding can be controlled, have greatly lessened the loss of life and damage to property. In this protection, upstream water-control and an efficient flood-forecasting service also play an important part.

Flood control on low-gradient rivers with a large sediment load becomes most difficult where the regime is violent (AM; AW). Abnormal daily spates as well as high seasonal floods occur throughout monsoon Asia, where a dense population, largely dependent on paddy, needs summer floods but also requires protection against abnormally high waters. Here flood-control structures dominate the landscape. The North China plain of about 324,000 km², exclusive of East Shantung, is dominated by the levées which stretch for 720 km along the Hwang Ho and 140 km along the Yungting Ho. The bed of the former, or Yellow River, near Kaifeng lies at a maximum of 15 m above the plain. As its drainage basin includes

288,234 km² of loess and is liable to winter frost, its silt content is notoriously high and its bed is being steadily raised (UNESCO, 1951a, pp. 309–4). During the last 4,242 years it has changed its lower course completely 7 times, breached its dykes more than 1,170 times, and overflowed them about 425 times. The annual flood damage is large and involves more than 29,600 km². Among the

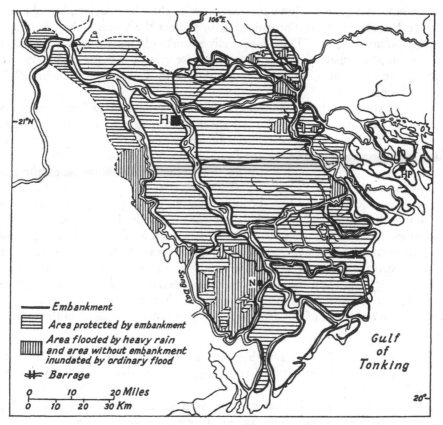

Fig. 10.III.1 Embankment system of the Red River delta, North Viet-Nam.
H. Hanoi   HP. Haiphong   N. Namdinh.
(Adapted from United Nations, 1966a, p. 15).

suggested remedies is a desilting basin to supply water free of coarse sediment (?silica particles) for irrigation. A great dam has been built across the river above Loyang, and for some years a scheme of 45 smaller dams and 26 reservoirs has been under construction higher upstream to regulate floods and reduce soil erosion.

Similar but smaller floodplains exist in most Asiatic countries, and only a few have as yet solved the flood-control problem. In North Viet-Nam the Red River and other streams form a delta that extends about 150 km inland (fig. 10.III.1). The Red River, which is 1,200 km long and rises in China, meanders for its last 220 km through a deltaic floodplain, sown almost entirely to paddy. Its annual discharge averages 3,900 m³/sec, but since recordings began has varied from 700 m³/sec to 35,000 m³/sec. The whole basin has a mean annual rainfall of 1,500 mm, and the lower course receives 1,800 mm. As the whole delta lies several metres below the highest flood-level, the people have fought a constant battle to contain floods in excess of those needed for paddy cultivation. In early times individual landowners strengthened their own dykes, but since the thirteenth century a central organization to protect the whole delta has evolved (United Nations, 1966a). This co-operative effort culminated between 1920 and 1944 in the strengthening of the complete dyke system. Levées now extend for 1,400 km and contain about 120 million m³ of earth. Only one serious breach has occurred since 1926.

Many of these basins in Monsoon Asia are so large and so fragmented politically that full flood control will need enormous capital and elaborate co-operation, and probably will not be realized for centuries. However, elsewhere many small basins and a few large basins (notably the Columbia, 772,000 km²) have already been controlled and instrumented sufficiently to allow scientific flood forecasting and streamflow routing (Rockwood, 1961; Lewis and Shoemaker, 1962). These ideas were first used comprehensively in 1943, and have progressed in value with the efficiency of the river control, with the knowledge of river hydraulics, and with the use of computers. Streamflow or flood routing provides a means of translating a hydrograph from an upstream to any downstream point so as to express the effects of the intervening channel and valley storage (Bruce and Clark, 1966, p. 197). It assumes that on a given reach of a river the outflow is equivalent to the inflow minus the amount of water stored in that reach. In practice, the technique of flood routing and forecasting is based on the assumption that:

1. rivers consist of reaches with relatively constant flow characteristics and that gauges or more detailed recordings exist at the entrance and exit of these reaches;
2. the routing period, or time interval used in the flood-routing formula, is the time taken for the flood wave to traverse the reach; and
3. the reaches are sufficiently short and self-contained to ensure that the storage in each is a function of the inflow and outflow of the river, or, in other words, is not unduly affected by local tributaries. If tributaries are present and are too small to warrant the establishment of a separate reach below their entrance their inflow should be added to that at the upstream end of the reach in question.

For flood routing the data required are the stage–discharge curve and the stage–storage curve for each reach of the river. The latter curve can be constructed from detailed flood recordings by reversing the routing procedure used. Thus, from past floods the difference between the inflow and outflow gives the amount of storage in the reach; the progressive sum of these storages plotted against the stage at the downstream end of the reach provides a stage–storage curve. It is necessary to begin the summation at a time when storage is about zero, that is when inflow and outflow of the reach are about equal. The methods and graphs are summarized by Bruce and Clark [1966, pp. 197–202]. Mechanical and electrical routing devices are installed on many rivers, and forecasts of flow and flood levels based on them are considered essential to make the optimum use of flow and to minimize flood damage. On uncontrolled rivers flood forecasts tend to be less accurate, but they still provide useful warnings, so that people and movable goods can be removed to safety and labour forces alerted. On partially controlled rivers reliable streamflow routing allows reservoir storage to be used most efficiently and economically. Thus flood control demands considerable accommodation space to be left inside reservoirs, whereas irrigation demands the maximum water storage. If a reservoir is kept too empty the irrigation may suffer from water shortage; if it is kept too full floods may damage the low-lying plains. Needless to say, the bigger the storage capacity, the better for all concerned, except those displaced by the reservoir. The phenomena of freeze and thaw which affects D regimes enters equally into streamflow routing. Scientific recordings and predictions now render obsolete the annual sweepstake at Yukon city, where a bell tied by a rope to a stake in the river ice gave the news and time of the break-up for the lucky winner. On tidal rivers, where for long distances the so-called tide is merely a freshwater wave moving upstream, the diurnal rise adds a further complication, as in East Pakistan, where over one-third of the total area ($140,000$ km^2) is liable to serious floods.

2. Irrigation

The application of water to the soil, or irrigation, is undertaken mainly to grow crops in dry climates and in humid climates with a marked dry season. It is, however, also widely practised in humid regions with a moderate rainfall in order to obtain higher yields and special crops. Thus, except in arid climates, irrigation is either supplemental or complementary to seasonal rainfall. The expenses involved are often offset by marked advantages of higher yields, better harvesting weather, and easier pest control. Also in some countries out-of-season products fetch extra prices on the national markets.

Most irrigation projects are on floodplains and valley floors, the main exceptions being where mountain springs are led overground to hillside terraces, as is common, for example, in Japan. Floodplains, if well drained, are often rich and easily cultivated; many are enriched almost annually by sediments; all are relatively flat and can be reached by simple water-lifting devices. In Eurasia the uncertainties and limitations of a seasonal flood with AM, AW, and CS river regimes led early to the building of earthen dams and to the invention of many water-lifting implements. On the Indus a wide variety of tanks and irrigation canals existed by about 3000 B.C.; on the Euphrates in Babylon earthen dams and an elaborate code of water laws existed in 2050 B.C. Most of the ancient methods of lifting water artificially are still in use. They were based on three elementary principles: the LEVER, as in the Egyptian *shadoof*, the Indian *picottah* and the Spanish *cigoñal*; the horizontal revolving WHEEL fitted with buckets round the rim, as in the Persian wheel, the Egyptian *sakiya*, Punjabian *harak*, and Spanish *noria*; and the SCREW, an invention of Archimedes of Syracuse in about 200 B.C. The last-named consists of a wooden cylinder containing a helix (corkscrew-shaped diaphragm) up which water is forced when the lower end is placed at a low angle in water and rotated.

Today irrigation is practised either seasonally or all the year, mainly where the river regimes are AM, AW, CS, HG, and HN. The seasonal form, often called basin irrigation, is commonest alongside uncontrolled rivers and is dependent on the flooding of embanked areas either direct or by means of artificially filled inundation canals. In contrast, perennial irrigation needs a partly controlled river with an almost constant water supply, thereby allowing two or three crops to be grown annually on the same plot. The basin method was, and still is, quite extensive, as many tropical and subtropical rivers rise 7 or 11 m in summer, and primitive lifting devices can raise the water farther to higher terraces. Thus in Egypt the Archimedes' screw raises water about 75 cm into a basin, whence a sequence of shadoofs, each with a rise of about 2·5 m, allows the upper terraces to be reached.

In some semi-arid regions with wadis that experienced a brief flow at one season (AW/BS regime) the peasants cultivated the wadi-floor after the rains. More commonly earthen dams were constructed across the valley to form tanks, such as survive in thousands in southern India and Ceylon. During floods much sediment was deposited in the reservoirs and in the floodplain. Recent analyses of the Nile 'mud' (UNESCO, 1951, pp. 279–99) show that the maximum concentration of silt in the river water reaches about 2·5 g per kg in August. If, as is usual, the basins are flooded to a depth of 1 m the layer of silt averages about 2·1 mm or about 10 tons per acre. Analyses of the chemical composition of the silt with regard to fertilizing properties revealed the following percentage (Table 10.III.1) of the chief plant nutrients and the approximate amount supplied per feddan or acre annually.

TABLE 10.III.1

	Percentage total deposition	Amount (kg)
Phosphorus (P_2O_5)	0·24	25·2
Potash (K_2O)	1·07	112·5
Lime (CaO)	4·16	430·6
Organic matter	2·42	254·1
Nitrogen (included in organic matter)	0·13	13·5

Thus the Nile mud is fairly rich in lime, potash, and phosphates and relatively deficient in nitrogen. However, the mechanical structure of the soil, when well drained, is very favourable to micro-organisms, especially nitrifying bacteria.

The main disadvantage of high-sediment content in rivers is the silting-up of reservoirs and of irrigation ditches. Probably the first dams with sluices near their base adequate to prevent excessive silting were the Periyar in southern India (1895) and the Aswan (1902). On the latter the sluices are kept open and flood-water allowed to pass through until the flood is lessening and its silt-content decreasing. The surest way of avoiding excessive sedimentation is to construct elaborate desilting basins, as in the Imperial valley schemes of the United States and at the new High Aswan dam. Without such desilting methods most of the world's great reservoirs on rivers will be largely silted-up within a few centuries, and many artificial reservoirs built in the twentieth century have already suffered appreciable sedimentation.

Irrigation practices and irrigated landscapes have evoked a large literature, including a global summary by Cantor [1967]. Of the numerous crops irrigated, paddy is supreme. It needs an almost constant water supply until towards harvest, and the more prolific varieties are usually grown where irrigation supplements a seasonal rainfall which helps to keep the basins filled and prevents problems of salinization.

Whereas Chinese, Japanese, and Arab (Moorish) irrigation methods dominated from early times in most of the Old World, in the New World the existing practices were greatly disrupted by the Spanish Conquest. In the United States irrigation did not revive until after about 1850. Here and elsewhere modern engineering techniques began to affect irrigation in the late nineteenth century. Up to 1902 the highest dam was at Furens in France, 170 ft (52 m). However, in the next two decades the increasing development of the petrol engine, of concrete construction, and of electrical equipment led to a great interest in dam building and water-pumping. By the 1940s dam erection was almost a national symbol of progress, especially as international concerns such as the World Bank made loans available for water-resource development. The colossal earth-shifting machines and cranes of some highly mechanized countries were offset in others by labour forces of up to 30,000 or 36,000 workers on a single dam. Richer nations vied and co-operated with each other to help and advise poorer nations in dam construction. During December 1962 there were at least 650 sizeable dams under construction, of which 217 were in the United States and 122 in Japan (United Nations, 1966b; Int. Comm., 1964; *World Almanac*, 1967, pp. 261–9). It is certain that the years 1930–70 will witness a greater in-crease in storage reservoirs and in irrigation acreage than has occurred in any previous century. Although many of the dams are multi-purpose and some are primarily for hydroelectric power, the post-1930 progress in water control for irrigation and other purposes may be judged on the growing size of barrages and reservoirs. In the 1930s the Hoover or Boulder Dam (221 m) was by far the tallest in the world. By the early 1970s it will be exceeded in height by at least thirteen others ranging up to 301 m (Inguri, U.S.S.R.). Most of these tall structures are of concrete, whereas the world's most massive dams are mainly earthfill or rockfill, usually with concrete sections. Of the twenty largest dams only Fort Peck, U.S.A. (125 million cubic yards; built in 1940) will pre-date 1950. Similarly, of the eleven huge barrages over 5 km long, only Fort Peck (6½ km) is more than thirty years old, and the barrage under construction at Kiev

R

will be 41 km in length. The world's greatest man-made lakes have become truly impressive. Whereas Lake Mead (Hoover Dam) led the way in 1936 with about 31 million acre-feet of water, by the early 1970s it will be surpassed by at least eleven reservoirs, irrespective of three other dams which already greatly increase the volume of natural lakes. The details in acre-feet are as follows: Kariba, Rhodesia, 130 million; Sadd-el-Aali (High Aswan), 127·3 million; Akosombo, Ghana, 120 million; Manicougan, No. 5, Canada, 115 million; Portage Mountain, Canada, 62 million; Krasnoyarsk, U.S.S.R., 59·4 million; Volga, V.I. Lenin, 47 million; Tabaga, Syria, 31·6 million. There are also eight other artificial reservoirs with a capacity of between 20 million and 28 million acre-feet. Such vast reservoirs allow interesting new possibilities of long-term storage.

No certain means exist of predicting at the start of a hydrological year the discharge for the coming year. The best that can be done from a long period of records, say of fifty or sixty years or more, is to find the mean annual trend, which varies appreciably with different decades (Moran, 1959; Hurst, Black, and Simaika, 1965 and 1966). However, the really large reservoir allows storage to meet most deficiencies. Thus the High Aswan or Sadd-el-Aali reservoir has four functions; to protect against dangerous floods; to produce hydroelectricity; to store sufficient water for use during the low-water stage; and to reserve excess water in abundant years to augment the supply in lean years. The total capacity at the maximum level of R.L. 182 m. is 157·4 milliard m^3 (1 million m^3 a day = 0·36 milliard m^3 a year). This is allotted as follows:

To silt trap	30 milliards	Reservoir Level 146 m
Over-year storage	90 milliards	Up to R.L. 175 m
To flood protection and annual storage	37·4 milliards	R.L. 182 m

As the silt-trap content remains dead storage, the over-year storage will be confined between R.L. 146 m and R.L. 175 m (content 120 milliards). The theory of this long-term storage is discussed by Hurst, Black, and Simaika [1965 and 1966]. The biblical seven fat years and seven lean years are now a legend; the only casualties may well be the marine life that thrived at the delta mouths.

With such storage expansion and river control it is not surprising that the irrigated acreage in the world has increased rapidly in the twentieth century. However, precise details are hard to obtain, especially as it is difficult to decide when the supplemental watering of crops causes them to be grouped under irrigation (Cantor, 1967; FAO, 1967, Table 2; Highsmith, 1965). In the late 1930s probably over 200 million acres were irrigated in the whole world. By the early 1960s this was estimated by some authors to have increased to about 370 million acres or 13% of the world's cultivated area. But, as Table 10.III.2 shows, the twenty-four leading countries for irrigation alone then had about 440 million acres irrigated, and if these returns were reasonably accurate the world total would be nearer 470 million acres. Irrespective of the accuracy of the global total, over 80% of all irrigated lands are in non-Soviet Asia, about 7·5% in the United States, and 5% in the U.S.S.R. The probable acreage under or available for irrigated crops in the early 1970s is also estimated in the Table. During the 1970s, if scheduled projects progress as expected, the world total under irrigation might well exceed 500 million acres. Remarkably little of this expansion will be in truly arid regions.

TABLE 10.III.2 Irrigable or irrigated land (millions of acres)

Country	Early 1960s	Early 1970s (estimated)*
Chinese Republic	183	190+
India	65	75
United States	35	37·5
U.S.S.R.	23·5	35
Pakistan	26·6	30
Indonesia	15·9	16
Iran	11·5	12
Iraq	9·1	9·5
Mexico	8·7	11
Japan	7·7	9
Italy	6·9	7·7
Egypt	6·1	7·5
France	6·2	6·5
Spain	4·6	5·8
Thailand	4·0	5·5
Turkey	3·2	4·5
Korea (Republic)	3·0	3·1
Argentina	2·8	3·8
Peru	2·8	3·2
Chile	2·7	3·1
Nepal	2·7	2·9
Australia	2·0	3·3
Sudan	2·0	3·0
Philippines	2·0	2·2

* Based on projects already planned or under construction. They may be rather conservative, for example, for Argentina.

By far the leading country for irrigation is the Chinese People's Republic, where large river-control schemes, much construction of hillside terracing, and rapid progress in mechanization, including electric pumping, have been, and still are being, undertaken. The reliability of the irrigation acreage given in the Table may be checked roughly by comparison with the area under paddy (for the most part irrigated). In 1966 the Chinese mainland had about 220 million acres under paddy.

The Indian subcontinent has well over 100 million acres under irrigated crops (cf. 115 million acres under paddy in 1965). Probably nearly a quarter of this is watered by wells or underground aquifers and, strictly speaking, should not be included in a discussion of the use of open channels. Here paddy (with jute in East Pakistan), millet, cotton, and corn are the leading hot season (*kharif*) crops, and wheat and vegetables the main cool season (*rabi*) crops. Several large new projects are under construction in the north-west, and a large scheme on the Mahanadi delta in Orissa. The Indus is probably the world's greatest river for irrigation. India and Pakistan have agreed upon a water-control scheme for it, and a large new dam has been built under this agreement. In Pakistan alone the river irrigates about 30 million acres, and the huge projects include the Sukkur barrage (completed in 1932), with 36,000 miles of main channels and distributaries supplying water to 2·6 million hectares (7·5 million acres); and, farther downstream, the Ghulam Mohammed barrage (headworks completed in 1955) that irrigates about 1,112,000 hectares (2·75 million acres), of which 40% is perennial (fig. 10.III.2).

In the United States, the third leading country for irrigation, the projects are mainly in the seventeen western states, the chief being California, with about 7·5 million acres. The crops include those of high dietary appeal or value, such as fruits and vegetables, for which there is an all-the-year-round demand by a nation endowed with a superabundance of basal foodstuffs. Cotton also is important, because of freedom from boll-weevil infestation and ease of machine picking. Fodder crops also play a large role in supplying safety and fattening for extensive grazing grounds (U.S. Dept. Agr., 1962). In 1939, when about 18 million acres were irrigated, the water withdrawal was about 63 million acre-feet, of which 83% came from surface streams. In 1959 the 34 million acres irrigated used 103 million acre-feet of water, of which only about 56% was derived from surface flow. The great modern interbasin water exchanges, especially on the Columbia and in California, have restored a wider and greater use of river water.

The fourth country for irrigation is the U.S.S.R., where large developments have progressed in the last thirty years, particularly in Soviet Central Asia near the Aral Sea. Here before 1960 the irrigators used mainly seasonal water from the smaller rivers fed by snow and rain (HN regime) and rising to flood stage in spring. Subsequently great projects have involved the larger rivers, the Amu-Dar'ya (or ancient Oxus) and Syr Dar'ya, which are fed largely by mountain snow and ice and maintain a high flood throughout the summer (HG and HN regime). It was calculated (Lewis, 1962; Vendrov *et al.*, 1964) that about 5 million hectares (12 million acres) could be irrigated here by 1965 and another 15 million hectares (37 million acres) by 1985. However, salinity and rapid seepage from distribution canals have been serious problems, and probably 'at the present level of efficiency, roughly 22 million acres is the ultimate amount of land that could be irrigated' in Soviet Central Asia (Lewis, 1962, p. 114).

Most of the countries with less than 20 million acres under irrigation have made considerable extensions since 1950. In Mexico, where cotton and sugar cane are important crops, modern methods have allowed some rehabilitation of older irrigated tracts rendered useless by excessive salinization. In Egypt the modern schemes include the Sadd-el-Aali dam, which is about 364 ft high,

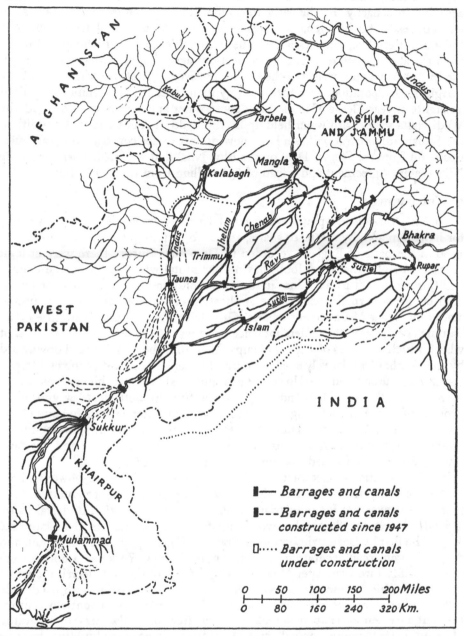

Fig. 10.III.2 Water Resource Development Projects in the Indus Basin in 1967.

The irrigated area extends within a few miles of the canal system. The territory was partitioned politically in 1947 and the international Indus Water Treaty was signed in 1960. The international boundaries are shown approximately by a pecked-dotted line. (Adapted from United Nations, 1966a, pp. 56 and 61.)

11,480 ft long, and holds back a reservoir 242 miles in length and 127 million acre-feet in capacity. By the early 1970s it will irrigate an extra 2 million acres and convert a further 700,000 acres from basin to perennial irrigation. The Sudanese Republic, with its large unused irrigable area, will benefit indirectly from the above scheme, as under reallocations agreed upon with Egypt in 1959, the Sudan Manaquil project of the Gezira receives 25% of the Nile water plus a margin for evaporation. In return the Egyptian (U.A.R.) Government acquired an eighteen years' lease of unused Sudanese water and permission to flood the Wadi Halfa valley with water held up by the Sadd-el-Aali barrage. In addition, the Sudan administration has a large new project under construction (132 miles upstream from the Sennar Dam on the Blue Nile) that was made possible by the Nile Waters Agreement and a loan of $32 million from the World Bank.

3. Water-power

From early times water-power was used to drive revolving wheels, such as *norias* for irrigation and, by means of shafts, stones for grinding cereals and pulses. In 1086 in the parts of England recorded in Domesday Survey there were at least 5,624 grist mills, many no doubt having horizontal wheels. During the twelfth and thirteenth centuries a minor industrial revolution was caused in some textile districts of western Europe by the application of water-power to drive fulling stocks by means of undershot wheels. Gradually where streams could be easily dammed or diverted the undershot wheel gave way to the more efficient overshot wheel, which consisted of a series of cups filled from above and forced downward by the weight of a relatively small amount of water. By the end of the eighteenth century in Europe and the United States quite elaborate machinery was water-driven, and the so-called Industrial Revolution in both textiles and iron-processing and manufacturing was based largely on water-power. In spite of the growing use of steam and later of internal-combustion engines, water-power remained significant in some areas well into the twentieth century and, for saw-mills and grist, is still used in a small way in some districts.

The introduction of electricity revolutionized the uses of water-power and led rapidly to the conversion of traditional wheels to electricity production. The first such conversion for industrial purposes was probably that undertaken by Aristide Bergès in 1869 in his woodpulp factory at Lancey in the Dauphiné Alps. The first hydroelectric station in France (of 900 h.p. on the Valserine), and the first commercial generating thermal stations in New York and London were built in 1882. Two years later Marcel Deprez began to transmit electrical current over appreciable distances from hydroelectric stations in France. Within a decade several other countries had small distribution networks. The twentieth century brought a tremendous expansion, particularly after 1918, when 24-hour working of plant became more general, and after 1945, when dam-building became extraordinarily popular. This expansion may be illustrated from Canada, where the hydroelectric generating capacity grew from 0·15 million kW in 1900 to 4·6 million in 1930, 7·8 million in 1949, 13 million in 1955, and 21·7 million kW in 1965 (Davis, 1957).

TABLE 10.III.3 Hydroelectricity production (million kWh) and installed capacity (in 1,000 kW) of leading producers

Country	Production 1965	1966	Hydro % of total electricity output (1965)	Installed hydro capacity
United States	197,001		17	44,492
Canada	117,063		81	21,711
U.S.S.R.	81,431		16	22,244
Japan	76,739		40	16,279
Norway	48,858		99·8	9,783
France	46,429	50,736	49*	12,683
Sweden	46,423		95	9,278
Italy	42,367	44,000	51*	13,955†
Brazil	25,515		85	5,391
Switzerland	24,015		98	8,120
Spain	19,550		62	8,141
Austria	16,083	17,327	73*	4,054
W. Germany	15,365	16,800	9*	4,072
India	14,807		41	3,331†
Finland	9,488	10,516	67*	1,857†
Yugoslavia	8,985		58	2,265
Mexico	8,609		50	2,327
New Zealand	8,588		81	1,910
Australia	8,367		23	2,092
United Kingdom	4,625		2	1,760
Czechoslovakia	4,456		13	1,540
Portugal	3,983		86	1,311†
Chile	3,954		64·5	710
S. Rhodesia	3,864		94	705
Colombia	3,218		63	793†
Peru	2,625		68	680
Turkey	2,167		44	510
Bulgaria	2,001		20	768

* For year 1966. † For year 1964.

In progressive societies living in countries deficient in coal and oil, hydroelectricity became the main motive force for factories as well as for domestic appliances and traction. Installation was cheapest where natural waterfalls occurred, as in glaciated uplands, where hanging valleys, ungraded rivers, and storage lakes abound. Thus in Norway about 60% of the main hydro sites have natural falls with drops ranging from 300 up to 1,008 m, while in Sweden and Finland, where the falls are lower, large natural reservoirs occur. Here and in many other mountainous and coal-deficient countries hydroelectricity has become the main source of power, and allowed the creation of manufacturing societies with a very high standard of living. In 1950 the percentage of total national electricity output provided by hydrogenerators was nearly 100 in Norway, 98 in Switzerland, 97 in Canada, 95 in Sweden and New Zealand, 83 in Italy, and 49 in France. On the other hand, in countries rich in other energy sources or lacking in waterfalls the hydro percentage of total national electricity output was 26 in the United States, 19 in West Germany, 3 in Great Britain, and zero in Denmark and the Netherlands. The situation in 1965 for all countries producing more than 2,000 million kWh of hydroelectricity annually is shown in Table 10.III.3 (United Nations, 1967, Tables 143 and 144).

Especially since 1918 electro-process industries have turned increasingly to large-scale hydro sources, such as the Niagara Falls. Certain industries are exceptionally greedy of power. Thus even in Canada, where large stations are favourable to cheap output, in the early 1950s the cost of energy was equal to nearly 11% of the value added by manufacture in paper products, 12·6% in non-ferrous metal products, and 14·6% in non-metallic mineral products. In the manufacture of newsprint the cost of energy amounted to 18% of the net value of the production; for aluminium and dissolving pulp the percentage was 27 and 30 respectively. Energy costs such as these attracted certain industries to hydro sites, and although their drawing power has weakened slightly today, they remain formidable, particularly for aluminium. In several countries two-thirds of the hydroelectric output is still used in the refining of non-ferrous metals, electro-chemicals, and pulp and paper manufacturing.

The world demand for electricity has grown so fast that it could be met only by greatly increasing the installed capacity of both thermal and hydro stations aided by advances in generating efficiency and in transmission. As modern hydroelectric plants already have a very high overall engineering efficiency, the future hope of improving the output per installed kW and the amount and cost per unit delivered to the customer lies in three methods.

First, in seeing that water flow or storage is adequate for capacity use all the year, especially where the regimes have a seasonal drop in discharge.

Second, in improving transmission. The long-distance bulk transmission of power at high voltages of 500 and 735–765 kV is being developed, especially in Canada, the United States, and Sweden. In the U.S.S.R. in 1963 the Bratsk hydro station on the Angara River in Siberia was brought up to 3·6 million kW installed capacity and was linked by a 500-kV line with Krasnoyarsk and the West Siberian power grids.

Third, in lowering the fixed costs, that is of initial constructions and installations. A large installation tends to be much cheaper than several smaller stations of an equivalent total capacity. Hence the tendency today to construct large storage reservoirs and/or long penstocks and tunnels. In Sweden, for example, it is common to utilize the head of a long stretch of rapids by building a reservoir and leading tunnels from it to a single large underground generating station. Since 1940 numerous investigations have been made into cheaper methods of dam construction and equipment installation, and certain companies and consortia have acquired a tremendous knowledge and experience (United Nations, 1957a). But the fixed costs remain high. In some of the huge Soviet schemes they

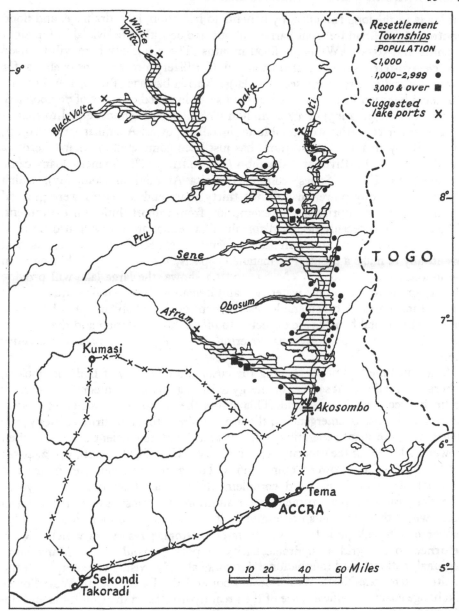

Fig. 10.III.3 The Akosombo Dam and the Volta River Project, Ghana, showing the main transmission lines (Adapted from Hilton, 1966).

have ranged from 12 up to 20% of the total outlay; in the small schemes of the Scottish Highlands the capital investment per hydro kW is two to four times that for conventional thermal generation; in Canada the large hydro units make the costs rather less unfavourable for water-power. However, the fact is that in many areas hydro power is either the only feasible local development or its high initial outlay is in the long run offset by benefits to irrigation, land drainage, and flood control. At the Sadd-el-Aali barrage about 210,000 kW are available for eight months and 50,000 kW for the flood months. The electricity here will be used for electro-ferrous and electro-chemical (fertilizer) processes as well as for pumping and general purposes. The large Kariba barrage (Reeve, 1960) on the Zambezi between Rhodesia and Zambia has an installed capacity of 705,000 kW and an output (in 1965) of 3,864 million kWh, which is almost equal to that of several other countries with double its installed power. A similar large project, also built by an Italian consortium, has just been completed in Ghana, across a gorge of the Volta River at Akosombo (fig. 10.III.3). The financial loans came mainly from United States' combines (Volta Aluminium Company), which guaranteed to buy power in bulk for thirty years and in return were granted freedom from expropriation and exemption from import duties on equipment. The power is especially designed for alumina reduction, but will also be used generally throughout southern Ghana, and is expected to cover costs and eventually to make a profit. The installed capacity is about 500,000 kW and can be increased to 768,000 kW. As fig. 10.III.3 shows, the large lake will provide fishing grounds, a 250-mile waterway, and irrigation for about 650 square miles around its shores. A considerable resettlement of the inhabitants will be necessary, but the total benefits are expected to offset this disruption and the inundation of an area of poor savanna covering about 3% of the national territory (Hilton, 1966).

The third method of cheapening the costs and efficiency of hydrostations is, where several different sources of energy exist in a region, to integrate all sources in an interconnected super-grid. This allows the economic exchange of power; mutual assistance in emergencies; the use of units with a low 'firm' capacity; the construction of giant generating stations; and the more efficient use of installed power by lessening the capacity kept in reserve (UNESCO, 1951b, pp. 224–54; OECD, 1961; Fed. Power Comm, 1965). Today in many countries hydro and thermal plants are considered complementary rather than rivals. As hydroelectricity can be rapidly turned on for addition to a transmission grid, even in areas where thermal plants predominate, it is becoming common to use surplus power of off-peak periods to pump water to storage reservoirs which can be returned to the grid as hydroelectricity at peak-demand hours. Many large regional, national, and international networks already exist. For example, in the United States the North West Power Pool and the Pacific South West Power Exchange ensure an efficient use of the great hydrosites on the Columbia and the Colorado. 'Power will be transmitted up to 870 miles . . . as far as southern California, at both 500-kilovolt alternating current and 750-kilovolt direct current' (Fed. Power Comm., 1964, p. 35). In the State of Washington the fifty largest hydrostations have a total installed capacity of over 10 million kW, and the Grand Coulée alone has 1,974,000 kW installed and a potential of 5,574,000 kW. The advantage of such large units are obvious when compared, for example, with those in the Scottish Highlands, where, owing to the small drainage basins, the building of more than fifty 'main' hydrostations has resulted in a total installed capacity of only 1,047,000 kW. In the latter area, a glaciated upland of 21,000 square miles and about 1 million inhabitants, the North of Scotland Hydro Electric Board was pledged to aid social and economic development generally. By 1966 it had supplied 96% of the premises, many of them remote and isolated dwellings, with electricity, but by then the total capital expenditure had risen to over £184 million. Thus the cost of firm power was several times that in countries with larger and more concentrated hydro resources (H.M.S.O.).

Sweden and France provide excellent examples of national integration schemes. In Sweden the bulk of the hydroelectricity is sent several hundred miles to the south, while in France the electricity production is predominantly

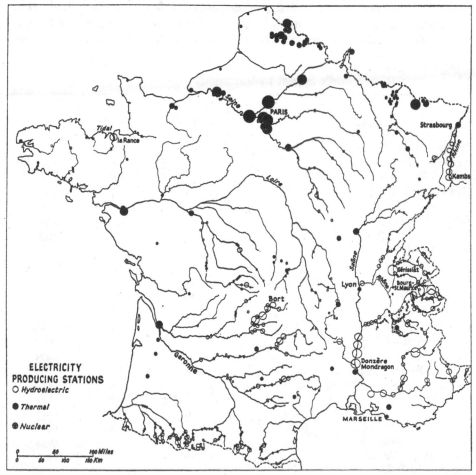

Fig. 10.III.4 Electricity Generation in France in the early 1960s, showing main thermal and hydrosites.

The circles are approximately proportional to the annual output, and may be judged from Kembs 840 million kWh and Donzère–Mondragon 1,890 million kWh. Small circles denote under 100 million kWh.

hydro in the south and thermal in the north (fig. 10.III.4). Still larger integration on an international scale is becoming common. All the Scandinavian countries are linked by transmission lines, and that between Sweden and Denmark, today of five cables, dates back forty years. In eastern Europe (Kish, 1968) the unified energy system or power grid of COMECON (the economic co-operation of the Soviet bloc in East Europe) has a total installed capacity of 35,000 MW.

Special interest in the world today centres on multiple-purpose dams, but irrespective of the variety of purposes, it is usually hydroelectricity that is the money earner, at least until other projects, such as elaborate irrigation, come into full operation. Many countries today finance construction schemes by loans from national and international agencies such as the World Bank, the International Monetary Fund, and the Inter-American Development Bank. This partly explains the unparalleled growth since 1950 of the number and global distribution of large hydrostations. Up to about 1950 the Hoover Dam (1,345,000

TABLE 10.III.4 World's largest hydrostations

Name	Capacity in 1,000 kW		Year of initial operation
	Installed	Ultimate	
Sayansk, U.S.S.R.		6,000	U.C.
Krasnoyarsk, U.S.S.R.		6,000	1967
Churchill Falls, Canada		6,000	U.C.
Grand Coulée, U.S.	1,974	5,574	1941
Sukhovo, U.S.S.R.		4,500	U.C.
Bratsk, U.S.S.R.	3,600	3,600	1961
Solteira Island, Brazil		3,200	1969
John Day, U.S.A.		2,700	1968
Nurek, U.S.A.		2,700	1970
Volgograd (22nd Congress) U.S.S.R.	2,543	2,560	1958
Kubyshev (V.I. Lenin), U.S.S.R.	2,100	2,300	1955
Portage Mountain, Canada		2,300	U.C.
Iron Gates, Romania-Yugoslavia		2,160	U.C.
Sadd-el-Aali, U.A.R.		2,100	1967
Mica, Canada		2,000	1975
Robert Moses-Niagara, U.S.A.	1,954	1,954	1961
St. Lawrence Power Dam, U.S.A.–Canada	1,880	1,880	1958
Guri, Venezuela		1,757	U.C.
The Dalles, U.S.A.	1,119	1,743	1957
Chief Joseph, U.S.A.	1,024	1,728	1956
Kemano, Canada	835	1,670	1954
Beauharnois, Canada	1,586	1,641	1951
Ingari, U.S.S.R.		1,600	1970
Kariba, Rhodesia	705	1,400	1959
Jupia, Brazil		1,400	1966
Sir Adam Beck, No. 2, Canada	900	1,370	1954
Hoover, U.S.A.	1,345	1,345	1936

kW) and the Grand Coulée Dam (1,947,000 kW installed) were unrivalled in magnitude. Since then at least twenty-five other stations have been started or completed with an ultimate capacity of over 1,345,000 kW and another fifteen of between 1 million and 1¼ million kW. In Table 10.III.4 blanks under the installed capacity indicate that the statistics were not available; U.C. denotes under construction (United Nations, 1966b).

The increasing demand for electricity makes the unused potential hydro power of considerable significance. Table 10.III.5 shows that enormous amounts of unharnessed water-power exist in Africa, South America, and Asia. Even in North America and in Europe, except in a few areas, the unused potential is large (Young, 1955; Hubbert, 1962, p. 99).

TABLE 10.III.5 Estimated world's water-power capacity (In 10^3 MW)

Region	Potential	% world total	Developed (1961)	% potential
Africa	780	27	2	0·3
South America	577	20	5	0·9
U.S.S.R., China, and Satellites	466	16	16	3·5
South East Asia	455	16	2	0·4
North America	313	11	59	19·0
Western Europe	158	6	47	30·0
Australasia	45	2	2	4·4
Far East	42	1	19	45·0
Middle East	21	1	—	—
WORLD	2,857	100	152	5·3

The future situation for water-power is largely a question of the cost per unit compared with that of thermal power sources, of the availability of capital for the initial installation, and of the possibilities of regional long-distance power exchanges. But hydrogenerators retain certain important advantages. They are non-pollutive, both of river water and of the atmosphere; they do not raise the water temperature; they have a relatively long life and are fairly easily maintained. In the United States the installed hydro capacity is expected to rise from about 40 million kW in 1963 to 80 million by 1981, but at the same time the proportion that water-power supplies to the total national electricity production will decline from 18 to about 15%. On the other hand, in many other parts of the world hydroelectricity will supply a steadily increasing percentage of the energy output.

4. Regional and inter-regional water-resource schemes

The majority of large water-resource schemes and dams are today multi-purpose and aim at controlling river flow for the best advantage of the economy of a drainage basin. The earliest comprehensive schemes were on the Rhône and

Tennessee. In 1921 the French Parliament laid down the principles of developing the Rhône for water-power, navigation, and irrigation. The expenses were to be met by the sale of electricity. But before the scheme got under way the Boulder Canyon Project Act in the United States (passed in 1928 and working under a multiple-purpose Federal Water Power Act of 1920) included in its terms of reference hydroelectricity, navigation, irrigation, flood and sediment control, and water supply for southern California. Here, too, electricity was the money earner, and indeed soon made the scheme viable. The trend towards multiple-purpose planning in the United States culminated in 1933 in the creation of the Tennessee Valley Authority, which soon produced a most highly developed drainage-basin scheme. Here twenty-eight multiple-purpose reservoirs are operated as a unit and the whole basin from watershed downward is under reasonable control (Patchett, 1943; T.V.A., 1963). A start had been made on this scheme when in 1934 the French *Compagnie Nationale du Rhône* was founded. The C.N.R. was a form of limited-liability company, with the Government acting as security for loans floated with the public. It lacked the broad powers and speed of decision enjoyed by the T.V.A., which consequently achieved its aims much more quickly. However, the C.N.R. scheme is nearing completion, and it and the T.V.A. remain the prototypes of similar and larger schemes elsewhere. The annual operation and maintenance costs and the benefit–cost relationships show that integrated basin development with large multi-purpose dams is in the long run a profitable investment that no riverine societies can afford to be without. By 1968 numerous countries were planning, or constructing, or had completed river-basin development schemes (United Nations, 1955–61, 1957b, 1958, 1960, 1962, and 1966a; White, 1962 and 1963). Among the many examples are the Damodar Valley Project in India; the Kitakami River Basin Project in Japan; and the Helmand Valley Authority in Afghanistan. The last-named includes most of southern Afghanistan and involves the reclamation and irrigation of a large area of desert and the provision of 'educational institutions, sanitation and public health centres, modern housing, resettlement, particularly for the nomadic tribes, cottage and small industries etc. in order to raise the levels of living of the people in general' (U.N., 1955–61, *Part 2D*, 1961, p. 8). The United States has since 1949 financed the building of eight dams on the Helmand River for hydroelectricity and irrigation. When finished the plan will restore to the area the rich agriculture that flourished centuries ago before over-cultivation and soil erosion turned it into desert.

These successful integrated socio-economic schemes have been for basins within a single national territory, whereas the basins of most great rivers, except in parts of the U.S.S.R., extend into several different political states, so that integrated basin projects need international agreements. The problem of international basin law is discussed on pp. 351–4. Here it may be noticed that development is much more difficult when only part of a basin can be controlled. The great schemes of ECAFE (Economic Commission for Asia and the Far East) began with flood control in 1949, and by 1952 had changed to multi-purpose river-basin development of water resources. The Lower Mekong River Project applies only to the lower course of a river which rises in China and drains 810,000 km² compared with the 99,000 km² of the Rhône. Yet even this restricted project involves national ownership by Thailand, Laos, South Viet-Nam, and Cambodia. Already over $100 million have been spent here on two large dams in Thailand and other improvements by twenty-one Western countries and the four riparian states. How successful international co-operation in drainage-basin development can be is also seen in the Columbia River agreement between the United States and Canada (U.S. Bur. Rec., 1941; Sewell, 1966), and the Indus agreement between India and West Pakistan. How unfortunate disagreement can be is seen in the Jordan basin, where Israel and the state of Jordan have constructed rival irrigation schemes which are extremely costly and mutually harmful (Smith, 1966).

5. Inter-basin water transferences

The twentieth century has popularized the technique of transferring water on a large scale from one drainage basin to another, with tremendous effects on the socio-economic geography at least of the receiving basin. In California about 70% of the water supplies are in the north, whereas 77% of the water needs are in the southern two-thirds of the state. Under the State Water Plan a 444-mile aqueduct leads water from the Sacramento–San Joaquin delta to the central and southern districts. More exciting is the transference of water across or through high watersheds, such as from the Colorado to Los Angeles. Frank Quinn (1968) has shown that the American West, which in the nineteenth century was won largely by the adjustment of rural communities to limitations of local water supply, now receives enormous quantities of water by interbasin transfers. 'One out of every five persons in the Western states is served by a water-supply system that imports from a source a hundred miles or more away. In total tonnage the amount exceeds that carried by all the region's railroads, trucks, and barge lines combined.'

Transference between basins draining to different oceans is becoming increasingly common. In the Andes at least one scheme in Chile and two schemes in Peru achieve a transference from the Atlantic to the Pacific. In New South Wales, Australia, the Snowy Mountains scheme leads water from the eastward Pacific slope to the westward Indian Ocean slope by means of two systems of tunnels. When completed in the early 1970s the nine power stations and seventeen dams will provide about 4 million kW at peak loads and about 2 million acre-feet of water, or enough to irrigate 640,000 acres of dry land in the Murrumbidgee–Murray basins.

Most of the water involved in these interbasin transference schemes is reservoir storage that would have run to waste in time of flood, and do not adversely affect the streams being used as feeders. But in at least one case, the Paraiba do Sul near Rio de Janeiro, a large water diversion (to feed a hydroelectric station outside the basin) has led to a serious loss of water in its lower channel.

Interbasin and inter-regional water transferences are in their infancy. Vast amounts of unharnessed hydroelectric power can in some countries probably be best used for pumping water long distances. Many nations have not yet realized what a desirable product water is to sell and what markets exist for distant pipe-line distribution. Water purchasers always have an assured exchange, at least in agricultural crops. The conjunction of Congo and Sahara offers more possi-bilities to mankind than journeys to the moon, and is altogether a much simpler and less costly project.

REFERENCES

BRUCE, J. P. and CLARK, R. H. [1966], *Introduction to Hydrometeorology* (Pergamon Press, London), 319 p.

CANTOR, L. M. [1967], *A World Geography of Irrigation* (Edinburgh), 252 p.

DAVIS, J. [1957], *Royal Commission on Canada's Economic Prospects: Canadian Energy Prospects* (Ottawa).

F.A.O. [1967], *Production Yearbook 1966*; Vol. 20 (Rome).

FEDERAL POWER COMMISSION (UNITED STATES) [1965], *44th Annual Report* (Washington, D.C.).

HIGHSMITH, R. M. [1965], Irrigated lands of the world; *Geographical Review*, 55, 384–7.

HILTON, T. E. [1966], The Akosombo Dam and the Volta River Project; *Geography*, 51, 251–4.

H.M.S.O., *Electricity in Scotland* and *North of Scotland Hydro-Electric Board Annual Reports* (Edinburgh).

HUBBERT, M. K., Editor [1962], *Energy Resources*; National Research Council Publica-tion 1000-D, National Academy of Science, Washington, D.C.

HURST, H. E., BLACK, R. P., and SIMAIKA, Y. M. [1965], *Long-Term Storage: An Experi-mental Study* (Constable, London).

HURST, H. E., BLACK, R. P., and SIMAIKA, Y. M. [1966], *The Major Nile Projects: The Nile Basin;* Vol. 10 (Cairo, U.A.R.) (see pp. 54–156).

INTERNATIONAL COMMISSION ON LARGE DAMS [1964], *World Register of Dams*.

KISH, G. [1968], Eastern Europe's power grid; *Geographical Review*, 58, 137–40.

LEWIS, D. J. and SHOEMAKER, L. A. [1962], Hydro-system power analysis by digital computer; *Proceedings of the American Society of Civil Engineers, Journal of the Hydraulics Division*, 88, 113–30.

LEWIS, R. A. [1962], The irrigation potential of Soviet Central Asia; *Annals of the Association of American Geographers*, 52, 99–114.

MICHEL, A. A. [1967], *The Indus Rivers* (Yale).

MORAN, P. A. P. [1959], *The Theory of Storage* (London), 110 p.

O.E.C.D. [1961], *Power System Operation in the U.S.A.*

PRITCHETT, C. H. [1943], *The Tennessee Valley Authority* (New York).

QUINN, F. [1968], Water transfers: Must the American West be won again?; *Geo-graphical Review*, 58, 108–16.

REEVE, W. H. [1960], The Kariba Dam; *Geographical Journal*, 126, 140–6.

ROCKWOOD, D. M. [1961], Columbia Basin streamflow routing by computer; *Trans-actions of the American Society of Civil Engineers*, 126, 32–56.

SEWELL, W. R. D. [1966], The Columbia River treaty; *Canadian Geographer*, 10, 145–56.

SMITH, C. G. [1966], The disputed waters of the Jordan; *Transactions of the Institute of British Geographers*, No. 40, 111–28.

T. V. A. [1963], *Nature's Constant Gift: A Report on the Water Resources of the Tennessee Valley* (Knoxville, Tennessee).

UNITED NATIONS [1955–61], *Multiple-Purpose River Basin Development* (ECAFE). (Parts 1, 2A, 2B, 2C, 2D being Flood Control Series Nos. 7, 8, 11, 14, 18.)

UNITED NATIONS [1957a], *Bibliographical Index of Works Published on Hydro-Electric Plant Construction* (Geneva).

UNITED NATIONS [1957b], *Development of Water Resources in the Lower Mekong Basin*; Flood Control Series No. 12 (New York).

UNITED NATIONS [1958], *Integrated River Basin Development* (New York), 60 p.

UNITED NATIONS [1960], *A Case Study of the Damodar Valley Corporation*; Flood Control Series No. 16 (New York).

UNITED NATIONS [1962], *A Case Study of the Development of the Kitakami River Basin*; Flood Control Series No. 20 (New York).

UNITED NATIONS [1966a], *Compendium of Major International Rivers in the ECAFE Region*; Water Resources Series 29 (New York).

UNITED NATIONS [1966b], *Fourth Biennial Report on Water Resources Development*; 40th Session, Supplement No. 3 (New York).

UNITED NATIONS [1967], *Statistical Year Book 1966* (New York).

U.N.E.S.C.O. [1951a], *Water Resources*; Vol. 4 (New York).

U.N.E.S.C.O. [1951b], Conservation and utilization of resources; Vol. 3 of *Fuel and Energy Resources* (New York).

U.S. DEPARTMENT OF AGRICULTURE [1962], *Land and Water Resources* (Washington, D.C.).

U.S. GOVERNMENT [1950], *A Water Policy for the American People;* 3 vols. (Washington, D.C.).

U.S. BUREAU OF RECLAMATION [1941], *Columbia Basin Joint Investigations* (Washington, D.C.).

VENDROV, S. L. *et al.* [1964], The problem of transformation and utilization of the water resources of the Volga River; *Soviet Geography*, 4, 23–34.

WHITE, G. F. *et al.* [1962], *Economic and Social Aspects of the Lower Mekong Development.*

WHITE, G. F. [1963], Contributions of geographical analysis to river basin development; *Geographical Journal*, 129, 421–32.

World Almanac [1967] (New York).

YOUNG, L. L. [1955], Developed and potential water power in the world; *U.S. Geological Survey Circular* 367 (Washington, D.C.).

CHAPTER 11

Long-term Trends

I. Long-term Precipitation Trends

R. G. BARRY

Department of Geography, University of Colorado

In Chapter 3.1(i) some indication was given of the variability of precipitation amounts, but the question of long-term trends was not explicitly discussed. This is clearly of relevance to our consideration of geomorphic processes and man's water requirements. Direct study of precipitation changes is unfortunately very restricted by the limited availability of precipitation records exceeding 100 or even 50 years duration, and it is vitally important, therefore, that such information as is available be analysed and interpreted correctly. For this reason we begin with a summary of the terminology of climatic change and of the simpler methods of analysing time series. Fuller details may be found in Mitchell *et al.* [1966], which provides the basis for the following account.

1. The terminology of climatic change

Three basic types of change can be distinguished. They are: a *discontinuity* – an abrupt and permanent change in the average value; a *trend* – a smooth increase or decrease, not necessarily linear, of the average; a *fluctuation* – a regular or irregular change characterized by at least two maxima (or minima) and one minimum (or maximum). Where a climatic fluctuation progresses smoothly and gradually between the maxima and minima it is termed an *oscillation*, and if the maxima and minima recur after approximately equal time intervals it is referred to as a *periodicity*.

It will be noted that these definitions do not incorporate any reference to time scale. Changes occurring during the instrumental period, on a scale of $10–10^2$ years, are termed '*secular*', those of the order of 10^3 years '*historical*'.

2. Time series

The year-to-year variability of precipitation totals may conceal long-term changes of one kind or another in a data series, and statistical techniques are necessary to suppress the short-term irregularities. The simplest method is the calculation of a *running mean* (or *moving average*), where mean values are determined for successive, overlapping periods of five, ten, or thirty years. For example, in the five-year case

$$\frac{P_1 + P_2 + P_3 + P_4 + P_5}{5} = \bar{P}_3$$

$$\frac{P_2 + P_3 + P_4 + P_5 + P_6}{5} = \bar{P}_4$$

where P_1 = precipitation in year 1 of the series;

$\bar{P}_3$ = running mean value for year 3.

More generally,

$$\bar{P}_x = \sum_{i=-n}^{n} P_{x+i}/2n + 1 \qquad (1)$$

where $\bar{P}_x$ = running mean value for x^{th} term of the series;

$2n + 1$ = number of terms in the running mean;

$\sum_{i=-n}^{n}$ = summation of the terms from $i = -n$ to $i = n$.

Figure 11.1.1 illustrates the smoothing effect of various running means for annual precipitation totals at Omaha, Nebraska. In the case of the five-year series there is a tendency for displacement and apparent inversion of the shorter fluctuations, but in general the effect is to clarify the long-term changes. There

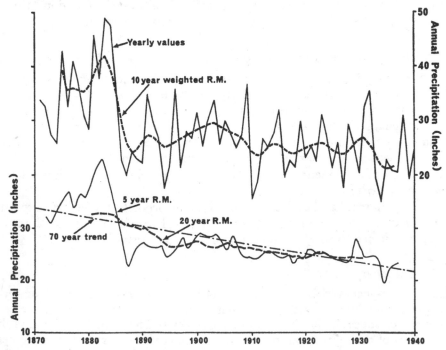

Fig. 11.1.1 Annual precipitation at Omaha, Nebraska, 1871–1940, and the smoothing effects of running means (Based on Foster, E. E., *Rainfall and Runoff*, 1949).

is a risk that some of the prominent longer fluctuations in the smoothed series arise from random variations, and indeed an original series of wholly random data can show quasi-cyclical rhythms after smoothing (Lewis, 1960). Simple statistical tests are available to assist the interpretation of smoothed series (Craddock, 1957).

Modification of this smoothing technique can give better results and avoid the inversion of the shorter fluctuations, noted above. For example, it is often useful to use a weighted running mean. This is termed a *low-pass filter*, since it retains changes of long wavelength and filters out the short ones. The procedure for determining appropriate weights cannot be dealt with here, but the following table illustrates a set of weights for studying variations with a wavelength longer

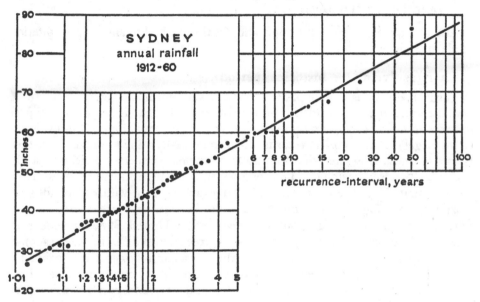

Fig. 11.1.2 Annual precipitation at Sydney, Australia, 1912–60, illustrating the calculated return periods (From Dury, 1964).

than 10 terms in the data series. The weights are based on the Gaussian (Normal) frequency distribution.

$i =$	-4	-3	-2	-1	0	1	2	3	4	Sum
Weight =	0·01	0·05	0·12	0·20	0·24	0·20	0·12	0·05	0·01	1·00

Figure 11.1.1 shows the application of this filter to the Omaha precipitation data. It is clear that this procedure gives more satisfactory smoothing than the simpler unweighted running means. The amplitude and time-phase of the fluctuations are accurately represented.

So far we have been concerned only with series of average values. Often, a more important question is the frequency of extreme values. A useful, basic technique is the analysis of *return period* (or recurrence interval); that is, the average time interval within which a rainfall, or flood, of specified amount or intensity can be expected to occur once. For example, a 10-year annual rainfall is the probable annual maximum once in a decade and the 100-year annual rainfall once in a century. This does not imply that the spacing is regular. A series of 70 years' data may well include the 100-year maximum and even the 500- or 1,000-year maximum. The method is as follows:

1. Rank the n observations in order of magnitude with the highest as 1 and the lowest as n.
2. The return period $= (n + 1)/r$, where r is the rank of a particular observation.
3. Plot the actual values against their computed return periods on semi-logarithmic graph paper.

Figure 11.1.2 demonstrates the application of the method to annual precipitation totals at Sydney 1912–60.

3. Changes during the historical period

Regular observations of precipitation amount were not made before about 1850, and therefore the only evidence of precipitation changes is for the most part indirect and of rather uncertain reliability. For the last millennium this evidence stems largely from historical records of natural disasters due to unusual weather events and also from tree-ring sequences of ancient timbers – a field of study known as dendroclimatology. Tree-ring width is partly a function of the tree's environment. Near the arctic tree-line, for example, ring width responds primarily to growing-season temperatures, but in semi-arid areas growth is a reflection of moisture conditions. In the western United States a wide ring usually indicates a cool, moist year. However, tree species differ in their response to particular climatic parameters and the seasonal regime. Dendroclimatic studies have made it possible to reconstruct changes in moisture conditions since the fifth century A.D. for locations in Arizona and Colorado, and Fritts has prepared maps of regional patterns of moist, cool or dry, warm decades since 1500 in western North America. More tentative indications of moister and drier periods during the last 10,000 years may be gleaned from analysis of peat-bog stratigraphy, especially pollen profiles preserved in the bogs. This evidence, like that of tree rings, is indicative of the moisture budget P-E, rather than precipitation alone. Moreover, the bog stratigraphy may reflect the local hydrological regime rather than climatic changes. For this reason we shall limit the discussion here to the surer documentary material.

Studies by Lamb [1965] of the historical records of good or bad harvests, of floods or droughts, and so on provide an index of 'summer-wetness' (July–August) in Europe since A.D. 1100 (fig. 11.1.3(a)). Each month scores 0 for drought evidence, $\frac{1}{2}$ if unremarkable, or 1 for indications of wetness. Decadal extremes of the index are 4–17, with a value of about 10 for unremarkable ones. A more tentative extrapolation has been made for the three centuries back to A.D. 800. Figure 11.1.3(b) shows the 50-year average rainfall as a percentage of

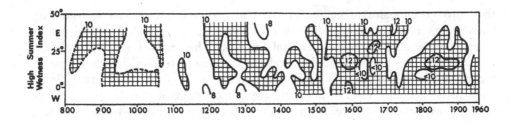

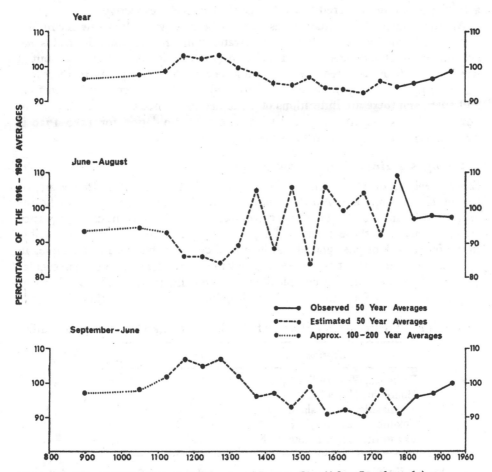

Fig. 11.1.3 Precipitation trends in Europe since A.D. 800 (After Lamb, 1965).
See text for details.

(*Top*). 'Summer-wetness' (July–August) in Europe near 50° N since A.D. 800.
(*Below*). Fifty-year averages of rainfall in England and Wales expressed as percentages of the mean for 1916–50.

the 1916–50 annual average for England and Wales. This series is based on correlations between decade values of rainfall since 1740 and of annual or winter temperature averages in central England. The reason for this correlation is the association between mild periods in the winter half-year with frequent, moist, south-west winds. Winter cold periods occur with drier anticyclonic conditions. Summer rainfall amounts are independent of annual precipitation totals, and have therefore been derived from the summer-wetness index. Summers appear to have been dry and fine between about 1150 and 1300 and wetter about 1600–1700, corresponding with an early medieval warm period and a cold epoch, often referred to as the 'Little Ice Age', respectively.

Another source of information has recently been provided by snow accumulation data from the South Pole. The stratigraphic record only indicates *net* accumulation – the effects of wind deflation or deposition are unknown – but at least the evidence is direct. Mean annual accumulation apparently increased significantly from $5 \cdot 4$ g cm^{-2} between 1760 and 1825 to $7 \cdot 5$ g cm^{-2} between 1892 and 1957, and there are indications of some decrease since the maximum about 1920–30. Deeper layers suggest that the mean accumulation for 1550–1750 is more or less the same as that for 1760–1957.

4. Changes during the instrumental period

Since precipitation measurements cover a period of about a hundred years, or less outside Europe and North America, only short-term changes can be investigated in any detail. There is, therefore, a risk that too much significance may be attached to these changes. They need to be interpreted as far as possible in the framework of post-glacial and historical climatic change. For example, an apparent trend in a short-period record may in fact be part of a long-term oscillation. At the same time the complexities of recent fluctuations (Veryard, 1963) help to prevent the formulation of over-simplified views about earlier changes.

TABLE 11.1.1 Percentage departure from the 1881–1940 mean annual rainfall

Station	1874–98	1907–31
Barbados, West Indies 13° N	14	−9
Honolulu, Hawaii, 21° N	13	−12
Townsville, Queensland, 19° S	17	−4
Colombo, Ceylon, 7° N	4	−8
Freetown, Sierra Leone, 8° N	11	−12

In the tropics there is evidence of a marked decrease of annual rainfall over extensive areas at the end of the nineteenth century. This is illustrated by Table 11.1.1 from Kraus [1955]. The greater aridity in the tropics in the early twentieth century was matched by rainfall decrease in south-east Australia and eastern North America, pointing to a general weakening of the moisture cycle in lower latitudes. Work in the Mediterranean area by Butzer [1957] strengthens this impression. Figure 11.1.4 shows that the Saharan–Arabian desert and the

Mediterranean region became much drier in the present century, whereas the North Atlantic, north-west Europe, and southern Russia experienced an increase of annual precipitation until about 1940. There is evidence that the trend towards drier conditions in the tropics was reversed during the 1930s, and in eastern North America and eastern Australia there has been a post-1940 increase in rainfall, apparently caused by more frequent tropical cyclones. It is significant that the changes have affected areas close to climatological boundaries, whereas the climate in 'core areas' such as Antarctica and the equatorial rain-forest seems to have remained more or less constant over long periods.

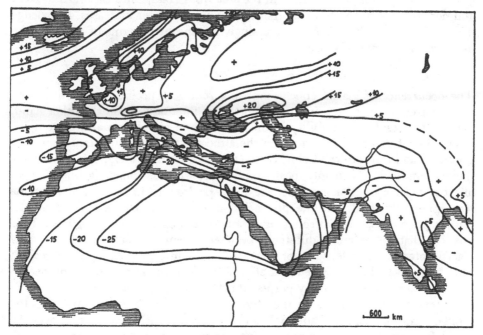

Fig. 11.1.4 Annual precipitation anomalies 1881–1910 to 1911–40, expressed as percentage deviations from the mean 1881–1910 (From Butzer, 1957).

Within the general trend of annual or seasonal averages less obvious, but none the less important, changes may occur. For example, the mean annual totals at Madras for 1813–80 and 1881–1940 are identical, but while the minor rains of May–July decreased after about 1890, the main October–December rains showed the opposite trend. Howe *et al.* [1966] demonstrate a marked increase in frequency of storm rainfall (daily totals of at least 2·5 in.) in mid-Wales during 1940–64 compared with 1911–40, while annual totals show only small changes. This caused an increase in flooding hazard. In New Mexico, Leopold [1951] found a steady increase in the frequency of daily rainfalls of less than 0·5 in. from 1850 to 1930 or 1940, and a subsequent decrease, without any variation in total rainfall. Such changes in semi-arid areas can be critical for plant growth, runoff, and erosion processes.

Precipitation fluctuations may assume economic significance, even in Britain. The period December 1963–February 1964 was the driest over England and Wales for more than 250 years, and if the September–February means are considered the years 1962–3, 1963 4, and 1964–5 had the lowest triple total, 45·1 in. (114·5 cm) for England and Wales, since 1757–60, when 44·6 in. (113·3 cm) was recorded. Dry winters lead to acute soil-moisture deficits in the following spring and summer, and to the likelihood of water restrictions in the south-eastern half of the country. Rodda [1965] estimates a 35–40% chance of a theoretical soil-moisture deficit (assuming that the potential evaporation rate is maintained) of 5 in. or more during the summer, in the area from Hampshire eastward to Kent and north-eastward to the Fens. As the demand for water increases, knowledge of such drought risks becomes increasingly significant in terms of planning the exploitation of water resources.

5. Causes of climatic change

The occurrence of glacial epochs probably requires specific causative factors, either terrestrial or extraterrestrial. Short-term fluctuations, on the other hand, may be the result of instability in the atmospheric circulation and of complex interactions between the oceans and the atmosphere. One important feature of the space and time scales of climatic fluctuations is the tendency for short-term fluctuations to be of mainly local significance, whereas longer-lasting changes affect a wider area. For example, the 1939–44 drought in south-east Australia was apparently local, but the protracted dry conditions in eastern Australia during 1896–1915 also affected other east-coast and tropical regimes. It is possible that the essential difference between recent fluctuations and those of glacial–interglacial scale is one of increased duration rather than greater magnitude of change, although 'feedback effects' between the atmosphere and oceans may well amplify and thereby perpetuate the initial change.

Important contributions to the study of recent drought conditions in the United States have been made by Namias [1960, 1966]. He shows that over the Great Plains a warm, dry spring tends to be followed by a warm, dry summer, whereas a cold spring is likely to be succeeded by a cold, wet summer. The physical mechanism involved is not yet clear. Drought over the north-eastern United States during 1962–5 seems to have been caused by below-normal sea surface temperatures offshore in spring and summer. This leads to increased cyclonic activity over the zone of maximum sea-surface temperature gradient, thereby strengthening dry, cool, north and north-westerly airflow over land. Corroboration of this idea is shown by the fact that three wet years, 1951–3, were associated with positive sea-temperature anomalies.

Precipitation changes with a time-scale of 50–100 years appear to involve the atmospheric circulation over at least the major part of one hemisphere. Kraus and Butzer suggest complementary explanations for the precipitation changes in the tropics and subtropics outlined earlier. In the tropics there seems to have been a shorter wet season associated with a narrowing of the intertropical convergence zone. The subtropical high-pressure cells apparently intensified and expanded both polewards and equatorwards, while cyclones in the westerlies of the northern hemisphere tended to shift northward. This represents a more zonal circulation, a tendency which has been drastically reversed since about 1940. In Mexico, however, Wallen [1955] found evidence of a significant increase in

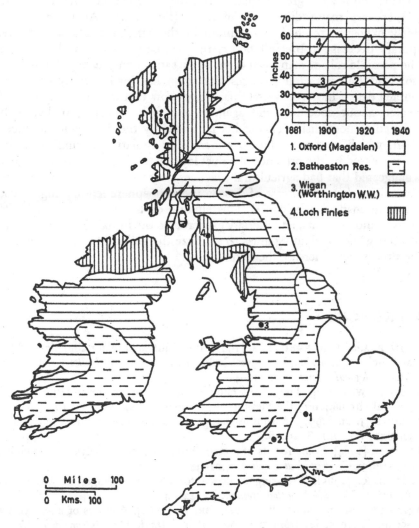

Fig. 11.1.5 Regions of annual rainfall fluctuations in the British Isles and their characteristic trends 1881–1940 (From Gregory, 1964).

annual, and especially July, precipitation from the 1900s to about 1930. This regional anomaly in the tropics need not invalidate the hypothesis of Kraus, but it does illustrate the necessity for cautious generalization. Similarly, Lamb [1967] demonstrates that regions exposed to prevailing winds from an adjacent ocean display a pattern of increased precipitation between about 1900 and 1940, corresponding to the period of strong wind circulation noted above, and a subsequent decline. Over England and Wales, for example, frequent westerly airflow, and therefore greater atmospheric transport of moisture, was experienced during 1910–40, with decadal means of precipitation for the 1910s and 1920s about 25% greater than in the 1880s. However, even in the British Isles there are major differences of trend (fig. 11.1.5). Areas with a north-westerly exposure reached a maximum in the early 1900s and underwent a temporary decrease of precipitation c. 1913–22 before rising again.

The possible role of man as an agent of climatic change cannot be overlooked. Bryson and Baerreis [1967] suggest that man has inadvertently extended the Rajputana desert in north-west India. They show that the area was successfully cultivated about 2000 B.C. and again in the fifth century A.D. before being abandoned to nomadism. The aridity is caused by the dominance of subsiding air aloft, which in turn is due to atmospheric radiative cooling. The cooling rate is increased by the presence of a dust layer extending up to 9 km (30,000 ft). It is suggested that bad agricultural practices may have facilitated wind removal of the soil, so that an increase of dust in the atmosphere led to a higher cooling rate and intensified subsidence. This sinking of the air reduces precipitation frequency and amounts, and thereby exacerbates the initial tendency. It remains to be seen whether planting grass to stabilize the soil might help to reverse the whole process and increase the rainfall.

REFERENCES

BRYSON, R. A. and BAERREIS, D. A. [1967], Possibilities of major climatic modification and their implications; Northwest India, a case for study; *Bulletin of the American Meteorological Society*, **48**, 136–42.

BUTZER, K. W. [1957], The recent climatic fluctuation in lower latitudes and the general circulation of the Pleistocene; *Geografiska Annaler*, **39**, 105–13.

CRADDOCK, J. M. [1957], A simple statistical test for use in the study of climatic change; *Weather*, **8**, 252–8.

DURY, G. H. [1964], Some results of a magnitude-frequency analysis of precipitation; *Australian Geographical Studies*, **2**, 21–34.

FRITTS, H. C. [1965], Tree-ring evidence for climatic changes in western North America; *Monthly Weather Review*, **93**, 421–43.

GIOVINETTO, M. B. and SCHWERDTFEGER, W. [1966], Analysis of a 200-year snow accumulation series from the South Pole; *Archiv für Meteorologie, Geophysik, und Bioklimatologie, A*, **15**, 227–50.

GREGORY, S. [1956], Regional variations in the trend of annual rainfall over the British Isles; *Geographical Journal*, **122**, 346–53.

HOWE, G. M., SLAYMAKER, H. O., and HARDING, D. M. [1966], Flood hazard in mid-Wales; *Nature*, **212**, 584–5.

JULIAN, P. R., and FRITTS, H. C. [1967], On the possibility of quantitatively extending precipitation records by means of dendroclimatological analysis; *International Association of Scientific Hydrology, General Assembly of Bern*, 243–50.

KRAUS, E. B. [1955], Secular changes of tropical rainfall regimes; *Quarterly Journal of the Royal Meteorological Society*, **81**, 198–210.

KRAUS, E. B. [1958], Recent climatic changes; *Nature*, **181**, 666–8.

LAMB, H. H. [1965], The early Medieval warm period and its sequel; *Palaeogeography, Palaeoclimatology, Palaeoecology*, **1**, 13–37.

LAMB, H. H. [1967], Britain's changing climate; *Geographical Journal*, **133**, 445–66.

LEOPOLD, L. B. [1951], Rainfall frequency: an aspect of climatic variation; *Transactions of the American Geophysical Union*, **32**, 347–57.

LEWIS, P. [1960], The use of moving averages in the analysis of time-series; *Weather*, **15**, 121–6.

MITCHELL, J. M., JR. *et al.* [1966], *Climatic Change;* World Meteorological Organization, Technical Note No. 79 (Geneva), 72 p.

NAMIAS, J. [1960], Factors in the initiation, perpetuation and termination of drought; *International Association of Scientific Hydrology, Publication No. 51*, 81–91.

NAMIAS, J. [1966], Nature and possible causes of the northeastern United States drought during 1962–65; *Monthly Weather Review*, **94**, 543–54.

RODDA, J. C. [1965], A drought study in south-east England; *Water and Water Engineering*, **69**, 316–21.

SMITH, M. P. [1965], Crisis looming on water shortage; *The Times*, 15 July, 11

VERYARD, R. G. [1963], A review of studies on climatic fluctuations during the period of meterological record; In *Changes of Climate*, Arid Zone Research, 20, UNESCO, (Paris), 3–15.

WALLÉN, C. C. [1955], Some characteristics of precipitation in Mexico, *Geografiska Annaler*, **37**, 51–85.

II. Geomorphic Implications of Climatic Changes

S. A. SCHUMM

Department of Geology, Colorado State University

As we view our familiar environment the question arises, did it always appear so? Major climatic changes are known to have occurred during the past million years of earth history, and over vast areas of the earth evidence of ice action dominates the landscape. In this chapter we concern ourselves only with the long-term effects of climate change on the hydrologic cycle and on the landscape, while omitting treatment of the more obvious effects of glacial and periglacial climates.

Any generalization concerning changes of climate can be very much in error for a given locality, but much evidence has been compiled to suggest that during the last million years average temperatures could have ranged from 10°·below to 5° F above present average temperatures. In most, although not all, regions higher average precipitation was associated with the lower temperatures of continental glaciation, and average precipitation was, at least for some presently arid and semi-arid regions of the United States, about 10 in. greater. During inter-glacial time and during a brief post-glacial episode higher temperatures prevailed, and average precipitation was about 5 in. less than that of today (Schwarzback, 1963).

The effects of climate changes of these magnitudes are not direct. Rather, with changing climate the relations between climate, runoff, and erosion are altered by significant changes of vegetation. Geomorphic evidence of a climate change is, in fact, evidence only of a change in the hydrologic variables of runoff and sediment yield. Therefore, the relations between climatic, phytologic, and hydrologic variables must be considered before the effect of a climatic change on the landscape can be evaluated.

1. Sediment movement

Modern hydrologic data from the United States have been used to demonstrate climatic influences on the quantity of runoff and sediment delivered from drainage basins. The family of curves of fig. 11.11.1 illustrate the general relation between climate and runoff (Langbein et al., 1949). The curves show that, as might be expected, annual runoff increases as annual precipitation increases. However, runoff decreases as temperature increases with constant precipitation because of increased evaporation and water use by plants.

The relations between annual sediment yield and annual precipitation and

s

temperature for drainage basins averaging about 1,500 square miles in the United States are presented in fig. 11.11.2. The 50° F curve of fig. 11.11.2 shows the relationship between sediment yield and precipitation adjusted to a mean annual temperature of 50° F (Langbcin and Schumm, 1958, p. 1076). Sediment

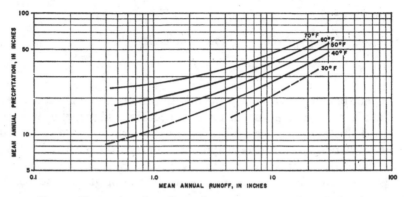

Fig. 11.11.1 Curves illustrating the effect of temperature on the relation between mean annual runoff and mean annual precipitation (After Langbein *et al.*, 1949 and Schumm, 1965).

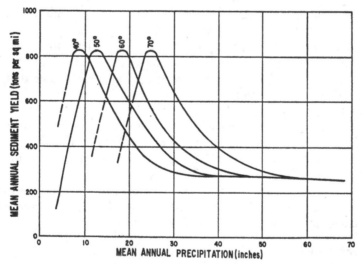

Fig. 11.11.2 Curves illustrating the effect of temperature on the relation between mean annual sediment yield and mean annual precipitation (From Schumm, 1965).

yield is a maximum at about 12 in. of precipitation, but it decreases to lower values with both lesser and greater amounts of precipitation. The variation in sediment yield with precipitation can be explained by the interaction of precipitation and vegetation on runoff and erosion. For example, as precipitation increases above zero, sediment yields increase at a rapid rate, because more runoff becomes available to move sediment. Opposing this tendency is the influence of vegetation, which increases in density as precipitation increases. At about 12 in. of precipitation on the 50° F curve the transition between desert shrubs and grass occurs. Above about 12 in. of precipitation on this curve sediment-yield rates decrease under the influence of the more effective grass and forest cover. Elsewhere in the world, where monsoonal climates prevail, sediment-yield rates may increase again above 40 in. of precipitation under the influence of highly seasonal rainfall (Fournier, 1960).

The sediment-yield curves for temperatures of 40°, 60°, and 70° F are dis-

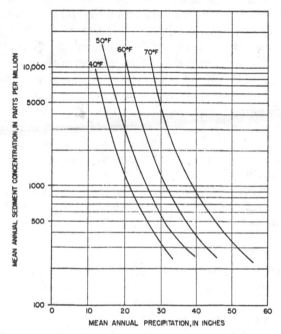

Fig. 11.11.3 Curves illustrating the effect of temperature on the relation between mean annual sediment concentration and mean annual precipitation (From Schumm, 1965).

placed laterally with respect to the 50° F curve (fig. 11.11.2). Together they indicate that, as annual temperature increases, maximum sediment yields should occur at higher amounts of annual precipitation. That is, higher annual temperatures cause higher rates of evaporation and transpiration, and less precipitation is available to support vegetation. Runoff is less, and so the maximum rate of sediment yield shifts to the right.

In addition to the amount of sediment moved, its concentration in the water by which it is moved is important. Curves were developed to show the relation between average sediment concentrations and average precipitation at different temperatures (fig. 11.11.3). For a given annual precipitation, sediment concentrations increase with annual temperature, whereas for a given annual temperature sediment concentrations decrease with an increase in annual precipitation.

One important point to be made with regard to figs. 11.11.2 and 11.11.3 is that, although more sediment is moved from a drainage system under semi-arid conditions, nevertheless sediment concentrations are greatest in arid regions. During a period of years the small number of high-concentration runoff events that occur in arid regions cannot transport the quantities of sediment that are moved by the greater number of lower concentration runoff events in semi-arid regions.

The curves of fig. 11.II.2 show that major changes in erosion rates will occur with relatively minor changes of climate if plant cover adjusts significantly to the climate change. That is, at a mean temperature of 50° F a small change of precipitation anywhere between 0 and 20 in. should elicit a significant hydrologic and geomorphic response, whereas this should not be the case in humid regions, because a major change of vegetational type and density will not accompany a small change of precipitation or temperature.

Changing type and density of vegetation exert a major control on landforms. For example, as the density of vegetation increases with annual precipitation (Langbein and Schumm, 1958, fig. 7), the rate of erosion of hill-slopes should decrease in semi-arid and sub-humid regions. However, in initially arid regions the major increase in runoff that occurs with increased precipitation (fig. 11.II.1) probably will more than compensate for any increase in vegetal cover. Therefore initially arid slopes should be subjected to more intense erosion. Although this last statement is conjectural, the peaks of the sediment–yield curves (fig. 11.II.2) demonstrate that more sediment is exported from semi-arid than from arid regions, and some of this increase must be derived from the hill-slopes.

2. Changes in the channel system

Moving down off the hill-slopes to valley floors, evidence concerning the response of river systems to climate change appears. Investigations into the relations among climate, runoff, and the character of channel systems indicate that within a given climatic region both the total length of channels per unit area (drainage density) and the number of channels per unit area (channel frequency) increases as annual runoff increases and as flood volumes increase. Therefore, increased runoff should cause lengthening of the drainage network and the addition of new tributaries to the system. This conclusion is based on data collected from within one climatic region, where, in fact, differences of soil type and geology exercise a dominant influence on both runoff and drainage density. Therefore, when the increase in runoff is accompanied by a major change in vegetational characteristics the results may be different. If a major climatic change causes a shift in vegetational type from shrubs and bunch grass to a continuous cover of grasses or from grasses to forested conditions it appears that the increased density of vegetation should prevent an accompanying increase in the length and number of drainage channels. In fact, worldwide measurements of terrain characteristics show that stream frequency is greatest in semi-arid regions, least in arid regions, and intermediate in humid regions (Peltier, 1959). It appears, then, that a major increase in precipitation can either increase or decrease the length and number of channels, and if drainage density were to be substituted for sediment yield on the ordinate of fig. 11.11.2 the curves should indicate in a very general manner the variation of drainage density with climate.

In summary, both drainage density and sediment yield should be greatest in semi-arid regions. The occurrence of maximum drainage densities and maximum sediment yields under a semi-arid climate suggests that high sediment yields are a reflection of increased channel development and a more efficiently drained system. Hence, a shift to a semi-arid climate from either an arid or a humid one should allow extension of the drainage network with increased channel and hillslope erosion.

Obviously the changing hydrology of the small upstream drainage areas and their hill-slopes will be reflected in the behaviour of the primary river channels that transport runoff and sediment to the sea. Again, the initial climate or the climate existing before a climate change is of major importance, for it determines the type of river to be considered. For example, in initially humid regions the rivers are perennial, and they remain perennial during a change to a wetter, cooler climate, although the channel will enlarge to accommodate the increased runoff.

The situation differs somewhat for an initially semi-arid region, for even the major rivers are initially either intermittent or characterized by long periods of low flow. The smaller tributaries are ephemeral, as are many in humid regions, but many more drainage channels should be present. With a shift to a wetter climate and greater runoff the flow in the major rivers and in many of the tributaries becomes perennial. Vegetation becomes denser, obliterating the smallest channels. As runoff increases, sediment yields and sediment concentrations decrease (fig. 11.11.1, 11.11.2, and 11.11.3). The result will be enlargement of the main channels. With a return to semi-arid conditions, the sediment yield increases, as both hill-slope and channel erosion increases, and runoff decreases. Deposition in and decrease in size of the main channels should result, as the tributaries again reach their maximum extent and number.

It has not always been recognized that the changes in tributary channels might not conform to changes along the larger rivers. This may not be important in humid regions, but it becomes of major importance in arid regions, where unfortunately few hydrologic data are available from which one may estimate river response to climate change. Nevertheless, increased precipitation should increase runoff in arid regions, but probably not to the extent that ephemeral rivers will be converted to perennial ones. Increased runoff should enlarge the tributary channels and extend the drainage network (arroyo cutting). Sediment will be flushed out of the tributary valleys into the main channels during local storms, and because the loss of water into the alluvium of the main channels is appreciable, aggradation of the main channels will result. In effect, increased runoff in arid regions will erode the tributary valleys, and this sediment will move downstream, where at least part of it will be deposited in the major river channels.

Without perennial flow, semi-arid and arid rivers probably are always characterized by a relative instability, and channel cutting can alternate with phases of aggradation. These events can be considered a natural part of the cycle of erosion of these ephemeral stream channels.

Deductions based on morphologic evidence have been made concerning channel adjustment to climate change, but what evidence exists that the postulated changes may be real? The evidence lies in studies of modern river behaviour in response to differing conditions of runoff and sediment yield. For example, many investigators have demonstrated that the width (w), depth (d), meander wavelength (l), and gradient (s) of rivers are related to the average quantity of water or the discharge (Qw) passing through a channel. As shown by equation (1), channel width, depth, and meander wavelength will increase with an increase in discharge, but gradient will decrease.

$$Qw \simeq \frac{w, d, l}{s} \tag{1}$$

Although the relation between these channel parameters and discharge are highly significant, nevertheless, for a given discharge a ten-fold variation in meander wavelength, width, and depth can occur. Recent work has demonstrated that with constant discharge an increase in the bedload (Qs), the quantity of sand and coarser sediment moved through a channel, will cause an increase in channel width, gradient, and meander wavelength but a decrease in channel depth, as follows:

$$Qs \simeq \frac{w, l, s}{d} \tag{2}$$

Channel shape is significantly influenced by sediment load, and as indicated by equation (2), an increase of bedload at constant discharge will cause an increase in the width–depth ratio, as a channel widens and shallows. Accompanying this change of shape is an increase in meander wavelength (l) and an increase in gradient (s). These changes are brought about by a decrease in the sinuosity, that is, a straightening of the course of the stream, which decreases the number of bends and steepens the gradient.

A glance at equations (1) and (2) reveals that changes in type of sediment load and discharge do not always reinforce one another, that is, although an increase in water discharge will increase depth, an increase in bedload will decrease it. The changes of river morphology that occur, therefore, depend on the magnitude of the changes of discharge and sediment load.

Evidence of river changes in response to climate fluctuations is difficult to obtain, because in most instances the adjustment destroys the pre-existing form, and the basis for comparison is lost. However, the Riverine Plain of New South Wales, Australia, is a unique area from which significant observations concerning river adjustment to changed climate can be obtained. Across this alluvial plain the Murrumbidgee River flows to the west, and evidence is visible on the surface of the plain of older, abandoned channels that functioned during different climatic episodes of the past (fig. 11.II.4). Oxbow lakes, which are remnants of an abandoned channel that was much larger than the modern river, are visible on the floodplain of the Murrumbidgee River. The channel shape and sinuosity of this channel is similar to that of the Murrumbidgee River, but its width, depth, and meander wavelength are larger. The abandoned channel is filled primarily with silts and clays, which must have comprised the sediment load of that channel and which is the predominant load of the modern river. Apparently, as a result of increased precipitation, much higher discharges moved out of the source area in the recent past; however, little change in the nature of the

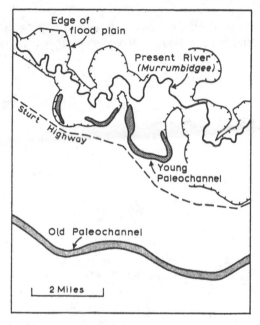

Fig. 11.11.4 Diagram made from an aerial photograph of a portion of the Riverine Plain near Darlington Point, New South Wales. The sinuous Murrumbidgee River, which is about 200 ft wide, flows to the left across the top of the figure (upper arrow). It is confined to an irregular floodplain on which a large oxbow lake (youngest paleochannel, middle arrow) is preserved. The oldest paleochannel (lower arrow) crosses the lower part of the figure.

sediment load occurred, and although the size of the channels are very different, they are in other respects morphologically similar. The contact between the Murrumbidgee River floodplain and the surface of the Riverine Plain proper (fig. 11.11.4) is a series of large meander scars, which are further evidence of the large size of this most recent palaeochannel. Another older set of abandoned channels can be detected in this area (fig. 11.11.4), and these palaeochannels are morphologically completely different from the palaeochannels, which once occupied the floodplain, and the modern river. The older palaeochannel (fig. 11.11.4) is straighter, wider, and shallower than both of the younger channels, and because of its straight course its gradient is twice that of the modern river. Its abandoned and aggraded channel is filled with sand and fine gravel, indicating that this channel was moving a very different type of sediment load from the source area, during presumably a more arid climate. The differences among these three channels can be explained by the changes in water discharge (Qw) and type of sediment load (Qs), as shown by equations (1) and (2).

Although direct evidence of the response of landforms to climate change is rare, the fact that it is the climatic effect on runoff and sediment yield that causes adjustment of river channels makes available other sources of information, as for example, downstream changes of river character with changing discharge and sediment load.

An excellent example of channel changes with changing discharge and sediment load is provided by the Kansas River system, which is tributary to the

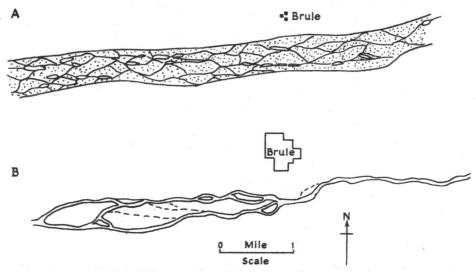

Fig. 11.11.5 South Platte River at Brule, Nebraska.
A. Sketch of channel from U.S. Geological Survey topographic map of the Ogallala Quadrangle, based on the field surveys of 1897.
B. Sketch of channel from U.S. Geological Survey topographic map of the Brule and Brule SE Quadrangles, prepared from aerial photographs taken in 1959 and field checked in 1961.

Missouri River in the western United States. Rising in eastern Colorado, the Smoky Hill River flows to the east. It drains a region of sandy sediments, and it is a relatively wide and shallow stream of steep gradient and straight course. In central Kansas two reservoirs retain much of the bedload of the Smoky Hill River, and below these reservoirs two major tributaries join the Smoky Hill River. These tributaries drain a region of fine-grained sedimentary rocks, and they introduce into the Smoky Hill River large quantities of suspended sediment (silt and clay). Although discharge is increasing in a downstream direction, the width of the channel decreases and its depth increases as a result of the influx of fine sediment and the decrease of bedload. The channel becomes more sinuous and gradient, width–depth ratio, and meander wavelength decrease. Farther downstream the Republican River, a major tributary, which drains an area about equal in size to that of the Smoky Hill River, joins the Smoky Hill to form the Kansas River. Below the junction of the two rivers, the major increase in discharge and the addition of large quantities of sand cause a major change in the channel morphology. The channel width increases markedly, as does the width–depth ratio and gradient, whereas sinuosity decreases and channel depth remains relatively unchanged.

The above changes in channel characteristics have occurred along one river as the nature of the sediment load was altered by changes in runoff and sediment yield from tributary basins. In addition, man has modified the flow patterns and sediment loads of rivers, and these changes often duplicate the effects of a climate change. Therefore, although geologic evidence is limited, the engineering literature is a fruitful source of information concerning river adjustment to man-induced changes of hydrologic variables. In fig. 11.11.5 a major change in the width of the South Platte River is illustrated. Both the North and South Platte Rivers in Nebraska were classic examples of braided rivers during the latter part of the nineteenth century, however, due to regulation of discharge, these channels have recently undergone major changes of dimensions and form. In the case of the South Platte River flood peaks have been reduced, and in response the river has changed from a wide braided channel to a narrow and somewhat more sinuous channel. Depth has probably also decreased. Similar changes have occurred along the North Platte River as a result of both a reduction in peak discharge and annual discharge. In these examples a major reduction in channel size has occurred as a result of reduced discharge, as indicated by equation (1).

Discussions of the adjustment of landforms to long-term climate changes are still highly speculative, but a more complete understanding of the possible changes that can occur is being developed. The ability to predict landscape changes resulting from a modification of the climatic or hydrologic characteristics of drainage systems has practical implications, for if man persists in his efforts to modify not only the hydrologic regime of river systems but also the climate of drainage basins, then he should be prepared to evaluate the consequences of these acts not only in terms of changed river discharge and sediment yields but also in changed channel and drainage basin morphology.

REFERENCES

FOURNIER, M. F. [1960], *Climat et erosion* (Presses Univ. France, Paris), 201 p.

LANGBEIN, W. B. *et al.* [1949], Annual runoff in the United States; *U.S. Geological Survey Circular* 52, 14 p.

LANGBEIN, W. B. and SCHUMM, S. A. [1958], Yield of sediment in relation to mean annual precipitation; *Transactions of the American Geophysical Union*, **39**, 1076–84.

PELTIER, L. C. [1959], Area sampling for terrain analysis; *Professional Geographer*, **14**, 24–8.

SCHUMM, S. A. [1965], Quaternary paleohydrology; In WRIGHT, H. E. and FREY, D. G., Editors, *The Quaternary of the United States* (Princeton Univ. Press), pp. 783–93.

SCHWARZBACK, M. [1963], *Climates of the Past*; (Van Nostrand Co., New York), 328 p.

III. Long-term Trends in Water Use

MARTIN SIMONS

Department of Education, University of Adelaide

1. The growing need for water

It has been reliably estimated that consumption of water in the United States alone will have risen before 1980 to an average of well over 400 billion (400,000,000,000) U.S. gallons per day, an increase of more than 30% over the present requirement, and there is every prospect of continued massive escalation of demand beyond that time. The total fresh water available in the United States from rain, rivers and ground water, even supposing all could be fully used, is about 3.5×10^{14} U.S. gallons per year. At the present rate of increase the United States will have outstripped available resources long before the middle of next century. The rising demand comes from all quarters (fig. 11.III.1). Domestic consumption is increasing, but is not the major problem. More water each year is used for irrigation, not only in arid regions but in more humid areas, where yields can be increased and made more reliable by supplemental irrigation. However, industrial uses of water are now becoming dominant, and by 1980 it is expected that more than half the water used in the United States will go to power stations and factories. Further than this, by the end of this century it is predicted that the most important single user of water will be the thermal electricity generating industry. This water is mostly used for cooling steam condensers for which fresh water is preferred and, in any case, central locations are superior to coastal ones to cut down electricity transmission costs. Nor does the increasing expansion of nuclear-power generation decrease the general water need, since up to the present time such power stations convert the energy of the atomic pile to steam, which is then used to drive the dynamos. In manufacturing industries water is employed in many ways, as a cleansing agent, as a constituent of some products, such as beverages, soaps and detergents, dyes, and various other chemical products, but cooling is still the major use. Already water shortage is hindering development of established industries in many places, and in these same areas the industrial pollution of rivers and lakes is creating its own particular problems.

The situation now facing Western Europe and the United States foreshadows developments to be expected in all industrial regions in the foreseeable future. Already in some parts of Western Europe the supplying of great conurbations like Manchester, Liverpool, London, and the Ruhr has proved much more troublesome than expected. Reservoirs constructed in Wales, the Lake District,

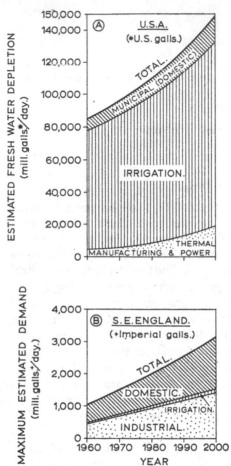

Fig. 11.III.1 Estimated fresh-water depletion for the United States (A), and maximum estimated demand for fresh water in South-East England (B) up to A.D. 2000. These totals do not include water used for hydroelectric generation which is immediately returned to surface storage. It should be emphasized that all values are only estimates, which may be greatly in error, particularly for the more remote dates.

and the Pennines have aroused strenuous opposition, and the cost of water, in cash and in loss of other amenities, is growing with the demand.

2. Geographical engineering

Over and above the accelerating expansion of water-using industries in the populated lands, the great deserts and semi-arid portions of the earth might sooner or later be developed, and water is unquestionably the first essential for this. These are continent-sized areas; if water can be made available they will hold continental populations and provide continental markets. This means water for new industries and new cities as well as for food production. To meet these needs, great new engineering works are required that will make the largest

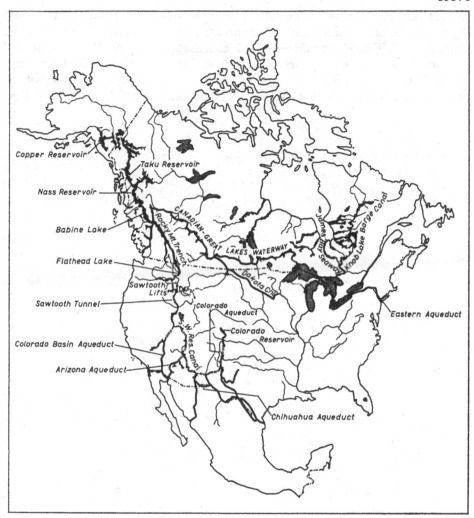

Fig. 11.III.2 The North American Water and Power Alliance proposals (Courtesy of the Ralph M. Parsons Company, Los Angeles and New York).

existing water-control schemes look puny. In retrospect it may turn out that projects like the NAWAPA scheme will be judged too small, rather than too ambitious.

The NAWAPA scheme has been described as 'plumbing on a continental scale', and envisages the control of all the major western rivers of the North American continent from Yukon to northern Mexico (fig. 11.III.2). The centre-piece of the scheme, upon which the rest largely depends, would be the creation of a 500-mile-long reservoir in the Rocky Mountain Trench from Flat-head Lake in Montana across the Canadia border into British Columbia. This would be a result of damming the upper reaches of the Columbia, Kootenay, and Fraser

Rivers. The lake floor would be 3,000 ft above sea-level, and from it water would flow in aqueducts, mainly leading southwards, into the arid west of the United States and, ultimately, to Mexico and aqueducts along the Rio Grande and Sonora Valleys. Associated with this mighty artificial lake would be another chain of reservoirs from Cathedral Rapids in Alaska through the valley of the Stewart in Yukon to the Taku, Liard, Mass, Skeena, and Salmon River valleys

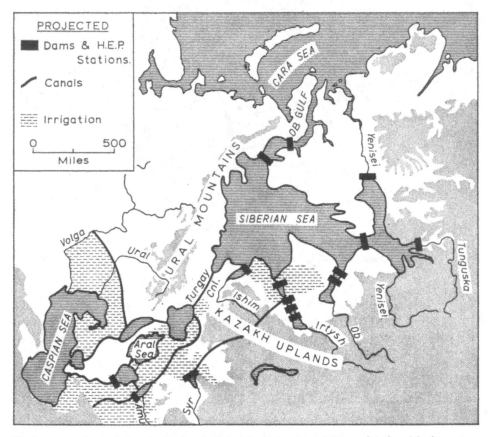

Fig. 11.III.3 One of the plans proposed for the irrigation of Central Asia with the waters of Siberian rivers.

in British Columbia. From these, some water would be pumped to the Peace River Reservoir, and this would feed a great navigable canal bringing water to the Canadian Prairies and, via a canal across the border, to the Dakotas and the upper Mississippi, while a Great Lakes Waterway would connect Lakes Winnipeg and Manitoba with Lake Superior, maintaining a constant water level in the Great Lakes as far as Buffalo, with an aqueduct to New York following the line of the Erie Canal. Associated projects in Labrador, Quebec, and Ontario, and the drawing into the scheme of many smaller local projects already in existence or under consideration are imagined. Compared with the large Californian Water Transfer Scheme, parts of which have been functioning successfully for many years and other parts of which are still debated, this plan seems colossal. There would be scores of dams, pumping houses, power stations, aqueducts, and navigable canals. The cost at 1966 prices was estimated by a sub-committee of the U.S. Senate at about $800 million; probably an underestimate. Thirty years, from the date of starting construction, would pass before the project's completion. Yet R. L. Nace pointed out in 1966 that even when finished only 13% of the potential water 'crop' of North America would be under the scheme's control. This, in the year 2000, may be too little and too late.

Comparable in scale is the proposed and long-debated Ob-Yenisei–Irtysh diversion scheme, the effect of which would be to bring water from the northward-flowing rivers of western Siberia to supply the Caspian Desert by creating an artificial sea in Asia with a surface area larger than England (fig. 11.III.3). Just as North American geologists have queried the ability of the Rocky Mountain Trench floor to withstand the great additional weight of water above, so Soviet climatologists have made gloomy prophesies concerning the probable deterioration of climate in the steppes south of the proposed flooded regions. A great reservoir on the Lower Ob, while doubtless aiding the arid lands and incidentally providing hydroelectric power for industries, would raise the West Siberian water-table, creating widespread swamps even greater in extent than those that would be drowned. Air masses forming above the lake would be cooled, and crops in the south would suffer. The more spectacular Ob-Yenisei plans have therefore been dropped, but others of equal importance have replaced them, and the debate continues.

Outside the U.S.S.R. and United States, nothing comparable in scale is under serious consideration at the present time, but schemes already in being or under construction, like the Australian Snowy Mountain Project and the Egyptian Aswan High Dam, are more ambitious than anything imagined in such areas fifty years ago, and it would be strange if still larger plans were not made during the next few decades. The Sahara Desert, for example, might one day be irrigated, although not as crudely envisaged by Sergel's scheme involving the flooding of about 10% of Africa (fig. 11.III.4).

Some of the above schemes seem like the wilder dreams of science-fiction, and yet they involve no important technical innovations. The engineering would be along well-proved lines, the dams would be larger and the aqueducts longer, but nothing new in principle would be involved. To some extent the same is true of the many schemes suggested for control and use of the sea. The Dutch are already admired for their ambitious land-reclamation projects, the entirely artificial polders which are being created in what used to be a tidal lagoon, the Zuyder Zee, and the Rhine Delta Scheme, which is currently under construction. It is sometimes forgotten by writers on these works that the Yssel Meer, the remnant of the Zuyder Zee, is now a freshwater lake behind the North Sea Dyke, and hence an invaluable source of water for the growing cities of the Netherlands. In the U.S.S.R. a project larger than this has been suggested for the Sea of Azov, at present more saline than the Black Sea, but capable of

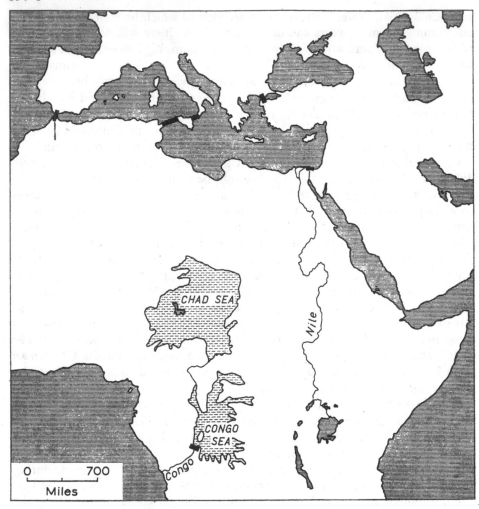

Fig. 11.III.4 The basis for H. Sergel's plan for the irrigation of the Sahara.

being enclosed by a relatively short dyke across the Kerch Strait. There are many other saltwater lagoons where similar projects could be suggested. On the smaller scale, in Britain the Wash and the Solway Firth might all physically lend themselves to such development, providing that the schemes were economically viable. In the United States similar freshwater reservoirs might be created in such inlets as Chesapeake and Delaware Bays, Long Island Sound, San Francisco Bay, and, by damming the Straits of Georgia and Juan de Fuca in Canada, a freshwater lake of great size could be made to supply the industrial northwest. Such plans, of course, cannot be considered in isolation. The enclosure of many of the world's major estuaries would have profound effects upon the cities and other communities around them which depended for their existence on access to the sea. Fishing villages in the Netherlands were forced to change their ways drastically when the Zuyder Zee was enclosed. A major port would insist on retaining access to the oceans, and might gain little from the impounding of its estuarine waters.

On the larger scale, even the enclosure of such vast bodies of water as the Mediterranean Sea, by damming the Straits of Gibraltar, has been proposed, though often, it seems, these ideas are deliberate pipe dreams or academic exercises rather than serious propositions. Nevertheless, under the sharp pressure of demand, even these notions will be taken seriously and will, at least, advance to the point of feasibility studies, if not further. The Mediterranean is somewhat more saline than the Atlantic, and shutting off its western end would if anything increase, rather than reduce, its salinity, which is caused by evaporation from its surface. It might even be expected that the level of the sea would fall and the lake become, like the Caspian, a shrinking reservoir of salt water. Effects on climate might also be considerable, for not only would conditions on the Mediterranean coasts be changed, possibly for the worse, but the cessation of currents through the straits might affect the larger pattern of water circulation in the Atlantic. Similar objections could be raised to some American projects, now abandoned, to divert the Gulf Stream by erecting a barrier between Florida and Cuba, with the intention of increasing the temperature of the Gulf of Mexico and thus preventing the damaging frosts that affect crops in the southern states. Another idea, to divert the cold Labrador current away from the American coast, by building a barrage into the Atlantic south-east of Newfoundland, would probably bring the Gulf Stream closer to American shores, but its likely corollary would be disaster for Europe, which would suffer a drastic fall of temperature. In the same dangerously speculative category must be placed the Soviet scheme to place a barrier across the Bering Strait and then, using a series of giant sluice gates, opened and closed alternately, allowing the flood tide to pass through but to restrain the ebb. In this way it would be possible technically to abstract large volumes of warmer water from the Pacific and pass them into the Arctic Ocean, where they might melt, or help to melt, the sea ice. Alternatively, by subtracting cold water from the Arctic, the Gulf Stream entering from the other side of the Ocean would be drawn farther into the basin, so having a similar effect. The climatic effects of any such scheme, put into operation, are at present quite unforeseeable.

3. Water purification

For any large-scale scheme innovation is necessary in the political and economic spheres, rather than in engineering. The Canadians are by no means so enthusiastic about NAWAPA as the Americans, for some see the project as robbing Canada of her water. Although at present she is not using more than a fraction of the available supply, there will certainly be a time when Canada's own development is restricted by lack of water. In Siberia, too, there are sectional interests opposed to the sacrifice of one region, by flooding or deprivation of water, to the convenience of another, though strong centralized government tends to conceal such divisions of opinion. In Africa political issues already loom very large in water control, and obviously no ambitious scheme for the Sahara can make headway unless the participating nations can agree. Political stability and trust are a primary requirement before any kind of financial arrangements can be made, and it is equally clear that however such enormous engineering works are to be capitalized, something new in the way of long-term investment and re-payment agreements will be necessary before it is worth any draughtsman's time to put pencil to paper. Very probably it will be difficulties in these realms, rather than in the technical field, that will compel men to turn to other sources of water, and to more careful control of existing supplies. There are two obvious lines that development will take. Water, unlike many other raw materials, is rarely destroyed. It is used and re-used by man at different stages of its journey seawards. However, it may enter the city or factory in a natural state and leave it more or less polluted. Rivers thus carry vast quantities of waste, and even the sea is polluted by sewage, by oil from the bilges of tankers, and by radioactive waste. Man could increase the amount of available usable water by large-scale purifica-tion and recirculation, just as in the closed environment of a space capsule a limited quantity of water is used again and again, purified, recirculated, re-purified, and so on for an indefinite period. A water-using industrial complex on earth could treat its water in exactly the same way, a necessary quantity being abstracted from the natural water cycle in the first place, and thereafter perpetu-ally recycled. Only solid waste would ever leave the site, and little further water would be needed unless some planned expansion of the industry were under-taken, when new water-treating plant would necessarily be included at the plan-ning stage. Although at the present time such closed water systems are not necessary, as a rule, they do exist on a small scale, particularly where water of greater purity than that found naturally is required. One example of an existing partial recycling system is at Bedford, England, where polluted water is taken from the Great Ouse, passed through a complex sequence of processes involving chemical precipitation, flocculation and softening, filtration, chlorination, and so on, before redistribution. Such schemes are expensive, but as demand rises, costs of supplying clean water from natural sources will increase and the political difficulties will loom ever larger, whereas further scientific advances and engineer-ing improvements can be expected to reduce the relative cost of recycling.

Similar arguments will prevail in future with regard to desalinization of sea-water and other mineralized sources. The cost per gallon of water obtained in this way, allowing for changes in the real value of money, is falling as improve-ments are made to equipment and as new sources of power are discovered. It is very probable that within a few decades it will become cheaper, as well as politically preferable, for cities like Los Angeles and London to augment their water supplies by large-scale distillation or other desalinization plant, powered by nuclear energy, or conceivably by cheap solid fuels. A large desalinization plant could be projected, designed, and constructed in a few years at most, and once built would remain under the control of the community which made the capital investment. Neither advantage accrues to the enormous water-transfer projects. Such engineering achievements as the Aswan High Dam and the Californian

Water Plan and the control of the Jordan have been handicapped by political troubles, and this must be a factor in the calculations of governments and investors as water shortage increases. A series of distillation or electrodialytic plants, while seeming less efficient overall, might be considered a more practicable investment, if locally demanded and locally financed, rather than schemes involving withdrawing resources from regions where their loss might later be deplored.

4. Conclusion

The past provides many examples of forecasts and projected trends which have proved wrong, but so far there are no signs of any tailing off of demand for water, and no suggestion of any important substitute being found. On the contrary, new industrial processes and new inventions frequently appear which increase the reliance of civilization upon this fundamental resource. Some of the expedients that will be necessary to meet the challenge will doubtless be extraordinary, others will be merely extensions and expansions of existing engineering methods. In the long run water supply may prove to be the ultimate limiting factor for world population. Synthetic food is a possibility, even a probability for the future feeding of the masses, but manufacture of new water by chemical means, though technically possible, can hardly be contemplated on a scale sufficient to make any substantial difference to the main problem. The total amount of water available to the race is virtually fixed. In the long term it is the rate at which the liquid can be passed through the biosystem that will set a limit to the development of human society. If all known sources of fresh water were fully used about $4 \cdot 5 \times 10^{15}$ (Imp.) gallons would be available, or 20,000 k^3. This would imply a world population of 20,000 millions at roughly the present *per capita* rate of consumption. Some estimates suggest that by A.D. 2100 world population will have reached or passed this figure if the food problem can be solved. After that, water must come from the sea.

REFERENCES

ACKERMAN, E. A. and LÖF, G. O. G. [1959], *Technology in American Water Development* (Resources for the Future, Inc., Baltimore), 710 p.

ADABASHEV, I. [1966], *Global Engineering* (Progress Publishers, Moscow), 237 p.

ANON [1960], New water treatment works of the Borough of Bedford Water Undertaking; *Water and Water Engineering*, 64, 293–9.

CAMPBELL, D. [1968], *Drought; Causes, Effects, Solutions* (Cheshire, Melbourne).

DIESENDORF, W., Editor [1961], *The Snowy Mountains Scheme: Phase I* (Horwitz Publications, Sydney).

FURON, R. [1967], *The Problem of Water; A World Study;* Translated by P. Barnes (Faber and Faber, London), 208 p.

INSTITUTION OF WATER ENGINEERS [1967], Symposium on conservation and use of water resources in the United Kingdom; *Journal of the Institution of Water Engineers*, 21, 203–330.

INTERNATIONAL UNION OF PURE AND APPLIED CHEMISTRY [1963], *Re-Use of*

Water in Industry; Committee of the Water, Sewage and Industrial Wastes Division, (Butterworth's, London), 256 p.

MILLER, D. G. [1962], *Desalination for Water Supply* (Water Research Association, Medmenham, Bucks).

NACE, R. L. [1966], Perspectives in water plans and projects; *Bulletin of the American Meteorological Society*, **47**, 850–6.

SYMPOSIUM [1964], *Water Resources Use and Management;* Australian Academy of Science (Melbourne University Press).

THE RALPH M. PARSONS CO. [1967], *North American Water and Power Alliance* (Los Angeles and New York), Brochure 606–2934–19.

CHAPTER 12

Choice in Water Use

I. Choice in Water Use

T. O'RIORDAN

Department of Geography, Simon Fraser University

and

ROSEMARY J. MORE

Formerly of the Department of Civil Engineering, Imperial College, London University

1. Traditional allocation

Water has been used by man in many ways. His first need was for a drinking supply for himself and his animals, then navigation on the waterways allowed him to move quickly and cheaply from place to place. In the drier climates water was early used for irrigation, in wetter, low-lying areas later drainage schemes made the land habitable. Man has long been the victim of disastrous floods, which today he is increasingly able to combat by flood-warning devices and reservoir storage. Currently water is also managed for a great variety of industrial purposes, for recreation and for its natural beauty in the landscape, all of which require maintenance of water quality and control of pollution.

The number of uses to which water is put varies in time and space, and the possibility that future new uses of water will be developed means that sufficient water should be left in long-term water inventories to accommodate them. In any specific region one or two water uses are often dominant, whereas others are subsidiary or unimportant. In many areas water uses conflict with one another (e.g. domestic water supply and irrigation), whereas some demands are complementary (e.g. pollution abatement, flood prevention, and recreation). Whether the many uses of water are conflicting or complementary, water management by allocation is necessary.

The classic means of allocating water between users has been by water rights. The oldest of these rights is the *riparian doctrine*, which was part of the civil law of Rome and later became incorporated in the common law of England. In brief, the riparian doctrine accords to the owner of riparian land, by or across which a stream flows, the right to use the water of that stream on or in connection with his contiguous land. It is not based upon use of water, and is not lost solely by disuse. Under English common law riparian rights may be altered by grant or prescription, for example, under the 'reasonable use' doctrine which states that riparian users must give consideration to reasonable demands by others with riparian rights. The *prescriptive right* to take water may be acquired through actual and uninterrupted use, generally for a long period, with acquiescence of the person from whom the right is acquired. All private industrial water supply from rivers in England and Wales, together with much irrigation, depends on prescription.

Whereas the land-based riparian and prescriptive rights were adequate in humid European conditions, they proved unmanageable when European immigrants began to farm and develop industries in the drier parts of the United States and Australia, and an *appropriative doctrine* was developed founded on the principle of temporal priority of diverting water and putting it to beneficial use upon land ('first come, first served'), regardless of the contiguity of the land to the source of supply. Appropriative rights, which are based upon use of water and are lost by non-use, allowed the spread of irrigation away from the river courses of the western United States.

Of more recent years, the allocation of natural water resources has rested largely upon the political process. In the United States, in particular, large-scale development schemes were the outcome of a combination of three factors: crises (usually natural catastrophes, though that preceding the New Deal 1933–9 was an economic one); a dominating political personality; and identification by vested-interest groups with the major federal agencies responsible for the implementation of large projects (dams, hydroelectric schemes, etc.). With the notable exception of the Tennessee Valley Authority, water-resources development during this period tended to be more project-orientated rather than conceived in terms of comprehensive resources management. Since dam construction meant employment, and stored water brought such varied benefits as flood protection, irrigation, and hydroelectric power, few politicians could afford to resist the temptation to attract federal capital into their constituencies, regardless of how uneconomic or non-integrated the project might turn out to be. The practice which developed whereby politicians helped each other to push water-resource projects through Congress was known as 'pork barrel legislation'.

2. The need for modern allocation techniques

As the demand for water grew both in relation to absolute requirements and with regard to the variety of uses to which it could be put, so unmanaged natural water-resources have become less able to supply these needs. This has been particularly the case in the drier western parts of the United States, where demands could be met only by large-scale transfers of water, and, even in the more humid regions, transfers of water resources, both in terms of space and function (e.g. change from irrigation to municipal use), are occurring, necessitating the introduction of rigorous techniques whereby water can be allocated both in space and in time in an optimal manner. Two major concepts in water-resources management have helped to emphasize the need for improved allocation techniques: those of multi-purpose use and of integrated river-basin development.

Multi-purpose use involves the simultaneous management of water resources (river basin, ground water, lake, reservoir, etc.) to produce a variety of functions, such as flood protection, public water supply, irrigation, hydroelectric power, navigation, pollution abatement, outdoor recreation, fishing and wildlife preservation, and the like. Each of these functions is spatially linked to a complementary use of land and economic activity. For example, water is consumed by activities on the land (e.g. irrigation) and urban complexes discharge waste products into the rivers, both of which are necessary to maintain the local economy. Therefore, the problem of allocating multi-purpose water resources also includes the important spatial element of land-use planning. *Integrated river-basin development* can be viewed in terms either of single-purpose development (e.g. flood protection, as in the Miami Conservancy in Ohio) or multi-purpose development, but in both cases the natural-resources base encompasses the total hydrological unit of the river basin. This concept again introduces very important spatial and temporal aspects, for any development of water resources in the headwaters of a river must undoubtedly affect the quantity and timing of downstream supply.

Present policy in relation to the need for large-scale spatial transfers is tending towards the concept of *multiple purpose re-use*. A 'use' or 'purpose' can be defined as the combination of specific productive inputs and the resulting outputs of goods and services. Traditionally, a distinction has been made between 'consumptive' uses, where water was actually used up (e.g. irrigation, public water supply), and 'non-consumptive' uses, where a certain volume of water was required but little or none was actually lost in the performance of the function (e.g. recreation, pollution abatement). In strict economic terms all uses are consumptive to a degree, whereas in physical terms most uses are largely non-consumptive, and a more accurate distinction can be made between those functions which involve a direct transfer of water outside the channels (e.g. public water supply and irrigation), which might be termed 'transferred' functions, and those which provide benefits through their presence in the channel (e.g. navigation, pollution abatement), which might be designated *in situ* functions. Multiple purpose re-use involves the spatial transfer of additional water supplies (e.g. ground water or surface water stored behind a dam) by means of the natural river channel. Not only is it much cheaper to utilize the natural channel as an aqueduct but by maximizing the use of water over space considerable economies may be achieved. Such a policy permits greater development of *in situ* functions, stimulates economic development throughout the river basin and its spread from one basin to another, yet still satisfies the demand made by 'transferred' functions. The 're-use' concept relates to the water abstraction by 'transferred-use' consumers from the river channel and its subsequent return after suitable treatment to the natural channel, though possibly at a different location (e.g. purified sewage effluent from the municipality).

The basin-water-resource base can be envisaged therefore as a system of interlinked multi-purpose demands (fig. 12.1.1), some conflicting and others complementary, which is managed by means of projects. A 'project' may be defined as a given set of inputs designed to produce the required combination of outputs from a resource system. The purpose of allocation techniques is, therefore, first to discover the optimum mix of functions for any given project in order to maximize net benefits (e.g. gross benefits minus gross costs), and secondly, to weigh-up the relative merits of alternative projects designed to achieve a similar set of productive outputs. In view of increasing demands upon both 'transferred' and *in situ* functions, and the effects upon both spatial linkages and economic activity, the need for a rigorous analysis of choice in water use is becoming more and more important.

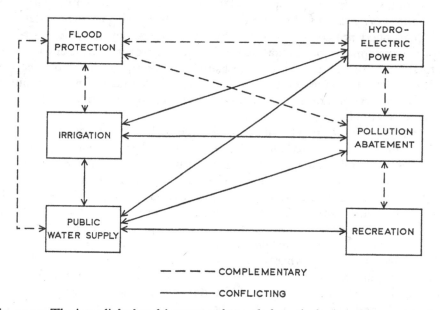

Fig. 12.1.1 The inter-linked multi-purpose demands from the basin water-resource base.

3. Economic and spatial problems in allocation techniques

In an interrelated multi-purpose water-resources system the evaluation of its component functions in precise economic terms is rendered more difficult on account of the highly intricate economic relationships and the spatial linkages involved. The latter are particularly concerned with 'collective benefits', 'externalities', and 'scale', whereas those of primarily economic interest are 'intangibles', 'opportunity costs', 'interest rate', and 'uncertainty and risk'.

Collective benefits

Certain benefits accruing from resources management are universal in character in the sense that they must be provided for a group regardless of individual preference or individual evaluation of those benefits (e.g. flood protection, pollution abatement, fishery protection, wildlife preservation, and the enhancement of aesthetic attractiveness). Since individuals cannot state their preferences or valuation in the market-place through the usual procedure of willingness-to-pay, collective benefits can be distributed and paid for only on a unified areal basis. One method of evaluating collective benefits on an economic basis is to calculate the least-cost alternative to provide the same supply.

Externalities

Water-resource uses are too closely interlinked to permit separate management of any one sector without impinging upon the production and development of at least one of the other sectors within the total complex. Such an interrelationship is called an externality, being positive where the relationship is complementary (e.g. the release of protectively-stored floodwater may dilute downstream pollution and be harnessed to generate hydroelectricity), and negative where the management of one function impinges upon the benefit outputs of another (e.g. the loss of aesthetic satisfaction in a recreation area due to the excessive discharge of polluting effluents). Externalities usually have a strong spatial component, for the effects on one function may be spatially removed from the management of the other. Thus a reservoir provides benefits both on site (recreation, power, water supply) and downstream (flood protection and pollution dilution).

Scale

The identification and calculation of benefits accruing from a project depend to a large extent upon its geographical and economic scale, and a project large enough to bring about repercussions throughout the entire economy must be treated differently from a regional resource development project, where secondary effects, particularly the extent to which they affect the local economy (i.e. the 'multiplier effect'), assume a greater meaning.

Intangibles

In any economic allocation process some common yardstick of evaluation is necessary by which to compare production inputs and outputs. The usual criterion is the tangible price of the goods in a free market. However, some goods (called 'intangibles') are essentially subjective or difficult to identify in monetary terms, such as the visual beauty of a landscape, the protection and preservation of a unique ecological habitat, or the maintenance of a clean river. Owing to the inherent difficulties of evaluation, intangibles are usually underemphasized or overemphasized, depending upon the personal attitudes of the decision-makers.

Opportunity Costs

Opportunity costs are of two kinds: the value of benefits forgone to one resource sector of the project owing to the selection of an incompatible alternative proposal (e.g. the loss of productive farmland when the headwaters of a river are dammed for the purpose of floodwater retention), and when a limited budget causes the diversion of capital away from other investment.

Interest Rate

The interest rate may be regarded as the opportunity cost of borrowed money on the assumption that increased capital investment in the present will result in a better return in the future. The interest rate at which money is borrowed to finance water-resource management projects is critical both in determining the economic justification of the project and in deciding upon the time distribution of expenditure. From fig. 12.1.2 it can be seen that at an interest rate of 7%, 90% of the benefits accruing from the project will be found within 34 years, while at

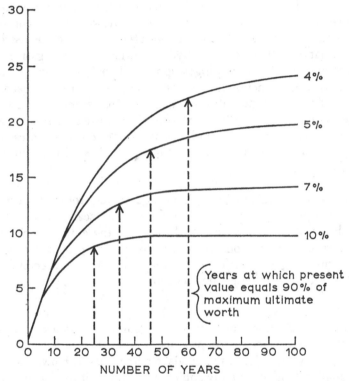

Fig. 12.1.2 Incomes and expenditures of dollars per year present value at various interest rates plotted against number of years. The vertical axis shows present annual value of costs and benefits (in dollars). (From Sewell et al., 1961.)

an interest rate of 4%, 90% of the project benefits will only be realized after 60 years. In the private sector of the economy, where the time horizon is short, interest rates are high, but in the public sector (where most resource-management projects are financed) a greater degree of attention is paid to benefits accruing to future generations, and interest rates are lower. In the United States the current interest rate for financing resource-management projects is approximately 3%.

Uncertainty and Risk

Both uncertainty and risk also reflected by the interest rate. Increasing risk, the gamble that is taken in the face of unknown and unpredictable supply and demand, usually results in the raising of the interest rate. Uncertainty involves such possibilities as the discovery or invention of new processes which might render an existing process uneconomic. For example, desalinization, coupled with nuclear-electric-power production, might seriously impair the long-term feasibility of projects designed to provide hydroelectric power and the long-distance transfer of fresh water.

For the reasons given above, multi-purpose resource-development projects are usually found in the public sector of the economy, it being generally the case that private interests do not take due consideration of all social (i.e. community welfare) costs, since they cannot recoup uncompensated benefits, and market incentives do not function ideally. However, it should not be thought necessarily that the public decision-making process is in any way more perfect than in the private sphere; and recent work in water-resource allocation techniques is concentrating on the role of the decision-maker and his attitudes to, and perception of, the problems that confront him. The decision to develop resources from a public or private standpoint is therefore by no means clear-cut, and the history of resources development has been marked by constant conflict between vested-interest groups whenever a proposal by private interests affects what is generally considered to be 'the public interest'. One of the most notable examples of this in American history was the decision to construct the Hetch Hetchy public-water-supply reservoir along with the production of hydroelectric power in the Yosemite National Park of California. This was treated as a test case against private development in public territory, but after more than a decade of struggle (1901–13) the dam was ultimately constructed, since there appeared to be no cheaper alternative that would provide the same benefits. A similar problem arose more recently in England when Manchester Corporation, facing a water shortage of 50 million gallons per day by 1970, unsuccessfully petitioned for an abstraction from two famous tourist-amenity lakes, Ullswater and Windermere. Such decisions on conflicting issues abound in resource-management history, and it is therefore not surprising that recent workers have become interested in such matters as basis of opinion, value judgement, policy, and perception, which represent the final phase of the allocation process.

4. Allocation techniques

There are two main technical approaches to the problem of water-resource allocation, benefit–cost analysis and systems analysis, the latter being in some respects a more sophisticated and dynamic development of the former.

A. Benefit–cost analysis

Benefit–cost (b–c) analysis is a technique for enumerating, evaluating, and comparing the benefits and costs that stem from the utilization of a productive resource base (a project). The economic theory upon which b–c analysis is based assumes that a desired objective is achieved through the lowest-cost alternative of a number of alternative projects designed to produce a similar objective or, in the case where the quantity of resource use is specified, the most

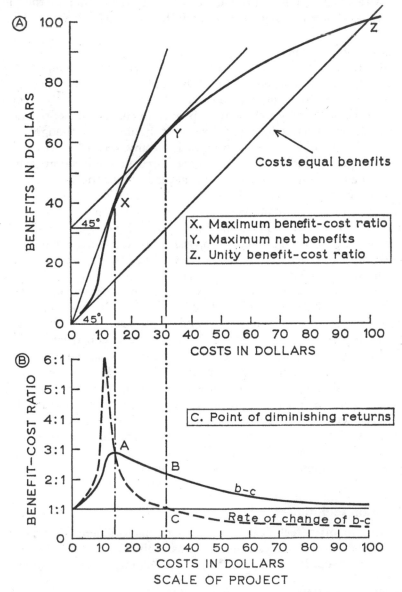

Fig. 12.1.3 Benefit–cost relationships for various scales of development (From Sewell, W. R. D., Davis, J., Scott, A. D. and Ross, D. W., Guide to benefit–cost analysis, In Burton and Kates, 1965, pp. 544–57).

efficient solution of a number of alternatives is that one which maximizes the desired objective. In terms of welfare, b–c analysis does attempt to maximize the net social benefits to a point of project implementation when the benefit accruing to any one set of consumers cannot be further increased without incurring losses upon another set of consumers.

According to b–c theory, the ideal project is the one where total net benefits (i.e. total benefits minus total costs) are maximized (i.e. fig. 12.1.3(a), point Y). At the point X on figure 12.1.3(a), the ratio between benefits and costs is at its greatest for the project, but the scale of the project is not maximized, since an increase of costs (even though costs are now rising in relation to benefits) will still produce sufficient benefits to increase the total net benefits. The optimal scale of development is reached when each component resource sector of the project is developed to the point of diminishing returns, i.e. where the last increment of cost invested just equals the value of benefit it produces (fig. 12.1.3(b), point C). Theoretically, the construction of such a project makes the best use of resources and capital in the sense that no greater net gain could be produced elsewhere in the economy.

The considerable complexity of economic and political issues that beset optimal allocation processes in the real world exert severe constraints upon the theoretical bases of b–c analysis which assume perfect knowledge of the market (i.e. no uncertainty), quantifiable benefits and costs as valued in the market, and an explicit statement of individual preferences. Furthermore, the objective of maximizing net social welfare relies upon equating individual preferences (utilities) and the total aggregation of inter-personal utilities (e.g. individuals' attitudes to water-based recreation). A weakness of conventional b–c analysis is that multi-purpose resource-development projects can be implemented for a variety of objectives, such as redistribution of national income, regional economic growth, and increased aesthetic satisfaction, for which the overriding principles of economic efficiency do not rule. When one asks, 'Whose welfare is being maximized?' and 'Who is paying for this?' it is found that the biggest beneficiaries usually exert the greatest vested interest in the decision-making process, such that these pertinent questions cannot properly be answered in a rigorous economic framework. B–c analysis, therefore, operates in the socio-political arena, where sub-optimum economic decisions must be partly based upon compromise and value judgements. Another difficulty is that b–c analysis often emphasizes the static view of the project, whereas objectives change both with the passage of time and as more data is collected – in other words, there is often considerable feedback (interaction) between policy making and programme design. Also, from a mechanical viewpoint, b–c analysis is such a laborious exercise that only a limited number of alternatives can be scrutinized with the necessary care, so that step-by-step analysis through time is virtually impossible.

Despite these considerable limitations, b–c analysis does have merits as an allocation technique and is still widely used in evaluating the feasibility of water-resource-management projects, because:

1. It is simple to understand and relatively easy to calculate in comparison with other resource-allocation techniques. Also, it does attempt to take the long view in the sense of analysing both short- and long-term benefits and costs, and the wide view by introducing the concepts of social (or collective) costs and benefits, and externalities.

2. By sifting out uneconomic proposals, it does provide a yardstick for measuring the *relative* feasibility of alternative resource-management programmes, whereby public money will achieve the most satisfying return.

3. Assuming that a common procedure for calculation is adopted, b–c analysis does help to establish priorities among a group of alternative resource-development proposals of a similar scale, and may also help in the sequence and timing of projects (e.g. a group of dams in a large river basin).

4. B–c analysis demands the detailed study and evaluation of each of the component resource sectors that constitute a project, and assists in identifying the location, intensity, and frequency of all benefits and costs, thereby helping to pin-point specific beneficiaries.

5. While the evaluation of intangibles is difficult to reconcile within the b–c framework, such costs and benefits are placed in juxtaposition with more tangible ones thereby providing at least some means for their inclusion in the decision-making process.

To sum up, b–c analysis is most relevant in a choice between resource-management projects within one context (e.g. various proposals to develop one river basin or group of basins), where the costs and benefits are relatively easily recognized and calculated (e.g. water-resource management in preference to education or health programmes), and are at a comprehensible scale (e.g. river basin, rather than on the scale of national resources).

The following Tables (12.1.1–3) give a simple example of benefit–cost analysis for a river basin which presents four possible dam sites for multi-purpose development (Spargo, 1961).

Assuming full employment, an interest rate of 5%, a project life of 50 years with no salvage value, and current price levels for all costs, which dam, or combination of dams, will achieve the maximum net benefits?

Obviously Plans 4, 6, and 8 can be quickly excluded, since their b–c ratios

TABLE 12.1.1

	Construction costs ($000)	Operation and maintenance ($000)
Dam A	10,000	30
Dam B	5,000	20
Dam C	3,000	20
Dam D	4,000	20

TABLE 12.1.2

	Benefits ($000)	Plan number
Dam A alone	1,000	1
„ B alone	400	2
„ C alone	200	3
„ D alone	100	4
Dams A + C	1,300	5
„ B + C	400	6
„ D + C	500	7
„ B + D	500	8
„ B + D + C	900	9

TABLE 12.1.3

Plan number	Construction costs ($000)	Present worth annual costs ($000)	Present worth total costs ($000)	Present worth total benefits ($000)	b–c ratio
1	10,000	549	10,549	18,300*	1·74
2	5,000	366	5,366	7,320	1·35
3	3,000	366	3,366	3,660	1·09
4	4,000	366	4,366	1,830	0·41
5	13,000	915	13,915	23,790	1·71
6	8,000	732	8,732	7,320	0·84
7	7,000	732	7,732	9,150	1·19
8	9,000	732	9,732	9,150	0·94
9	12,000	1,098	13,098	16,470	1·27

* The factor 18·3 used to generate the total benefits is derived from the 5% interest rate over a fifty-year time span (Kuiper, 1965, p. 408).

fail to reach unity. Although Plan 1 appears to have the most favourable b–c ratio, its scale of development will only be optimal if its *marginal* benefits equal *marginal* costs (fig. 12.1.3(*b*), point *C*). The method of testing for this is to compare the benefits and costs of the project with the lowest total costs with that having the next lowest total costs (because economic efficiency aims at minimizing total costs). Thus, for example, comparing Plan 2 with Plan 3:

$$\frac{\text{Benefits of Plan 2} - \text{Benefits of Plan 3}}{\text{Costs of Plan 2} - \text{Costs of Plan 3}}$$

$$= \frac{7{,}320{,}000 - 3{,}660{,}000}{5{,}366{,}000 - 3{,}366{,}000} = \frac{3{,}660{,}000}{2{,}000{,}000} = 1 \cdot 83$$

The ratio 1·83 represents the marginal benefit–marginal cost ratio comparing Plan 2 with Plan 3.

Plan 2 is therefore superior to Plan 3, since, despite greater costs, the incremental benefits thereby achieved are higher, and hence net benefits are greater. Table 12.1.4 continues this analysis.

T

TABLE 12.1.4

Plan comparison	Incremental costs ($000)	Incremental benefits ($000)	Incremental b–c
3 alone	—	—	1·09
2 over 3	2,000	3,660	1·83
7 over 2	2,366	1,830	0·78
1 over 2	5,183	10,980	2·12
9 over 1	2,549	1,830	0·72
5 over 1	3,366	5,490	1·67

Thus, by elimination, Plan 5 (i.e. the joint construction of dams A and C) is superior to Plan 1 (i.e. dam A alone) in terms of maximizing net benefits.

B. Systems analysis

A system is a unit composed of interacting parts, whose qualities and relationships can often be conveniently expressed in mathematical terms (by means of a model) and analysed as a whole. River basins obviously can be viewed as open, self-regulating physical systems maintained by a throughput of water. However, each river basin is hydrologically unique and demands a fundamentally different systems model for its analysis. Within this varied hydrological framework of channel geometry, hydrological events (e.g. storm rainfall) and basin responses (e.g. surface runoff), there is often a great variety of possible dam sites, the crucial building blocks of most water-resource systems, their respective sizes (i.e. a few big or many little ones), whether they operate in series or in parallel, the positions and sizes of associated hydroelectric plants, the maintenance of sufficient reservoir capacity to combat possible floods, the maintenance of sufficient channel flow for irrigation, navigation, pollution control, etc. The major problems in basin planning are not so much where to send the water but how much and at what time to do it.

Where economic development involves the control of a hydrological system a more complex hydro-economic system results, wherein economic and social variables interact with, and constrain, the purely hydrological ones, giving rise to the now-classic geographical notion, exemplified by T.V.A., of the fusion of a hydrological system with a coherent socio-economic system. The description of such a system involves the specification of:

1. The nature of the hydrologic and socio-economic inputs (e.g. rainfall, runoff, population statistics, demands for water from various users, etc.);
2. The nature of the outputs (e.g. water supply, electrical energy, etc.);
3. The variables involved (e.g. reservoir capacity, size of power plants, etc.); and
4. Some mathematical statement (model) relating inputs, outputs, and system states in time. The state of the system is its instantaneous condition, which is characterized by its composition, organization, and energy flows, and

defined by system parameters. Hydro-economic systems are thus highly complex and contain many elements that are difficult to quantify (e.g. constraints on the system arising from political and institutional factors). Nevertheless, the development of electronic computers has enabled quite complex systems of variables and their relationships to be treated.

There are four main steps in the design of a systems analysis for an integrated river-basin development:

1. Specification of objectives.
2. The translation of the objectives into design criteria.
3. Field-level planning.
4. Comparison of the results with the objectives.

Specification of the objectives

As has been seen, economic and social objectives may vary widely and involve such broad considerations as increasing the overall economic and social efficiency of the region, the income *per capita*, and the employment opportunities. Systems analysis differs from benefit–cost analysis, therefore, in that its aims may not be narrowly financial (e.g. the net profit to be made from engineering structures), but in some respects its accounting procedures are similar to, and represent an extension of, b–c analysis. Thus, where possible, the collective objectives are quantified and specified as an *objective function* whose value is to be maximized, a simple example of which can be expressed as follows:

$$\sum_{t=1}^{T} \frac{E_t(y_t) - M_t(x_t)}{(1+r)^t} - K(x) \tag{1}$$

where $E_t(y_t)$ = gross efficiency benefits in the tth year from the output (y) of the system (x) in that year;

$M_t(x_t)$ = operation, maintenance, and replacement (OMR) costs in the year (t) for the constructed system;

$K(x)$ = capital costs for the constructed system;

r = discount rate;

T = length, in years, of the economic time-horizon.

The summation yields the present value of the continuing stream of gross benefits, less OMR costs, for a period of t years, discounted at an interest rate of r. Subtraction of the undiscounted capital costs (assumed to be incurred in year 1) yields the present value of net efficiency benefits, which is the quantity to be maximized.

The translation of the objectives into design criteria

This involves such processes as mathematically specifying the relative importance of the objectives, the relationships between the various attributes of the completed system, any limiting constraints on these attributes (e.g. maximum physical size of the possible dams, maximum capital available at a given stage of the development, minimum dry-season river flows, etc.), together with discount rates and opportunity costs. The most important relationships are the *cost–input*, *benefit–output*, *production*, and *capital cost–output* functions (fig. 12.1.4).

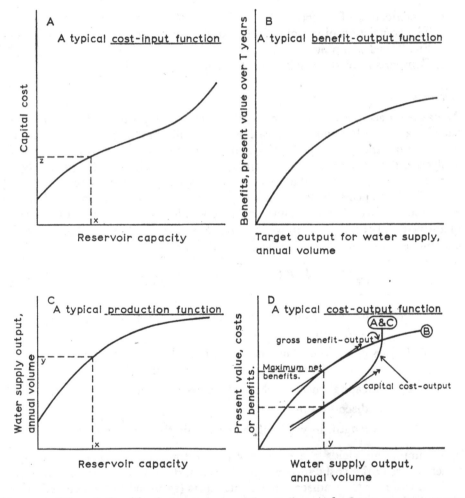

Fig. 12.1.4 The relationships between some of the attributes of a water-resource system (After Hufschmidt, M. M., The methodology of water-resource system design. In Burton and Kates, 1965, pp. 558–70).

A. A typical cost–input function.
B. A typical benefit–output function.
C. A typical production function.
D. A typical cost–output function.

Field-level planning

This aims at maximizing the value of the objective function, subject to the above relationships and constraints. At this stage the computer juggles with all the variables in the objective function, maintaining the relationships and constraints expressed by the design criteria, to produce one or more alternative allocations of water to fulfil the initial objectives. Obviously the large number of variables involved, their range of hydrological and economic character, and the complexity of their superimposed relationships and constraints makes their mutual optimization a highly complex matter, even with the aid of computers. This optimization can be attempted either by constructing mathematical *analytical models* or through *simulations* of the system. These are both *paper experiments* wherein the specimen inputs, storages, states, relationships, and outputs of the system are specified and analysed mathematically. This allows for the simultaneous consideration of such variables as monthly averages and probable variations of streamflow, the capacities of the contemplated reservoirs, the amount of HEP needed month by month, the changing probable needs for irrigation and industrial water, etc.

Analytical models grossly simplify the system down to a few parameters in an attempt to discover optimum values for the most important relationships within the system. Two valuable techniques in this connection are linear and dynamic programming. Linear programming is an optimization procedure to find the maximum or minimum value of a combination of linearly-related variables subject to a number of constraints on the values which they may take. Dynamic programming is a more time-oriented mathematical approach to problems involving an optimum sequence of decisions. It assumes that whatever the initial state and initial decision may be, the remaining integrated decisions constitute an optimum policy with regard to the state resulting from the first decision. The chief value of analytical models in water-resources-system analysis is that they can be used to develop a set of manageable mathematical relationships which can be solved to indicate the range of variation within which a reasonably optimum solution can be obtained by exploration, often by a sequence of simulations.

It is difficult to build the time element satisfactorily into analytical models, although less so than into b–c analysis. In contrast, the *simulation model* can accurately express the step-by-step changes in element states and interactions during each small segment of time of operation by sets of Markov type, or differential, equations. This allows that the states at the end of one time interval can form the input, under differing conditions (e.g. of investment, demand for water, etc.) if necessary, into the start of the next time interval. In such a flexible simulation model the events occur in the same temporal sequence as their counterparts in real life, and in practice each time interval represents a dynamically programmed analytical model providing an output which is the basis for the input for the next time interval, and so on. Thus an initial set of conditions can be investigated in conjunction with a variety of design parameters proposing many alternative relationships and changing conditions through time. The simulation can be run again and again on the computer with different combinations of design parameters, approaching optimization by whittling away at the possible combinations of states, variables, changes, etc. Hufschmidt and Fiering [1966] have shown how a simulation analysis was used in a water-resource-system development plan for the *Lehigh River*, a tributary of the Delaware in Pennsylvania. The design criteria involved the construction of up to six dam sites to provide regulated flows for water supply (domestic and industrial) in the Bethlehem area; recreation at the reservoir sites; dam storage for flood-damage protection, and storage and head for hydroelectric power generation. There were 42 major design variables, 16 dealing with possible physical facility components (i.e. sizes of 6 reservoirs, of 9 power plants, and the construction of a diversion channel), together with 24 variables relating to the allocation of reservoir capacity. In addition there were 12 monthly values of

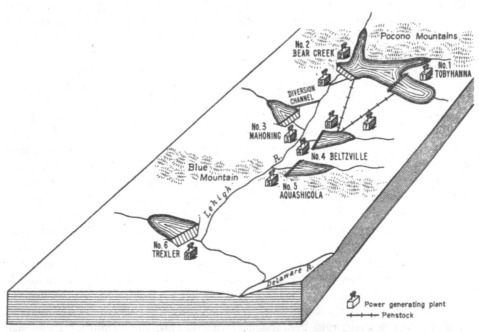

Fig. 12.1.5 The possible physical facility components for the development of the Lehigh basin (Reservoirs 1 and 2 are alternatives) developed by computer simulation. Optimum amounts of reservoir storage, channel flow, hydroelectric output, etc., are calculated by systems analysis (From Hufschmidt and Fiering, 1966).

flood storage and recreation allocation, and 2 output design variables – the target volumes of water supply and electrical energy. The model was programmed so that investments in reservoirs, power plants, and recreational facilities could be progressively scheduled to assumed changing levels of demand. Up to 10 changes in investment and target-output levels could be accommodated, adding 10 time-associated variables, making a total of 52 major design variables in all.

Comparison of the results with the objectives

Even allowing each of the 52 major design variables to have only 3 states ('high', 'middling', and 'small'), this yielded 6 million, billion, billion possible design combinations for the Lehigh River programme, so that a decision was reached after rigorous sampling and testing of the more likely solutions (involving mean and standard deviations of benefits from water supply, energy, recreation, and the prevention of flood losses), together with the examination of each in terms of the initial objectives. Figure 12.1.5 is a sketch of the 16 possible physical facility components.

5. Institutional aspects of water-resource allocation

Allocation techniques suggest how it might be possible to maximize the net social benefits stemming from any given set of inputs, and provide a common economic basis for choosing between various alternative proposals for developing the water-resource complex. However, the decision as to the nature and timing of any particular project rests with what can loosely be described as the political institutional process – that complex of vested-interest group lobbying, public emotion, personality clash, and protested compromise that takes place in committee rooms and council chambers throughout the world. The decision-making process itself is still only imperfectly understood, but it is certainly dependent upon the political institution, for although the allocation of multi-purpose water and land resources involves certain common problems, the final outcome (i.e. implementation of a plan) will differ according to the general policies and attitudes of the nation in question.

Most democratic countries attempt to emphasize the economic significance of the allocation process, although their decisions are often affected by poorly co-ordinated executive powers of governmental agencies and by the pre-existing social and legal institutional arrangements. The presence of these 'non-optimal' factors in the real world introduces practical difficulties which inhibit the full realization of any optimizing allocation technique and the achievement of complete economic efficiency. Water-resources policy, therefore, will not only vary significantly from nation to nation but even between regions because of the differing political, social, economic, and psychological stresses that are placed upon decision-makers. In reality, therefore, decision-makers are faced with a large number of constraints which must be considered in reaching any conclusion regarding water-resources development:

1. *Physical constraints.* Any resource base must have a physical limit of development, regardless of the total economic inputs.
2. *Fiscal constraints.* Total development will depend upon the amount of public money available in the light of other demands.
3. *Policy constraints.* Certain water-resource functions may not be permitted, either for social or political reasons, to drop below a clearly defined quantity of use (e.g. public water-supply requirements and pollution control).

4. *Legal constraints.* Water-resource legislation is usually the outcome of a compromise between the demands of vested-interest groups. An example of the favouring of vested interest would be the financial encouragement of irrigation in the western United States.

5. *Administrative constraints.* The formulation of an optimum water-resources allocation policy depends upon the willing coordination between existing local authorities.

6. *Ownership constraints.* The management of water resources inevitably involves the management of associated land resources, and opposition by private owners may inhibit or prevent project development.

7. *Quantification constraints.* The constraints outlined above assume greater importance in view of the inherent difficulties of quantifying many intangible benefits and costs, particularly of a social or aesthetic character.

8. *Perception constraints.* The behavioural outcome of choice in water use is affected by the perception of the range of choice available to the decision-maker – be he a direct beneficiary, engineer, or resource planner. Perception of alternative uses in water-resource management is still imperfectly understood, but is now considered to be of great importance in the allocation process.

Three national examples serve to illustrate some different institutional approaches to water-resource-allocation decision-making.

Policy-making and the allocation process in the *United States* is divided between the theorists and the pragmatists. The theorists adhere to the economic principle of *marginal productivity*, which states that the cost of providing the last unit of water should be equal to that which the customer is prepared to pay, so that the different users will adjust their various demands, and for all uses marginal productivity will become equal. Theoretically all water would then be used efficiently, the aggregate marginal valuation of the total-water-resource base would be maximized, and the charge for each use would reflect the loss of water to the system and the opportunity costs of its consumption. Such pure economic theory is not implemented in practice. For example, the marginal productivity of water in the western United States is being artificially distorted in favour of irrigation, and it has been estimated that if 10% of the water at present used in large-scale irrigation were to be transferred to municipal and industrial uses, then output in these latter sectors would triple before the end of the century and would constitute a much more efficient allocation of water. However, a change in policy will depend, in part, upon the extent to which the distortion is perceived by decision-makers and more efficient alternatives incorporated. The pragmatists, on the other hand, recognize the importance of ethical and institutional considerations in influencing allocation policy-making towards preserving and/or enhancing the quality of the environment through multi-purpose resource-management programmes. They hold that what is evaluated as economically optimum by a computer may not necessarily be so visualized by the human user, whose attitude towards water-resource use is related to his *perception* of the problem in the light of the information available to him regarding the various alternatives that might produce the same benefits. Increasing attention is being paid, therefore, to the perception of alternative choices and to the attitudes held by decision-makers and the public, together with the considerable influence that these have upon behaviour and action in multi-purpose water-resources management.

Present *British* water-resources policy centres upon the 1963 Water Resources Act, before which water-resource legislation had been fragmentary and ill co-ordinated and had largely dealt with local individual water-resource functions as they became apparent. The 1963 Act recognized the need for a complete water inventory and constituted a system of twenty-nine River Authorities in England and Wales charged with conserving, redistributing, and otherwise augmenting the water resources of their area in all their economic and social aspects. These River Authorities are responsible not only for a detailed account of water supply through hydrometric schemes but also for the evaluation of demand by instituting a comprehensive system of licensing all water abstractors. The licensing scheme, which is related to a system of charges, has superseded the common law (riparian) doctrine of rights. The Water Resources Board is the national co-ordinating body, advisory in capacity, but responsible for liaison, research, and policy-making, particularly in relation to transfers of water between River Authorities. The Act also saw the need for some overall criterion which would relate the varying seasonal and spatial demands placed upon regional water resources, in proposing the concept of a 'minimum acceptable flow' (MAF) for each main river. The MAF is extremely difficult to establish, since its calculation depends upon the licensing system. Its chief advantage is that it can be used flexibly to set seasonal standards for local amenity and river quality in relation to fluctuating demands. The system of charging for water use is at present being developed and will take into account the amount of water used, the extent to which it is removed from the system (with an adjustment for the location, quantity, and quality of returned water), the nature of the source, the season of the year, and the rate of abstraction. The aim is to weight user charges in relation to the effect of his abstraction upon the total water-resources system and not simply upon initial direct removal. For example, irrigators, who return no water to the system, might be subject to a weighting factor of 100%, but water abstracted for industrial cooling processes may only be levied at 1% of the total amount abstracted, since almost all the initial abstraction is returned to the river close to the point of removal. The charging scheme could thus be used as a form of marginal-cost pricing, in that additional demands for water will contribute indirectly to the capital investment required for providing it, with the weighting functions reflecting the opportunity costs in use set by transferred demands (e.g. irrigation, domestic and industrial water supply). The revenue from charges levied on beneficiaries could prove a very positive factor in financing British water-resources development.

In sharp contrast to both the above institutional bases for water-resource allocation is the *Israeli* legislation, the product of a harsh physical, but even more harsh political, environment. Here all water is considered to be the property of the State under the jurisdiction of a Water Commissioner, who can pre-scribe norms for the quantity, quality, price, conditions of supply, and use of water, which is regulated by licence for specific purposes only. During periods of shortage Rationing Areas are declared in which uses of water are ranked in the priority – domestic and public purposes, industrial and irrigation.

6. Water-resource systems

One has seen how water-resources management can be visualized in terms of an interrelating complex of functions that constitute a system. Water-resources management is, however, only one such system, and the bonds which unite its component functions are complemented by the links between it and other rural and urban resource-management systems, each of which in turn is composed of interconnecting multi-purpose functions. Regional development plans are often designed to co-ordinate such a complex of resource-management projects (fig. 12.1.6). Water-resources development is essentially spatial in concept, for while the initial programme often relates to a specific drainage system which is usually clearly delimited in physical terms, the benefits may relate to wider areal units, which may be regional, national, or even international in character.

A water-resource *system* is thus an integrated complex of interlinked hydro-logic and socio-economic variables operating together within a well-defined area, commonly a drainage-basin unit. Such systems are composed of interlinked *sub-systems*, which are simpler on account of the smaller number and greater simplicity of components, their more restricted areal coverage, and their more restricted aims (i.e. water spreading, ground-water recharge, and human use by wells on an alluvial fan). Water-resource systems are more complex, in that they represent larger-scale integrated systems, in which the hydrological and socio-economic variables are interlocked more complexly within a well-defined areal unit of some magnitude. The large drainage basin being developed in an inte-grated socio-economic manner as a unit forms the ideal water-resource system. However, modern water needs often demand the creation of *super-systems*, in which, for example, a given socio-economic policy is imposed on a variety of drainage-basin units (each perhaps having very different hydrological characteris-tics), or in which one very large and uniform hydrological unit is being de-veloped under a number of differing socio-economic bases (usually international). In the former the super-system complexity stems from the need to operate a varying hydrological reality towards some integrated socio-economic goal (e.g. the California Water Plan), whereas in the latter the complexity arises from attempts to operate a large hydrological unit under differing institutional bases (e.g. the Mekong River Plan). The future will obviously see a great expansion in scale of these international super-systems.

The Beech River Watershed, one of fourteen tributary basins of the Tennessee Valley Authority, provides an example of a small water-resources system. On the basis of an Inventory Report on resources and future needs, it is proposed to construct seven multi-purpose dams and one dry dam for flood-

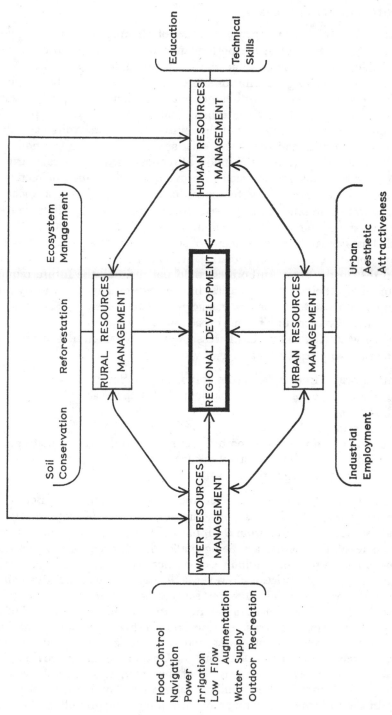

Fig. 12.1.6 The complex of resource-management projects involved in regional water-resource development.

control purposes only, and to improve 80 miles of channel. The resulting flood protection of 17,000 acres of fertile bottomlands will stimulate more intensive agricultural development there, while the steep erodable valley sides will be returned to soil-conserving pasture and forests. More stress is being placed upon a mixed pastoral economy, with larger, more efficient and capital-intensive agricultural units encouraged by controlled rural depopulation. The provision of an assured water supply from one of the multi-purpose reservoirs has helped to attract new industries and to expand existing firms. In all, 2,300 non-farm jobs have been created to absorb the inflow of dispossessed small-scale farmers. The multi-purpose dams are designed to provide a varied range of recreation facilities, some zoned exclusively for specialized activities, such as boating, fishing, and group camping. An improved road network will facilitate convenience and accessibility for the whole region.

In South-East England recent estimates of short- (until 1985) and longer-term (until 2001) domestic, industrial, and irrigation water needs,[1] together with the recognition that many River Authorities could not meet all the future requirements from within their own boundaries, has caused ten River Authority areas to be grouped into a water-resources management super-system. The existence of a future deficiency zone running in a broad sweep from Northampton through London to Ipswich (fig. 12.1.7), has necessitated the consideration of six alternative plans involving regional transfer of water:

1. Importing water from outside (possibly from the Severn or Wye).
2. Estuarinal barrage construction, particularly for the Wash.
3. Desalinization.
4. Artificial recharge of aquifers.
5. Surface storage in reservoirs (both direct supply and river-regulating).
6. Controlled ground-water abstraction from the Chalk (either by direct abstraction or river-regulating).

To meet short-term demands (i.e. over the next ten years) the Board has adopted a 'progressive development programme' which is designed to tap the surface and, particularly, underground supplies of the Great Ouse and Thames. Longer-term feasibility studies are being made of a barrage across the Wash which may provide up to 620 million gallons per day at an estimated capital cost of £300 million (1966 prices). However, a decision of this expensive nature, while it would largely free inland water sources from meeting the considerable future demand, must also be made in relation to its effect on pollution, navigation, existing tidal currents, marine and freshwater ecology, the local economy, etc. Improved ground-water management is another longer-term possibility, whereby water is pumped from the Chalk aquifers to provide additional *in situ* water supply during periods of low flow. One possible scheme, designed to yield an additional 270 million gallons per day, is expected to cost only an estimated £3 million, compared with reservoirs providing the same

[1] Total estimated *deficiencies* in South-East England are 100 million gallons per day in 1971, rising to 400 by 1981, and to 1,100 by 2001.

supply costing between £30 million and £45 million and using up to 10 square miles of land. A technically more sophisticated variant of ground-water abstraction is the artificial recharge of an aquifer whereby water is pumped into the ground-water reservoir during winter periods of excess supply and pumped out again in the spring and summer. However, considerable research is still needed in assessing the hydro-geologic capabilities of the Chalk in southern England

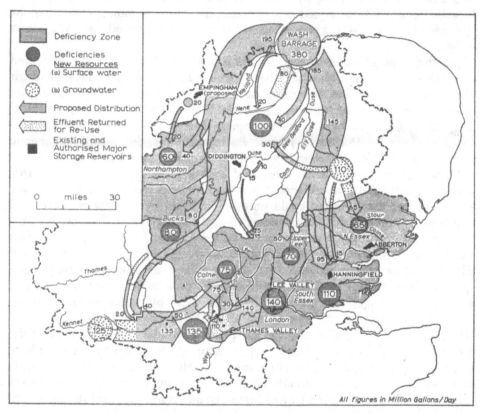

Fig. 12.1.7 One proposed scheme for the supply of water to South-East England in A.D. 2001 involving a Wash barrage, management of the Chalk ground water and surface reservoirs (After Water Resources Board, 1966. Crown Copyright Reserved).

before any major artificial recharge scheme can be put into operation. Of the two remaining proposals, neither inter-regional transfer of water nor desalinization are considered economically feasible in competition with these alternative measures, although technological improvements may render nuclear-powered desaliniz-ation, linked to electricity supply, competitive, at least on a small scale. Figure 12.1.7 shows one proposed long-term scheme involving the Wash barrage and increased ground-water use.

An equivalent super-system is being operated under the California Water Plan, which was initiated in 1960 in response to the astronomical rise in State water consumption. The population had increased by 250% in 25 years to 18·5 million in 1965, consuming some 30 million acre-feet of water per year. The Water Plan (fig. 12.1.8) faces the problem that although 75% of the State's water occurs north of San Francisco, 75% of the use is south of it. The basis of the scheme is the construction of a large number of dams, including the Oroville dam (3·5 million acre-feet, generating 644,000 kW of hydro-electricity) and those on the Feather River, linked to the coastal and other cities by a massive series of huge aqueducts, operated by costly pumping plants (e.g. at Tehachapi, where over 4,000 ft³/sec of water will be raised over a 2,000-ft ridge).

Another type of super-system, in which one river basin is being developed internationally, is the Columbia River covering 282,000 square miles, of which the 33,000 square miles of headwaters lie in Canada and the 259,000 square miles of the middle and lower reaches in the United States. Over the past thirty years the Americans have been developing their portion of the Columbia to provide cheap hydroelectric power, associated with the positive economic externalities of flood protection, recreation, irrigation, and water supply. However, any river composes an integrated hydrological system that disregards artificial international boundaries, and it became apparent that comprehensive multi-purpose management could only be achieved by pooling resources and combining proposals for development so as to satisfy mutual objectives at a lower overall cost than the sum of the two national parts. This international aspect of the development raises many problems, including those concerned with differing water laws on either side of the boundary, differing goals by the two governments involved and hence two distinct policies and attitudes towards river-basin management, differing attitudes held by water-resources bene-ficiaries, and, the most thorny problem of all, the division of benefits and costs. The Americans recognized that the full power potential of their portion of the Columbia could only be realized by providing upstream storage in Canada to produce a more regulated discharge, even though, as there would be too little head to provide much hydroelectric power in Canada, most of the power benefits would accrue to the United States. In addition, upstream storage would provide increased flood protection, and cheap power would encourage local employment downstream, whereas Canada would forgo the real costs of loss of flooded land and the opportunity costs of the future unavailability of the Columbia basin water. It is little wonder therefore that the Columbia River Treaty Protocol Agreement, which was signed by Canada and the United States in 1963, took nineteen years to ratify. The Americans agreed to pay cash to the Canadians as compensation for real costs and to share their down-stream benefits of cheap power and increased flood protection. The latter in-volves payments to Canada of a total of $254·5 million for power benefits and $64·4 million for flood-protection benefits over the first thirty years of the sixty-year treaty (after which Canada can elect to sell her share of U.S. benefits). In return, Canada has agreed to provide 15·5 million acre-feet of storage (in-cluding 8·5 million acre-feet for flood-protection purposes) by means of three

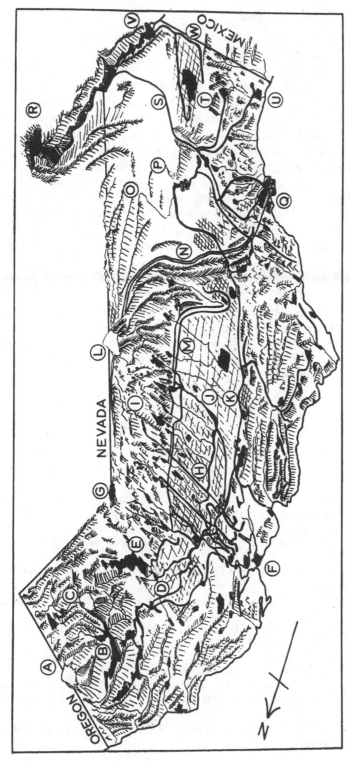

Fig. 12.1.8 California water resources development (from the map issued by the Irrigation Districts Association of California). The small 'boxes' are pumping-stations. (A) Mt. Shasta; (B) Shasta Dam; (C) Mt. Lassen; (D) Sacramento River; (E) Oroville Dam; (F) San Francisco; (G) Lake Tahoe; (H) Hetch Hetchy Aqueduct; (I) Yosemite Valley; (J) San Joaquin River; (K) California Aqueduct; (L) Mt. Whitney; (M) Friant Kern Canal; (N) Los Angeles Aqueducts; (O) Death Valley; (P) Mohave Desert; (Q) Los Angeles; (R) Hoover Dam; (S) Colorado River Aqueduct; (T) Salton Sea; (U) San Diego; (V) Imperial Dam; (W) All American Canal.

projects, although operation of the existing twenty-four dams and all future projects will rest with the United States.

Although the planning of such super-systems may seem impressive, in the long term the greatest future problem associated with the exercising of a conscious choice in water use is the paradox that, whereas on the one hand, significant and economically viable schemes must of necessity be on a large scale and involve complex long-term planning, on the other hand, the increasing tempo of technological, economic, and social changes imply that the vehicles and even the objectives of such schemes may constantly change over comparatively short time intervals. Thus, whereas the vehicles of water-resources planning are becoming more massive and complex, the requirement for their manoeuvrability is also increasing, and the aim of all future planning is to produce a large-scale and completely integrated scheme capable of constant re-evaluation. Until recently no satisfactory methods for such re-evaluation existed, but the advent of computer-based dynamic programming goes some way towards resolving the paradox. Buras [1966], in a very interesting paper on this subject, outlines the possibilities for greater use and development of dynamic programming, particularly with reference to conjunctive management of surface and ground-water storage. Although this approach faces many, as yet, unsolved problems (e.g. in flow probabilities and analysis of large systems), it seems that the most efficient choice in water use will only be made when spatial and dynamic aspects of the problem can be considered in detail simultaneously.

REFERENCES

BURAS, N. [1966], Dynamic programming in water resources development; in Chow, V. T., Editor, *Advances in Hydroscience*, Vol. 3 (Academic Press, New York and London).

BURTON, I. and KATES, R. W., Editors [1965], *Readings in Resource Management* (Chicago Univ. Press). (See especially chapters by Hufschmidt, M. M. and by Sewell, W. R. D. *et al.*)

DAHL, R. A. and LINDBLOM, C. E. [1953], *Politics, Economics and Welfare* (New York), 557 p.

ECKSTEIN, O. [1958], *Water Resource Development: The economics of project evaluation* (Cambridge, Mass.)

FOX, I. K. [1966], We can solve our water problems; *Water Resources Research*, 2, 617–23.

HIRSCHLEIFER, J., DE HAVEN, J. C., and MILLIMAN, J. W. [1960], *Water Supply; Economics, Technology and Policy* (Chicago), 378 p.

HUFSCHMIDT, M. M. and FIERING, M. B. [1966], *Simulation Techniques for Design of Water Resource Systems* (Harvard University Press), 212 p.

KATES, R. W. [1962], *Hazard and Choice Perception in Flood Plain Management*; University of Chicago, Department of Geography Research Paper No. 78, 157 p.

KRUTILLA, J. V. and ECKSTEIN, O. [1958], *Multiple Purpose River Development* (Baltimore), 301 p.

KRUTILLA, J. V. [1967], *The Columbia River Treaty – The Economics of an International River Basin Development* (Baltimore), 211 p.

KUIPER, E. [1965], *Water Resources Development, Planning, Engineering and Economics* (London), 483 p.

LOWENTHAL, D., Editor [1966], *Environmental Perception and Behavior;* University of Chicago, Department of Geography, Research Paper No. 109, 88 p.

MAASS, A., HUFSCHMIDT, M. M., DORFMAN, R., THOMAS, H. A., MARGLAN, S. A. and FAIR, G. M. [1962], *Design of Water-Resource Systems* (Cambridge, Mass.), 620 p.

MCKEAN, R. N. [1958], *Efficiency in Government through Systems Analysis: With Emphasis on Water Resources Development* (New York), 336 p.

MORE, R. J. [1967], Hydrological models and geography; In Chorley, R. J. and Haggett, P., editors, *Models in Geography* (London), pp. 145–85.

NATIONAL ACADEMY OF SCIENCES [1966], Alternatives in Water Management; *National Research Council Committee on Water, Publication No.* 1408, (Washington, D.C.).

O'RIORDAN, T. and MORE, R. J. [1969], Choice in water use; *Transactions of the Institution of Water Engineers.*

PREST, A. R. and TURVEY, R. [1966], Cost–Benefit Analysis – a Survey; *Economic Journal*, 16, 683–735.

SAARINEN, T. F. [1966], *Perception of the Drought Hazard in the Great Plains*; University of Chicago, Department of Geography Research Paper No. 106, 198 p.

SEWELL, W. R. D., DAVIS, J., SCOTT, A. D. and ROSS, D. W. [1961], Guide to benefit-cost analysis; *Resources for Tomorrow Conference, Background Papers I*, (Queen's Printer, Ottawa), p. 17.

SIMON, H. A. [1957], *Models of Man* (New York), 287 p.

SPARGO, R. A. [1961], Benefit Cost Analysis and Project Evaluation; *Resources for Tomorrow Conference, Background Papers* 1 (Queen's Printer, Ottawa), pp. 299–310.

UDALL, S. C. [1963], *The Quiet Crisis* (New York), 224 p. (See especially pp. 132–4.)

WATER RESOURCES BOARD [1966], *Water Supplies in South-East England* (HMSO, London).

WENGERT, N. [1955], *Natural Resources and the Political Struggle* (New York), 71 p.

WHITE, G. F. [1957], Trends in River Basin Development; *Law and Contemporary Affairs*, 22, 157–87.

WHITE, G. F. [1964], *Choice of Adjustment to Floods*; University of Chicago, Department of Geography Research Paper No. 93, 164 p.

Index

This index, as with the volume, deals mainly with principles, processes, and practices. Reference to persons is restricted to those who contribute important formulae.

D. M. BECKINSALE

Index